Sustainable Cities and Communities Design Handbook

Sustainable Cities and Communities Design Handbook

Green Engineering, Architecture, and Technology

Second Edition

Edited by

Woodrow W. Clark
Qualitative Economist, Managing Director
Clark Strategic Partners, Beverly Hills
CA, United States

Butterworth-Heinemann is an imprint of Elsevier
The Boulevard, Langford Lane, Kidlington, Oxford OX5 1GB, United Kingdom
50 Hampshire Street, 5th Floor, Cambridge, MA 02139, United States

First edition 2010
Second edition 2018

Library of Congress Cataloging-in-Publication Data
A catalog record for this book is available from the Library of Congress

British Library Cataloguing-in-Publication Data
A catalogue record for this book is available from the British Library

ISBN: 978-0-12-813964-6

For information on all Butterworth-Heinemann publications visit our website at https://www.elsevier.com/books-and-journals

Publisher: Matthew Deans
Acquisition Editor: Ken McCombs
Editorial Project Manager: Serena Castelnovo
Production Project Manager: Paul Prasad Chandramohan
Designer: Victoria Pearson

Typeset by TNQ Books and Journals

Contents

17. Energy Economics in China's Policy-Making Plan: From Self-Reliance and Market Dependence to Green Energy Independence

Xing Li, Woodrow W. Clark, II

18. Energy Strategy for Inner Mongolia Autonomous Region

Woodrow W. Clark, II, William Isherwood

19. Business Ventures and Financial Sector in the United Arab Emirates

Robert Rumiński

20. ECO-GEN Energy Solutions: Tahiti Story

Cheryl Stephens

21. Japanese Smart Communities as Industrial Policy

Andrew DeWit

List of Contributors

Talia Arnow, Coalesce Accelerator, Boston, MA, United States

Samantha Bobo, Rice University, Houston, TX, United States

Danilo Bonato, Remedia Consortium, Milano, Italy

Woodrow W. Clark, II, Aalborg University, Aalborg, Denmark; Clark Strategic Partners, Beverly Hills, CA, United States

Attilio Coletta, Università degli Studi della Tuscia, Viterbo, Italy

Andrew DeWit, Rikkyo University, Tokyo, Japan

M.K. Elahee, University of Mauritius, Reduit, Mauritius

Michael Fast, Aalborg University, Aalborg, Denmark

Sierra Flanigan, Coalesce Accelerator, Boston, MA, United States

Julian Gresser, Alliances for Discovery, Santa Barbara, CA, United States

Andrew Hoffmann, Independent Consultant, Chicago, IL, United States

Emily Howald, Ohio Wesleyan University, Delaware, OH, United States

William Isherwood, Asian Development Bank, Manila, Philippines

Anjun J. Jin, Huaneng Clean Energy Research Institute, Beijing, China

Joseph Kantenbacher, University of Surrey, Guildford, United Kingdom

A. Khoodaruth, University of Mauritius, Reduit, Mauritius

John Krygier, Ohio Wesleyan University, Delaware, OH, United States

Calvin Lee Kwan, University of California, Los Angeles, CA, United States

Benjamin Leffel, University of California Irvine, Irvine, CA, United States

Xing Li, Aalborg University, Aalborg, Denmark

Henrik Lund, Aalborg University, Aalborg, Denmark

Myra Moss, Ohio State University Extension, Columbus, OH, United States

Wil Nagel, University of California, Irvine, CA, United States

V. Oree, University of Mauritius, Reduit, Mauritius

Raimondo Orsini, Sustainable Development Foundation, United Kingdom

Poul A. Østergaard, Aalborg University, Aalborg, Denmark

Tom Pastore, Sanli Pastore & Hill, Los Angeles, CA, United States

Wenbo Peng, Huaneng Clean Energy Research Institute, Beijing, China

Robert Rumiński, University of Szczecin, Szczecin, Poland

Don Schultz, California Rate Payer Commission, Sacramento, CA, United States

Rebekah Shirley, Power for All, Berkeley, CA, United States; University of California, Berkeley, CA, United States

Arnie Sowell, California State Assembly Office, Sacramento, CA, United States

Cheryl Stephens, ECO-GEN Energy Inc., Van Nuys, CA, United States

Douglas N. Yeoman, Parker and Covert LLP, Sacromento, CA, United States

Gerardo Zamora, Anthem Software Solutions & Services (S^3), San Jose, Costa Rica

Tor Zipkin, Aalborg University, Denmark

About the Editor

Woodrow W. Clark, II, MA, Ph.D.

Dr. Clark was Senior Policy Advisor to California Governor Gray Davis (2000–03) for renewable energy reliability. He was formerly Manager of Strategic Planning/Implementation at Lawrence Livermore National Laboratory. He is a qualitative economist who is the Managing Director of Clark Strategic Partners, in Los Angeles, California.

Chapter 1

Introduction

Woodrow W. Clark, II[a]
Clark Strategic Partners, Beverly Hills, CA, United States

Chapter Outline

An earlier book on sustainable communities (Clark, 2009a) was a series of a dozen sustainable case studies that have actually been implemented. They ranged from public schools and community colleges in the nonprofit world to city governments to corporations and business around the world. The idea was to give people cases that are real and models for them to understand the breadth and depth of sustainability as both a concept and mechanism for action.

Sustainable development thus has become an acceptable policy initiative and programmatic implementation strategy. However, the "devil is still in the details" as to what sustainable development means. On the one hand, there are businesses and governments that see sustainable development as being a strategy for building more homes, office buildings, and large communities. In order to acknowledge that communities were sustainable, groups formed to provide scores and credit points. The most popular one in the United States is the US Green Building Council (USGBC) that developed a scoring mechanism called Leadership in Energy and Environmental Design whose basic score starts with being "accredited" to its highest as being "platinum."

Without going into details (see that book The Next Economics, Springer Press, 2009), the USGBC has become a world leader in establishing sustainable buildings. There are now efforts and pilot programs to do the same for communities and even cities. A number of other organizations in the last decade have done something similar in the United States, including the Climate Action Registry, originally founded to be California centric but now both national and global. Because of the energy crisis that hit California at the

a. Woodrow W. Clark II, Ph.D. is a Qualitative Economist who is Managing Director of Clark Strategic Partners at www.clarkstrategicpartners.net.

Sustainable Cities and Communities Design Handbook. https://doi.org/10.1016/B978-0-12-813964-6.00001-X

turn of the 21st century, a number of energy efficient and conservation programs were created including Flex Your Power, which still exists. My first book, *Agile Energy Systems* (AES; Elsevier Press, 2004), was written with Professor Ted Bradshaw, while I served Governor Davis as his Renewable Energy, Emerging Technologies, and Finance Advisor.

The updated AES is now with a global perspective (not just California) and was published in August 2017. Chapter 3 in this book covers AES briefly, which is the combination of central power grids along with on-site distributed energy systems (such as solar, wind, and geothermal) that were the answer to the energy crisis in California plus much more. The California government programs and subsequent actions were led by the California Energy Commission and the California Public Utility Commission (CPUC), which were able to establish the state as a world leader in energy conservation and renewable power generation then and now even more so today. In the cases presented in this book, there is one from CPUC with specific areas of focus in order to define what sustainable communities are. My last three books helped do this with one the *Green Industrial Revolution* (GIR) from Elsevier Press in 2014 and an earlier shorter version of that book in Mandarin that was published in 2015. The last one in 2016 was *Smart Green Cities* from Routledge Press. However, there have been very few cases in these books. Now there are many cases in this second Sustainable Cities and Communities Design Handbook (SCCDH). And hopefully there will be more, even annual, editions of the book.

Meanwhile, the US government had established programs (e.g., Energy Star) that both ranked the energy output (hence savings from conservation) of appliances and equipment and the national energy laboratories (e.g., National Renewable Energy Laboratory) for establishing the ratings or electric power derived from sunshine, wind, geothermal, and other renewable energy resources. The Lawrence Livermore National Laboratory and Lawrence Berkeley National Laboratories along with another national research labs, universities, and companies have moved ahead with other national research labs like Sandia, Oakridge and NASA (https://www.nasa.gov/ames/facilities/sustainabilitybase).

Chinese President Jinping Xi said repeatedly since his talk at the UN G20 Summit, sponsored by China in Hangzhou, China (September 2016) as "Green Development for China and other nations in order to reverse climate change" (http://g20executivetalkseries.com).

Today China has the largest solar farm in the world, which covers a coal mine, as part of China's 13th Five Year Plan (see Case on this and China communities too) which removes it from depending on fossil fuels and nuclear power into what is known as "green energy" (https://www.dailydot.com/debug/largest-floating-solar-farm-china/).

International organizations have been formed for doing similar rankings and scores for the EU nations and globally. See the cases in the book as well as

Sustainia from Denmark at: www.sustainia.me. The most significant global group leader is the United Nations, whose Paris Conference in December 2015 created the G20 nations including the United States, China, Russia. the UK, Germany, Turkey, Brazil, and others. The B20 was created to support the G20 nations in issues ranging from finance to labor to infrastructures that in 2017 was sponsored by Germany and focused on Africa. See the Case on B20 for 2017 from Germany.

Meanwhile, unfortunately the federal US government currently in power does not see the need to continuing these policies, plans, and financing. The US federal government denies there is climate change and therefore is reducing its climate preservation funding, science, and technological solutions (https://www.commondreams.org/views/2017/05/26/if-theyre-wrong-planet-dies).

The result is that the state of California (and now other US states) has been and will continue to be the leader in climate preservation. California has the sixth largest GDP in the world and continues to take aggressive actions, technologies, and financing for climate preservation.

This is what the rest of the United States and world need to do. Germany, Japan, Korea and China have already been doing this for over 2 decades now as some cases demonstrate in the book (http://www.eseoulpost.com/news/articleView.html?idxno=9432).

It will not be easy. Japan as Chapter 24 demonstrates has been a leader for decades in part due to its reliance on energy from fossil fuels and even nuclear power plants despite the atomic bombs that were dropped on the nation at the end of World War II.

SCCDH (second edition) is a book long in the making and overdue. Basically the book maps out what communities need to do when thinking about how to protect their environment while repairing, building, or expanding themselves. (At this point, there is no need to review all the chapters and cases but only to refer to some.)

What the book does is basically outline the need for specific tools or mechanisms that allow people and companies to create sustainable communities. The best way to tell the story about sustainable as green, smart, and healthy development is by way of examples. Therefore, communities are defined and illustrated by the cases since they are often self-contained communities with all the attributes therein from residential to recreational areas that include the use of basic infrastructures that need to be integrated ranging from water and energy to transportation and telecommunication systems.

My 2014 book on the GIR is a concept that puts these issues into a broader picture about society and its industrial growth. The GIR is an industrial revolution that uses renewable energy power and fuel generation along with storage and new technology devices including the interconnection of communities into "smart grids." As Chapter 2 on the GIR stated, the GIR is now in the EU and Asia and has only recently come to the United States in the last few years. And it is remarkably different from the Second Industrial Revolution (2IR) of fossil fuels and nuclear power (http://www.apo-tokyo.org/publications/apo_news/september-october-2016/).

Jeremy Rifkin, the environmental economist saw this GIR coming but called it the Third Industrial Revolution (2005) and started a series of groups to support and implement it. Now communities throughout the world are participating and receiving advice and plans to implement. The 2IR was an era dominated by the fossil fuel and now nuclear power industries. Yet they all knew of the dangers that their systems caused to climate and environment. Among all of them Shell Oil in the European Union (EU) was very aware: https://www.theguardian.com/global/video/2017/feb/28/what-shell-knew-about-climate-change-in-1991-video-explainer.

And as noted, President Xi of China calls the GIR "green development," which is now implemented. Public Policy and Leadership is the title of Chapter 4 by Scott McNall, which notes how significant it is for communities to have a plan or programs prior to making public policy decisions. Be they elected officials or appointed, leadership over government programs is critical. However, today leadership among corporate decision makers is equally significant. Most companies will make decisions based upon profit and loss, but the political arena and economic concerns over how to handle climate preservation is critical for most corporate leaders. The bottom line tends to always be there and provide the base line or bar to prevent sustainable development programs. In short, there are many problems with the theory(s) of economics as they are not scientific and therefore cannot predict issues, problems, or the solutions therein. See Chapter 5 on Cross-Disciplinary Science with Economics. And also Qualitative Economics (Clark and Fast, 2008) does just that and will be updated to publish in 2018.

However, even with decision makers providing leadership and public policy, the bottom line remains for the need to make money. What new Technologies and Scientific Planning (Chapters 5 and 6) need is to be in place before change occurs. And even more significant, how does an organization or company know when is the right time to "try something new" from the GIR. The answer is that there is no perfect time. Most construction contractors and builders will use the same tried and true technologies or whatever is in their self-interest from their last client. All they want to do is make money and influence (lobby) publically elected officials: https://www.theguardian.com/commentisfree/2016/nov/30/donald-trump-george-monbiot-misinformation?CMP=fb_gu.

Often these traditional or conventional technologies are all from the 2IR and therefore inappropriate for sustainable development. The "stranded costs" alone will be long lasting (20–30 years) and delay the GIR for another 2–3 decades. This is unacceptable given the need to mitigate global warming and climate change, which is now "climate preservation." Hence as Chapters 4 and 5 outline, there needs to be a constant technology review and management program to investigate and perform "due diligence" on the technologies that are or could be associated with the GIR. State laws often require objective "oversight committees" to check on contracts and costs. China, in the Appendix, has its 13th Five Year Plan out now (March 2016), which stipulates these plans, goals, measurements, and reviews for ethical and moral reasons.

There are communities and regions that have implemented green development agencies such as those outlined in Chapters 2 and 3 with several cases. California's path toward leading the United States and global sustainability movement has begun and provides some interesting models. While some would credit this statewide effort to Governor Schwarzenegger, the reality is that he took the programs and basic public policy leadership from Governor Davis who preceded him. The difference is that Davis left state office with the California finances on an upswing, while Schwarzenegger during his terms as governor "bankrupted" the state by not providing government regulations and rules that return funds to the state of California. The point is that sustainable development does cost money and resources. But it need not be totally dependent upon the neoclassical theories of finance and accounting for its existence and performance. The government is not an invisible hand (Clark and Fast, 2008).

That topic of finance and accounting gets into life cycle project models in Chapters 7–10 that go through the process as noted in several cases. In Chapter 7 by Tom Pastore, for example, the actual economic analyses for the accounting are provided for sustainable development costs. Life cycle and cost benefit analyses are critical areas for any organization to study in order to make decisions. Even more significantly today in the GIR there needs to be factored in the rebates, incentives, and tax benefits for technologies and programs from the GIR.

The big issue is often the bottom line for businesses. And this is increasingly so for government and nonprofits since the 2008 economic "deep recession." What is important about Chapters 7, 9–11 (as well as appendices and cases) on economics and accounting are that they point out and demonstrate the shift from cost benefit analysis to "life cycle analysis" and how this economic change has helped bring innovations, emerging technologies, and the GIR into the market sooner than normally expected in the conventional neoclassical "market economy" theory.

The economic change has, however, also come with new economic and legal programs that have helped provide longer-term financing for innovation and the GIR. One of the key areas is the use of power purchase agreements (PPA), which provide financing to an organization for 15 to 25 years. Such long-term or "life cycle" legal and financial commitments have helped solar systems to be installed at a faster rate, along with government tax and grant incentives, than in the last decade. Since 2010, solar energy has moved rapidly into the key renewable energy spot than in prior years. The big change was from solar "farms" to onsite and distributed systems for buildings and communities. This process has also reduced the price of solar systems due in part to a growing number of new solar manufacturing companies.

One reason was the creation and then implementation of the Property Assessment Clean Energy (PACE) that the city of Berkeley, California, enacted in 2008. The program was especially useful for getting solar energy into homes, buildings, and complexes. However, the city got the loans or other funds should the building owner go bankrupt or foreclosed. Hence, the US Federal Housing Authorities (Fannie Mae and Freddie Mac) declared the PACE illegal since the lenders would get no funds. PACE then only become possible for business and corporation funding.

This funding process and then others since 2010 resulted in solar systems becoming cost competitive with regular energy generation systems. The same economic phenomena came with wind when a combination of long-term financing mechanisms and government tax and incentive programs reduced the costs of wind turbines to below that of natural gas power generation by the end of the first decade of the 21st century. On-site and distributed power has become a key part of "green development" for both individual homes as well as businesses, resorts, and other complexes.

Hence, the financial and accounting economics for renewable energy power generation for the GIR is moving away from a PPA long-term finance mechanism. This does not mean that the cost benefit analysis (short term in quarters or even 2–3 years) works now. But soon the shift might be to a more traditional economic and accounting model. Now, however, there are two models that have become part of the GIR that Chapter 7 refers to and several remaining chapters mention. The two emerging models are: (1) feed-in-tariffs (FiT), which charge a higher energy purchase rate to consumers but also allow consumers to sell their power to the central grid supplier or other energy customers; and (2) regular leases.

The FiT was started and very successful in Germany in the early 1990s. It has since gone under some expansion and revision. Then Spain started a national program in the mid-2000s that appeared to be too aggressive with overbuilding of solar plants and systems with some facilities. The net result in both Germany and Spain was higher employment and job creation in the solar sector. And for Germany, it became the number one nation in solar manufacturing. Now other nations and communities are starting FiT programs.

The other emerging economic model is a regular lease. However, the costs are high although for a shorter time. The costs of solar and other renewables are coming down, so that one "old economic" model is such that all renewable power generation might be factored into the regular operational costs (like heating, air conditioning, electrical and plumbing) for buildings (homes, offices, storage, etc.). These systems can then be part of the total costs for a mortgage of any building and then applied to different or more expansive areas like college campuses, shopping malls, etc., which have clusters of buildings.

Chapter 8 on Legal Contracts, Lease and Power Purchase Finance by Douglas Yeoman explores the PPA and also some of the newer models. The legal section, however, also covers the need for construction contracts, liabilities, warrantees, and insurance. What is important is to know that such legal documents can be templates and models for other programs, buildings, and clusters.

Chapters 10 and 11 then get into the actual cases or examples of how the GIR works. Some of the information that had used the same database for the technical and financial optimization for the same colleges is rooted in Chapter 4.

Chapter 11 then looks into sustainability as to how economics needs to combine with public policy known as "civic capitalism." In the end, however, it is important to identify and find communities that have been sustainable and represent what the GIR is about. Denmark (where I was a Fulbright Fellow in 1994 and then a visiting professor the next 6 years) has been a leader in this regard in terms of national policies, financing, and operational programs (Clark, 2009a,b; Clark and Lund, 2007, 2008). Lund in particular has been tracking renewable energy, especially the wind manufacturing industry that went global in Denmark through a merger of several Danish wind turbine companies (Lund and Clark, 2002); Vestas is today the largest, most dominant wind turbine manufacturing company in the world.

Hence Chapter 12 concludes on climate preservation from the bottom-up community approach: a case in Denmark by Lund and Poul Alberg Østergaard (academic practitioners from Aalborg University in Denmark) looks at Frederikshavn, a small city in the northern Jutland region of Denmark. This city is a main transportation and shipping hub for northern Demark with western Sweden and southern Norway. Because of that, the city is aware and very aggressive in becoming sustainable due to the environmental needs but also the control over energy and fuels from the 2IR.

There are appendices in the book that are well worth using as references. One appendix presents the *California Standard Practices Manuel* (CSPM) of which I was a coauthor while working in state government. Basically, it is an economic model for doing life cycle analysis on projects. There are guidelines and formulas. The CSPM was published in 2002 but was originally created in the mid-1980s for doing cost analyses on projects for the CPUC. However, it was not revised until 2002. The CSPM is now used extensively in California government project finance. It remains the guiding model for doing economic data projections, analyses, and evaluative outcomes.

Another appendix on the use of an Optimization Energy Plan for Los Angeles Community College Campuses is even more useful in that it provides a working model on how to understand, apply, and analyze technologies for use in applied areas like college campuses. The same optimization code can be used for other nonprofit organizations as well as government. And with some modifications, the optimization may be very useful for companies and businesses, especially those that have clusters or groups of buildings in one area.

Finally, the last chapter has conclusions for the next generation of people, companies, and governments about sustainability. I wrote this chapter because the problems today with global warming and climate change are directly the fault of my generation. We "baby boomers" lived off our land and exploited others for decades. We are now paying the price because the world is indeed "round"—not "flat" as some "populist" economic commentators would argue—hence the atmospheric and other pollutions that we cause in one part of the world travel to our part of the globe too. In fact, this group's current defense of the flat world idea in economic and business terms is equally as wrong and short sighted.

Consider the economic crisis that did not hit just the United States or even the EU and Asia. The "d-recession" impacted everyone. Money has been misplaced and even stolen from New York City to London to Paris to Rome and around the world again to Tokyo, Seoul, Sydney, New Delhi, and now to Beijing, Shanghai, and Hong Kong. Everyone must be careful.

The point is that the last chapter tries to provide the details from the green industrial revolution perspective in terms of what this means for jobs, careers, and the future of our planer that we have handed over to young people. Without doubt, the problems are enormous and will need time, money, and people to solve them. The clock is ticking and the environment becoming worse daily around the world. With the United States withdrawing (June 1, 2017) from the UN Paris Accord from December 2015, there will be even more "drama" for the next few years.

These are not statements from one more of those crazy scientists, academics, or economists on the far "left." No! The facts exist. Insurance companies are the bottom line here and their rates and even reluctance to insure certain communities, regions, or areas of the United States has started and expanded. Some countries and their cities as well as industries can no longer get insurance. In short, our world is at risk. We do not have any time left to debate it. We *all* must act *now*.

REFERENCES

Clark II, W.W., 2009a. Sustainable Communities: Case Studies. Springer Press.

Clark II, W.W., 2009b. Analysis: 100% renewable energy. In: Lund, H. (Ed.), Renewable Energy Systems. Elsevier Press, pp. 129–159 (Chapter 6).

Clark, W.W., Fast, M., 2008. Qualitative Economics. Coxmoor Press, Oxford, UK.

Clark II, W.W., Lund, H., 2008. Integrated technologies for sustainable stationary and mobile energy infrastructures. Utility Policy Journal 16 (2), 130–140.

Clark II, W.W., Lund, H., December 06, 2007. Sustainable development in practice. Journal of Cleaner Production 15, 253–258. Elesvier Press.

Lund, H., Clark II, W.W., Nov. 2002. Management of fluctuations in wind power and CHP: comparing two possible Danish strategies. Energy Policy 27 (5), 471–483. Elsevier Press.

FURTHER READING

Clark II, W.W., Kuhn, L., Summer 2017. Violence in Schools, Colleges and Universities. NOVA Press.

Clark II, W.W., Summer 2017. Agile Energy Systems: Global On-site and Distributed Energy. Elsevier Press.

Clark II, W.W., Cooke, G., March 2016. Smart Green Cities. Routledge Press.

Clark II, W.W., Cooke, G., JIN, A.J., LIN, C.-F., August 2015. Green Industrial Revolution in China (Mandarin). Ashgate and China Electric Power Press.

Clark II, W.W., 2014. Global Sustainable Communities Handbook. Elsevier Press.

Clark II, W.W., Fall, 2012. The Next Economics: Global Cases in Energy, Environment, and Climate Change. Springer Press.

Clark II, W.W., Cooke, G., 2014. Green Industrial Revolution. Elsevier Press.

Clark II, W.W., Cooke, G., Fall, 2011. Global Energy Innovations. Praeger Press.

Rifkin, J., 2011. Third Industrial Revolution. Barnes and Noble.

Chapter 2

The Green Industrial Revolution

Woodrow W. Clark, II
Clark Strategic Partners, Beverly Hills, CA, United States

Chapter Outline

Time is passing quickly for United States and other nations, while Europe and Asia now and the People's Republic of China (PRC) have been developing sustainable, energy-independent communities for the last 2 decades. As a nation and the leader of democracy for two centuries, the United States must examine its own "roots" and provide the future direction for humanity. The Green Industrial Revolution (GIR) is now strongly embedded in other nations so that these countries are no longer dependent on fossil fuels or nuclear power, which defined the Second Industrial Revolution (2IR). The GIR primarily generates stationary power and creates fuel from renewable energy sources.

Nations must take action now to create and implement GIRs for themselves. They need to reduce the energy dependency on the Mideast, a geopolitical region whose instability constantly threatens national security and keeps them from focusing on crucial domestic issues such as health care, financial reform, and innovation as well as the global planetary environmental crisis. Becoming

Sustainable Cities and Communities Design Handbook. https://doi.org/10.1016/B978-0-12-813964-6.00002-1

involved and part of the GIR must be recognized and implemented by nations and communities sooner rather than later. The United States, for example, must go beyond the 2IR, with its massive and inefficient fossil fuel generation, environmental degradation and move rapidly into the GIR, with its community-centric and environmentally friendly renewable energy generation. Europe and Japan have already done so for the last 2 decades. Where is America?

Social and economic forces are coming together as the nation ponders its sustainable future. Now with global warming and climate change impacting everyone's daily lives, can anyone wait any longer? On an economic front, the world is battling the most severe economic turndown since the Great Depression of the 1930s. Nationwide, states are reeling with the loss of tax and real estate development revenue. California has been "bankrupted" by its governor, whose efforts to balance a shattered budget are subject to serious questions.

California is the world's eighth largest economy. Nine years ago, the state was the world's sixth largest economy and held the distinction as number seven from 2003 to 2008. However, in mid-2008, the recession started. The basic result of the California budget signed in September 2009 was to handicap the entire state from its public education and welfare systems to basic needs such as fire, police, water, energy, waste, transportation, and prisons. On the other coast, the American auto industry, once the nation's pride as the leader of the global manufacturing sector, is on life support from the federal government. The era of the V8 and the megaton SUV is fading in the review mirror as it should have a decade ago. Now General Motors (GM) is renamed by the general public and federal government decision makers as Government Motors.

Americans are wondering what their vital interests in any international arena should be. The world's oil and natural gas supplies have peaked and are rapidly declining. As the Shell Oil geophysicist M. King Hubbert observed in his startling prediction, first made in 1949, the fossil fuel era would be of very short duration in "Energy from Fossil Fuels, Science" [February 4, 1949]. In 1956, Hubert predicted that US oil production would peak in about 1970 and then decline. At the time he was scoffed at, but in hindsight he proved remarkably accurate.

Just as the world's oil and natural gas supplies have peaked, there is renewed interest in nuclear power. This too is a false hope. The US Department of Energy has reported a key set of figures documenting the declining and limited supplies of gas, oil, and coal. One surprising statistic, which bodes badly for the nuclear industry, is that, there are only 61 years left of uranium.

While America's domestic oil supplies peaked in the 1970s, and international oil supplies sometime around the early part of the 21st century (estimates are now at 2030), and with demand rising from newly developed nations, pushing for more oil and gas with tax breaks or even land options is the wrong policy and certainly not part of the GIR. When these measures and others related to "balancing a budget" for the short term are then implemented, it means future generations will be paying taxes for years at triple or quadruple the original costs. This 1-year "fix" to balance budgets or justify expenses is misguided, wasteful, and economically crippling on our children and grandchildren. Any new funds and resources must be focused on renewable energy generation and related technologies for storage and waste, transportation, and related areas. If not, then global political and social tensions will mount since fossil fuels will become scarce and more expensive.

Our nation can no longer afford more "oil wars"; nor can it continue to deny that it needs to take a new path. As a nation, Americans must come up with a national energy policy that makes sense, as the entire country must move rapidly from the 2IR that dominated the 20th century to the GIR. This transition has already started in the Europe and Asia, and it may be the "new world order" of the 21st and 22nd centuries. The 2IR was dependent on fossil fuels, internal combustion engines along with massive infrastructures to support energy and transportation. The GIR is focused on using renewable energy to power "smart" local communities where onsite building-by-building renewable power and smart grids can monitor to conserve power and increase efficiencies.

Europe, Japan, and South Korea have been in the GIR for the last 2 decades. A large-scale effort is now underway in China. A recent report by the international think tank the Climate Group finds that China is rapidly gaining in the race to become the leader in development of energy technologies. America definitely has some catching up to do. The sooner it starts the faster it can achieve the inherent benefits of a sustainable and localized-energy-generated lifestyle that focuses on sustainable communities while creating new companies, careers, and areas for employment.

In the 19th century, the United States started to be the leader in the 2IR. By the end of the 20th century, America was the world leader in innovation and entrepreneurship, so that by the new millennium (21st century) it was creating historic advances in computerization and information technology. Now that distinction as innovator and entrepreneurial dynamo is challenged as the world seeks leadership in the battle to stop global warming and reverse climate change.

Germany is now the number one producer and installer of solar panels for homes, offices, and large open areas. Japan is now leading the world in auto manufacturing since it began making vehicles that are not damaging the environment and atmosphere. Other nations in the European Union (EU), such as the Nordic countries and Spain, have been aggressively implementing policies and programs to become energy independent in 4 decades. And they are succeeding. See Appendix A for Denmark's accomplishments. However, unlike other EU nations, the Danish government is focused both on national policy and plans and local distributed systems as they move ahead to implement the GIR.

HOW COMMUNITIES AND NATIONS MOVE AHEAD

The place to start is to recognize that there is confusion in the national dialog as the nation waves good-bye to 2IR and cast its eyes and focus toward GIR.

Why? The confusion is exacerbated by the American media with its digital communication systems that are ubiquitous and instantaneous, yet are shallow and politically biased when dealing with significant issues. The public is besieged by the latest "buzz words" and concepts like sustainability, renewable energy, green jobs and careers along with energy efficiency, conservation, greenhouse gases, global warming, and climate change. All these words are without definitions.

Basically, as Clark and Fast (2008) argued in their book *Qualitative Economics*, there must be definitions of concepts and ideas such as "clean." For example, there is a qualitative and quantitative difference between "clean" and "green." By 2010 with the success of Al Gore's film "An Inconvenient Truth" making the public and policy makers aware of the problem of global warming, too many concepts were "green washed" and passed off as something they are not. The terms get tossed back and forth by scientists and politicians so that everyone thinks they know what they mean, until they try and use them in a sentence and then the conversation quickly becomes as painful as that of the 2007 South Carolina's Miss Teen Contestant YouTube video incident.

Even *The Economist* admitted that "modern economic theory" (July 2009) had failed along these lines. As they put it, "economics is not a science." To help sort this out, Clark helped to create the field of "qualitative economics" (2008) to make distinctions between words, concepts, and even numbers that are often misused (see Appendix B for an example of how audited data was misused by ENRON during the California energy crisis). The issue is that numbers, words, and ideas are all too often not defined or even discussed. The public and decision makers just use them. So do companies and lobbyists to whom "clean" energy means the use of energy and fuels such as natural gas and diesel. These are fossil fuels and emanate gases and particulates that pollute the environment. These chemical wastes cause massive health and environmental problems. "Green" on the other hand, in the context of energy and fuel, means renewable energy from natural resources like wind, sun, geothermal, ocean, and tidal waves as well as the flow of water in rivers.

Whether America is ready or not, GIR is at its doorstep, now. The huge amounts of federal stimulus money in 2009—about $250 billion earmarked for energy conservation and renewable generation—coupled with crashing local government budgets (particularly in California and New York) will propel Americans to look in the direction of energy independence, sustainable activities and communities. In the small town of Benicia, California (population of 30,000), the city's $2 million annual energy bill represents about 5% of the budget. Eliminating that expense would allow the city to beef up safety personnel and community services, or give the city a buffer for the leaner days ahead. Unfortunately, most of the federal stimulus funds for energy are focused only on efficiency and conservation.

While a start, renewable power generation is the core need for GIR. Again, will America fall further behind? There is that distinct possibility as the EU,

Japan, and now China become ever more aggressive in renewable energy generation and technologies.

Energy independence and subsequent elimination of energy bills are part of the potential benefits waiting as we transition into GIR. As soon as possible, America needs to give up freebasing fossil fuels and embrace a healthier community with intelligent development and greater community connectivity. What is crucial is that Americans, starting in local communities, must see the vision and take action. Almost every community has the renewable resources to make itself energy independent and carbon neutral. The United States must get started. Americans must come to an understanding and develop a national energy policy, then get out of the way and let America's historic innovation and entrepreneurship take over and "leapfrog" to what other nations have done and are doing. Clark notes some of this in his study for Asian Development on Inner Mongolia in China (2007), which has been published globally in the *Utility Policy Journal* (2010).

Clark and Bradshaw (2004), in their pioneering book on the future of energy policy due to the "global lessons from the California energy crisis," *Agile Energy Systems*, concluded by noting that the "new localized energy (read: distributed energy systems) market place will redefine how integrated resource management (read: renewable energy power generation and storage that is combined or integrated into "smart grids") is implemented in a public market (read: regulations and standards must exist and be adhered to) where private companies can compete in a socially responsible manner (read: basic infrastructures like energy, water, waste, transportation etc. must be provided for everyone)" (2004, p. 459).

WHAT IS A RENEWABLE ENERGY POWER SOURCE?

"Renewable" energy generation is part of being sustainable, one of those terms that everyone thinks they understand until forced to use it in conversation. Basically, it's a source of energy that is not carbon based and would not diminish—that is, it's the "gift that keeps on giving." For example, the sun is always shinning and the wind blows fairly consistently. Each needs some form of storage or feedback when the wind is not blowing or at nighttime when there is no sunshine. That is why these forms of energy generation are called "intermittent" and need technologies to provide for base load (round-the-clock power availability) energy generation.

The ocean is always there with tides and water. The most common renewable energy sources are systems that make use of the wind, the sun, water, or a digestive process that changes waste into biomass and waste recycling for fuel generation. Other renewable sources include geothermal, "run of the river streams," and now increasingly, bacteria and algae.

Wind generation is fairly straightforward and has been used as a power source for hundreds of years. A large propeller is placed in the path of the wind, the force of the wind turns it, a gear coupling interacts with a turbine, and electricity is generated and captured. While ancient in form, there have been significant technological advances. The new generation of wind turbines is stronger, more efficient, quieter, and less expensive. Today, wind turbines are being installed in small communities and even smaller systems on rooftops as part of the natural flow of air over buildings.

Solar generation systems capture sunlight including ultraviolet radiation via solar cells (silicon). This process of passing sunlight through silicon creates a chemical reaction that generates a small amount of electricity. This process is described as a photovoltaic or PV reaction and is at the core of solar panel systems. A second process uses sunlight to heat liquid (oil or water), which is then converted to electricity. A number of communities are now looking into more and more solar "concentrated" systems where the sun is captured in heat tubes and used for heating and cooling. This is a great renewable technology for use in water systems and buildings that have swimming pools.

Biomass is a remarkable chemical process that converts plant sugars (like corn) into gases (ethanol or methane), which are then burned or used to generate electricity. The process is referred to as "digestive" and it's not unlike an animal's digestive system. The ever-appealing feature of this process is that it can use abundant and seemingly unusable plant debris, such as rye grass, wood chips, weeds, grape sludge, almond hulls, etc.

Geothermal is power extracted from heat stored in the earth, which originates from the formation of the planet, from radioactive decay of minerals, and from solar energy absorbed at the surface. It has been used for space heating and bathing since ancient Roman times, but is now better known for generating electricity. Worldwide, geothermal plants have the capacity to generate about 10 GW as of 2007, and in practice generates 0.3% of global electricity demand. In the last few years, engineers have developed several remarkable devices called geothermal heat pumps, ground source heat pumps, and geoexchangers that gather ground heat to provide heating for buildings in cold climates. Through a similar process they can use ground sources for building cooling in hot climates. More and more communities with concentrations of buildings, like colleges, government centers, and shopping malls are turning to geothermal systems.

Ocean and tidal waves generate power, the use of which was been pioneered by the French and the Irish with their revolutionary SeaGen tidal power system. The French have been generating power from the tides since 1966, and now Électricité de France has announced a large commercial-scale tidal power system that will be enough to generate 10 MW per year. America, particularly the Pacific coastline, is equally capable of producing massive amounts of energy with the right technology. Ocean power technologies vary, but the primary types are: **wave power** conversion devices, which bob up and down with passing swells; **tidal power** devices, which use strong tidal variations to produce power; **ocean current** devices, which look like wind turbines and are placed below the water surface to take advantage of the power of ocean currents; and **ocean thermal energy conversion devices**, which extract energy from the differences in temperature between the ocean's shallow and deep waters.

Bacterial, or microbial fuel cell energy generation, sounds too far out there, but Better Products (or British Petroleum) made a $500 million investment in the process, which is now being developed by researchers at the University of California, Berkeley, and the University of Illinois, Urbana. The process uses living, nonhazardous-use microbial fuel cells bacteria to generate electricity. The researcher envisions small household power generators that look like aquariums but are filled with water and microscopic bacteria instead of fish. When the bacteria inside are fed, the power generator—referred to as a "biogenerator"—would produce electricity.

While all these power generation systems result in electricity, none are as cheap as current fossil fuels such as coal, oil, and natural gas. Nor were fossil fuels cheap when they started in the late 1890s, forming the basis for the 2IR. To maximize renewable power efficiency, renewables need to be integrated as linked or bundled supply sources according to the natural physical characteristics of the area where they exist. Further, these intermittent power generation resources are greatly enhanced with storage devices since the sun is not always shining (especially at night) or the wind blowing.

Thus there is a need for storage devices either natural, like a salt formation, or artificial, like a battery—new advance batteries and fuel cell programs are now coming out in California and through the US Department of Energy. Once you collect the electricity, the storage device lets you regulate the distribution so that you can optimize the process. The government support for the 2IR in terms of tax incentives, funds, and even land must be repeated for the GIR. The incentives for the 2IR must be reduced and applied for the GIR. This is called "tax shifting" and has been very successful in other areas so that there is little or no additional tax upon the consumers. There is no need for further debate or delay.

As people become familiar with the concepts and the fact that renewable energy technology is not as cheap today as fossil-fuel systems, Americans will begin to understand the economic move and change for local renewable energy generation and distribution of power (see Appendix C) from central grid power plants. The industrial revolution that developed central grid power plants was significant at the time for the coordination and costs for generating power supplies to communities. Nonetheless, this 2IR meant that the price of fossil fuels for power plants was reduced over time. At least 3 or 4 decades were needed to achieve that goal. There is substantial evidence and the series of laws at the turn of the 20th century to document how central power plants along with fuel supplies became economic monopolies that then controlled the fuel (primarily turning to fossil fuels such as coal, oil, and gas) supplies and hence large manufacturing and industrial markets. Despite litigation over the next 4 to 5 decades, these large fuel suppliers and power generators remain the dominant global economic business organizations.

While some fossil fuels like coal are still cheap today, they are the major American and global atmospheric polluters. If the human and environmental impacts of coal were calculated into its costs, then the real cost of coal energy generation for power would soar. The GIR needs the same sort of economic, tax, and funding support or incentives that the 2IR received over a century ago. This is a key action point for all American communities, regions, and states. And these financial actions will enhance and reduce global warming as a result.

The result of 2IR was the creation, operation, and maintenance of big centralized fossil-based power plants as Appendix C illustrates. They had to be powerful to withstand the degradations over the vast distribution of a central-powered grid system. At each conversion from AC to DC, the electricity loses some of its capacity. However, there is so much of it at the beginning that it does not matter several thousand miles away at the end. This results in the loss of efficiency in transmission over power lines as well as the constant need for repairs and upgrades.

Not so in the case of environmentally sound renewable systems. For best results, they need local renewable power generation and distribution systems, "smart local and onsite grids" that do not travel far and do not lose any of the electricity to inefficiencies. The other way to do it is to hook into a transmission line. This way it's additive to existing energy distribution or so that the transmission line acts as a "battery" for the renewable energy that needs storage. Some have equated this to a model of the Internet where there is no one area for control over data (or in this case power) but it is spread out and localized.

Energy independence will not happen tomorrow, just like the SUV and the carbon-intensive economy did not become social and political realities overnight. America spent a trillion dollars on the Iraq war, and it will probably cost at least that much to turn America into the leader of the GIR. However, national survival and international political leadership are compelling us to quickly surpass what has begun in parts of Europe, Japan, and China.

Fortunately, some in America are taking the first step. Consider California, where in the early part of the 21st century the world's largest energy efficiency program was implemented. The state is taxing the utilities' ratepayers and pushing that money back into making business and facilities more efficient. California is putting about $3 billion into the 2010–12 energy efficiency cycle with energy savings targets for the years 2012–20 of over 4500 MW, the equivalent of nine major power plants.

New York City, which is struggling to hold onto its leadership in the financial world, is facing severe capacity issues, particularly in Manhattan. Taking a page from California, New York has embarked on a similar state policy–directed energy efficiency effort. Other states like Pennsylvania, New Jersey, Illinois, and Missouri are coming along. But the heavy coal-burning states in the mid-West are in denial mode and refuse to give up burning coal, probably since the rancid and toxic residue is blowing east and not spoiling their own environments.

While energy efficiency is a first and important step, complete energy independence is within our technological grasp. A third generation of renewable technologies is coming and it is much better—lighter, thinner, stronger, and cheaper. Wind and solar power, coupled with highly efficient storage devices, smart grids, and local distribution systems are coming together (see Appendix D for a graphic example). These independent power systems need to be integrated. What is lacking is the large national financing and political leadership to make the commitment and push America past the threshold into the GIR. Sustainable development as a key component in the GIR, like its predecessor, depends on this leadership and financial support.

DEREGULATION BENEFITS: MYTHS ABOUT ECONOMIC EFFICIENCY AS CONSERVATION IS NEEDED TOO

The key issue for the economics of utility deregulation is that there be opportunities for market response that will increase production efficiency and reduce prices to consumers—in short, keep the neoclassical balance between production and demand. There are many economic arguments that became articles of faith supporting the premise that benefits would follow from deregulation and more competition in electricity. For the most part economists did not follow their own early warnings that it would be complex and difficult to actually deregulate the electricity sector. Instead, over the subsequent years economists started to believe that they could actually deregulate and that benefits would follow.

The following list of benefits are derived from Joskow's (2000, p. 119–24) examination of deregulation in the late 1990s in California and other states, though the article was written before the economic collapse in California. Joskow notes that in spite of general efficient operation of the electrical system and the comparative success of American efforts compared to those in most other countries, the pressures for deregulation increase. There are several reasons, with our analysis of each one.

Better Investment Decisions

First, deregulation in the long run will provide better guidance in deciding what generation plants to build and will exert pressures to control construction costs. The largest costs in the utility industry are in capital expenditures, and this fixed long-term investment is around half of the cost of generation of power. However, the cost of generation even with the same fuel and size of plant varies widely across regions of the nation and world, and the risk of bad decisions needs to be placed on the suppliers of these generation services. For economists this is a fundamental variable in all industries, where efficient capital investment provides competitive advantages. Can competition induce better investment decisions?

On the positive side, accountability for mistakes under the traditional system is largely shifted to ratepayers with little recourse to investors. The example of Diablo Canyon nuclear plant is instructive. Cost overruns make the power too expensive to sell under normal expense recovery calculations, even though PG&E wanted to shift all the burden of cost to the ratepayers. The regulator, the CPUC, wanted the utility to absorb most of the burden. Finally a compromise was reached in which the power was sold to customers at a rate well above the average of other sources of power (12 cents/KWh), based on performance criteria.

If performance was below a set level the balance would be absorbed by the company. In practice, before deregulation, the plant performed reliably above the minimum, and costs were passed on to consumers. Under deregulation, presumably, cost overruns on plants like Diablo Canyon would not be competitive and would lead to huge financial losses to the utility.

From a conventional economic point of view, under regulation the public regulator protects the utility from some risk for the investment decisions that are made. The fact that the utility can pass on costs shifts some of the risk and allows for poor investment decisions. Under deregulation, this risk-buffering function will be removed, subjecting these decisions to competitive markets.

A counterargument is needed, however. California's experience has repeatedly shown how private companies would have made horrible decisions if it were not for the regulatory system to enforce restraint. If Diablo Canyon was a problem, one can only imagine the financial crisis of the utilities and the state if the 12 nuclear plants that were planned for construction were actually built. While the regulatory system is not capable of micromanaging or of assuring that all investment decisions are wise, the fact that there is a second opinion may lead on average to fewer mistakes.

Good decisions are benefits for which there should be a public benefit, not just private gain, especially for those low-cost sources of power that result from the collaboration between the public through the regulator and the private utility. The benefits of the shared risk between the utility and the guaranteed pass through of the costs to the customer has made some projects feasible and has assured low costs of capital because investors sense a guaranteed return on their investment. Indeed the whole utility industry is more of a public trust than is an industry such as oil refining. Many of the resources used in the generation of power are derived from public lands, especially hydro, and the power lines are typically acquired through the use of eminent domain proceedings to force access or sale of land; the fact that power cannot be easily stored means that the entire integrated system is critical.

Politicized Priorities Excluded

The economists point to the fact that regulation reflects politicized priorities that counter the best economic interests of efficiency and low costs. The economists do not like the fact that some of the decisions in the regulatory process reflect political priorities that turn out to have higher costs and add nonmarket factors into production decisions. This includes for California the contracts with QF producers that were priorities of the regulators and the QF industry, not the regulated utilities. The utilities would have preferred to remain in control of all of their generation, but PURPA opened up the supply to other providers, and this political decision led to inefficiencies. Contracts based on faulty assumptions led to problems, but also to system advantages.

Clearly, the fact that regulatory commissions are public bodies means that they are under pressure from the public to reflect political priorities rather than just private priorities. From a regulatory perspective, one can see the danger that too many public good priorities for the power sector can divert from its primary mission and cause costly dysfunction. For example, in India, subsidies to farmers (virtually free power) was a political promise one political party offered to gain support for public investment in rural electrification, and now the farmers getting the subsidy are such a large political force that any reform of the power system is politically impossible. The goal of helping farmers by policies that affect the electricity industry was a shortsighted solution to their real needs.

However, most regulatory bodies do not have as much of an overtly political aim that would undermine the electricity industry as they did in India. The examples of regulatory involvement generally have to do with natural resource issues such as air pollution, water management, nuclear waste, and other sitting issues, or with equity issues such as assuring low rates for poor customers, reliable service in rural areas, and safety. In addition, regulators may be concerned with economic development and the use of power for public purposes such as street lighting and streetcar service in urban areas. In developed countries it is true that regulators have political agendas for shaping the efficiency and cost of power, and to some extent this is true in all countries.

Critics of the high price of power in California have been unhappy with the decision of California following PURPA to open generation to small qualified producers by offering long-term standard offer contracts to facilitate new suppliers' entry into the power market. Without long-term contracts, QF producers could not get the long-term financing they required to enter the market. The opening of the closed vertically integrated utility system to other producers was a first break in the utility monopoly, and the utilities were distrustful. Although the QF contracts were supposed to be neutral to consumer and utility, all parties were involved in miscalculating fuel costs, which led to overly expensive long-term contracts with the QF providers.

The QF standard offers were negotiated between the utilities and the independent producers, and the rates were based on data provided by the utilities, presumably the same data on which they would make internal facility sitting decisions. In addition, none of the parties anticipated that there would be much interest by QF providers in the standard offer contracts. While all parties may have been less than fully honest and cooperative, the process was designed with only one premise—that the consumer would be indifferent to the fact that power was generated by QFs and that this rate would be set to match the utilities' avoided cost (Summerton and Bradshaw, 1991).

These contracts were not just created in a political process, they were based on utility provided numbers reflecting what the utility would have spent if it had gone ahead and built another plant—in this case a gas-fired steam plant that was the best technology at the time. Parenthetically, the problem with QF contracts came with technologies where there were no direct comparable fuel costs, such as wind, and so the contract negotiators took estimates for future oil and gas costs as a basis—but they made a mistake and calculated these costs way too high.

The real issue was not with the concept of avoided cost. Another problem came in that neither the regulators nor utilities accurately estimated the amount of interest by independent producers, which turned out to be much more than expected. No cap or recalculation was included in the rate design to be instituted after a certain amount of power was contracted.

The problem from an economist's point of view is that the system is regulated to achieve political goals, and the inclusion of these political goals into the rate system leads to inefficiencies that the market is supposed to correct. There are two sides to this argument. First, the regulated system is not perfect, and as Joskow points out, it can do better without being thrown out. Mistakes are made in the regulatory process, and neither party built in adequate protection in case the policy did not work out as well as desired. Two redundant voices in planning seem to be better than one, and the benefits of the public input are not balanced against the losses.

The other side of this issue is that in the interest of the public good many of the values the economists define as political interference need to be included by regulation or else they would not be included at all. For example, increasing reliance on renewable sources of energy is a social value that would probably not be realized under pure market forces. Yet the development of adequate renewable industry is essential for reducing greenhouse gases, for promoting the technological innovation in solar and wind that will arguably be the technologies for the future. The capital-intensive nature of the electricity industry means that long-term industrial values need to be introduced by regulation rather than the market, and it is good for the whole industry in the long run to enforce change. The economists miss the value of the latter.

THE ISSUE OF DEREGULATING AGILE ENERGY SYSTEMS

The deregulation structure and mechanisms in California were designed by generally well-meaning people. Most of the government officials and industry leaders who made decisions leading to the California crisis were not personally incompetent, malicious, or corrupt, but they simply designed a system that assumed idealistic market mechanisms based upon one set of economic theories. The basic problem with these theories is that they ignored reality and hence were so flawed that it led to a $40 billion catastrophe and the recall of the California governor. The deregulated system left open market opportunities for corporate officials and their companies, who were mostly out-of-state energy market players. California, as it has historically, beckoned to these companies and allowed them to engage in illegal and unethical behaviors. Without regulations and by manipulating a system that was more complex than the ability of those in charge to understand, monitor, control, or even dampen the extent and nature of the problems that would emerge, energy companies could make quick profits.

The deregulation plan more or less seemed reasonable at the time it was designed. The last piece of legislation, for example, needed to enact deregulation was approved without one negative vote. Bipartisan support was overwhelming, yet naïve. Woo (2001, p. 753) notes that this apparent reasonable quality among all the decision makers was fairly persuasive:

> *A cursory review may lead one to believe that the California market reform should be able to provide reliable electricity at low and stable prices. After all, the PX-CAISO market design marries the economic model of market consumption and the engineering model of optimal dispatch. Competitive bidding among PX buyers and sellers should theoretically yield an efficient allocation of electrical energy. Competitive auctions of ancillary services should enable the CAISO to perform least cost dispatch. Zonal pricing of transmission should efficiently allocate the limited transmission capacity. Surging zonal MCP and the associated excess profits should attract new generation at the locations with dwindling supply or rising demand. But the reality of the events that have occurred since Summer 2000 obliterates such a belief.*

The plan that was adopted in California had many flaws and pointing them out is easy picking for a critic. In fact, since late 2001, most experts and economists now refer to the deregulation public policy as "restructuring." Most analysts agree that aspects of the deregulation scheme have proven faulty and should not be repeated by any future deregulation scheme. However, it is far more compelling to question the very premise on which deregulation was proposed—the premise of competition. In the words of Governor Pete Wilson, who pushed the policy and claimed at its signing that the policy would combat the 40% extra that consumers in California were paying every time they flicked on the light switch, competition and dismantling the electricity monopoly would "guarantee lower rates, provide customer choice and offer reliable service, so no one is literally left in the dark."

Of course, the public now knows more (but will probably never know all) about how brokers, traders, and generators used illegal and unconscionable gaming of the California electric system with such well-named schemes such as "Death Star, Get Shorty, Fat Boy, and Ricochet" to create artificial scarcity and to spike prices. For example, it is now known how reluctant the FERC was to respond to price crises in California for largely political reasons. However, most of the people initially responsible for making the decisions to deregulate believed the economic analysis and assumed these illegal schemes could not happen.

At the time of deregulation decisions, many analysts were concerned with the deregulation proposal and expressed concern that an unregulated market would not invest in enough generation capacity, or that competition would not be strong enough to control market power (Clark and Lund, 2001; Clark and Morris, 2002; Clark, 2002). Energy regulators were very familiar with schemes to manipulate the market, and they understood that the task of regulation is to control expected industry efforts to make more profit in monopoly industries than is fair and reasonable.

The problem was that the voices of caution and warning were not heeded, and the overwhelming enthusiasm for the deregulation experiment got out of control. When the prices for power were lower than expected for the first 2 years after the plan went into effect and stranded costs were being paid off at a faster rate than expected, the enthusiasts said "not to worry, we told you so." Eventually, however, controls on excessive prices simply did not work because gaming was perfected in a tight market and events happened that were not anticipated, though in hindsight they should have been expected.

There is much that cannot be properly known and charted in advance when public interests intersect with complex large-scale technical systems such as the electricity grid. In spite of the vast amount of information that was available and that helped to run the state power system for several years, the planners forgot that large electricity grids are so complex that it is impossible to know enough in advance to avoid mistakes. It is expected that power providers have backup plans if a part of their system fails; policy makers had no backup plan in case deregulation did not work as expected.

The purpose of this chapter is to explore how power system information was used and misused. More importantly, complex systems have a language and organization all their own. The electrical grid is no exception, and as subsequent events since 2000 have well documented, public policy makers were sold a "bill of goods" that promised lower prices and efficiency from economic "mythical models" based on ideal types, situations, and market forces. Senator Dunn (2002, p. 8), who later headed the California senate investigation of the energy crisis, tells what he thinks happened:

> *Everyone in Sacramento was driven by just one thing, across political lines. They were driven by the desire to deliver to their constituents lower prices; that's it. From that point on there was little additional analysis other than for a handful of folks like Steve Peace. Ninety-eight percent of the legislators just said: "Folks are telling me it's going to get lower prices"...*

The subsequent vulnerability of energy system decision makers to unprecedented misinterpretation and market manipulation cascaded into the energy crisis in California and subsequently into other regions of the world.

Rapid swings due to market forces are important to examine in energy availability and how price caused large effects and how undeterminable consequential forces led to surprising outcomes of vulnerability, reliability, and price increases. The rapidly changing political agendas aggravated the conditions that they were supposed to stabilize into competitive markets. The misinformation, the poor timing in which public policy was made, and the market control exercised by private sector firms led to the rapid escalation of chaos and complex systems out of control. For example, the rapid increase and fluctuation in prices was nowhere near linear to the reduction in supply of energy. Furthermore, even when demand decreased, prices did not come down to previous levels. The political process necessary to control a complexity crisis in uncharted territory is different from what is needed to manage a system that is well understood.

In order to not lose sight of the scale of the problem in the electricity system, it is essential not to forget that the California price/cost problems were not experienced for the first time in the 2000–01 energy crisis. Under the historical public structured energy markets, the unknown had led to previous policy failures:

- The consensus was that nuclear power would be inexpensive, ultimately too cheap to even meter. The mistaken enthusiasm in the 1960–70s over nuclear power led utilities to build some of the most expensive plants instead of the cheapest.
- Utility forecasts for power demand during the 1970s expected demand to increase, whereas per capita demand fell instead of increased. As already noted, California avoided problems of overconstruction of plants through a timely review of the methodology for demand projection.
- Prices for power paid to qualified facilities and independent producers under PURPA were appropriately linked to oil and gas prices, and projections of these prices were for rising prices and limited availability. Instead, prices fell and gas supplies were abundant during most of the 1980–90s. Most of these contracts ended up being priced way too high because they were based on faulty price assumptions with no correction mechanism.
- Restructuring proponents assumed that spot market prices would remain low and that they would provide the foundation for lowering prices following deregulation. Instead, they became unstable.

In this chapter, complexity theory in public policy is discussed in order to offer a theoretical perspective on the information availability, influence, and problems for practical interpretation that affect the management of large technical systems. Then we give some examples of how these misinterpretations, miscalculations, and manipulation have plagued the electrical power system in California and elsewhere. In conclusion, suggestions are given for managing large complex technical systems under conditions of uncertainty and intense public interest stake in avoiding crises.

CONCLUSION: ECONOMIC DEVELOPMENT IN AN AGILE ENERGY SYSTEM

One of the most significant studies of the impacts of moving to an agile energy system that rests on efficiency, conservation, and renewable power assumptions and standards is a series of simulations. Comparing a business-as-usual model with a "clean energy scenario," Geller shows that a series of 10 clean energy strategies would in combination produce a decline in total energy consumption in the United States starting in 2010.

In addition such renewable energy assumptions would reduce the national energy intensity from 7.4 kBtu per dollar (1999 value) based on business as usual to 5.5 kBtu per dollar with these and other clean energy strategies implemented. These policies would cost by 2020 $674 billion in the United States, while saving a total of $1229 billion over the same period. Hence, a net gain of $554 billion would be achieved in addition to reducing American dependency on international oil and gas supplies. In other words, the investment would not only pay for itself but would produce a gain of 82%. These savings would keep on growing into the future.

Such savings would reduce per-household costs for energy from all sources (household, industrial, transportation, etc.) from about $5500 in 2000 to $3800 in 2020 under the clean energy scenario, compared to an increase per household to about $6250 in annual energy costs with the business-as-usual scenario. In short, the average household would reduce their energy bill by $2400, or 40%.

Jobs under the clean energy scenario would increase rather than decrease, multiplying the economic value of the strategies. While not a part of the Geller study, one estimate is that the clean energy strategy would create 870,000 jobs. This is a clear example of the puzzling technology implementation question—if people could gain so much and it would not cost them anything, why are we not doing all these good projects that would also clean the environment and create jobs?

We have argued here that public policy directed toward advanced, renewable technologies with agile energy systems, which are disbursed and distributed energy generation on the regional and local level, mean the creation of local businesses and jobs. The evidence is mounting in California that such is the case with the enactment of measures to stem the California energy crisis, legislation to set new standards, codes, and protocols. Above all, sunset government programs are needed to stimulate initially the clean energy sector while other programs limit the growth of fossil energy power generation.

What California has learned and its citizens embraced is the basic concept of "social capitalism": clean environment and renewable energy can be profitable. The citizens of the state do not want to see more oil wells off their coast or in their parks. Nor do they want to import fossil energy fuels from other states, which must violate their environments as well. Instead, the citizens of California have embraced what the Europeans call "social constructions" whereby they believe that both can happen—a clean environment and profitable businesses.

The pathway has been set with government and the public sector taking the lead in "greening public buildings" as well as procurement of "environmentally sound technologies." The private sector has rapidly taken the lead and appears poised to implement "civic capitalism." A win-win scenario is not far behind for all Californians.

APPENDIX A: THE DANISH CASE FOR THE GREEN INDUSTRIAL REVOLUTION

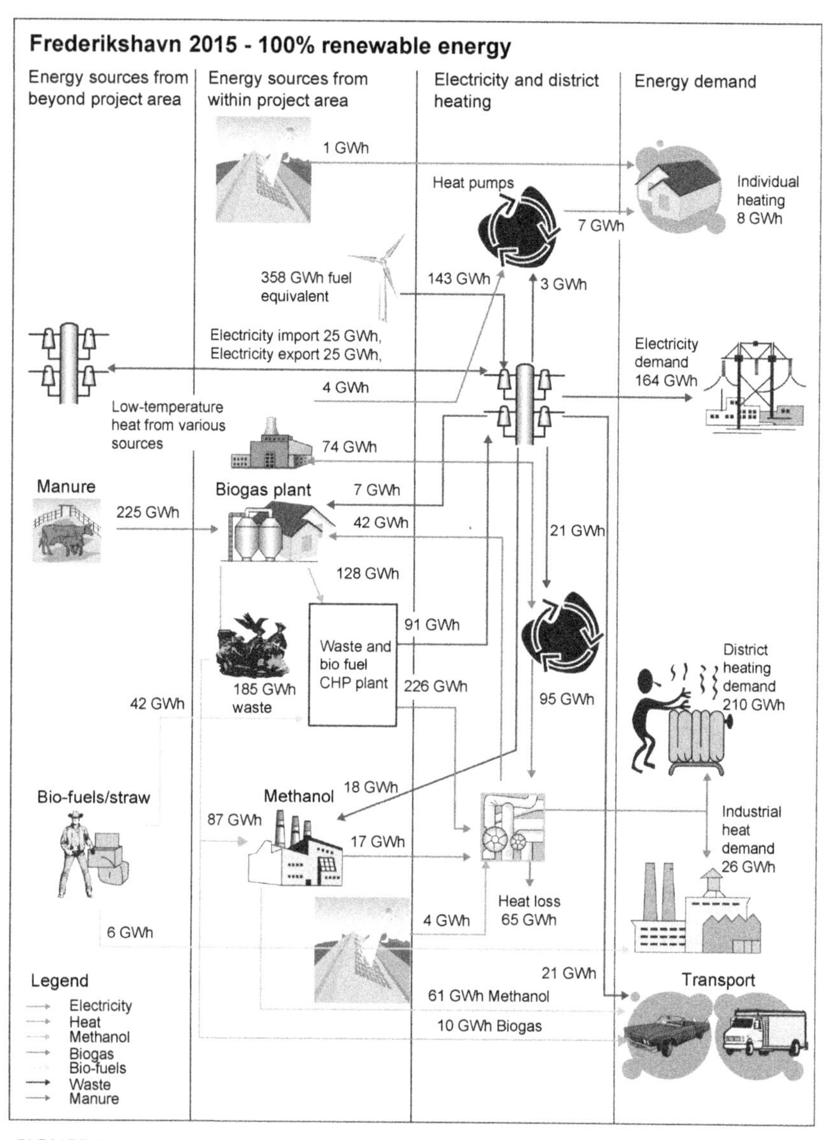

FIGURE 2.1 Flowchart of the proposed energy system in Frederikshavn, Denmark, in 2015.

APPENDIX B: E-MAIL SURVEILLANCE DATA ON ENRON

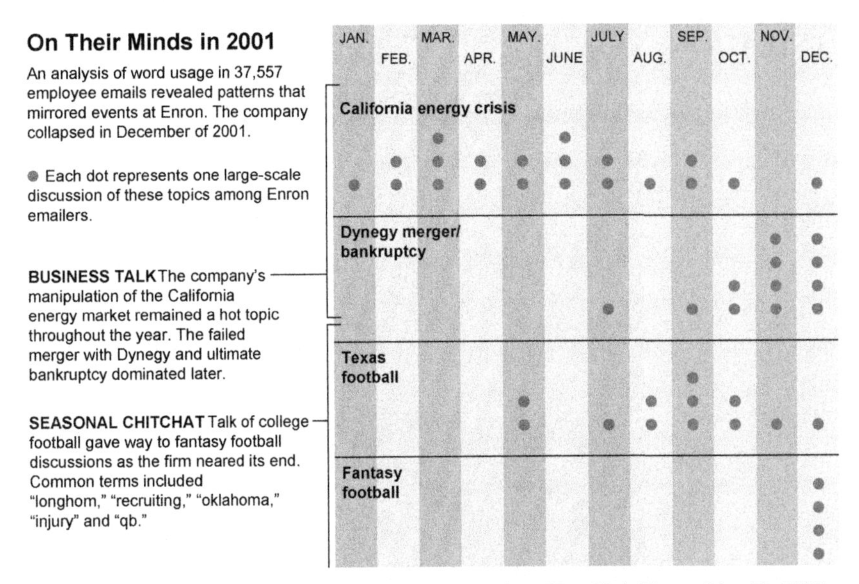

FIGURE 2.2 Email surveillance data on ENRON. *Kolata,* New York Times *(May 22, 2005).*

APPENDIX C: FROM CENTRAL POWER GRID TO LOCAL DISTRIBUTED POWER SYSTEMS

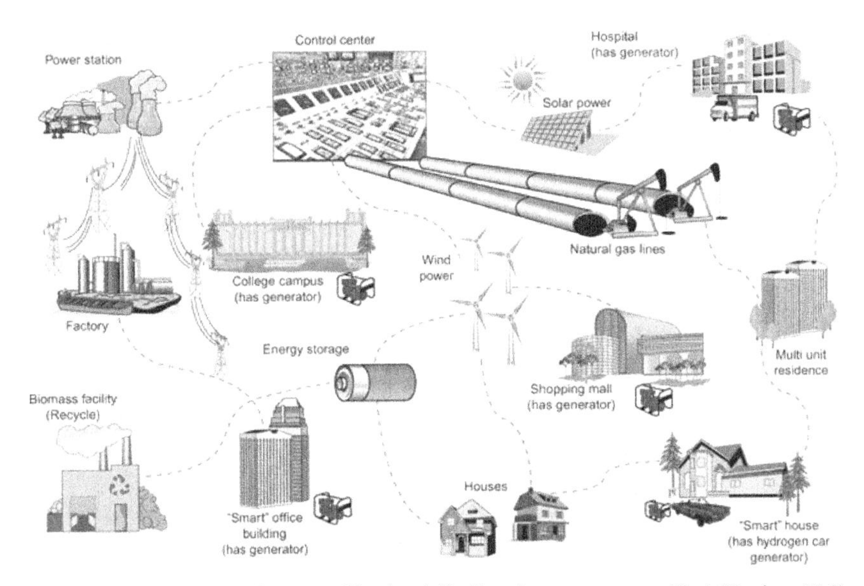

FIGURE 2.3 From central power grid to local distributed power systems. *Clark Woodrow W. II and Larry Eisenberg, "Agile sustainable communities: On-site renewable energy generation, in Journal of Utility Policies" 16(4) 2008: pp. 262–274.*

APPENDIX D: SMART GRID THAT INCLUDES LOCAL DISTRIBUTED POWER AND RENEWABLE ENERGY GENERATION

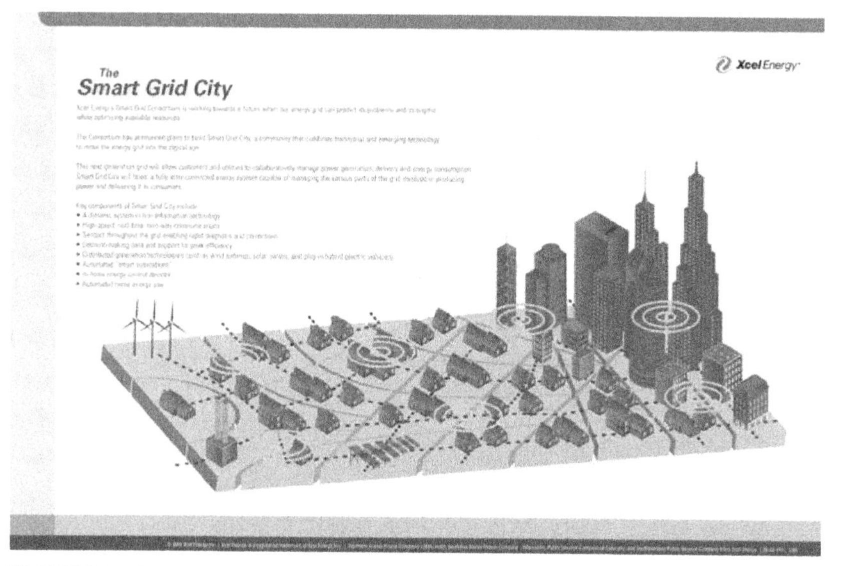

FIGURE 2.4 Smart grid that includes local distributed power and renewable energy generation. *Xcel Corporation (2009).*

REFERENCES

Clark II, W.W., Bradshaw, T., October 2004. Agile Energy Systems: Global Lessons from the California Energy Crisis. Elsevier Press, London, UK.

Clark II, W.W., Fast, M., 2008. Qualitative Economics: Toward a Science of Economics. Coxmoor Press, London, UK.

Clark II, W.W., Isherwood, W. (Eds.), 2010. Utility Policy Journal. Energy Instructions in the West: Lessons Learned for Inner Mongolia, PRC. (Asian Development Bank, 2007 Report). Elsevier Press.

The Economist, Special Issue and Cover on Modern Economic Theory Failing, July 17, 2009.

Hubbert, M., King, Chief Consultant (General Geology), June 1956. Exploration and production research division, nuclear energy and the fossil fuels. Publication Number 95. In: Presented at Spring Meeting of the Southern District, American Petroleum Institute, San Antonio, Texas, March 7-8-9, 1956. Shell Development Company, Houston, Texas.

Clark II, W.W., Lund, H., December 2001. Civic markets in the California energy crisis. International Journal of Global Energy Issues 16 (4), 328–344. Inderscience, UK.

Clark, W.W., Morris, Gregg, August 2002. Policy Making and Implementation Process: The Case of Intermittent Power. International Energy Electrical Engineers (IEEE). http://grouper.ieee.org/groups/scc21/1547/index.html.

Clark II, W.W., 2002. The California challenge: energy and environmental consequences for public utilities. Utilities Policy 5 (28), 587–645. Elsevier, UK.

FURTHER READING

Clark II, W.W., 2009. Sustainable Communities. Springer Press, New York, NY.
Rifkin, J., 2004. The European Dream. Tarcher/Penguin, New York, NY.

Chapter 3

Cross-Disciplinary Scientific Foundation for Sustainability: Qualitative Economics

Michael Fast[1], Woodrow W. Clark, II[2]
[1]*Aalborg University, Aalborg, Denmark;* [2]*Clark Strategic Partners, Beverly Hills, CA, United States*

Chapter Outline

We need to understand everyday interactions as "qualitative economics" as they have their roots in the historical, subjectivist, and philosophical tradition. As a starting point, we use the philosophy of science perspective and analysis. As the 21st century moves rapidly into a new economic era, there is a need to examine the ontological roots of economics as people, organizations, environment, and business to understand the local and global economies of

Sustainable Cities and Communities Design Handbook. https://doi.org/10.1016/B978-0-12-813964-6.00003-3

today and tomorrow. In fact, our forefathers in economics were themselves philosophers before the field became an academic discipline.

It is in this context that we name the perspective and the tradition from an old philosophical tradition "Lifeworld," which is a school of philosophy. Lifeworld comes from the German *die Lebenswelt*, with its roots in the 18th century philosophy of Kant, and later on Husserl, Heidegger, Schutz, and Gadamer. The theoretical development from this philosophical tradition is seen in different schools of contemporary social science thought ranging from phenomenology, hermeneutic, ethnomethodology, linguistics, and symbolic interactionism. For us, the Lifeworld tradition and its interactionism theoretical development is an approach to theorizing, hypothecating, describing, understanding, and explaining everyday life. In short, the Lifeworld and interactionism represent the elements of any natural, physical, or engineering science. This is discussed in the following paragraphs.

Overall, the Lifeworld school of thought is placed in the paradigmatic world of science as the "subjectivist" perspective. Lifeworld is the individual whose thoughts and daily interactions between people and larger groups are essential in understanding social reality, everyday life, and social construction with the world. Today in Europe, much of the discussion focuses on phenomenology and "social constructionism," which we view as part of the Lifeworld tradition. In the following discussion, we also apply more US-based theories (particularly interactionism and Chomsky's linguistics) in the tradition of sociology and philosophy to economics, thus creating the science of qualitative economics.

This discussion takes a different philosophical paradigm and applies it to the creation of the science of economics. We see the parallel in linguistics today with the need to understand the meaning behind ideas, words, sentences, and meanings by following the scientific method. Linguistics is remarkably similar to economics in its need to become a science. The field needs to define and explain the definition(s) behind numbers and statistics. Finally, linguists like any science, offers hypothesis testing along with universal rules and formalism in predicting future action and interactions.

The philosophical roots of the Lifeworld tradition are primarily European. Lifeworld can be traced to set the stage for the interactionism subjectivist theoretical perspective exemplified by Herbert Blumer, the 20th century American sociologist at the University of California, Berkeley. Blumer and his mentor George Herbert Mead, the early 20th century philosopher at the University of Chicago, are the American roots for a Lifeworld tradition.

We take the subjectivist approach as practiced in interactionism for describing and analyzing business. What is even more significant, however, is the application of linguists combined with symbolic interactionism and phenomenology. As a science, this perspective provides theoretical and analytical constructions that are appropriate to the science of economics. When such qualitative data are combined with quantitative economics, a comprehensive science emerges as based on a qualitative perspective.

To get the debate focused, the Lifeworld tradition will be applied initially herein to business economics. And to add some perspective, purpose, and planning to it, the focus will be primarily upon the business of energy. This particular perspective reflects one of the authors' area of expertise and recent experiences and also a very contemporary issue that merges both economics and science—energy is viewed as science and engineering. Moreover, the business economics of the energy sector is placed in the context of global warming and climate change.

We will discuss how to understand the very concept of organizations and how organizations are constructed and developed. We need to have an understanding of what people are and what they bring to the organizational economic context by interacting with one another and in groups.

We will also discuss how to understand organizations from a methodological viewpoint. It is no mere coincidence that qualitative methods have been developed within a Lifeworld tradition. Furthermore, we will show how to understand this subjective and scientific paradigm along with the theorizing within this tradition that transforms economics into a science along the lines of natural and physical sciences. The aim is therefore—through the everyday life sociological tradition—to discuss the central issues and basic concepts put forth by Herbert Blumer's symbolic interactionism, phenomenology, and Chomsky's linguistics to understand and develop a qualitative economic perspective.

Today there are many insightful criticisms of the objectivist paradigm or as it is better known, neoclassical economics, which is the mainstream theory in economics. We believe, however, that most of these critiques fall within the same philosophical tradition or objectivist paradigm of economics and hence are really "revisionist." An excellent example that has gained considerable attention is "Freakonomics: A Rogue Economist Explores the Hidden Side of Everything" (Levitt and Dubner, 2005). In the book and several lectures, the authors present qualitative insights but in quantitative methods. In short, they rework neoclassical economics around qualitative ideas. Although this is a start at being scientific, it lacks explanation and prediction hypotheses must be made, tested, and then retested with the purpose of seeking understanding.

This chapter presents an entirely different route to understanding economic and business phenomena. As Erik Reinert, the economic historian, put it there are three significant issues in addressing economic theory:

> *One: how economic growth is 'created'; two, the alternative mechanisms through which growth and welfare are 'diffused' between and within the nation-states, and to the individual; and three, how this alternative understanding is based on a different philosophical basis.*
>
> Reinert (1997, p. 9).

We follow a historical but yet a "new paradigm" today to economic theory. In other words, our approach to the science of economics is an entirely different paradigm. Our approach argues that the current conventional neoclassical paradigm including economic revisionism with its focus on perfect information in a balanced equilibrium system is the basis of economic theory today and is fundamentally wrong. Even some of the more popular books on the earth being flat, global or different ("Freakonomics"), are all based on the conventional neoclassical economic paradigm that is nonscientific.

Adam Smith, under the objectivist paradigm along with his followers to this day notes that the basis for business is barter and exchange. In fact, they are adamantly opposed to a Lifeworld view of economics. The objectivist's paradigm in a pre–mass production globalized world provides little guidance in terms of uneven economic growth and clearly substantive problems with understanding the needs of developing nations. Instead, it argues that "market forces" and business in general can innovate and hence meet new challenges. Yet that neoclassical economic perspective is part of the problem as it only considers supply and demand in quantifiable statistics.

For example, how does "climate change" enter into the conventional economic paradigm? It does not because climate change at the United Nations or the US, and the EU international levels require extensive government involvement, public policy, standards, regulations, and controls that can be set and monitored. Indeed, that is the message gaining prominence globally, as all but three nations (one of which was the United States) signed the Kyoto Accords. What is interesting about this issue is that the conventional neoclassical perspective would rather deny the science of global warming and ignore the economic needs for combating climate change. This has been the position of most economists such that their answer is that the "market forces" will correct the problem. The code concept for neoclassic economics is to label the global energy crises as a "perfect storm" (sic), which makes it ironic in the context of global warming and now the climate and weather changes experienced globally. Only since science has almost unanimously confirmed global warming, have economist sought other ideas.

THE PARADIGM SHIFT: ECONOMICS AS A SCIENCE

Business activities are a part of our everyday world, and the phenomenon of business presents itself in many ways, forms, and shapes. Business is a large part of our everyday life, and business is life itself, often helping to define who we are and what we do. We interact with people not only in daily conversations but also in just living, that is, going to work, driving on a highway, studying in school, attending a lecture or movie, and going on a holiday. Life for us as individuals and everyone around us is interaction with others in one form or another.

The Internet, email, as well as wireless and new technologies heighten our daily interactions as never before, now on a 24-h, 7-day-a-week, and global basis. Business and economics are interaction and "exchanges" of anything (information, knowledge, goods and services, or whatever) that define and provide meaning and understanding in our everyday lives. Almost everything people and groups do has some sort of business interaction attached to it.

The *reality* of business economics has been investigated and explained in many ways. And the *science of business economics* is, as it has "always" been, is central in the social discussion. However, understanding business research and the method of research along with the (ontological and epistemological) assumptions lying behind the research and its reality in everyday life is rarely discussed. Business economics, and its close cousin economics, are not sciences, but "art forms," as C.P. Snow (1959) would call it, disguised in the aura of scientism.

The problem with the dominant objectivist paradigm in the philosophical tradition of neoclassical economics in business and economics today stems from its historical roots. Many Western philosophers, theologians, scientists, and laymen have, more or less successfully, contributed to the discussion and understanding of reality and science. They have agreed or disagreed about thinking in various traditions about how the broad social sciences have been developed.

Some traditions have been dominant for some time and have then been replaced by others. Others have been rediscovered and developed. Yet the basic philosophy of business economics has continued to follow two particular traditions. Most of what we encounter in the scientific world today is an expression of a certain tradition or another that is not presented as one of many but as *the only one* giving evidence of reality. A particular tradition, or as Kuhn (1962) noted as a "paradigm," has a long history and to a great extent is dominating the social discussion and the organization of society, including science itself.

The dominant business economic tradition has come from the objectivist tradition and manifested itself as *positivism* and *rationalism* as contained in the prevailing schools of thought like structural functionalism, system theory, and game theory. As Reinert (1994, p. 80) puts it, "Neo-classical economics is essentially a theory of the exchange of goods already produced, taking no account of the diversity of conditions of production and their influence on pricing behavior. Neo-classical theory is, it seems, a theory which cannot accommodate for the existence of fixed costs, since these create increasing returns."

Discussions of philosophy in science and methodology are important for understanding reality and theorizing on its applications in everyday life. It is precisely these connections between ontology and epistemology in philosophy of science that theorizing and methodologies arise to capture the reality, which must be in the center of any scientific discussion. Furthermore, openness and a specific discussion of an alternative philosophical approach to the established traditional way of seeing science and reality are necessary. Thinking and reflection are critical in the scientific investigation of reality together with and related to the basic philosophical assumptions. It is only in this connection that we can talk about something being true (e.g., correct) or false.

There are economists and business academics who discuss philosophy of science as it relates to business economics, but only a rare minority make an impact on the research practice of social science and business or economics in particular. The problem with most methodological discussions and theoretical works is that they only discuss "choice" of methods in the context of methodological considerations and techniques of investigation and analysis.

It is difficult to find an explicit discussion on connections between philosophy/philosophy of science and methodology, especially focused on opposing philosophical traditions. In particular, the scholarly discussions lack epistemological considerations and make assumptions that underlie the very choice of those methods.

Another problem in traditional social scientific research is the lack of in-depth discussion of the background of the qualitative methods, especially of the Lifeworld ontology and everyday life epistemologies. The current conventional objectivism tradition has established positivist and rationalist theories in the functionalistic paradigm, which lack understanding about why the qualitative methods exist and for which epistemological grounds they are significant in understanding business actors, actions, and situations.

Furthermore, the objectivist fails to understand in which contexts these methods appear, and to which contexts they relate or underscore and support quantitative and statistical methods. Finally, the issue is rarely raised in business economics as to how do we incorporate qualitative theories and methods in the production of knowledge to which they can contribute in-depth understanding?

To a great extent, much of the problems exist because there is a basic lack of historical consciousness, debate, or concern over the progress of science in relation to social science (i.e., the theoretical discussions and background of science) and understanding of different traditions of philosophical thinking. Hence it must be recognized that the meta-theory of social science must be philosophy if it is to have any significance. Some scholars may say that the "quantitative" researchers are not especially conscious of or need to be directed to the history of science and ideas in their work. They argue that they do not need to think in terms of philosophical traditions, even though they are part of a particular tradition themselves and reflect its biases, beliefs, and assumptions (see Gadamer, 1993). Hence these scholars reflect historical traditions that presuppose and bias their results under the guise of "well-established" (usually) quantifiable "facts." Here we challenge that basic traditional assumption and argue instead for a new perspective and paradigm in the study of business economics based on an old tradition to understand, appreciate or criticize, and apply the results of business research to understand everyday business activities.

When the conventional paradigm continues to dominate both researchers and students alike, then their study of social science phenomena, in the objectivistic tradition, is "allowed" to prevail. In short, their perspective (rooted in this one particular paradigm) directs their work, influences the results, and skews the applications. To some extent, objectivism is now a restrained and an ambiguous concept, which according to many researchers covers not only different but also often conflicting relations.

In research there has, however, been a slow shift over the years toward a more "qualitative" approach. Empiricism as reflected in structuralism, logical positivism, or quantification of data has not shown that it can understand business and economics. And while we argue that more qualitative science must be a part of business economics, we fully acknowledge that the same concern to probe and ask basic scientific questions about perspectives of direction of work and skewed results must apply herein as well.

Today we see more "case studies"[1] (but with widely different approaches and perspectives) and qualitative investigations than we have experienced previously as graduate students and even in more recent years as professors. This social constructionism or the so-called softening partly originates from the *anomalies*, which Kuhn (1962) explains is part of the continuing development of science. Researchers must experience the problems and explanations in the existing frame of theory to start looking for alternative paradigms. However, we shall demonstrate how "qualitative" research or ethnography can still be associated with the positivist paradigm, and therefore be subject to the same basic problems as its quantitative equivalent. The key is to "map or link" the subjectivist tradition with the objectivist techniques (numerical, mathematics, etc.) when appropriate and useful.

Our approach has its roots (ontological and epistemological) in the Lifeworld tradition, with the theoretical perspectives reflected in the subjectivism paradigm. We will discuss a different philosophical tradition that presents an alternative to the objectivist tradition in social science and business economics. The Lifeworld tradition and subjectivist paradigm reflect the conception of science as reflected in linguistics, which is an established science directed to the everyday life reality of human interactionism.

This subjectivist paradigm is not an alternative philosophy of science, as it simply follows a different philosophical heritage than the current dominant paradigm of objectivism. The subjectivism paradigm deals with a different

1. Albeit primarily used as exploitative investigations, and often used before "the real investigation is started," i.e., a quantitative questionnaire/survey investigation.

strain of theory from the conventional objectivism philosophy of science in vogue today. Subjectivism discusses understanding of reality that combines everyday life in a sociological frame of reference wherein business economics is viewed as everyday business actions and interactions that can be seen, heard, recorded, and communicated. In short, this discussion represents a different philosophical perspective to understanding everyday business economics and business reality in a more scientific manner.

SUBJECTIVISM PARADIGM: LIFEWORLD PERSPECTIVE AND SYMBOLIC INTERACTIONISM

The *Lifeworld* tradition as reflected in subjectivism and everyday life perspectives is described by others as an *interpretive paradigm, symbolic interactionism, hermeneutic approach,* or a *qualitative perspective*. The tradition provides a different set of theories and methodologies within which we picture reality and our conception of science. In short, qualitative economics is a science in the philosophical tradition of the natural and physical sciences because it seeks to create understanding, define terms and assumptions, and use logic and empirical data to draw both conclusions and further verifiable hypotheses. This process, as seen in modern linguistics, is the basis of a scientific inquiry.

Our discussion deals with how everyday reality and economic science are understood. Besides discussing the philosophical assumptions and the history behind the tradition, we also discuss some of the traditions related to our perspective. They are primarily hermeneutics, philosophical hermeneutics, sociological phenomenology, and symbolic interactionism.

PHILOSOPHY OF SCIENCE

Mainstream philosophical traditions that contribute to the understanding of everyday business realities today are positivist and rationalist traditions, known as part of the objectivist paradigm. We purposely pay attention to philosophical arguments that are counter to the current economic paradigm prevalent in American and European business programs, schools, textbooks, and administrations. Curiously enough, however, these positivist and rationalist traditions are not underlying Asian and transitional economies. Nor are the predominant economic paradigms today used in practice in most businesses.

Instead, in Asia much of the economic and business traditions are based on a different paradigm and tradition. However, actual business practices globally are neither locked into a specific rigid model nor follow one set pattern of economic actions. In fact, it is this "disconnection" between the everyday of life of business activity that inspired us to seek different understandings of the situations that actors' experience in their daily business lives. We believe that our discussion hits directly at the heart or core of what Western industrialized capitalism has become today through an in-depth analysis of the philosophical and historical roots of science.

Our argument is that the objectivism paradigm in science became the dominant one with its theories and arguments for rationalism, empiricism, and ultimately quantification of social phenomena. In short, numbers without meaning and definition became economics. The objectivist paradigm has prevailed for over a century throughout other social sciences as well. The impact on business economics has led to a reductionist approach in theory and methods that dictates how economic and business analyses are conducted.

Another paradigm that is an alternative in philosophy is *subjectivism*. Through the Lifeworld tradition, this paradigm forms the basis for qualitative economics. Herein emanates the focus on Lifeworld, everyday life, symbolic interactionism, and the use of interpretative theories and qualitative methods that are missing today in business economics and business studies. Moreover, economics becomes a science when qualitative theories, data, and methods are used because actors, events, groups, surroundings, and their interactions can be constructed into hypotheses, which are tested, described, explained, and analyzed. The outcome can be laws and rules used for forecast and prediction. In short, economics becomes a science.

Phenomenology: The Tradition of the Lifeworld Perspective

Central in our discussion of a new perspective on economics is phenomenology. The connection of individuals in interaction is one of the cornerstones to understand the development of the firm. The present and past actions and interactions of people within any situation define an organization. As Fast (1993) argues, the very definition of an organization, firm, or company is the sum of all its *past, present, and future actions* interpreted by the actors and attached meanings. Consequently, understanding a company can be seen in the actors' actions and interactions among the people who comprise it. To understand how organizations operate in a regional, national, or international context, it is critical to analyze its *interactions* within itself and with other organizations.

Any organization can be understood through the actors who by their actions and knowledge create the everyday life of any firm. The actions and interactions exist in a context that is created by the actor through his actions. The action is related to the actors' concrete understanding of the situation and his context of meanings (Blumer, 1969; Schutz, 1972; Mead, 1934; Brown, 1978; Jehenson, 1978). Actors have motives and definitions of situations that make their social world into an inner logic.

Knowledge can be understood as motion picture of reality: experiences and information are produced through actions and transformed (by interpretation and retrospection) to the knowledge that the actors experience as useful and relevant. Reality is not a snapshot or still photograph. The world constantly moves in which actors are confronted with and have various experiences in which the process of consciousness develops or simply shifts toward different paths (or structures), which can be transformed into further actions. In this process, the actor uses and develops a scheme for interpretation to connect episodes or situations of social action in a sensible and thoughtful way of behavior. Hence situations are schemes that can be understood as active information-seeking pictures that accept information and orient actions continuously (Weick, 1979; Bartunek, 1984).

Knowledge is constructed by the actors in their "environmental" situations and events. Precisely because knowledge is a relation to and an orientation toward the "environment" through action, the environment itself can be defined as *the experiential space* and as *the interpretation space*. The experiential space is what is close and concrete, where, e.g., the actors travel and interact. This can be seen as the consciousness of any human being in "the natural attitude" first of all is interested in that part of his everyday of life world that is in his reach and that in time and space are centered around him (see also Schutz, 1973, p. 73).

Actors construct their own reality, individually and collectively, but they do not experience it in the same way. Moreover, actors see reality as if they live in an external world independent of themselves. Through their language, behavior, and typifications, actors often understand events, situations, and actions of others as being natural and that society is something "out there" that cannot change. They are wrong. The reason for this view of stability is that from the actor's knowledge, human beings "know" the world and that other actions confirm this in their given understanding of the world (Hennestad, 1986; Silverman, 1983). However, the international experiential space is not something that exists independent of the actors, and through the action–knowledge process, actors create their internationalization and the international experiential space. Therefore it is problematic to talk about borders between the firm and the environment. The norm in everyday life is change and not stability.

Experiential space exists "inside the firm." The experiential space is the actors' moving picture of events and everyday life, which is constituted by the interaction and knowledge processes of others. On the other hand, actors are confronted with everyday circumstances in the experiential space that one cannot claim that they have invented and that they cannot disregard. Actors exist in a society outside which they cannot place themselves. However, the firm cannot be seen as a reaction to things that happen "out there." What is "out there" is still an item for a subjective with an intersubjective interpretation and understanding. In other words, organizational actions influence and change the experiential space directly.

The central point is not only the product, the marketing, or the economic development of the firms in which actors talk and act but also the way in which they talk or communicate about behaviors and the way in which this talking creates a situation and interactions. A moving picture of reality is created. The actors have to understand how they create their experiential space and in which way they can act sensibly in it. Actors who are conscious about their experiential space will be less orientated toward rigid views of what is true or false and more oriented toward what is a flexible, creative, sensible, and fluid depiction of everyday life. Therefore no interaction represents truth or falsehood, but only versions that are more or less sensible and explain everyday life.

SYMBOLIC INTERACTIONISM: IN THE SUBJECTIVIST THEORETICAL PARADIGM

The primary mode for understanding organizational or collective interactions is through the *symbols* (or *meaning*) involved in the situations and events. *Symbolic interactionism* is the study of collective action between groups or organizations from the actors' Lifeworld perspective. The analysis of organizational actions must be seen within the context that helps define the interactions. However, each context has a history of events that frame it. And the interactions themselves redefine and create a new set of circumstances from which the organization operates.

Contextual analysis, therefore, can be limited and static since it only reflects the status quo and on-dimensional perspective of the past. To understand the present actions of an organization, and even attempt to predict its future actions, specific situations must be studied. Therefore transformational grammar provides the framework of scientific analyses and rule making. From human interactionism through language, scientific hypotheses can be created with explanations and predictive models.

George Herbert Mead (1962; originally 1932) at the turn of the 20th century from the University of Chicago formulated the philosophical basis for the symbolic interactionist perspective upon which Herbert Blumer (1962) expounded. The symbolic interactionist perspective discusses how human beings act and interact in everyday life. Mead, with his student and subsequent chief proponent Blumer, laid the groundwork for much of the theory behind today's "qualitative theory" in sociology. Mead rejected the classical English and American traditions and drew instead upon philosophical elements in both continental European and Far Eastern philosophy to counteract the empiricist and positivist determinists who were beginning to dominate the development of the social sciences.

Mead and Blumer argued that individuals are actors who alone or in groups interact in a variety of daily situations, be they personal, business, social, or whatever. Since human beings are thinking and reflecting, these interactions and the study of them are the basis of all human behavior. Language is used between actors as they interact and communicate. The ability of humans to create symbols (language and gestures, etc.) distinguishes the human species from all others. Bugs, animals, and fish (even dolphins) do not communicate to understand, gain and restore knowledge, or act. Understanding and explaining everyday actions, however, is the extension and essence of human interactions.

Blumer refined Mead's theories into a practical and straightforward approach to understanding how people act and interact in everyday life. Blumer assumes that since humans think, they must reflect before they act. In short, humans create and take action in various situations through the thinking process based on their reflective ideas and thoughts. To theorize as to how this is done, Blumer used Mead's concept of the "generalized other" or the fact that people think and reflect to themselves before they take action.

Human behavior is unpredictable, full of uncertainty, and therefore not rational. When scientists study and theorize about normative behavior, they have focused on some set of elements that compose human behavior. Because people are human beings, their everyday lives are made up of uncertainties and nonlinear actions. Human beings have an infinite set of behaviors and possible patterns to follow. Everyday life may be composed of sets and regular routines, but these are neither normal nor indicative of the creative potential of individual actors. They simply signify what people follow for convenience or expediency sake. They certainly are not the situations from which to draw significant conclusions about actors, situations, groups, or collective behavior.

In short, human interaction is by definition "abnormal." The essence of abnormal behavior, however, is that it constitutes its own processes and orderliness for individual actors and groups. The understanding of "abnormal" behavior is really the knowledge of what is "normal" for actors. And can best be seen in conflictual situations in which actors will display underlying emotions, feelings, and thoughts. Thus everyday life in business is not predictable or even normal. It is more often than not composed of change and irregularities. In short, the abnormal is the normal from which rules can be constructed.

Key to collecting data on actors, situations, groups, and collective behavior are the methodologies employed. Here the qualitative methods[2] from anthropology and sociology play a significant role. For decades, anthropologists and sociologists have conducted research studies using qualitative methods. The results of these studies are often case studies that describe the "static state" of a culture or a group of people. As described in this volume, one of the basic methods used in qualitative economics is "participant observation."

2. That is, field work, participant observation, action research, interaction and dialogue analyses, etc.

Most qualitative studies require that the researcher live and work in a particular environment to understand the people in their everyday life activities. In addition to the traditional anthropological methods, other related qualitative methods are used. Collection of prior data is always a method used by a qualitative researcher before going into the "field." In the context of business, this method would be called "market assessment," whereby the researcher/business person would want to know what information is already available about the culture. However, the researcher needs to be "objective" about these materials since previous studies may be biased. As outlined, field or site visits are used to gather data, analyze it, and then draw conclusions. Today, the use of electronic data is more and more common. However, in particular, legal data such as depositions, court testimony, and the like are a large part of qualitative data collection since it is legally factual and variable.

What remain critical in the qualitative economic perspective are the interactive methods utilized in the actual data collection process along with the interpretation itself. These methods are akin to the scientific methods used in physics, chemistry, biology, or engineering among the natural sciences. Actual verbal data from linguistic interactions are similar to the data collected in physics or chemistry.

The "site or field visit" through extensive interaction with the actors such as action research as participant observation and use of interviews is the single most important method in qualitative research. Here the researcher/business person must actually go and see what the culture looks like. The researcher/business person must live in another place and experience the culture and people. It is not enough to visit or tour. There are a number of subtle, but often critical facts learned from such visits. Among others, is the local infrastructure of a community or region such as transportation, communications, housing, and commerce, which may influence a business and economic development.

However, observing and recording data are not enough. And in many cases, observations can be wrong or misleading. In-depth interviews are necessary from a variety of people. Usually, anthropologists identify key individuals or "opinion leaders," that is, those whose information is variable as consistently "correct" and "adequate." Such people are not always the leaders in the culture, but they have rare insights into how the culture operates. Opinion leaders can verify and correct observations. More importantly, through interactions with them, a deeper understanding of the everyday phenomenon, situation, event, and local markets can be made.

Sørensen and Nedergaard (1993) have described how "intuition" plays a role in business, which can be seen as that of the visionary. In other words, business persons who uses their own sense of what product or market works may also be those with a vision. Often that concept (vision) is applied to business persons. These individuals have an idea and then pursue it. Although others may have had the same idea or vision, it is the business person who actually does something about it.

Research results are analyses that check the results and provide verification to provide useful written descriptions. This entire process is what Blumer (1969) describes as the "symbolic interactionist perspective." The basic assumption for the symbolic interactionism is that actors interact with one another and form relationship with others. However, each actor also interacts with himself or herself. In short, the actor reflects and contemplates his or her actions. Blumer calls this thinking process, the "generalized other," because actors do this all the time: they think, reflect, think again, act, think, and continue to move ahead. In other words, Blumer provides a theoretical framework for understanding intuition when seen as part of an actor's interaction with others.

For the business person, the result of qualitative research can be a strategic plan of action. Even though anthropologists rarely do any analysis or prediction with their data, the business person can. In particular, anthropologists never forecast, predict, or explain situations and cultures. They try not to influence the local culture in any way. This nonaction model has come under considerable criticism but is considered by anthropologists as following the natural scientific method: objectivity. A business person would develop an "action plan" and move on it immediately. The business person would want to see the needs of the culture and fulfill them. In many cases, the business person has the vision of a concept for the future economic development of the culture and will act upon it. Typically, the business persons will "carry through" or "follow-up" on their analysis of the culture because they see a business opportunity, can set goals, and create performance objectives.

TRANSFORMATIONAL LINGUISTICS: ECONOMIC RULES OF FORMALISM IN BUSINESS PRACTICES

The entire qualitative approach is the process of symbolic interaction at work. It also sets the stage for understanding how actors interact and create universal concepts that can be applied in a variety of situations. The underlying rules that explain the action of the actors can then follow the linguistic paradigm outlined by Chomsky (1957) so that the explanation of interaction is seen in the formation of rules.

Linguistics uses a qualitative methodology to identify sources of data such as native speakers/hearers of a language. Sentences are created and repeatedly tested against those of native speakers. Underlying the transformational grammar approach to linguistics is the assumption that languages have universal characteristics. The task of linguistics is to identify and derive the grammar for a language. Data are collected and comparisons are made to other languages. However, linguistics (and now cognitive psychologists) have found that native speakers/hearers do know rules and representations of their own language. They know what sounds right and correct.

Some evidence may bear on process models that incorporate a characterization of grammatical competence, while other evidence seems to bear on competence more directly, in abstraction from conditions of language use.

Chomsky (1957, p. 201).

In order words, qualitative methods for language usage, are the basic data collection procedure for linguists. They use discovery and description of everyday language as the basic core for their analyses and theories. Clark and Sørensen (1994a,b, 1995, summarized in Clark and Sørensen, 2002) have applied Chomsky's linguistic theory(s) to understanding business and economics. Clark and Fast (2008) have now refined qualitative economics in terms of theory and methodology. In particular, the direct application of the theories and methods in this volume to specific business cases has proven invaluable. Of particular interest for the business community has been the plethora of legal cases involving corporate governance, scandals, and bankruptcies in the Unites States. Qualitative economics is both useful and scientific in analyzing, understanding, and predicting future corporate actions (Clark and Demirag, 2006).

QUALITATIVE ECONOMICS: TOWARD A SCIENCE OF ECONOMICS

The *subjectivist* paradigm is reflected in an interactionism perspective to specific concepts in business economics. We want to explore a few areas in which some of the concepts in the book might be applicable to everyday business. Although our concern is primarily with organizations (organizing) and international business, we do present other basic concepts of business in new theoretical perspectives.

In particular, we take both real business situations and cases as examples of how the science of qualitative economics works. We discuss situations wherein research organizations seek to commercialize their discoveries. Also, we discuss companies that appear to have exposed conventional economic models and annual reports whereby their statistics and accounting lead to false and criminal acts.

Moreover, we look at the science of economics from the public policy perspective wherein decision makers have created deregulation or privatization schemes for certain vital infrastructure sectors to the determinant of the general public, while enhancing the enormously false profits of their executives. A qualitative economics approach to business exposes such financial schemes.

Organizing: Fitting Together of Lines of Activities and Actions

Through the processes of interaction, the actors construct some results: the interaction means organizing and creating the firm, and the actors create a moving picture of and a relation to the experiential space. The actors create intersubjective moving pictures of the reality, which is an organizational paradigm.

Organizing: Dynamism of the Firm

The actors create over time something we define and call a "firm." The processes that occur can be understood as organizing, which not only focuses upon action and interaction but also on moving pictures of reality and intersubjectivity.

Essentially, the firm can be understood as overlapping interactions. The actors create the firm through interactions, but "it" has also an influence upon them through their interpretation of "it." This dialectical perspective appears from the view that the firm only exists through the interactions between the actors and thus is viewed as a corollary of these interactions. Simultaneously, the organization is historically to the individual member, but the individual is confronted with an already existing organizational everyday of life, which sets the institutional parameters for his self-development. Self and organization thus develop together and because of each other in a dialectical process of mutual transformation (Singelmann, 1972, p. 415; see also Mead, 1934; Berger and Luckmann, 1966; Benson, 1977; Arbnor and Bjerke, 1981/1997).

The actors have to live with and exist with uncertainty and ambiguity. In other words, the way in which the actors handle themselves is in itself uncertain and exposed to many different interpretations and understandings. To reach security, the actors attempt to organize their activities.

Organizing means assembling the actions and should be seen in relation to interpretation and understanding by the actors. The actors form their actions so as obtain information and experiences that give meanings to the organizational world. This is organized by the actors in an attempt to construct an understanding. In the organizing the dependent actions are oriented toward removing contradictions and uncertainty: the actors seek to define and make sense in their situation, and thus they both create the firm and the experiential space. Organizing is to be seen as a social, meaning-making process in which order and disorder are in constant tension with one another, and wherein unpredictability is shaped and "managed." The raw materials of organizing are people, their beliefs, actions, and their shared meanings that are in constant motion (see Sims et al., 1993, p. 9).

There is a similarity between the phenomenological meanings of the practical activity of organizing and theorizing—the act of sense-making is in fact the central feature of both. Theorizing is most fundamentally an activity of making systematic as well as simplified sense of complex phenomena that often defy understanding by everyday, common-sense means. Theorizing might also be seen as a means by which people in organizations make their own and others' actions intelligible by reflective observations of organizing processes; through these processes, novel meanings are created and possibilities for action are revealed. Theorizing becomes an act of organizing, first, when it is a cooperative activity shared in by several or even all of the actors in an organizational setting, and second, when its purpose is to reveal hidden or novel possibilities for acting cooperatively. Organizing is cooperative theorizing, and vice versa (see Hummel, 1990, p. 11). In short, the firm is a social construction and a collective phenomenon.

This discussion of organizing can be seen as the beginning of an understanding of changes and innovation in the "firm."

Intersubjectivity and the Organizational Approach

The actors act and develop knowledge, and in the same time, they create a moving picture of the firm and the experiential space, which over time, through interaction processes, becomes the actors' intersubjective moving picture of reality (or paradigm). Among the actors moving pictures are created of a reality that contains specific actions (routines, traditions, procedures, politics, myths etc.), mental maps of the experiential space, norms, and values (as symbols). This is related to the actors' interpretations and expectations of each other in the organization or firm.

In the social subsystem constituted by a formal organization, the assignment of meanings is not left to the discretion of the members alone. The organization presents the individual member with a number of anonymous, functional typifying schemes that will help him orient his behavior toward the incumbents of other positions, especially hierarchical positions. These types are furnished to the newcomer in the organizational chart or the nomenclature of organizational titles. They underlie job descriptions and expose rights and duties attached to each organizational position, rules of conduct, customs, etc. By such standardization of the scheme of typifications, the organization attempts to establish a congruency between the typified scheme used by each actor as a scheme of orientation and that of his organizational fellowmen as a scheme of interpretation. This standardization is supposed to promote the smooth flow of authority relationships required for the efficient functioning of the organization (see Jehenson, 1978, p. 226; Benson, 1977).

Silverman's (1983) understanding of action and development can be seen in relation to the discussion of organizing and the organizational paradigm. His understanding of the organizational connections is that the path in interactions and in related meanings is built up over time. This reflects the consequences of the actions of the different actors and their knowledge. There are institutionalized expectations of possible actions by others—the foundation for social life—"the rules of the game." In the group there is acceptance because the actors do not themselves feel that they can change "the rules," and at the same time, stabile group relations give some advantages.

The organizing should be seen in relation to the extent of the actors' involvement in more or less involvement to keep or to change the rules. Organizational changes are the actors' change of the rules or their change of attitudes toward them. The actors solve problems with developed definitions and take actions in relation to the dominant views of reality. They act rationally and logically according to their understanding and interpretation. Rationality should therefore be understood as a social construction in itself and as a social product rather than action guiding rules for organizational life. It is a symbolic product, constructed by actions depending on the actors' moving picture of reality and interaction. That is, the structuring of organizational interactions requires members to rely upon shared but largely tacit background knowledge that is embodied in an organizational paradigm (Brown, 1978, p. 374; Garfinkel, 1967). Rationality as well as the definition of "problems," "situation," "leadership," and so on, are afforded by the dominant moving picture of reality.

Interaction between actors in a situation allows for many different interpretations whereby the actors are facing multiple realities. The interaction between different opinions means that new conceptions may arise. The reality is seen differently, which produces changes. Brown states that the organizational change could be seen as an analogy with scientific change (see also Imershein, 1977):

> *...most of what goes on in organizations, involves practical as well as formal knowledge. That is, the relevant knowledge is often a matter of application, such as how to employ the official procedures and when to invoke the formal description of those procedures, rather than abstract knowledge of the formal procedures themselves. Paradigms, in other words, may be understood not only as formal rules of thought, but also as rhetoric and practices in use.*
>
> Brown (1978, p. 373).

Bartunek (1984, p. 355) talks about an organizational paradigm as interpretive schemes (with references to Schutz and Giddens), which describes the cognitive schemata that map our experience of the world through identifying both its relevant aspects and how we are to understand them. Interpretive schemes operate as shared, fundamental (though often implicit) assumptions about why events happen as they do and how people are to act in different situations.

The essence of all this is that the meaning people create in their everyday reality gives the understanding of why people are like they are, which can be seen in their interaction and intersubjectivity, including their common interpretations, expectations, and typifications. As long as organizational actors act as typical members, they tend to take the official system of typification for granted as well as the accompanying set of recipes that help them define their situation in an organizationally approved way. The emergence of other, non-organizationally defined typifying schemes results from the breaking down of the taken-for-granted world when the actors enter into face-to-face relationships.

The Actors' Experiential Space: Organizational Lifeworld

The actors construct their reality, individually and collectively, but they do not experience it in this way. Moreover, they see reality as if they live in an external world independent of themselves. Through the language and typifications, we understand things as being natural and that society is something "out there" that we cannot change. The reason for this stability is that from our knowledge we "know" the world and that actions confirm us in a given understanding of the world (see Hennestad, 1986; Silverman, 1983). However, the experiential space is not something that exists independent of the actors and, as it is argued, it is through the action–knowledge process that the actors create their organizational activities and the experiential space. Therefore it is problematic to talk about borders between the firm and the environment:

> *While the categories external/internal or outside/inside exist logically, they do not exist empirically. The "outside" and "external" world cannot be known. There is no methodological process by which one can confirm the existence of an object independent of the confirmatory process involving oneself. The outside is a void, there is only the inside. A person's world, the inside or internal view is all that can be known. The rest can only be the object of speculation.*
>
> Weick (1977, p. 273).

The experiential space exists "inside the firm": The experiential space is the actor's moving picture as constituted by the interaction and knowledge processes. On the other hand, the actors are confronted with circumstances in the experiential space that one cannot claim that they have invented and that they cannot disregard. The actors exist in a society outside which they cannot place themselves. However, the firm cannot be seen as a reaction on things that happen "out there": What is "out there," is still an item for a subjective and an intersubjective interpretation and understanding. In other words, the organizational actions will influence and change the experiential space directly. The central point is not only the product, marketing, or the economy but also everything that the actors see and talk about: it is the way in which they talk about it, and the way in which this talking creates a situation, actions, and moving pictures of reality.

The actors have to understand how they create their experiential space and how they can act sensibly. So actors who are conscious about the fact that they create their experiential space will be less orientated toward what is true or false and more oriented toward what is sensible in the situation. Therefore nothing represents true or false, but only versions that are more or less commonsensible and exhibit their own logic.

Through interaction the actors have constructed a moving picture of the experiential space. This contains not only their orientation toward and relation to well-known actors (customers, suppliers, agents, competitors) but also moving pictures of less known actors (other firms, other competitors, public authorities, and institutions). There are happenings in the world that are not directly connected with the actors, but which they may notice and relate themselves to their interpretation space and with which they may be confronted in an interaction (such as the reconstruction of USSR, The Gulf War, the situation in the former Yugoslavia, among others).

The experiential space that one has chosen and formed does not directly influence actions and senses. However, when the experiential space exists, it works as a possible guide for actions and interpretations. Thus a newly created experiential space is a historical document, formed as knowledge. An important characteristic is therefore that the experiential space is a social construction. It is through interpretation that the actors create a moving picture of the experiential space and from this act and interact.

The discussion so far is the basis for understanding organizational changes and activities. Through the discussion of action, knowledge, and interaction, and its consequences (i.e., organizing, the organizational paradigm and experiential space) we can begin to understand all about what is the "firm" and how it changes.

Constituting of the Organizational Activities and of the "Firm"

If we relate this to organizing and intersubjectivity, Fig. 3.1 can be presented. The idea is to illustrate that the actors' knowledge and actions should be seen in relation to interaction and intersubjectivity in their everyday of life.

Actions and knowledge are to be understood in relation to the interaction processes by the actors. The experiential space is what the actors interpret as their situation, market, and surroundings, understood in relation to situations in which the actors interact with their product. However, the experiential space is

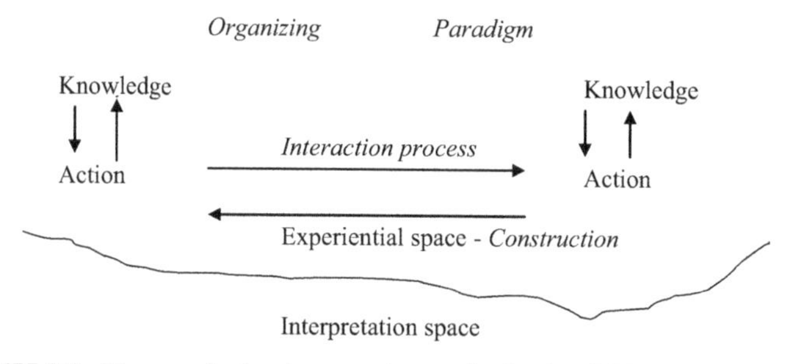

FIGURE 3.1 The organizational context in organizational activities and development.

also the social situations and the relations that the actor has with other actors "from outside the firm." This is understood as a broad spectrum of relations and is the actors' involvement in interaction connected with knowledge and understanding of reality such as friendship, family, etc., or seeing reality as a "multiple reality." In other words, the actors not only create their understanding of reality at work between nine and five but also have a history and life beside their involvement in the firm.

The organizational context is the actors. Their organizing and intersubjectivity are the organizational paradigm or their organizational identity. This includes the circumstances that involve the actors, their situation, problems, and possible solutions. Interpretations and understanding in this context influence the way in which the actors construct the experiential space and interact, and also the way in which the development of knowledge occurs, and to what extent it influences actions. It is the actors who interpret and have influence on the organizational development through their knowledge and experiences from different situations. The actors interpret others' actions in the experiential space or changes in it and construct an understanding, a moving picture of the connections between them and the experiential space.

In the process, actions and knowledge are seen in the actors' attempt to improve their capability to act in the experiential space, in other words, in the actors' construction and extension of the experiential space. This is constituted in the organizing and in the actors' attempt to thinking-in-future. The attempt to implement the thoughts depends on the action and interaction processes, and "success" in the situation is seen in the organizational actions or as an attempt to act in relation to new situations and moving pictures of reality. The background for this is in the actors' intentions with the actions and the development of the actor's knowledge. Success should also be understood as the actors' motives and expectations, and in their experiences of the results of the actions. The important issue in the actors' attempt to improve their situation is therefore the connection between the organizing and thinking-in-future, and the foundation for development; the connection between knowledge and action. Thinking-in-future should be understood in relation to processes of interaction and to intersubjectivity, and as a formulation of the project and the future actions. Therefore internationalization can be seen as construction and extension of the experiential space and development of knowledge in the experiential space.

The differences between the actors' interaction on the home market and interaction on the international market are exactly the confrontation of different moving pictures of reality and typifications between the actors involved. The actor's knowledge, typifications, and scheme to interpretation are confronted with foreign actors with another moving picture of reality deriving from the fact that they have a different context and everyday of life, and thus have developed their understanding and interpretations. This is, for instance, what some regard as differences in culture

(Bate, 1984; Meyerson and Martin, 1987; Gullestrup, 1992) and what others discuss in relation to international communication as an important dimension in internationalization.

When the actors interact with the foreign actors, they have to interpret and form an understanding of those actors' moving pictures of reality and their way of interpreting reality. They have to construct intersubjectivity, which is the foundation for further interaction. Their approach is comparative to researchers' approach quite in accordance with hermeneutic and qualitative methodology: the way in which we understand others' actions and the meaning of those actions (see Weber, 1964; Goffman, 1959; Garfinkel, 1967; Giddens, 1976; Schutz, 1978; Morgan, 1983).

When the actors act and interpret the situation they try to be sensible. There is logic in their attempt to be sensible, and they use earlier interpretations that have worked. When a new uncertainty is confronted with those earlier enactment's, not everything is noticed, and others are noticed as well known. In the same time, the actors experience the world as changing and as unpredictable and therefore that it is something more than what they already know. The actors try to reach out for those changes through interaction and enactment and act in such a way that those changes could be recognized and that they could be able to relate themselves to them.

The understanding of action and knowledge processes can be discussed in a frame of reference based on three areas: the actor's development capability, organizing and the organizational paradigm, and the actor's extension of the experiential space.

The Actors' Development Capability

This should be understood as a process and as a result of the actions. The development is connected with processes of interaction, but where the state of development depends on the interpretations and the actors' creation of interaction. The actors, with their specific qualifications, experiences, and personality, are not passive participants in a prefabricated reality; they are contributors and creators of meanings (see Allaire and Firsirotu, 1984).

Interaction and development are results and process: a situation and an involvement of something (not necessary a product) in a situation. However, the process of development does not start with the situation or the involvement but with a combination of both, based on the actors' interpretation and understanding of experiences used in future activities and in processes of interaction. The firm's development can thereby be understood as the actors' development and readjustment capability to change moving pictures of reality and interpretations and new actions. This can be seen as the capability to develop knowledge and transform this to actions, realized through the process of interaction in the organizing space and with actors in the experiential space.

Organizing and the Organizational Paradigm

When actual changes in a family, community, group, or community of any kind are organized, they become a change in the paradigm as seen in the actors' understanding of formal structures, goals, politics, management processes, and resources, and are all in all a result of the organizational reality and the way of functioning in the everyday of life. They contain a living dimension in a connection of shared and meaningful symbols manifested in interactions, interpretations, myths, ideology, values, and multiples cultural forms of expressions as a result of interaction processes and interpretations. An important part of this everyday of life is the business area that is based on previous actions oriented toward what to produce, which resources and technology one should use, and an understanding of the existing knowledge and needs for knowledge development in relation to the business area. This could also be understood as what Diamond (1990, p. 34) named as "the organizational identity." Organizational identity is the product of the group's intersubjective organization of experience at a given point of time: the "story" they share about what is real to them. It is a picture of the meaning, purpose, and intention, collectively and unconsciously assigned to common experience and behavior of organizational members especially during critical incidents, or what Benson (1977) calls "organizational morphology," which refers to the officially enforced and conventionally accepted view of the organization. It refers to the organization as abstracted from its concrete, intricate relations with other aspects of social life. This is the administrators' vision of the organization, the form that they try to impose upon events. Since they are partly successful, the morphology may also be somewhat accurate as a description of the organization.

It is the connection between the actors' knowledge, their understanding of the business area, and their organizational actions that create the orientation toward and the mode of handling organizational activities. This should be seen in the light of the actors' attitudes of and intentions with the activities. The actors' development capability is constituted by the interaction process and by the actors' interpretation and knowledge of the business area. The actors' change of moving reality picture is therefore important in a situation in which a contradiction exists between the business area and the experiential space. This can be seen in situations in which one shifts to new technology and new products, enters new market, changes services, etc., or an extension of the experiential space. In each of these situations there is a need to change the knowledge. Therefore a recognition among the actors of the problem, new actions, or actions in a new way are important. An experienced need among the actors to develop knowledge should be seen in relation to the organizing and the organizational paradigm.

Actor's Extension of the Experiential Space

Through the interaction process and interpretation *the actor's experiential space* influences the organizing and the development capability. There is a confrontation with other actors' external to the firm, but the organizing and the development of knowledge are to a less extent activated when the actors experience the experiential space as stable. This experience means less possibility for changes in the actors' organizing and development capability due to intersubjectivity, and the interaction is satisfied. On the other hand, activation of interaction is more probable when the experiential space is experienced as uncertain or ambiguous. In other words, it is the experiences in the interaction with other actors and the way in which the actors interpret and act accordingly that is crucial to their development. The foundation for experiences and interpretation of the experiential space, for knowledge development and for changes of organizational activities, are the actors' interaction and involvement. In situations with changes, the central issue is the actors' knowledge and change of interpretations that can be related to and transformed into actions.

SUMMARY

Volumes have been written about the future. Economics is frequently cited. However, unlike in science, economists make predictions that conflict and often fail. For the most part, economists take the present and past data to reformulate them into predictions. Although the future must be rooted in the past, it will need to be based upon thinking "outside the box," as Toffler calls it. Nonetheless, from his work on the future (*Future Shock* and *The Third Wave* in particular), he admits and advocates that "the intellectual framework that might unify management theory and economics is not yet in place. The task of creating that framework still lies ahead" (in "Foreward" to Gibson, 1997, p. X). In 2005, Jeremy Rifkin published a book, *The European Dream* (Rifkin, 2005) whose subtitle states "How Europe's vision of the future is quietly eclipsing the American Dream" by moving into a "Third Industrial Revolution" that is fossil fuel free, and has localized renewable energy for travel and power through public and private partnerships. The key is in Rifkin's definition of the European dream since it

> *Emphasizes community relationships over individual autonomy, cultural diversity over assimilation, quality of life over the accumulation of wealth, sustainable development over unlimited material growth, deep play over unrelenting toil, universal human rights and the rights of nature over property rights, and global cooperation over the unilateral exercise of power.*
>
> Rifkin (2005, p. 3).

Rifkin and others need economics to provide the data and understanding for the new European dream let alone those new dreams coming in Asia, Africa, and the Americas. Many of the economic policies advocated in Rifkin's book are becoming a reality in the European Union as the Parliament passed a three-tier program for implementing the "Third Industrial Revolution" (May 2007). Meanwhile, America cannot resist its reliance on statistics, quantitative reports, and the "art" of economics. Economic data are historical (or from the past) by their very nature and definition but open for interpretation and understanding, and not all "data" are gathered as it is impossible. Future economic trends or predictions are impossible and baseless. If any established, new, or emerging nation is to advance, it must understand its past and present and strategically interact for its future based on the application of scientific know-how, ideas, and wants, thereby creating its future.

Indeed, economics has been toted as the science for explaining or even predicting the future. However, its analyses are all based on the numbers from the past. What this discussion has done is provide a scientific framework rooted in philosophy, theory, and empirical data to create a science of economics. Furthermore, applying qualitative economics to business and management through a Lifeworld perspective and the use of interactionism is a scientific perspective that actually can handle the contradictions and paradoxes that society meets today in the discussion of sustainability.

This discussion presents not only a "think-outside-the-box" perspective on conventional and dominant functionalist economics paradigm, but also a dynamic, interactive, and new perspective based on the philosophy of science. Furthermore, Gibson and others in his volume (1997) argue that the world of the future is uncertain and even chaotic. It is a dramatic departure from the post—World War II world. New thinking in economics, and other fields, are nonlinear and dynamic to account for a world in which the future is "terra incognita" as Toffler puts it. The landscape of tomorrow is both unchartered and accelerated. Economists can think of it as a speeding uncontrollable automobile or as a challenge for making some sense from it.

Some economists have put forth a different approach to economics, such as Levitt and Dubner in *Freakonomics* (2005) to signal that there is something radically wrong with the conventional paradigm. What Levitt, a young and increasingly well-established economist, has described as being of interest is "the stuff and riddles of everyday life" (Levitt and Dubner, 2005, p. xi). As Levitt says later in the book,

> *There is at least a common thread running through the everyday application of Freakonomics. It has to do with thinking sensibly about how people behave in the real world.*
>
> Levitt and Dubner (2005, p. 205).

However, the authors fail to probe into the depth the interaction between people in situations and events. They do not "ask the second" or follow-on question. However, their book does raise questions about economics and provokes scholars to think outside the box.

Other economists argue that it is time to reexamine the accepted economic norms and popularization of globalism (Saul, 2005). What is obvious about the future of economics is that, as Gibson notes (1997, p. 3) "economics will be based not on land, money, or raw materials but on intellectual capital." When the case was first written in the late 1990s fuel cells were not known to economists let alone the general public. Today, fuel cells are widely known. New innovations and commercialization into the marketplace are the economic drivers for the future. However, the technological innovation case study exemplifies another significant issue: current economic theories and business management practices cannot describe or explain the commercialization phenomenon for business development.

Conventional economic development theories of today are inadequate by defining innovation commercialization as due to "externalities" of one sort or another. Or more insignificantly, new business products or services depend on "market study" and demands. There are, for example, simply not "five forces" at work. And to attempt to explain the commercialization of the fuel cells in this manner is fallacious. Business economics is simply not a phenomenon that occurs in a "black box" subject to external influences.

Still other economic theorists have attempted to create new frameworks for understanding and explaining the commercialization and deployment of innovations into the business community. Yet these attempts are drawn from the same neoclassical functionalist paradigm and therefore suffer from the same problems. Resource-based theory is a good attempt in the investigation of those elements (both tangible and intangible) which a firm may then turn into various business opportunities. This approach clearly recognizes "intellectual" or knowledge capital, for example, as a valuable resource beyond the conventional economic definition of the firm. With its focus on capabilities as central to business development, there is a growing concern that any business or "firm" still remains a mystery. In other words, no economist knows how companies, businesses, or firms develop, grow, and sustainable themselves. The basic problem remains and is recognized by forward-thinking economic theorists: What is the firm? Is there a theory of the firm?

Nonetheless, as argued indeed there is a "theory of the firm" when the scholar uses as the starting point, looking inside the firm first. An understanding of the everyday business life of the firm (in our case, research that turns innovation into a new commercial venture) provides the opportunity for theory building, which has never before been considered in economics. In other words, when economists start from a Lifeworld perspective that is rooted inside the firm and builds from the ground up, then the results are likely to be more significant, accurate, and scientific regarding how a business operates, grows, and is managed.

Furthermore, through a growing body of literature, the firm, interactionism, and its linguistic framework for analysis provides hypotheses, description, explanation, and rules that become generalizable. In short, qualitative economics is science. However, for our purposes now, the consideration of the firm, and particularly the commercialization of an innovation, provides an understanding on the building of economic theory from the Lifeworld paradigm in everyday business life.

Interactionism is the cornerstone for moving from the microeconomics or case study perspective to the larger more "universal" macroeconomic one. In this context, research on more cases about innovation in light of business economics must be gathered and analyzed. Transformational linguistic theory provides the framework for scientific rule making. The definition of economics as exchanges and therefore as engagement allows for an understanding of economic phenomena, events, and situations when the data are drawn from a knowledge base to construct the economic reality of everyday life. Since action among actors is predicated upon their knowledge and understanding of others, the description of the cases presented earlier in such terms should be considered.

To summarize the interactionist framework from Blumer, symbolic interactionism draws its roots from the subjectivists' philosophical tradition in Europe and the United States, which is now conceptualization in the Lifeworld paradigm. When we look at symbolic interactionist theory and the interactionist perspective applied to business economics, cases can be described, understood, and explained for business activities in a scientific manner. When transformational linguistics is included, then there is a very powerful and robust theoretical model from which to formalize (not quantify) business economics into sets of rules, interactions, and predictive analyses.

REFERENCES

Allaire, Y., Firsirotu, M.E., 1984. Theories of organizational culture. Organization Studies 5 (3).

Arbnor, I., Bjerke, B., 1981/1997. Methodology for Creating Business Knowledge. SAGE, USA.

Bartunek, J.M., 1984. Changing interpretive schemes and organizational restructuring: the example of a religious order. Administrative Science Quarterly 29.

Bate, P., 1984. The impact of organizational culture on approaches to organizational problem-solving. Organization Studies 5 (1).

Benson, J.K., March 1977. Organizations: a dialectical view. Administrative Science Quarterly 22.

Berger, P.L., Luckmann, T., 1966. The Social Construction of Reality – A Treatise in the Sociology of Knowledge. Doubleday & Company, New York.

Blumer, H., 1962 (1969, 1986). Symbolic Interaction – Perspective and Method. Prentice-Hall, Englewood Cliffs, NJ.

Brown, R.H., 1978. Bureaucracy as praxis: towards a political phenomenology of formal organizations. Administrative Science Quarterly 23.

Chomsky, N., 1957. Syntactic Structures. Mouton & Co., The Hague.

Clark, W.W., Fast, M., 2008. Qualitative Economics: Toward a Science of Economics. Coxmoor Publishing Company, Oxford updated Springer Press 2018.

Clark Jr., W.W., Demirag, I., 2006. US financial regulatory change: the case of the California energy crisis. Journal of Banking Regulation (Special Issue) 7 (1/2), 75–93.

Clark Jr., W.W., Sørensen, O.J., 1994a. Toward a theory of entrepreneurship. In: Paper Presented at the 39th World Congress of Small Business, Strasbourg, France, 1994. Aalborg University Press, Aalborg, Denmark.

Clark Jr., W.W., Sørensen, O.J., 1994b. Linguistics in Constructing a Theory of Business/Economics: Some Empirical Applications. Aalborg University, Aalborg, DK (paper in preparation).

Clark Jr., W.W., Sørensen, O.J., 2002. Entrepreneurship: theoretical considerations. International Journal of Entrepreneurship and Innovation Management 2 (1) (Inderscience, London, UK).

Fast, M., September 1993. Internationalization as a social construction. In: Proceedings From the 9th IMP – Conference, Bath, UK.

Gadamer, H.-G., 1993 (1975). Truth and Method. Sheed & Ward, London.

Garfinkel, H., 1967. Studies in Ethnomethodology. Prentice-Hall, Englewood Cliffs, NJ.

Gibson, R. (Ed.), 1997. Rethinking the Future: Rethinking Business, Principles, Competition, Control and Complexity, Leadership, Markets and the World. Nicholas Brealey, London.

Giddens, A., 1976. Hermeneutics, ethnomethodology, and problems of interpretative analysis. In: Coser, L.A., Larsen, O.N. (Eds.), The Uses of Controversy in Sociology. The Free Press, New York.

Goffman, E., 1959. The Presentation of Self in Everyday Life. Doubleday, New York.

Gullestrup, H., 1992. Kultur, Kulturanalyse og Kulturetik – eller hvad adskiller og forener os? Akademisk Forlag, København.

Hennestad, B.W., August 1986. Organizations: frameworks or frame work? Scandinavian Journal of Management Studies 3 (1).

Hummel, R.P., Spring, 1990. Applied Phenomenology and Organization. Administrative Science Quarterly 14.

Imershein, A.W., 1977. Organizational change as a paradigm shift. London. In: Benson, J.K. (Ed.), Organizational Analysis – Critique and Innovation.

Jehenson, R., 1978. A phenomenological approach to the study of the formal organization. In: Psathas, G. (Ed.), Phenomenological Sociology – Issues and Applications. John Wiley & Sons, New York.

Kuhn, T., 1962. The Structure of Scientific Revolution. University of Chicago Press, Chicago.

Levitt, S.D., Dubner, S.J., 2005. Freakonomics: A Rogue Economist Explores the Hidden Side of Everything. William Morrow, New York, NY.

Mead, G.H., 1962 (1934). Mind, Self, & Society – from the Standpoint of a Social Behaviorist. The University of Chicago Press, Chicago, IL.

Meyerson, D., Martin, J., November 1987. Cultural change: an integration of three different views. Journal of Management Studies 24 (6).

Morgan, G. (Ed.), 1983. Beyond Method – Strategies for Social Research. Sage Publications, Beverly Hills.

Reinert, E., 1994. Symptoms and Causes of Poverty: Underdevelopment in a Schumpeterian System. Forum for Development Studies, No. 1–2, pp. 73–109.

Reinert, E., 1997. The Role of the State in Economic Growth. Working Paper #1997.5. University of Oslo, Centre for Development and the Environment, pp. 1–58.

Rifkin, J., 2005. The European Dream: How Europe's Vision of the Future is Quietly Eclipsing the American Dream. Penguin, New York, NY.
Schutz, A., 1978. Concepts and theory formation in the social sciences. In: Emmet, D., Macintyre, A. (Eds.), Sociological Theory and Philosophical Analysis. The Macmillan Company, New York.
Schutz, A., 1972. The Phenomenology of the Social World. Heinemann Educational Books, London.
Saul, J.R., 2005. The Collapse of Globalism: And the Reinvention of the World. Viking, Toronto, Canada.
Schutz, A., 1973. Hverdagslivets sociologi. Hans Reitzel, København.
Sims, D., Fineman, S., Gabriel, Y., 1993. Organizing & Organisations. Sage Publications, Great Britain.
Silverman, D., 1983. The Theory of Organisations. Heinemann, London.
Singelmann, P., August 1972. Exchange as symbolic interaction: convergences between two theoretical perspectives. American Sociological Review 37.
Snow, C.P., 1959. The Two Cultures and the Scientific Revolution.
Sørensen, O.J., Nedergaard, A., 1993. Management Decision Making in an International Context: The Case for Intuition and Action Learning. University of Aalborg, International Business Economics, Denmark. No. #2 Reprint Series.
Weber, M., 1964. Social and Economic Organization. The Free Press, New York.
Weick, K.E., 1979. Cognitive processes in organizations. Research in Organizational behavior vol 1.
Weick, K.E., 1977. Enactment processes in organizations. In: Staw, B., Salancik, G.R. (Eds.), New Directions in Organizational Behavior, Chicago.

FURTHER READING

Buchholz, T.G., 1989. New Ideas from Dead Economists. Penguin Press, New York.
California Energy Commission, February 2002. Report on Summer 01. Sacramento, CA.
Clark Jr., W.W., Bline, D., Demirag, I., 1998. Financial markets, corporate governance and management of research and development: reflections on US managers' perceptions. In: Demirag, I. (Ed.), Comparative Capital Systems. Pinter, London. Chapter in Book.
Clark Jr., W.W., Lund, H., November 2001. Civic markets. International Journal of Global Energy Issues. https://doi.org/10.1504/IJGEI.2001.000927.
Clark, W.W., Bradshaw, T., October 2004. Agile Energy Systems: Global Lessons from the California Energy Crisis. Elsevier Press, Oxford, UK. Updated in 2017.
Clark, W.W. (Ed.), 2012. The Next Economics. Springer Press.
Clark, W.W., Cooke, G., 2014. The Green Industrial Revolution. Elsevier Press. Edited Mandarin, 2015.
Clark, W.W., Cooke, G., 2016. Smart Green Cities. Rutledge Press.
Davis, G., January 31, 2002. Letter to FERC Chairman Wood.
Economist, January 4, 1992. Examining the Mystery, p. 17.
Economist, August 23, 1997a. The Puzzling Failure of Economics, p. 11.
Economist, August 23, 1997b. Europe: Remodeling Scandinavia, pp. 37–39.
Economist, September 20, 1997c. The Visible Hand, p. 17.
Economist, December 6, 2001. The Amazing Disintegrating Firm.
Economist, February–March, 2002a. RE: Enron.
Economist, March 4, 2002b. Social Science Comes of Age, pp. 42–43.

Economist, January 19, 2002c. The Real Scandal.

Fortune, January 16, 2002. The Letter to Ken Lay.

Freeman, C., March 2–7, 1994. Lecture Series: "Interfirm Cooperation and Innovation; Technical Change and Unemployment: The Links between Macro-economic Policy and Innovation Policy; Schumpeterianism: between History and Economic Theory". University of Aalborg, Denmark.

Freeman, C., Perez, C., 1989. Structural Crisis of adjustment, business cycles and investment behavior. In: Dosi, G., Freeman, C., Nelson, R., Silverberg, G., Soete, L. (Eds.), Technological Change and Economic Theory. Pinters Publ., London, pp. 38–66.

Freeman, C., Hagedoorn, J., 1995. Convergence and divergence in the interantonalization of technology. In: Hagedoorn, J. (Ed.), Technical Change and the World Economy: Convergence and Divergence in Technology Strategies. Edward Elgar, Aldershot, UK.

Grabher, G. (Ed.), 1993. The Embedded Firm: On the Socioeconomics of Industrial Networks. Routledge, London, UK.

Hakansson, H., Snehota, I., March 1994. Developing Relationships in Business Networks. University of Uppsala, Sweden.

Hakansson, H., Johanson, Jan, 1993. The network as a governance structure: interfirm cooperation beyond markets and hierarchies. In: Grabher, G. (Ed.), The Embedded Firm: On the Socioeconomics of Industrial Networks. Routledge, London, UK, pp. 35–51.

Heilbroner, R., 1989. The Making of Economic Society: Revised for the 1990s. Prentice-Hall International, London, UK.

Heilbroner, R., Thurow, L., 1994. Economics Explained (rev.). Touchstone Press, New York.

Hirschman, E.C., Holbrook, M.B., 1992. Postmodern Consumer Research: The Study of Consumption as Text. Sage, New York, NY.

Holland, J.H., 1992. Adaptation in Natural and Artificial Systems. MIT Press, Cambridge, MA.

Husserl, E., 1962. Ideas. Macmillan, New York.

Kahn, J., January 7, 2002. One plus one makes what? Fortune.

Kuttner, R., March 25, 2002. The road to enron. Am. Prospect 13 (6).

Kuttner, R., January 1–14, 2001. The enron economy. The American Prospect 13 (1).

Lundvall, B.-A., 1993. Explaining interfirm cooperation and innovation: limits of the transaction-cost approach. In: Grabher, G. (Ed.), The Embedded Firm: On the Socioeconomics of Industrial Networks. Routledge, London, UK, pp. 52–64.

McGraw Hill, PowerWeb, 2002. Entrepreneurship. http://register.dushkin.com.

McNeill, D., Freiberger, P., 1993. Fuzzy Logic: The Discovery of a Revolutionary Computer Technology–and How it Is Changing Our World. Simon & Schuster, New York.

Mowery, D.C., Oxley, J.E., 1995. Inward technology transfer and competiveness: the role of national innovation systems. Cambridge Journal of Economics 19 (1).

Nelson, R.R., 1990. Capitalism as an engine of progress. Research Policy 19. Elsevier Science Publishers B.V., North Holland.

Nelson, R.R., Winter, S.G., 1982. An Evolutionary Theory of Economic Change. Harvard University Press, Cambridge, MA.

Perkins, L.J., 1996. What is physics and why is it a "science"?. In: Lecture Presented at Graduate Physics Seminar, University of California, Berkeley. Lawrence Livermore National Laboratory, University of California.

Porter, M.E., 1980. Competitive Strategy: Techniques for Analyzing Industries and Competitors. Free Press, New York.

Porter, M.E., 1990. Competitive Advantage. Free Press, New York.

Rand Report, Bernstein, Mark, January 2001. California Energy Market Report. Winter 2000–2001,Unplublished paper presented to Staff in California State Legislature, Sacramento, CA. Rand Corporation, Santa Monica, CA.

Reinert, E., 1995. Competitiveness and its predecessors – a 500 year cross-national perspective. Structure Change and Economic Dynamics 6, 23–42.

Reinert, E., 1996. Economics: 'the dismal science' or 'the never-ending frontier of knowledge?: on technology, energy, and economic welfare. Norwegian Oil Review (7), 18–31.

Reinert, E., 1998. Raw materials in the history of economic policy. In: Cook, G. (Ed.), The Economics and Politics of International Trade. Rutledge, London & NY, pp. 275–300.

Reinert, E., Daastol, A.M., 1997. Exploring the Genesis of Economic Innovations: the religious gestalt-switch and the duty to invent as preconditions for economic growth. The European Journal of Law and Economics 4 (3/4), 1–58.

Sacramento Bee, April 2001. Gorging Profits. California Energy Commission, Sacramento, CA.

Schumpeter, J., 1934. The Theory of Economic Development. Harvard Un. Press, Cambridge, MA.

Schumpeter, J., 1942. Capitalism, Socialism and Democracy. Harper & Brothers, New York, NY.

Davis, G., March 21, 2002. Talking Points: Enron. Sacramento, CA.

State of California, Jan. 31, 2002. Governor Davis Calls for Federal Investigation of Enron's Role on California Energy Crisis. Press Release.

Teece, D.J., Bercovitz, J.E.L., De Figueiredo, J.M., 1994. Firm Capabilities and Managerial Decision-making: A Theory of Innovation Biases. Unpublished. University of California, Berkeley, USA. Haas School of Business.

Chapter 4

Political–Economic Governance of Renewable Energy Systems: The Key to Create Sustainable Communities

Woodrow W. Clark, II[1,a], Xing Li[2]
[1]*Clark Strategic Partners, Beverly Hills, CA, United States;* [2]*Aalborg University, Aalborg, Denmark*

Chapter Outline

a. The authors thank Jon McCarthy for both economic details and edits as well as Lucas Adams, KentaroFunaki, Claus Habermeier, and Russell Vare for international perspectives and data. Namrita Singh and Jerry Ji checked on data contributions.

Sustainable Cities and Communities Design Handbook. https://doi.org/10.1016/B978-0-12-813964-6.00004-5

CORPORATE AND BUSINESS INFLUENCES AND POWER

Corporate interests and impacts on public policy are extremely significant around the world. In the United States, the US Supreme Court ruled in 2012, about 6 months before the US National election for president and members of the Congress, that anyone (defined now as even companies and corporations) could contribute any amount of money to people running for election to any office. This put the US national election in the hands of the wealthy and those with corporate interests. In total over USD $1 billion was put into the presidential and congressional elections. About 70%–80% of the money went to Republican candidates. The results were different from what was predicted as the public voted for Democrats, reelecting President Obama, increasing the Democratic majority in the US Senate and increasing its minority position in the US House.

While the antilarge funds from individuals and corporations did not do well in the last US national elections, the future is uncertain with specific needs to finance and support programs such as renewable energy. One case in point for the US national election concerned the Obama administration's support of a solar company for over USD$500 million in debt loans. In the year before the national election and less than 2 years after receiving the US debt funds, the company declared bankruptcy. There was no investigation (at the time of writing this chapter) into the solar company or its supporters, some of whom were associated with the Obama administration through the US Department of Energy.

The US national government is not alone. Such cases arise and are common not only in developing countries but also in other Western developed countries. For example, even in California on the local level, where the public votes for funds to rebuild and modernize its schools, a superintendent and his facilities manager in a wealthy Southern California School District were both convicted and jailed on fraud for taking public money for themselves. Furthermore, a large national American corporation settled out of court for over USD$6.5 million in payments back to the school district.

INTERNATIONAL CASES

For example, in Ukraine, the former Prime Minister accepted 10% of natural gas funds from Russia that were piped through Ukraine. She was voted out of office, went on trial and was convicted, and was then put in jail. Similar situations occur all over the world. Consider China, whose last Central Government administration had a high-level official whose wife was convicted of supporting a Western business person and then killing him. This resulted in a dramatic change in the Central Government leadership that took office in early 2013 after the Chinese New Year in February.

CHINA LEAPFROGS AHEAD

China, however, has developed its own form of non-Western government and economics (Clark, 2012). Clark and Li made this point for over a decade, calling China's government "social capitalism" (Clark and Li, 2004, 2010, 2013), which means that governments must be responsible for social issues including the environment and renewable energy, as well as health and retirement issues for its people. Western economics looks at what is best for the corporation, shareholders, and employees first and foremost. This economic ideology lacks both ethics and honest use of economics, calling into question the fact that economics is not a science but simply a way to use numbers to justify political positions and decisions. In *Agile Energy Systems* (Clark and Bradshaw, 2004) the point was made with a review of what happened in the California energy crisis from 2000 to 2003 and how companies manipulated "markets" to their own economic advantages while allowing brownout and blackouts throughout the state.

The Chinese governments have led in the role of government providing direction with 11Five-Year Plans and funding to support these plans. Renewable energy is a key component of the 12th and then the 13th Five-Year Plans that started in March 2011. Each plan provides clear and formulated policies, and their intended budgets, to address environmental issues and their solutions. Denmark (and other Nordic nations, including Germany) also has national plans including renewable energy. Their current one (2012) calls for Denmark to be 100% energy independent through renewable energy by 2025. Denmark is already 45% there today.

Meanwhile, the United States has no such plans, especially in energy—and renewable energy in particular. Instead, the United States leaves the decisions on energy to "market forces," which are focused primarily on their past business models that are usually rooted in fossil fuels from the Second Industrial Revolution (2IR). It is not surprising that the United States remains behind in the Green Industrial Revolution (GIR). Clark and Cooke (2011) document the problem with the United Stateswhen compared with the rest of the developed and developing world. Their new book on the GIR will be published initially in Chinese as a wake-up call to the West, and particularly the United States.

CHINA HAS "LEAPFROGGED" INTO THE GREEN INDUSTRIAL REVOLUTION

To avoid the mistakes of the Western developed nations, China has moved ahead in a variety of infrastructure areas (Clark and Isherwood, 2007, 2010). Also, the United States must look comprehensively into the corporate and political reactions to the 2011 Japanese tsunami and ensuing nuclear power plant explosions, as well as the 2010 BP oil spill in the Gulf of Mexico off

Louisiana. The United States and other countries cannot ignore the environmental consequences and economic costs of the 2IR that have handicapped its moving into the GIR. The end result is not good for the Americans, let alone the rest of the world.

The deregulation of industries starting in the Reagan and Thatcher eras was a mistake and a completely naïve view of reality from neoclassical economics by Adam Smith. There has never been a society or area in the world in which the principles of capitalism have been proved to work in reality. Instead just the opposite has been the reality. Chomsky (2012) looks at the history of economics in a far more concrete manner.

THE WESTERN ECONOMIC PARADIGM MUST CHANGE

Even *The Economist* in two special issues labels modern economics as "state capitalism" (January 23, 2012b) and in another issue soon after that as The Third Industrial Revolution (3IR) (April, 2012a), a theme from Jeremy Rifkin (2004) and his book with that title in 2012. Clark has published several articles and given numerous talks about the 3IR (2008, 2009, 2010, and 2011) but prefers to think of it as the GIR (Clark and Cooke, 2011). Basically, the GIR concerns renewable energy, smart green communities, and advanced technologies that produce, store, and transmit energy for infrastructures, while saving the environment.

The point is that the development of the United States into a powerful world leader had a lot to do with its military strength and also its economic development of fossil fuels for over a century in the 2IR including the technologies that support them, such as combustion engines and related technologies such as atom bomb and nuclear power (Chomsky, 2012). The growth of the United States started with businesses and their owners who control the economy today. There is little or no competition. However, even more significant is that the basis for this wealth is in fossil fuels and continues to be there. Hence the environment is continuing to be damaged to produce more oil and natural gas causing climate change. However, this 2IR retards and places the United States decades back when compared with emerging economics and even other Western developed nations.

As historians have documented, the development of the 2IR in the United States was primarily based on "state capitalism" since oil companies got land grants, funding, and even trains or pipelines for transporting their fossil fuels. That governmental support continues today. Consider the issue of the United States getting shale oil from Alberta, Canada, and the massive pipelines that were installed through the United States to get the oil to the country. Furthermore, these same companies get tax breaks and credits such that their economic responsibility to the United States is minimal. The argument that the United States will be "energy independent" with these fossil fuels is false. The United States needs to stop getting its energy from fossil fuels anywhere in the world, including domestically or from its neighbors.

Hence the argument is that China will buy the oil from Canada. Basically Canada (and the United States) should not even extract oil from the ground, which permanently destroys thousands of acres of land, making them impossible to repair or restore. There are far more and better resources from renewable energy like sun, wind, geothermal, run of the river and ocean, or wave power.

INTRODUCTION AND BACKGROUND

The GIR emerged at the end of the 20th century due in large part to the end of the Cold War that dominated the globe since the end of World War II (WWII). The 2IR had dominated the 20th century because it was primarily based on fossil fuels and technologies that used primarily mechanical and combustion technologies. On the other hand, the GIR is one of renewable energy power and fuel systems and smart "green" sustainable communities that use more wireless, virtual communications and advanced storage devices like fuel cells (Clark and Cooke, 2011). The GIR is a major philosophical paradigm change in both thinking and implementing environmentally sound technologies, which requires a new and different approach to economics (Clark, 2011).

The United States lived in denial during the 1970s and then again since the early 1990s, which became apparent for both Democrat and Republican presidential administrations in their lack of proactive polices globally through the Kyoto Accords and most recently the UN Intergovernmental Panel on Climate Change Conference in Kopenhaven (December 2009) and Cancun (2010). On the other hand, in the early 1990s, economic changes in Europe and Asia were made due to the end of the Cold War to meet the new global economy. The Asian and EU conversions from military and defense programs to peacetime business activities were much smoother than that of the United States. Environmental economist Jeremy Rifkin recognized this change and developed the concept of a "Third Industrial Revolution" in his book, *The European Dream* (2004). According to Rifkin the GIR took place a decade earlier in some EU countries. He did not recognize that Japan and South Korea had been in a GIR even decades before that (Clark and Li, 2004).

At the same time, Clark et al. (2006) published an article on "Green Hydrogen Economy" that made the distinction between "clean" and "green" technologies when related to hydrogen and other energy sources. The former was often used to describe fossil fuels in an environment-friendly manner, such as "natural gas" and "clean coal." Green, on the other hand, means specifically renewable sources such as the sun, wind, water, wave, and ocean power. In short, the article drew a dividing line between what technologies were a part of the 2IR (that is clean technologies such as clean coal and natural gas) and the GIR (solar, wind, ocean and wave power as well as geothermal). The GIR focused on climate change and changing the technologies and fuels that caused it, or could at least mitigate and change the negative pollution and emission problems that impacted the earth.

Clark and Fast (2008) in founding the science of "qualitative economics" made the point about economics since it needed to define ideas, numbers, words, symbols, and even sentences due to the misuse of "clean" to really mean fossil fuels and technologies that were not good for the environment. The documentary film, *Fuel* (Tickell, 2009) made these points too as it told the history about how "clean" was used to describe fossil fuels like natural gas to placate and actually deceive the public, politicians, and decision makers. For example, Henry Ford was a farmer and used biofuels in his cars until the early 1920s when the oil and gas industries forced him to change to fossil fuels.

Hawkins et al. (1999) refer to the environmental changes as the beginning of "The Next Industrial Revolution." This observation only touched the surface of what the world is facing in the context of climate change. Moreover, the irony is that China has already "leapfrogged" and moved ahead of the United States into the GIR (Clark and Isherwood, 2008, 2010). While China leads the United States now in energy demand and CO_2 emissions, it is also one of the leading nations with new environmental programs, money to pay for them, and installation of advanced infrastructures from water to high-speed rail systems.

These economic changes came first from Japan, South Korea, and the European Union. Rebuilding after WWII from the total destruction of both Asia and Europe meant an opportunity to develop and re-create businesses and industrials and commercialize new technologies. The historical key in Japan and then later in the European Union was the dependency on fossil fuels for industrial development, production, and transportation. For Japan, as an island nation, this was a critical transformation in the mid-19th century with the American "Black Ships" demanding that Japan open itself to international, especially American, trade. However, as recent events testify, Japan made the mistake of bending to the political and corporate pressures of the United States to install nuclear power plants despite the atomic bombings of two of its major cities in WWII. The final results of tragedies from the 9.0 earthquake are not final yet in terms of the nuclear power plants in Fukushima and its global impact on the environment, let alone in Japan and the immediate region of northern Asia.

Soon after the end of the Cold War in the early 1990s, the GIR become dominant in Japan and spread rapidly to South Korea as well as Taiwan and somewhat to India. China came later when it leapfrogged into the 21st century through the GIR. Germany, Japan, and South Korea took the lead in producing vehicles that required less amounts of fossil fuels and were more environment-"friendly," often again called "clean tech." Hence their industrial development of cars, high-tech appliances, and consumer goods dominated global markets.

The United States ignored the fledging technological and economic efforts in the European Union, South Korea, and Japan as the nation tilted into a long period of self-absorption, bubble-driven economic vitality. The nation had a history of cheap fossil fuels primarily from inside the country and the given high-tax breaks and incentives (Tickell, 2009). The 2IR also had survived

WWII successfully. Furthermore, the end of Cold War meant to the Americans that their 2IR was to dominate and be in control of global economic markets. The Soviet Union had failed to challenge them. Then came 9/11 and its aftermath along the longest continuous war in American history as well as the continuing battle with fundamental Islamic terrorists. With its own unique and fractured political debate and power struggles, the United States labored to make sense of a post–Cold War era in which special interests replaced reason and any movement toward a sound domestic economic policy.

Instead, the American ideological belief in a "market economy," entrenched in the late 1960s to mid-1970s, replaced the reality of how government and industry must collaborate and work together. The evidence of the problems and hardships from "market forces" came initially from a convergence of events in the early part of the 21st century including a global energy crisis, the dotcomcollapse, and terrorist attacks. Spending and leveraging money into the market caused the global economic collapse almost a decade later in October 2008.

The Economist characterized the basic economic problem the best when in mid-2009, a special issue was published under the title "Modern Economic Theory," superimposed on the Bible melting (Economist, July 18–24, 2009). Basically the case was made that economics is "not a science" in large part because its theories and resultant data did not predict the global economic recession that started in the fall of 2008 and continued3 years later. From that special issue of *The Economist* in the summer of 2009, an international debate about conventional modern economics started and continues today.

The GIR impacts the United States at the local level in a completely different perspective and rational than at the regional, state, or national level. Infrastructures of energy, water, waste, transportation, and information technology among others and how they are integrated are the core of the GIR (Clark and Cooke, 2011). These infrastructure systems need to be compatible yet integrated with one another. For example, renewable energy power generation must be used for homes, businesses, hospitals, and nonprofit organizations (government, education, and others) that are metered and monitored as "smart on-site grids" and also used for the energy in vehicles, mass trains, and buses, among other transportation infrastructures (Knakmuhs, 2011). Such "agile energy" or "flexible systems" (Clark and Bradshaw, 2004) allow people to generate their own power while also being connected to a central power grid. However, both the local power and central power in the GIR need to be generated from renewable energy sources, with standby and backup storage capacity.

There are five key elements for the GIR: (1) energy efficiency and conservation; (2) renewable power generation systems; (3) smart grid-connected sustainable communities; (4) advanced technologies like fuel cells, flywheels, and high-speed rail; and (5) education, training, and certification of professionals and programs. First, communities and individuals all need to conserve and be efficient in the use of energy as well as other natural resources like land, water, oceans, and the atmosphere. Second, renewable energy generated from wind, sun, ocean waves, geothermal, water, and biowaste must be the top priority for power onsite and for central plants.

The third element is the need for smart girds on the local and regional levels in which both the monitoring and controlling of energy can be done in real time. Meters need to establish base load use so that conservation can be done (systems put on hold or turned off if not used) and then renewable energy power is generated when demand is needed. The fourth element needs to be advanced storage technologies such as fuel cells, batteries, regenerative brakes, and ultracapacitors. These devices can store energy from renewable sources, like wind and solar that produced electricity intermittently, unlike the constant supply of carbon-based fuel sources. Finally, the fifth element is education and training for a workforce and entrepreneurial and business sector that is growing and providing employment opportunities in the GIR.

In general, the GIR must provide support and systems for smart and "green" communities so that homes, businesses, government, and large office and shopping areas can all monitor their use of natural resources like energy and water. For example, communities need devices that capture unused water and that can transform waste into energy so that they can send any excess power that is generated to other homes or neighbors. Best cases from around the world of sustainable communities that follow these elements of the GIR exist today (Clark, 2009).

Essentially the GIR was started by governments that were concerned about the current and near-future societal impact of businesses and industries in their countries. The European Union and Asian nations in particular have had long cultural and historical concerns over environmental issues. The Nordic nations, for example, have started programs on ecocities as well as reuse of waste for more than three decades. Sweden, Denmark, and Norway have all either eliminated dependency on fossil fuels now for power generation or will be eliminating the dependency in the near future. All but Finland have shut down nuclear power plants and their supply of energy as well. The same since the 1980s has been true in most other EU nations except France.

However, the key factor in the European Union and Asia has been their respective government leadership in terms of public policy and economics. Consumer costs for oil and gas consumption are at least four times those of the United States due to the higher taxes (or elimination of tax benefits) to oil and gas companies in these other nations. The European Union has implemented such a policy for two decades, which has also completed people to ride more in trains and mass transit rather in their individual cars. The United States, on the other hand, continues to subsidize fossil fuels and nuclear power though tax incentives and government grants. It is not so in the European Union and Asia. The impact of fossil fuels on climate change was the basis for changing these policies and financial structures over two decades ago. Today the impact on the environment has become severe and thus even more significant for future generations around the world.

The historical difference has been that the American contemporary economic ideology of marketforces is simply a balance of supply and demand. This neoclassical economic model has failed for many reasons, but especially due to one of the two key issues presented in *The Economist* special issue (July 2009) that points out that economics is not a science. This is important for a number of reasons, but the basic one, which pertains to the GIR, is that contemporary economics does not apply to major industrial changes, such as the GIR, let alone the beginning of the 2IR. For most economists to be confronted with a challenge to their field being a science or not is disturbing. The "dismal science" may be boring with its statistics, but to not be a science brings questions to the entire contemporary field of economics and its future.

The debate is over how does a community or nation change? Economics is one of the key factors. The issue is, are "market forces" the key economic change factor? The 2IR discovered that market forces or businesses by themselves could not get fossil fuels and other sources of energy into the economy at reasonable costs. Time, government support, and policies provided the market with capital and incentives. Additionally, the GIR economics includes economic externalities such as the environmental and health costs.

In short, the "market force" neoclassical paradigm represented the American economic policies (and also the United Kingdom) for over the last four decades. Prime Minister Thatcher and then President Reagan were the embodiment and champions of this economic paradigm derived from Adam Smith (Clark and Fast, 2008). Market force economics had some influence on the European Union and Asia but then demonstrated its failure in October 2008 with the global economic collapse that started in the United States on Wall Street. This failure meant that some of the government programs in the European Union and Asia, which had succeeded, now needed to be given more economic attention because they basically differed greatly from the US and UK economic models.

These other nations have been in the GIR now themselves for several decades, which succeeded and continued to do so with a different economic model. Northern European Union, Japan, South Korea, and China are clear documented examples of a different economic model. For example, a key economic government program representing the GIR in the European Union is the Feed-in-Tariff (FiT), which started in Germany during the early 1990s and was successfully taking route in Italy, Spain, and Canada as well as nations in the European Union and Asia. There are problems in Spain and Germany has decided to cut it back, whereas the United States has not started a FiT in any significant, long-term planned policy program on a national, let alone a state, level. Some US communities and states have started very restrictive and modest FiT programs.

EU Policies

Germany jumped out in the lead of the GIR in the European Union with its FiT legislation in 1990 (Clark and Cooke, 2013). Basically the FiT is an incentive economic and financial structure to encourage the adoption of renewable energy through government legislation. The FiT policy obligates regional or national electricity utilities to buy renewable electricity at above-market rates. Successful models like that exist such as the EU tax on fuels and the California cigarette tax, both of which cut smoking dramatically in California and for people to use mass transit and trains rather than drive their cars as much in the European Union. However, they also provide incentives and metering mechanisms to sell excess power generated back to the power grid. Other EU nations, especially Spain, followed, and the policy is slowly being developed in Canada and some US states and cities. Chart A shows the economic impact of the FiTs. Over 250,000 "green" jobs have been created in Germany alone. The graphs in Chart A also show the growth in Germany of the solar and wind industries and how this expansion is becoming global.

Germany is now the world's leading producer of solar systems because it has more solar systems installed than any other nation based on the creation of world-leading solar manufacturing companies, solar units sold and installed are measured by sales, amount of kilowatts per site, and records keep by the local and national governments (Gipe, 2011). China took the lead at the end of 2011. The extensive use of solar systems by Germany is despite the fact that the nation has many cloudy and rainy days along with significant snow in the winter, which is common in northern Europe. Japan implemented in 2010 a similar aggressive FiT system to stimulate its renewable energy sector and regain renewable energy technological (solar and system companies and installations) leadership that it held in the early part of the 21st century. Measurements were kept by the solar companies as well as local and national governments. MITI, the Japanese national research organization, measures the use of renewable energy systems on a quarterly basis. However, the aftermath of the Japanese earthquake and destruction of the nuclear power plants in April 2011 could actually expedite renewable energy growth and installation through a number of government programs and incentives that are being proposed.

Other European countries have similar GIR programs as well. Denmark, for example, will be generating 100% of its energy from renewable power sources by 2050. While trying to meet that goal, the country has created new industries, educational programs, and therefore careers. One good example of where this policy has accomplished dramatic results is the City of Frederikshavn in the Northern Jutland region of Denmark. The city has 45% renewable energy power now, and by 2015, it will have 100% power from renewable energy sources (Lund, 2009). In terms of corporate development in the renewable energy sector, for example, one Danish company, Vestas is now the world's leading wind power turbine manufacturer with partner companies all over the world because of its partnership and joint ventures in China. Vestas continues to introduce improved third-generation turbines that are lighter, stronger, and more efficient and reliable. They also continue to design new systems, like those that can be installed offshore away from impacted urban areas.

Germany, Spain, Finland, France, United Kingdom, Luxembourg, Norway, Denmark, and Sweden are on track to achieve their renewable energy generation goals. Italy is fast approaching the same goals when in 2010, it gained the distinction of having the most megawatts of solar energy installed from Germany. However, Denmark is one of the most aggressive countries due to its seeking 100% renewable energy power generation by 2050. Already Denmark has a goal of 50% renewable energy generation by 2015 (Clark, 2009). Other EU countries are lagging behind, especially in Central and Eastern Europe. The European Union has required all its member nations to implement programs like those in Western European Union to be energy independent from getting oil and gas, especially now since most of these supplies come from North Africa, the Middle East, and Russia.

Various EU nations have widely different starting positions in terms of resource availability and energy policy stipulations. France, for example, is a stronger supporter of nuclear energy. Finland, recently has installed a nuclear power plant due its desire to be less dependent on natural gas from Russia. However, Sweden is shutting down its nuclear power plants. The United Kingdom and the Netherlands have offshore gas deposits, although with reduced output predictions. In Germany, lignite offers a competitive foundation for baseload power generation, although hard coal from German deposits is not internationally competitive. In Austria, hydropower is the dominating energy source for generating power, although expansion is limited.

Other European Union directives toward energy efficiency improvement and greenhouse gas (GHG) emission reductions also impact electricity generation demand. Many EU members have taken additional measures to limit GHG emissions at the national level. Since the EU-15 is likely to miss its pledged reduction target without the inclusion of additional tools, the European Parliament and the Council enacted a system for trading GHG emission allowances in the community under the terms of Directive 2003/87/EC dated October 13, 2003. CO_2 emissions trading started in January 2005 but have not produced the desired results due to limitations of "cap-and-trade" economic measures and the use of auctions over credits given for climate reduction.

After being established for 3 years, by 2007 the results are not good, however, as the economics and "markets" are not performing as predicted. Basically the carbon exchanges have performed poorly and not as promised to either buyers or sellers of carbon credits (or other exchange mechanisms). The initial issues are emission caps not being tight enough and lack of significant EU or local government oversight (EU, 2009). By 2010, many of the exchanges have closed or combined with others. The problem is often cited as the lack of supporting governmental (European Union or by nation) policies, but the real issue is that the economics does not work as well as the control over carbon emissions. The trading and auction mechanisms furthermore, do not provide direct and measurable solutions to the problem of emissions and its impact on climate change. A far more direct finance and economic mechanism as proposed by several EU nations and China would be to have a "carbon tax."

An important lesson from the FiT policies in Germany came from two - decades of the policies from 1990 through 2007. As Chart B shows, Germany learned that a moderate or small FiT was not sufficient to push renewable energy systems like solar systems into the mainstream of its economy. In short, far more aggressive use of the FiT type of financing and/or direct carbon taxes needs to be made. On its own the solar industry would not move fast enough into the GIR. In many ways, this is the lesson for other nations. In fact, the reality of the 2IR historically has been to have strong and continuous government incentives from the late 19th century to the present day. The definition and model of economics of a market remains critical in understanding how the United States can move into the GIR. Consider now how Japan and South Korea did just that: moved into the GIR with strong government leadership and financial support.

Japan and South Korea are Leaders in the Green Industrial Revolution

It took an extraordinary political transition to prompt Europe to open the door to the GIR, whereas Japan and South Korea in particular have taken a completely different path. And now China is moving aggressively ahead in the GIR. Most of the information and data below will be focused on China (Clark, 2009). For example, China led the United States and the other G-20 nations in 2009 for annual "clean energy investments and finance," according to a new study by "The Pew Charitable Trusts" (Lillian et al., 2010, p.4):

> *Living in a country with limited natural resources and high population density, the people of Japan had to work on sustainability throughout their history as a matter of necessity. With arable land scarce—some 70-80% of the land is mountainous or forested and thus unsuitable for agricultural or residential use—people clustered in the habitable areas, and farmers had to make each acre as productive as possible. The concept of "no waste" was developed early on; as a particularly telling, literal example, the lack of large livestock meant each bit of human waste in a village had to be recycled for use as fertilizer.*
>
> *Along with creating this general need for conservation, living in close proximity to others inspired a culture in which individuals take special care in the effect their actions have on both the surrounding people and environment. As such, a*

desire for harmony with others went hand in hand with a traditional desire for harmony with nature. Nature came to be thought of as sacred and to come into contact with nature was to experience the divine. Centuries-old customs of cherry blossom or moon viewing attest to the special place nature has traditionally held in the Japanese heart.

However, in April 2011, China became the world leader of financial investment in "clean tech" with $54 billion invested, which was over $10 billion than Germany, which took the second place, and almost double that of the United States, which was at the third place (San Jose, 2011, p. 8). Wind was the favorite sector of renewable energy with $79 billion invested globally. This article noted in particular a comment by a senior partner in a venture capital firm, "a lot of the clean technologies are dependent on policy and government support to scale up. In some other parts of the world (not USA), you have more consistency in the way these types of funds are appropriated" (San Jose, 2011, p. 8).

The Japanese have had a long cultural and business history in commercializing environmental technologies. The 2011 earthquake made Japan focus back on that historical tradition. The future has yet to become clear and will not be defined for some months and years ahead. However, in Japan, the environment took a backseat to industrial development during the drive toward modernization and economic development that began in the latter half of the 19th century. After nearly 300 years of self-imposed isolation from the world, Japan was determined to catch up with the industrialized West in a fraction of the time it took Europe and the United States to make their transitions, eventually emerging as a great power in the beginning of the 20th century.

Economic development continued unabated until WWII, when its capacity was destroyed by American bombings. Economic growth restarted again in the postwar period at a rapid pace but with a distinctive orientation and concern for the limited nature resources of the island nation. By the 1970's, based on the strength of its industry and manufacturing capabilities, Japan had attained its present status as an economic powerhouse. Companies like TOTO (concerned with water and waste conservation and technologies) along with the auto makers concerned with atmospheric pollution emerged as global leaders. A large part of this success was the need for the government to invest in research and development organizations (e.g., METI) to support companies and business growth, what would now be called the GIR. For example, high-speed rail was started in Japan in the mid-1980s (see Chart C) and expanded. The transportation systems were economically efficient as well as environmentally sound and provided for the public at reasonable rates.

Although this incredibly successful period of development left many parts of the country wealthy, it also resulted in serious environmental problems. In addition, the oil crisis had hit Japan particularly hard because of its lack of natural resources, making it difficult for the industrial and manufacturing sectors to keep working at full capacity. To respond to the effects of pollution, municipalities began working earnestly on ways to reduce emissions and clean up the environment, while the Japanese industry responded to the oil crisis by pushing for an increase in energy efficiency.

At the same time, Japan's economy was evolving more toward information processing and high technology, which held the promise of further increases in energy efficiency. Japan created new innovative management "team" systems that were copied in the United States and the European Union. Many manufacturing firms saw value in establishing plants in other developed countries in part to create a market for their products, employ local workers, and establish firm and solid roots. For example, Toyota and Honda established their Western Hemisphere Headquarters in Torrance, California. Other high-tech companies established large operations throughout the US. In this way, Japanese in government, industry, and academia have worked collaboratively with local and regional communities to reincorporate traditional Japanese ideas about conservation and respect for the environment to create sustainable lifestyles compatible with modern living.

Community-level government efforts in Japan, supported by national government initiatives, have led to unique advancements in energy efficiency and sustainable lifestyles, including novel ways of preventing and eliminating pollution. As it stands, Japan is responsible for some 4% of global CO_2 emissions from fuel combustion, and although this is the lowest percentage among major industrialized nations, it is still something the country intends to reduce, with a long-term goal of reducing emissions by 60%–80% by 2050. With the majority of energy still coming from coal, Japan is also attempting a large shift toward renewable energy.

As of November 2008, residential-use solar power generation systems have been put in place in around 380,000 homes in Japan. A close examination of the data on shipments domestically in Japan shows that 80%–90% are intended for residential use, and such shipments are likely to increase, as the government aims to have solar panel equipment installed in more than 70% of newly built houses by 2020 to meet its long-term goals for reductions in emissions. Current goals for solar power generation in Japan are to increase its use 10-fold by 2020 and 40-fold by 2030, and large proposed subsidies for the installation of solar power generation systems—9 billion yen or $99.6 million total in the first quarter of 2009—along with tax breaks for consumers will continue the acceleration of solar adoption by Japanese households.

In recent years, places like Europe, China, Southeast Asia, and Taiwan, saw tremendous growth in energy generation almost entirely from solar power installations. However, these have mostly involved large-scale solar concentrated power facilities not fit for individual households. In Japan, however, as solar power generation systems for residential use become increasingly commonplace, they have become concentrated by creating sustainable communities through use on roofs of local homes and businesses.

The same is true with the light-emitting diode (LED) light bulbs, and now solar panels. Today LED bulbs may cost a few pennies more, but they last far longer than a regular light bulb and can be recycled without issues of mercury and other waste contamination. The result is better lighting for homes and offices but significantly less costs in terms of the systems and the environment. Some LED bulbs are guaranteed to last from 6 to 8 years (Nularis, 2011). While energy demands in homes and offices continues to rise due to the Internet, computers, and video systems, the installation of energy-efficient and now cost-saving systems if very much in demand. Some states are even being required by law the to change over from the older light bulbs to the newer LED ones.

Distributed Renewable Energy Generation for Sustainable Communities

Adding more complications to the policy decisions of the European Union, Japan, and South Korea is the reality of an aging grid and undercapacity. The European Union must crank up investment in new generation. Estimates are coming in that indicate that to meet demand in the next 25 years, they will need to generate half-again as much electricity as they are now generating. According to the International Energy Outlook 2010, conducted by the US Energy Information Administration (USEIA, 2011), world's total consumption of energy will increase by 49% from 2007 to 2035. This could result in a profound change in the European Union's power generation portfolio, with options under consideration for new plants including nuclear energy, coal, natural gas, and renewables.

Originally, when nations electrified their cities and built large-scale electrical grids, the systems were designed to transmit from a few large-scale power plants. However, these systems are inefficient for smaller scale distributed power from renewable sources (Clark, 2006). Although some systems will allow for individual households to either buy power or sell power back to the grid, the redistribution of power from numerous small-scale sources are not yet managed well economically (Sullivan and Schellenberg, 2011). As Isherwood et al. (1998 and then in 2000) document in the studies of remote villages, renewable energy for central power can meet and even exceed the entire demand for a village, hence making it energy independent and not needing to import any fossil or other kinds of fuels. This model and program has worked in remote villages, but it can also be applied to island nations and even larger urban communities or their subsets.

The grid of the future has to be "smart" and flexible and based on the principles of sustainable development (Clark, 2009). As the Brundtland Report said in 1987"as a minimum, sustainable development must not endanger the natural systems that support life on Earth: the atmosphere, the waters, the soils and the living beings" (Bruntland, 1987: Introduction). With that definition in mind, a number of communities sought to become sustainable over the last three decades.

Integrated "agile" (flexible) strategies applied to infrastructures are needed for creating and implementing "on-site" power systems in all urban areas that often contain systems in common with small rural systems (Clark and Bradshaw, 2004). The difference in scale and size of central power plants (the utility size for thousands of customers) with on-site or distributed power can be seen in the economic costs to produce and sell energy. Historically the larger systems could produce power and sell it for far less than the local power generated locally for buildings. Those economic factors have changed in the last decade (Li and Clark, 2009). Now on-site power particularly from renewable energy power (e.g., solar, wind, geothermal, and biomass) has become far more competitive and is often better for the environment. Large-scale wind farms and solar-concentrated systems are costly and lose efficiency due to transmission of power over long distances (Martinot and Droege, 2011).

Developing World Leaders in Energy Development and Sustainable Technologies

Some of the major benefits of The GIR are job creation, entrepreneurship, and new business ventures (Clark and Cooke, 2011). Considerable evidence of these benefits (Next 10, 2011) can be seen in the European Union, especially Germany and Spain (Rifkin, 2004). Many studies in the United States have documented how the shift to renewable energy requires basic labor skills and also a more educated workforce, but one that is also locally based and where businesses stay for the long term. This is a typical business model for almost any kind of business and is what has motivated EU universities to create "science parks," which take the intellectual capital from a local university and build new business across nearby to the campus (Clark, 2003a,b).

Asia's shift to renewable energy will require extensive retraining. Consider the case of wind power generation in China. In the early 1990s, Vestas saw Asia and China as the new emerging big market. Vestas agreed to China's "social capitalist" business model (Clark and Li, 2004; Clark and Jensen, 2002), where the central government sets a national plan, provides financing, and gives companies direction for business projects over 5-year time frames, which are then repeated and updated. Business plans are critical to any company, especially when set and followed by national governments.

A major part of the Chinese economic model required that foreign businesses be colocated in China with at least a 50% Chinese ownership. This meant that in the late 20th and early 21stcenturies, the Chinese government owned companies or were the majority owners of the new spin-off government-owned ventures, established international companies, or started businesses in China. Additionally, China required that the "profits" or money made by the new ventures be kept in China for reinvestments.

Additionally the results, such as with the renewable energy companies like wind and solar industries, were that all the ancillary supporting businesses also needed to support the companies from mechanics, software, plumbing, and electricity to installation, repair and maintenance, and other areas. Supporting industries were also needed such as law, economics, accounting, and planning, especially since the Chinese government began to create sustainable communities that required all these skill sets (Clark, 2009, 2010). Hence these businesses grew and became located in China.

However, the Chinese social capitalism model is not rigid with the government owning controlling percentage (over 50%) of a company. Many businesses were started by the Chinese government with its holding of 25%–33% of shares, whereas other firms were owned by the former government employees, until the companies went public (Li and Clark, 2009). Yet in almost all cases, the companies are competitive globally and are performing remarkably well as demonstrated again in the renewable energy sector where in early 2011, SunTech, a China-based publically traded company, became the world's largest manufacturer and seller of solar panels (Chan, 2011). According to a press release by the company in February 2011, it has delivered more than 13 million PV panels to customers in more than 80 countries.

Today China is a (if not the) world leader in wind energy production and manufacturing with over 3000 MW installed in China alone (Vestas, 2011). The Chinese are now following a similar business model in the solar industry (Martinot and Droege, 2007–2010). As such, China and Inner Mongolia (IMAR) have contracted Vestas to install 50 MW in IMAR (Vestas, 2011), according to a report from the Asian Development Bank (Clark and Isherwood, 2008, 2010), which argues for targeted needs to:

- Create international collaborations between universities and industry
- Conduct research and development of renewable energy technologies
- Build and operate science parks to commercialize new technologies into businesses
- Provide and promote international exchanges and partnerships in public education, government, and private sector businesses

The end results for the European Union are smart homes and communities. The GIR starts in the home so that energy efficiency and conservation are a significant part of everyone's daily life. The home is the place to start. However, it is also the place to start with the other elements of the GIR, such as renewable energy generation, storage devices, smart green communities, and new fuel sources for the home and transportation.

Costs, Finances, and Return on Investment

Government policy(s) and finance are critical for economic growth especially concerning the environment and climate change. The basis of the GIR in the

European Union, South Korea, and Japan can be seen in their articulation of a vision and financial programs. Most of these countries also had established government energy plans. China in fact has had national plans since the People's Republic of China was established in 1949. Having a plan is in fact the basic program and purpose of most business educational programs. Governments need to have plans, as most businesses do. Business plans are for themselves and their clients. Yet the United States continues without any national energy or environment plans. Most US states do not have them either, whereas an increasing number of cities and communities are developing them to plan for becoming sustainable.

This lack of planning has both long-term and short-term impacts. The finance of new energy technologies and systems (like any new technology) often depends on government leadership through programs in public policy and finance (Clark and Lund, 2001). Fossil fuel energy systems in the 2IR have been funded and supported by the governments of Western nations through tax reductions and rebates that continue today. For the GIR, it is only logical and equitable that such economic and financial support continues. That means the US national government should provide competitive long-term tax incentives, grants, and purchase orders for renewable energy sources rather than just fossil fuels.

Meanwhile the European Union, South Korea, and Japan took the leadership in the planning, finance, and creation of renewable energy companies, whereas other nations including the United States did not (Li and Clark, 2009). For example, because of the national policy on energy demand and use, Japan has one of the lowest energy consumption measurements in the developed world. This has been made possible by its continued investment in long-term energy conservation while developing renewable sources of energy and companies that make these products. Japan's per capita energy consumption is 172.2 million Btu versus 341.8 million Btu in the United States.

One critical feature of a long-term economic plan is the need for life cycle analysis (LCA) versus cost–benefit analysis (CBA). Although not discussed much in this chapter, Clark and Sowell (2002) discuss these two very different accounting processes in-depth as the systems apply to government spending. Each approach is critical in how businesses learn what their cash flow is and their return on investment (ROI). The CBA model only provides for 2- to 3-year ROI since that is what most companies (public or government) require for quarterly and annual reports. However, for new technologies (like renewable energy, and also wireless and Wi–Fi technologies) more than a few years are needed on the ROI. The same was true in the 2IR when oil and gas were first discovered and sold. Now in the GIR, longer economic and financial ROIs are needed.

LCA covers longer time periods, such as 3–6 years, and within renewable energy systems some are as long as 10–20 years depending on the product and/or service. Furthermore, LCA includes externalities such as environment, health, and climate change factors, all of which have financial and economic information associated with them. The point is CBAs are limited. The basic concept is that the LCA consists of one long-term finance model in the United States today for solar systems called a Power Purchase Agreement (PPA) that contracts with the solar installer or manufacturer for 20–30 years. The PPA is a financial arrangement between the user "host customer" of solar energy and a third-party developer, owner, and operator of the photovoltaic system (Clark, 2010).

The customer purchases the solar energy generated by the contractor's system at or below the retail electric rate from the owner, who in turn along with the investor receives federal and state tax benefits for which the system is eligible on an annual basis. These LCA financial agreements can range from 6 months to 25 years and hence allow for a longer ROI. However, there are other ways to finance new technologies especially if they are installed on homes, office, and apartment buildings. Today financial institutions and investors can see an ROI that is attractive when the solar system on a home, for example, is financed as a lease, part of tax on the home, or included in the mortgage itself like plumbing, lighting, and air-conditioning.

Some newer economic ideas on how to finance technologies that reduce "global climate change" are interesting. One way to describe the GIR financial mechanisms is by looking at the analytical economic models that financed the 2IR. For example, the 2IR was based on the theory of abundance. The earth had abundant water and ability to treat waste. Hence buildings, businesses, homes, and shopping complexes all had plumbing for fresh water and drainage for waste. The same scenario occurred in electrical systems that took power from a central grid for use in the local community buildings. Locally and globally, people have found that systems work, but that now with climate change, they need to conserve resources and be more efficient.

When these economic considerations are factored into even the CBA rather than an LCA financial methodology, the numbers do not work (Sullivan and Schellenberg, 2011). Financial consideration for energy transmission and then monitoring by smart systems are needed, but costly. Long distances make them even more costly today because then the impact of the climate (storms, tornadoes, floods, etc.) with required operation and maintenance are added with security factors. The actual "smart" grid is at the local level where these and other uncontrolled costs can be eliminated and monitored.

The financing of water, waste, electrical, and other systems for buildings was over time incorporated into the basic mortgage for that building. In short, modern 2IR infrastructure systems were no longer outside (e.g., the outhouse or water faucet) but inside the building. What this 2IR financial model does is set the stage for the GIR financial model. Much of the 2IR financing for fossil fuels and their technologies came about as leases or building mortgages. A variation of the 2IR model, which is a bridge to the GIR, is the PACE (Property Assessed Clean Energy) program started in 2008 in Berkeley, California, whereby home owners can install solar systems on their buildings, for example, but pay for them from a long-term supplemental tax that is transferred with the sale of the property assessment on their property taxes. The financing is secured with a lien on the property taxes, which acquires a priority lien over existing mortgages. The program was put on hold in July 2010 when the Federal Housing Finance Agency expressed concerns about the regulatory challenge and risk posed by the priority lien established by PACE loans. Nevertheless, the US Department of Energy continues to support PACE.

The dramatic change to the GIR, however, moves past that financial barrier. Mortgages are part of the long-term cost for owning a property. Therefore in the GIR, the conservation and efficiency for the 2IR technologies in buildings can be enhanced with the renewable energy power, smart green grids, storage devices, and other technologies through mortgages that can be financed from one owner to another over decades (20–30 years or more). This sustainable finance mortgage model is a long-term or a LCA framework and provides for technologies and installation costs to the consumer that makes the GIR attainable with a short time. Changes, updates,and new technologies can easily be substituted and replace the earlier ones. It is needed that the banking and lending industries try this GIR finance model on selected areas. After some case studies the model can be replicated or changed as needed.

CONCLUSIONS AND FUTURE RESEARCH RECOMMENDATIONS

The basic point of this chapter is to highlight the need for economics to be more scientific in its hypothesis and data collection. Furthermore, the economics of the 2IR and the GIR are very similar if not parallel, for example, the role of government since it must often take the first steps in directing, creating, and financing technologies. As the 2IR needed government to help drill for oil and gas as well as mine for coal, the government needed to build rail and road transportation systems to transport the fuels from one place to another.

The GIR is very much in the same economic situation. The evidence can be seen in Asia and the European Union. And especially now in China, the central government plans for environment and related technologies help the nation move into the GIR. Moreover, there is a strong need for financial support, not tax breaks or incentives, but investments, grants, and purchasing, for GIR technologies, such as renewable energy. This can be seen in the United States today with the debate over smart grids. What are they? And who pays for them? When the smart grid is defined as a utility, then the government must pay for it since it is part of the transmission of energy, for example, over long distances, which must be secure and dependable.

However, as the GIR moves much more into local on-site power, the costs of the smart grid are at the home, office buildings, schools and colleges, shopping malls, and entertainment centers. Local governments are also involved as they are often one of the largest consumers of energy in any region and hence emitters of carbon and pollution. Within any building, a smart grid must know when to regulate and control meters and measurement of power usage and conservation. The consumer needs the new advanced technologies, but the government must support these additional costs and their use of energy as they impact the local community and larger regions' residential and business needs.

Economics has changed in the GIR. And yet, economics has a basis of success in the 2IR. Historically, 2IR economics was successful because the government was needed to support its technologies along with goods and services. The evolution into the neoclassical form of economics was far more a political strategy backed by companies that wanted control of infrastructure sectors. However, the reality was that "greed" took over and has now forced a rethinking of economics as nations now move into the GIR.

CHART A: THE GERMANY FEED-IN-TARIFF POLICY AND RESULTS (1990–2010)

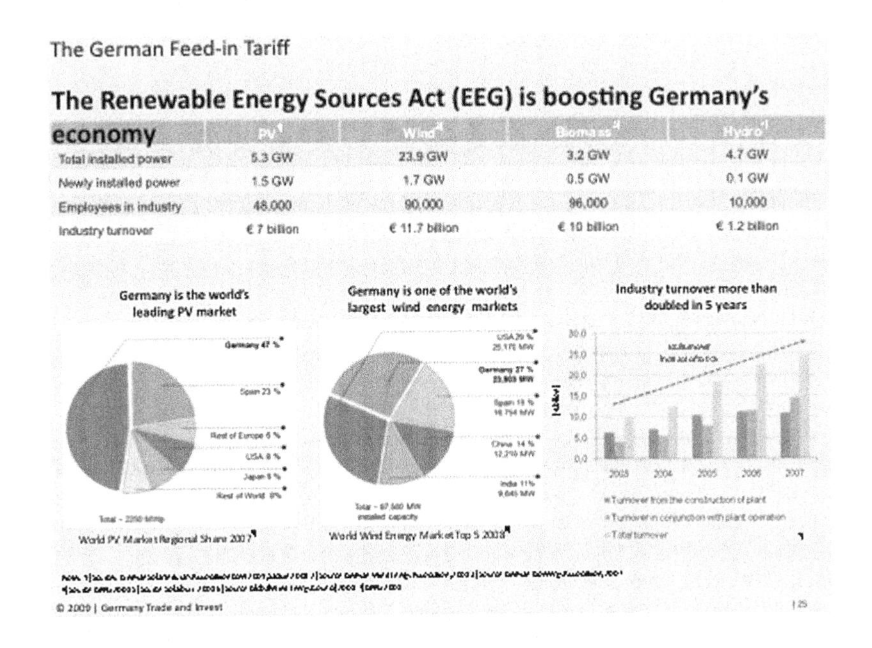

CHART B: THE GERMANY FEED-IN-TARIFF POLICY ECONOMIC RESULTS (1990–2007)

The German PV Market

The PV market in Germany boomed in 2004 following amendment of EEG

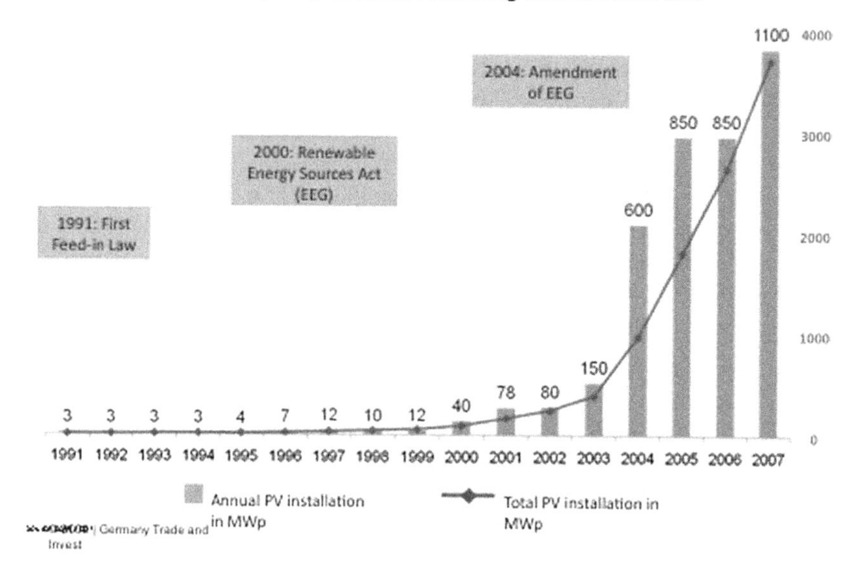

REFERENCES

Bruntland Commission Report, 1987. Our Common Future. UN Commission General Assembly Resolution #38/161 for Process of Preparation of the Environmental Perspective to the Year 2000 and beyond, 1983. Oxford University Press.

Chan, S., March 2011. Global solar industry prospects in 2011. In: SolarTech Conference. California, Santa Clara.

Chomsky, N., May 8, 2012. Plutonomy and the Precariat: On the History of the US Economy in Decline, pp. 1–5. Tompatch.com. http://truth-out.org/news/item/8986-plutonomy-and-the-precariat-on-the-history-of-the-us-economy-in-decline76.

Clark II, W.W., 2012. The Next Economics. Springer Press.

Clark II, W.W., Cooke, G., Fall 2011. Global Energy Innovations. Praeger Press.

Clark II, W.W. (Ed.), June 2010. Sustainable Communities Design Handbook. Elsevier Press, NY.

Clark II, W.W., Isherwood, W., January 2010. Inner Mongolia Autonomous (IMAR) Region Report. Asian Development Bank, 2008 and reprinted and expanded on in Utility Policy Journal. Special Issue. China: environmental and energy sustainable Development.

Clark II, W.W. (Ed.), 2009. Sustainable Communities. Springer Press.

Clark II, W.W., Fast, M., 2008. Qualitative Economics: Toward a Science of Economics. Coxmoor Press, London, UK.

Clark II, W.W., 2008. The green hydrogen paradigm shift: energy generation for stations to vehicles. Utility Policy Journal 16. Elsevier Press.

Clark II, W.W., Rifkin, J., et al., 2006. A green hydrogen economy. Special Issue on Hydrogen Energy Policy 34, 2630–2639. Elsevier.

Clark II, W.W., Bradshaw, T., 2004. Agile Energy Systems: Global Lessons from the California Energy Crisis. Elsevier Press, London, UK.

Clark II, W.W., Li, X., 2004. Social Capitalism: transfer of technology for developing nations. International Journal of Technology Transfer, Interscience 3 (1). London, UK.

Clark II, W.W., Dan Jensen, J., 2002. Capitalization of Environmental Technologies in Companies: economic schemes in a business perspective. International Journal of Energy Technology and Policy 1 (1/2). Interscience, London, UK.

Clark II, W.W., 2003a. Science parks (1): the theory. International Journal of Technology Transfer and Commercialization 2 (2), 179–206. Interscience, London, UK.

Clark II, W.W., 2003b. Science parks (2): the practice. International Journal of Technology Transfer and Commercialization 2 (2), 179–206. Interscience, London, UK.

Clark II, W.W., Sowell, A., Nov 2002. Standard Economic Practices Manual: life cycle analysis for project/program finance. International Journal of Revenue Management. London, UK: Interscience Press.

Clark II, W.W., Lund, H., December 2001. Civic markets in the California Energy crisis. International Journal of Global Energy Issues 16 (4), 328–344. Interscience, UK.

Economist, April 21, 2012a. The Third Industrial Revolution. Special Issue.

Economist, January 23, 2012b. The Rise of State Capitalism: The Emerging World's New Model. Special Issue with Lenin smoking a cigar on the Cover. Two articles.

Economist, July 18, 2009. Modern Economic Theory: Where it Went Wrong – and How the Crisis Is Changing it. Special Issue with Bible melting on Cover. Editorial and three articles.

EU, CCC (October 12, 2009). Meeting Carbon Budgets – the need for a step change. Progress report to Parliament Committee on Climate Change. Presented to Parliament pursuant to section 36(1) of the Climate Change Act 2008. The Stationery Office (TSO). ISBN:9789999100076. http://www.official-documents.gov.uk/document/other/9789999100076/9789999100076.pdf.

Federal Housing Finance Agency. http://www.fhfa.gov/webfiles/15884/PACESTMT7610.pdf.

Gipe, P., March 2011. Feed-in-Tariff Monthly Reports.

Hawkins, P., Lovins, A., Hunter Lovins, L., 1999. Natural Capitalism: Creating the Next Industrial Revolution. Little, Brown and Company, Boston, NYC.

Isherwood, W., Ray Smith, J., Aceves, S., Berry, G., Clark II, W.W., Johnson, R., Das, D., Goering, D., Seifert, R., 2000. Remote power systems with advanced storage technologies for Alaskan Village. University of Calif., Lawrence Livermore National Laboratory, UCRL-ID-129289 Energy Policy 24, 1005–1020. January, 1998.

Knakmuhs, H., April 2011. Smart transmission: making the pieces fit. RenGrid 2 (3), 1. www.renewgridmag.

Li, X., Clark, W.W., 2009. Crises, opportunities and alternatives globalization and the next economy: a theoretical and critical review. In: Li, X., Winther, G. (Eds.), Globalization and Transnational Capitalism. Aalborg University Press, Denmark. Chapter #4.

Lillian, J., May 2010. New & noteworthy. Sun Dial, Solar Industry 3–4.

Lund, H., 2009. Sustainable towns: the case of Frederikshavn. 100 percent renewable energy. In: Sustainable Communities. Springer Press. Chapter #10.

Martinot, E., Droege, P. Renewable energy for cities: opportunities, policies, and visions. Series of Reports from 2007–2010. http://www.martinot.info/Martinot_Otago_Apr01_cities_excerpt.pdf.

Next 10, January 19, 2011. Many Shades of Green, 2011. Silicon Valley and Sacramento, CA. www.next10.org.

Nularis, 2011. Data and Information. www.nularis.com.

Rifkin, J., 2004. The European Dream. Penguin Putnam, New York, NY.

Clean energy financing jumps to record $243B. San Jose Business Journal, April 1, 2011 8.

Sullivan, M., Schellenberg, J., April 2011. Smart grid economics: the cost-benefit analysis. RenGrid 2 (3), 12–13. www.renewgridmag.

Tickell, J., 2009. Fuel. Independent Documentary Film. La Cinema Libra, Los Angeles, CA.

Vestas, 2011. www.vestas.com.

FURTHER READING

Chanlett-Avery, E., February 9, 2005. CRS Report for Congress, Rising Energy Competition and Energy Security in Northeast Asia: Issues for U.S. Policy. www.fas.org/sgp/crs/row/RL32466.pdf.

Clark II, W.W., Cooke, G., Jin, J., Lin, C.F., 2013. The Green Industrial Revolution. Published in Chinese.

Clark II, W.W., Winter 2013. The Economics of the Green Industrial Revolution: energy independence through renewable energy and sustainable communities. Special Issue Contemporary Economic Policy Journal (CEP). Western Economic Association International.

Clark II, W.W., Li, X., 2012. Social capitalism: China's economic rise. In: Clark II, W.W. (Ed.), The Next Economics. Springer Press. Chapter #7.

Sustainable Energy and Transportation. Special Issue. In: Clark II, W.W., Lund, H. (Eds.), Utility Policy Journal 16.

Clark II, W.W., December 06, 2007. Partnerships in creating agile sustainable development communities. Journal of Cleaner Production 15, 294–302. Elsevier Press.

Environmental Protection Agency. http://www.epa.gov/greenpower/buygp/solarpower.htm.

Lo, V., April 25, 2011. China's role in global economic development. In: Speech by Chairman of Shui on Land, Given at Asian Society, Los Angeles, CA.

Malthus [(1978/1970)] who argued that conflict over natural resources would arise as a consequence of population growth and environmental degradation – in the economic literature on conflict.

Office of the Comptroller of the Currency (OCC). http://www.occ.gov/news-issuances/bulletins/2010/bulletin-2010-25.html.

Rifkin, J., 2012. The Third Industrial Revolution. Penguin Putnam, New York, NY.

Sanders, B., April 30, 2011. US Senator, Speech before California State Democratic Convention. Sacramento, CA.

U.S. Department of Energy. http://www1.eere.energy.gov/wip/solutioncenter/financialproducts/pace.html.

U.S. Energy Information Administration (US EIA). See Annual Reports at: http://www.eia.doe.gov/oiaf/ieo/world.html.

Chapter 5

Renewable Energy: Scaling Deployment in the United States and in Developing Economies

Joseph Kantenbacher[1], Rebekah Shirley[2,3]
[1]*University of Surrey, Guildford, United Kingdom;* [2]*Power for All, Berkeley, CA, United States;* [3]*University of California, Berkeley, CA, United States*

Chapter Outline

INTRODUCTION TO SOURCES AND USES OF ENERGY IN THE UNITED STATES

The United States is the largest consumer of energy in the world. In 2016, the United States consumed nearly 100 quadrillion Btus (quads) of primary energy, a bit more than one-sixth of the global total for that year. A full 81% of

Sustainable Cities and Communities Design Handbook. https://doi.org/10.1016/B978-0-12-813964-6.00005-7

that energy was supplied by fossil fuels (coal, petroleum, and natural gas). Nuclear and renewable power sources provide the balance of the US energy mix.

American homes and businesses consumed 3800 TWh of electricity in 2015, an average of 11,800 kWh per capita (Electric Power Annual, 2015). Although coal provided as much as half of the United States' electricity as recently as 2008, it now provides the fuel for less than one-third of the total electricity consumed. Coal's dominance of the US electricity market has ebbed in large part due to competition with low-cost natural gas. Between 2000 and 2010, natural gas power plants accounted for 81% of new power production capacity, and now natural gas is the fuel for 32% of US electricity. Nuclear power provides a further 19%, maintaining its market share for years, even though only one new reactor unit has come online since 1996 (in large part due to the substantially improved capacity factors for existing plants). Large-scale hydroelectric facilities form the other major, low-carbon electricity source, contributing about 6% of the electricity consumed. A mix of conventional renewable energy sources like wind, sunlight, and biomass contribute roughly 7.4%.

Residential and commercial buildings are large energy consumers in the United States, accounting for more than a quarter of end-use energy consumption (Estimated U.S. Energy Consumption, 2016). Natural gas and electricity are the primary energy sources in the built environment, with temperature regulation, lighting, and water heating being the chief energy-consuming activities.

The transportation sector, which comprehends personal, public, and commercial ground travel as well as aircraft and marine vessels, is almost exclusively powered by fossil fuels. Petroleum products account for 92% of the primary energy input in the transportation sector in 2016, and natural gas contributed an additional 2.7%. Biomass-derived ethanol and biodiesel are the chief sources of renewable transportation fuel, providing 5%. At present, electricity is a negligible component of the transportation energy system, a mere 0.1%. Significant growth in electric vehicle use in the coming years is forecast, however.

WIND

Wind turbines translate the kinetic energy of moving air into mechanical energy, which in turns powers a generator that produces electrical energy. Conventional turbines have a horizontal-axis design, in which two or three rotor blades are mounted atop a tower and arrayed such that they resemble airplane propellers. When oriented into the wind, the movement of air across the blades generates lift, spinning the shaft connected to an electric generator. Wind power output from this conventional turbine design is a function of two factors: swept area and wind speed. As the rotor area (determined by the blade length) of the turbine doubles, the power output quadruples. A doubling of incoming wind speed translates to an eightfold increase in power output.

Since 1998, the size of wind turbines has substantially increased, with rotor diameter more than doubling and tower height increasing by nearly 50% (Wind Technologies Market Report, 2015). This has contributed to a 180% increase in power output and has contributed to the price of wind-generated electricity dropping from about 40 cents/kWh in the early 1980s to around an average of 2–5 cents/kWh for newly built projects today (Wind Technologies Market Report, 2015; Cost of Wind Energy Review, 2015). Modern turbines can reach peak power outputs in the megawatt range, meaning that utility-scale aggregations of turbines (i.e., wind farms) can readily scale up to several gigawatts in size. Although arraying turbines in wind farms can help achieve economies of scale (particularly with respect to transmission costs), the use of stand-alone turbines can be an economically viable means of providing power to systems in remote locations, such as communications towers or rural irrigation networks.

Also in operation are various vertical-axis designs, which have a vertically oriented main rotor shaft and are usually situated on the ground or rooftops. Because wind speeds tend to be lower closer to the ground and these vertical-axis systems tend to be small in size, the power output from this category or turbines is usually low (in the watt or kilowatt range). Vertical-axis turbines represent about 2% of the generation capacity of small wind systems but are increasingly popular installations for distributed, renewable energy generation (Distributed Wind Market Report, 2015).

Although wind power has been employed in the United States since the mid-19th century, the US modern wind industry did not develop until the 1970s, when it was launched in response to the increasing cost of oil-based electricity generation. By the mid-1980s, California had 1.2 GW of installed wind capacity, which accounted for more than 90% of the global total. California's dominance in this area is commonly attributed to the federal and state investment tax credits that were in place, as well as state-mandated utility contracts for wind power. After the expiration of state and federal investment incentives, the US wind industry stagnated until the late 1990s, at which time the first in a series of production tax credits (giving renewable power producers a rebate for each kWh generated) sparked renewed growth in domestic installations. In recent years, wind power has been a major component of nationwide generation capacity addition, growing from 10% of new capacity additions in 2005 to more than 40% of new additions in 2015 (Wind Technologies Market Report, 2015).

In 2015, nearly 8600 MW of new wind generation capacity was brought online in the United States, bringing the cumulative installed capacity to nearly 74,000 MW. At 17,700 MW, Texas leads the country in terms of total wind capacity, trailed by Iowa, California, and Oklahoma. Seventeen states have greater than 1 GW of installed capacity, and in 12 states, wind energy accounts for at least 10% of in-state power production (Wind Technologies Market Report, 2015).

Although the wind power resource in the United States is substantial, it is also, significantly, distributed across the continent with substantial variation. Although several areas, particularly in the East and West coasts, feature a coincidence of both strong continental wind resources and high-density population centers, the broadest swath of wind resource is located in the plains states, whose low population levels and densities decrease the economic efficiency of employing those wind resources locally. Capitalizing on the resources of the plains will require the build-out of expensive transmission lines, increasing a given project's expense and challenging its viability. Offshore wind resources, which are generally stronger and steadier than their continental counterparts, have also attracted the attention of wind power developers. However, the cost of electricity produced by offshore wind plants is thought to be about three to four times greater than that produced by terrestrial systems, due to higher costs across the breadth of construction, including turbine manufacturing and installation (Wind Technologies Market Report, 2015; National Offshore). In addition, the added expense of undersea transmission lines further diminishes the attractiveness of investment in offshore systems.

Although utility-scale turbines dominate the US wind market, distributed wind installations can be found in all 50 states. Although large installations (greater than 1 MW) are the primary source of distributed capacity addition in the last decade, small wind installations (under 100 kW) are an active portion of the sector. In 2015, domestic small turbine installations grew by 16%, placing an additional 4.3 MW in service (Distributed Wind Market Report, 2015). By number of projects, the most common contexts of distributed systems built in recent years are agricultural and residential, whereas by total power capacity, the most common applications are industrial and commercial. Uncertainty about state and federal subsidies and competition with solar photovoltaics and natural gas may inhibit the expansion of wind power in distributed systems.

SOLAR

The sun is the primary energy input for virtually all processes, biological and otherwise, that occur on the earth's surface. While in the broadest sense several of the energy sources currently employed today—wind, biomass, and fossil fuels—have a solar genesis, there is a class of technologies that directly converts electromagnetic energy from the sun into useful energy. The two main categories of such technology are solar photovoltaics and solar thermal.

Solar Photovoltaics

Solar photovoltaic (PV) technologies harness the photoelectric effect, using sunlight to excite electrons in a semiconductor and generate direct current electricity. Two main categories of PV technologies that currently exist are silicon crystal and thin film. Today, silicon is the most commonly used semiconductor in the PV industry, on account of both the relatively high efficiencies it can achieve as well as its relative abundance as a raw material, although the process of refining silicon to a sufficient level of purity is expensive. Thin-film technologies, such as those made from cadmium telluride or copper indium gallium selenide, promise to address the high materials costs of silicon PV, combining low-cost materials (such as glass or plastic) with thinly spread semiconductors. Although thin-film technologies have limited efficiencies, they have the potential to provide solar power at a lower cost per watt than silicon crystals.

In the United States, PV originally had a niche application in the space program, providing power to shuttles and orbiting satellites. With improved efficiencies (from laboratory-best figures of 15% in the early 1980s to better than 40% today (New world record, 2014)) and decreasing production costs (by about a factor 100 since the 1950s), PV came to be an economical power source of a variety of Earth-bound applications, including off-grid living and remote communication devices. Prior to 2005, the majority of cumulative US PV installations were dedicated to off-grid applications, even though utility-scale PV arrays started to be deployed in the early 1980s. Since 2005, however, most of the added PV capacity has been grid-tied, with the sizeable majority taking the form of utility-scale developments. [In 2016, more than 70% of the added solar capacity came from systems larger than 2 MW (Utility-Scale Solar, 2015).] The cost of installing solar panels has fallen dramatically in the last decade, at about $1.50 per Watt for a utility-scale system today from a cost of about $4.50 per Watt in 2009 (U.S. Solar Market Insight, 2016). Electricity is produced by current solar PV systems at a cost of about 5–7 cents per kWh, which is competitive with many other sources of power.

Solar Thermal

Solar thermal technologies harness sunlight to produce thermal energy. This heat is then used directly or to generate electricity.

In contrast with solar PV, electricity generation by solar thermal technology is almost exclusively the province of large, utility-scale generators. Concentrated solar power (CSP) systems use mirrors or lenses to focus solar energy, heating a working fluid (such as water or oil) that drives an electricity-generating turbine. Based on how they collect solar energy, there are four mainstream CSP designs—trough, linear Fresnel reflector, tower, and dish—although one (trough) has dominated the US CSP market to date. These systems employ parabolic mirrors that concentrate solar heat on fluid-filled receiver that runs the length of each trough. There is currently 1800 MW installed CSP capacity in the United States, 1300 MW of which is located in California (Concentrating Solar Power).

The operation of concentrated solar plants requires both large tracts of contiguous land for siting and substantial volumes of water to provide a cooling reservoir for the steam turbine. Many of the prime locations suitable for CSP siting are remote, requiring additional transmission to connect to the grid, and arid, placing a burden on scarce local water resources. These factors, combined with long construction times and higher cost relative to PV projects, mean that CSP systems are unlikely to achieve greater penetration market than photovoltaics. However, unlike PV, CSP systems can be readily and economically fitted with thermal storage systems such as molten salt. Adding storage allows for operating at night or during cloudy conditions and turns CSP into a dispatchable power source. CSP systems can also be paired with other thermal energy-based systems, such as natural gas power plants, to increase reliability and efficiency.

On a megawattage basis, small-scale distributed applications of solar thermal energy have the biggest market share. As of 2013, US consumers had installed nearly 17,000 MW-th (thermal equivalent) of solar thermal systems, which include pool heaters and water heating systems (Solar Heat Worldwide, 2013). In 2015, the investment tax credit for solar water heaters was extended, which may foster continued investment in these systems. However, cost reductions for PV systems and natural gas may still harm the economic viability of solar hot water systems.

GEOTHERMAL

Geothermal energy systems tap into underground heat reservoirs, utilizing the stored thermal energy directly or as a feedstock for electricity production.

Hydrothermal resources exist where magma comes close enough to the surface to transfer heat to groundwater reservoirs, producing steam or high-pressure hot water. When hydrothermal resources are sufficiently hot (several hundred degrees Fahrenheit) and close to the surface (within a few miles), it can be economically sensible to drill a well and use the steam or hot water either as a direct power input into a turbine (as with dry and flash steam plants) or as a heat source to produce steam with a secondary fluid (as with binary-cycle plants). Shallow hydrothermal resources of more moderate temperature, which in the United States are located primarily in Alaska, Hawaii, and many western states, are commonly used directly to provide heat for buildings, agriculture, and industrial processes.

Even in the absence of hydrothermal resources, geothermal energy can be harnessed for use. Geothermal heat pumps for buildings take advantage of the constant temperature nature of subsurface Earth, cooling warm summer air in the underground or, in the winter, drawing heat up from the relatively warm ground. As this technology does not require the presence of hydrothermal resources, its adoption is feasible in all regions of the country. In the domain of electricity production, next-generation "enhanced geothermal systems" are being designed to access the hot, dry rocket at several miles' depth and, through the injection of water, artificially create a hydrothermal resource (Annual U.S. and Global Geothermal, 2016).

One key feature of geothermal energy is that it is continuously available, as opposed to intermittent resources like sunshine and wind. As such, geothermal power is one of the leading options for renewable resource-based base load electricity generation and provides a technically viable option for supplanting coal power plants, which are arguably the most environmentally pernicious class of generators currently operating.

Although in the late 19th century numerous communities in the United States made use of surface-level hydrothermal resources for residential and commercial heating services, it was not until 1922 that the first geothermal electricity generator was brought online at The Geysers near San Francisco, California. In 1960, the first US utility-scale geothermal plant, an 11-MW facility, was completed. After rapid expansion in the 1980s, spurred in part by California's Geothermal Grant and Loan Program, growth in total capacity and output has slowed. In the past decade, the average growth rate of electrical output from the geothermal industry was about 1% per year (Annual U.S. and Global Geothermal, 2016).

In 2014, geothermal energy resources provided about 5% of US renewable electricity generation, and as of 2009, there were more than 3700 MW of geothermal electricity generators online. Owing to its location on a series of tectonic plate conjunctions, California enjoys considerable hydrothermal resources and, at 2700 MW of installed capacity, is the premier geothermal power producer in the country. With 600 MW of installed capacity and more projects under development than other states, Nevada's fleet is also a significant component of the US geothermal power plant stock. As of 2016, reported projects currently under development are set to add 1200 MW of generation capacity. In addition, it is estimated that about 50,000 homes have geothermal heat pumps installed each year (Geothermal Heat Pumps).

BIOPOWER

Biomass resources exist where solar energy is stored in an organic form as plant matter or other biological material. Biopower technologies utilize biomass to generate electricity.

Direct-combustion steam production is the most common mode for generating biopower, although the scale and thermal efficiency of those boilers tend to be smaller and lower, respectively, than their coal counterparts. Cofiring biomass in conventional power plants can simultaneously scale up the use of biopower and mitigate the environmental impacts of fossil fuel combustion. Gasification systems and organic, anaerobic digestion can also be used to convert biomass into syngas (a mixture of carbon monoxide and hydrogen gas) and methane, respectively. Biomass-derived gases can be used in high-efficiency combined cycle generators and other modular systems, making them prime distributed generation devices.

The broad distribution of resources and diversity for utilizing them makes biomass a viable renewable energy source for much of the contiguous United States. It is a particularly important renewable resource in the southeast, which has a modest endowment of solar, wind, and geothermal resources. A survey of available biomass resources from forest and agricultural lands determined that the technically and economically exploitable volume of biomass is more than seven times greater than the current usage rate (Estimating Renewable Energy). Transportation costs and seasonal availabilities diminish the attractiveness of biomass as an energy resource, particularly in electricity markets, where it competes with coal's low costs and all-year availability.

In 2015, Americans consumed 3.9 quads of biomass-based energy (Short-term Energy Outlook, 2017), about 4% of the total energy used in the United States that year. Woody biomass and wood by-products account for 52% of the energy content of biomass feedstocks and is used primarily for electricity production and industrial heating. Ethanol and biodiesel make up 35% of the biomass total, and combusted agricultural and municipal solid wastes supply most of the biomass balance. Taken together, these three sources of energy provide two-fifths of the renewable energy consumed in the United States. Some 17,300 MW of currently installed electricity-generating capacity has some form of biomass as its primary feedstock (Preliminary Monthly, 2017). In 2015, 64 TWh of biopower electricity were produced, half of which was produced outside of the electric power sector (Electric Power Annual, 2015).

MARINE AND HYDROKINETIC

In oceans and rivers exist large, regular, and untapped ebbs and flows of water. The technical resource potential of these energy sources is substantial, summing to as much as 750 TWh per year (Quadrennial Technology Review, 2015).

There are three potential categories of marine and hydrokinetic energy (MHK) resources: waves, tides, and river and ocean currents. As a whole, the MHK energy industry is in the early stages of development, and although there is a broad array of concepts and designs for capturing marine energy, there has been relatively little in the way of deployment or standardization, particularly at the commercial level. As of this writing, there are no commercial MHK systems operating in the United States (Quadrennial Technology Review, 2015).

ADVANCED RENEWABLES DEPLOYMENT

Through the experience of deployment and the engineering breakthroughs of the laboratory, renewable energy technologies have over the last decades steadily improved in terms of performance (efficiencies, capacity factors) and economics (capital costs). These iterative advances have brought renewable energy technologies ever closer to shedding the long-held perception that they are "alternative" energy sources.

Moving forward, in addition to the regular technological advance of renewable energy devices, there are several measures and developments that can further enhance the utility of renewable energy systems. Broadly, these categories include building integration, vehicle integration, and hybrid systems.

Renewables and Buildings

One of the fastest-growing segments of the solar industry is that of building-integrated photovoltaics (BIPV). BIPV designs seek to replace and enhance certain elements of a building, such as the roof, window overhangs, or walls, with solar panels. This reduces both the materials cost of building construction and the installation cost of the PV panels, and ensures that the PV panels will be optimally situated on the finished structure. Passive solar building design can also take advantage of solar energy, using windows and interior surfaces to regulate indoor air temperatures.

Building-integrated wind designs have also been proposed and implemented, but such arrangements have so far featured substandard turbine performance and unappealing impacts on building inhabitants.

Vehicle-to-Grid Systems

Plug-in hybrid vehicles (PHEVs), which feature both internal combustion engines and electric motors, have the potential not only to reduce the consumption of petroleum products but also to facilitate greater penetration of renewable power sources. The electric battery in PHEVs can both be charged by and discharged into the electric grid, turning the car into a mobile, distributed electricity storage device. This storage capability is thought to be of particular benefit to wind turbines, the power output of which is generally greatest at night when the demand for electricity is the lowest. PHEVs would allow for higher penetrations of wind power than might otherwise be economically viable, storing excess generation at night and dispatching that electricity to meet greater loads during the day. An intermediate step in the vehicle-to-grid system might be a vehicle-to-home approach, wherein electricity is delivered to a household through a direct connection with a PHEV.

Hybrid Systems

Pairing renewable energy systems has the potential to improve economics or performance over what could be achieved by each system working in isolation. For example, sitting solar and wind systems together can reduce the overall transmission costs as there is a day/night complementarity in peak output that could reduce grid congestion and allow for smaller transmission lines. Similarly, combining offshore wind turbines with marine power stations has been proposed as a means of reducing the construction and maintenance costs for each. Hydroelectric systems can serve as energy storage units for wind farms, making use during the day of water pumped uphill by the excess generation of turbines at night. Such hybrid systems have yet to achieve substantial deployment but may have promise as a means to overcome the limitations of individual renewable systems.

SUMMARY OF SCALING RENEWABLES IN THE UNITED STATES

Fossil fuels are the dominant energy source in the United States, providing over four-fifths of all primary energy consumed. However, renewable energy sources are abundant, and when harnessed they can contribute significantly to the satisfaction of US heat and power demands. Over the past several decades, renewable energy technologies have advanced significantly and have achieved ever-greater levels of deployment, both by utilities and by individual consumers:

- Wind turbines currently produce electricity at costs competitive with other fossil-fuel-based generators, leading annual capacity additions of wind power to be among the highest in the electric power sector.
- Solar technologies can efficiently utilize sunlight to provide both heat and electricity. PV and thermal systems both have become popular modes of distributed energy provision, with rapidly falling prices for PV systems making solar power particularly attractive.
- Geothermal deposits provide consistent, clean energy for both electricity generation and heating. The number of hydrothermal power plants is rapidly increasing, particularly in the western United States, through which heat pumps bring geothermal energy to households in every region.
- Biomass is one of the largest sources of nonfossil power in the United States, especially for industrial applications. The quantity and variety of feedstocks available for combustion are substantial, providing for multiple avenues for expanding the biopower sector.
- Marine energy resources are vast, but the technologies to exploit them are still immature.

Owing to its size and patterns of energy consumption, the United States presents an informative study of the obstacles and opportunities associated with scaling renewable energy within an established sociotechnical system that is already capable of providing a full set of "modern" energy services. For many countries, however, the rate of provision of essential energy services falls well short of the desirable levels. In such cases, distributed renewable energy (DRE) systems can be instrumental in achieving goals for sustainable developments.

THE ENERGY ACCESS GAP: REMOTE AND UNDER-GRID POPULATIONS NOT BEING REACHED

Providing universal energy access is one of the greatest development challenges of our time. With almost 16% of the global population lacking access to electricity, the United Nation's Sustainable Development Goal 7 (SDG7), to ensure affordable and clean energy to all people by 2030, is a welcome vehicle by which to unlock the underlying benefits of energy access, which include choice, opportunity, and freedom. Universal energy access is critical to the achievement of many other SDG goals, such as ending poverty, achieving food security and promoting sustainable agriculture, ensuring equitable quality education, achieving gender equality, promoting sustainable economies and productive employment for all, reducing inequality, and responding with urgency to climate change and its impacts.

However, all authoritative sources tracking global electrification rates (such as the World Bank's Global Tracking Framework) find that the absolute number of unelectrified people globally is changing slowly and that more concerted global action is needed to close the access gap and achieve universal electricity access by 2030. At the current electrification rates we are not on track to reach SDG7 goals (Global Tracking Framework, 2017). More specifically:

- 1.06 billion people lacked access to electricity in 2014, an improvement of only 2 million over 2012. Although 86.5 million people gained electricity access each year, this was offset by population growth of 85.5 million per year.
- The energy access gap is decreasing everywhere except sub-Saharan Africa (SSA), where, over 2012–2014, energy access growth of 19 million people per year was outpaced by population growth of 25 million people per year. In SSA, 600 million people lack access to electricity.
- Based on International Energy Agency (IEA) World Energy Outlook projections, with the current trajectory 784 million people will continue to lack access to electricity in 2030.

Most importantly, over 80% of unelectrified people live in rural areas, where connections to the central grid are often economically prohibitive and can take many years to realize (World Energy Outlook (WEO)). Unelectrified rural populations in low energy access (LEA) countries—where 50% or less of the total rural population has access to electricity—represent more than a third of the total global unelectrified population alone (WEO). LEA countries have traditionally relied on grid expansion as the pathway to increase energy access. This conventional approach, however, has seen limited success. Remote communities continue to experience persistent and pervasive "energy isolation barriers" in the context of centralized, grid-based electrification as a result of the multiple dimensions of geographic, economic, and political remoteness (Alstone et al., 2015).

For instance, complex geography and diffuse population inflates the marginal cost of grid extension in many rural areas for poor nations, whereas the economic limitations of such communities are reflected in low energy consumption and the inability to afford central grid connection fees. The political currency often required for central grid expansion often represents a further social barrier for marginalized or opposition communities. As large-scale public investment schemes, grid expansion has the potential to be subject to domestic political disputes and divisions. Such divisions, corruption, and mismanagement may lead to inconsistent priorities in investment or for entire pockets of the population to be overlooked.

Furthermore, many of the energy-poor people also live in periurban areas where connection to central grid can be difficult (Lee et al., 2016). The financial cost of grid connection is a particularly heavy burden, and as a result, many households remain outside the grid despite close proximity to grid infrastructure. In Kenya, for example, the cost for connection is estimated at about $400 per household, representing a significant cost barrier in a country where the per capita income ranges around $1300 (Lee et al., 2016). More than 31 million Nigerians (40% of total unelectrified) are estimated to be "under the grid"—living within 10 km of a transmission line but with no connection (New Estimates). In Kenya, the "under the grid" population is estimated at around 21 million (50% of total unelectrified) (Shedding New Light).

DISTRIBUTED RENEWABLE ENERGY SOLUTIONS: PIVOTAL TO UNIVERSAL ENERGY ACCESS

As such, achieving universal energy access and unlocking its subsequent benefits will require a concerted focus on serving rural and off-grid communities with clean, affordable, and fast electricity solutions. DRE technologies, which generate, distribute, and/or store energy services independently of a centralized system, offer an unprecedented opportunity to accelerate the transition to modern energy services in such remote areas by complementing or substituting centralized systems while also offering significant cobenefits such as improved health, positive impacts on income growth, women's empowerment, distributive equity, and climate change mitigation (REN21, 2016).

For instance, solar lanterns can help displace the need for kerosene lamps, which contribute to an indoor pollution that is responsible for more than 4 million deaths per year globally, while solar-powered water filtration and desalination technologies support the delivery of safe drinking water to millions without access in remote regions. Minigrids are expanding opportunities for rural enterprise, job creation, and increased agricultural productivity through services such as irrigation, milling, and drying. Furthermore, DRE can provide critical energy for health services and medical equipment needed by the 1 billion people worldwide served by health centers and hospitals that lack access to electricity, a burden disproportionately borne by women. DRE solutions can not only empower women through health but also can benefit women directly by proactively integrating women across the value chain of the technologies, as designers, educators, trainers, managers, and microenterprise entrepreneurs.

These decentralized solutions—renewable energy–based minigrids, solar lights and home systems, behind-the-meter storage, and a host of associated appliances—are increasingly considered a more affordable option than grid extension for communities living far from the electricity grid. According to the World Bank's Global Tracking Framework, in 2015, 717 MW of off-grid renewable energy in Africa provided electricity access to 60 million people, or about 10% of the total off-grid population [1, p. 43]. Indeed, DRE systems already provide energy services to millions of people, with market penetration increasing annually.

The Global Off-Grid Lighting Association (GOGLA) estimates that over 20 million quality verified lighting products have been sold in the last ten years by over 100 companies actively focused on solar lanterns and solar home kits [25, p. 2]. In fact according to GOGLA's forecasts, about one in three off-grid households globally will use small-scale off-grid solar by 2020 [8]. The DRE sector as a whole thus ranks with some of the largest utilities in the world in terms of household reach. The magnitude of DRE's global market penetration over such a short period of time highlights its role as a cheap and fast solution to basic energy access.

Well-designed and well-implemented policy is one of the most important factors in enabling energy access, as is well documented in the literature (Desjardins et al., 2014; Walters et al., 2015; Franz et al., 2014; Murali et al., 2015). Yet specific focus on solutions for rural electrification, including specific inclusion of DRE, is not commonly reflected in national policy. In fact, 73% of LEA countries do not have DRE targets and more than a third have no national energy access targets at all (Fig. 5.1) (Bloomberg New Energy Finance, 2016). So if LEA countries desire to establish frameworks that quickly catalyze development of the DRE sector in-country, what are the most critical elements of an early-stage policy portfolio?

Here we will briefly explore the success factors for policy best practice to catalyze DRE market growth based on policy trends observed in high growth

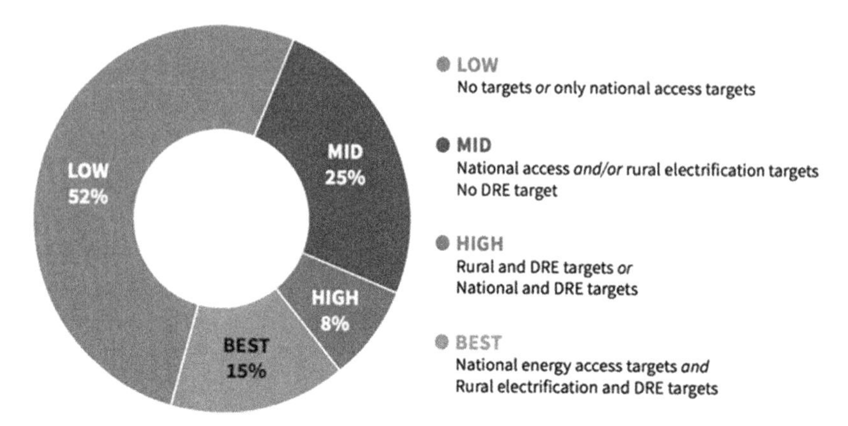

FIGURE 5.1 Energy Access target performance of low energy access countries.

markets and provide insight based on practitioner survey and analysis of policy performance indicators.

THE HABITS OF HIGHLY EFFECTIVE MARKETS: TRENDS ACROSS HIGH-PERFORMING COUNTRIES

A simple exercise allows us to observe the correlation between policy performance indicators and DRE market growth. Let us focus on the five largest DRE markets globally: Ethiopia, Tanzania, Kenya, India, and Bangladesh. India and Bangladesh account for 97% of DRE adoption in developing Asia, and Kenya, Tanzania, and Ethiopia account for 67% of DRE adoption in SSA. We can take cumulative sales of quality-verified solar lighting products normalized by unelectrified population as an estimate for DRE market development (a conservative estimator, as solar lighting product sales does not incorporate energy service provision from minigrids or other DRE systems) (Bloomberg New Energy Finance, 2016; Lighting Global, 2015; Quddus) and policy indicators from a well-established clean energy investment climate index (Methodology, 2015).

Observing correlation and deconstructing the five policy performance indicators with strongest correlation highlights important market accelerators for catalyzing DRE market growth in LEA countries (Fig. 5.2):

- Reduction of import duties and tariffs on DRE-related products
- Supporting the availability of local finance through loans and grants and microfinance
- Establishment of energy access targets or national commitments to electrification
- Establishment of rural electrification plans or programs that incorporate DRE

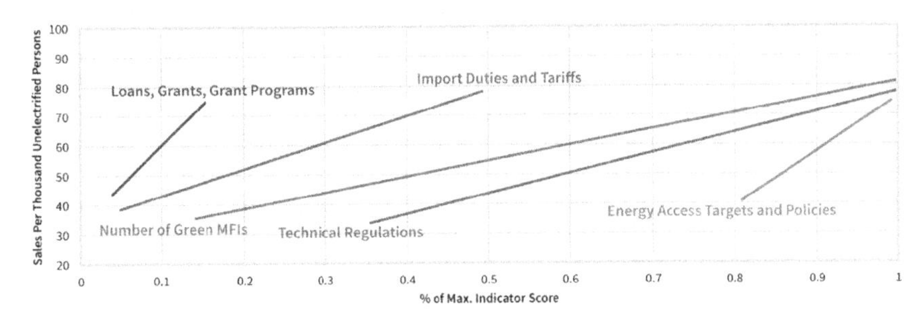

FIGURE 5.2 Regression analyses for distributed renewable energy sales against policy performance.

- Technical regulation through established licensing procedures for minigrid operators and through adoption of quality standards for products and services.

These empirical findings strongly corroborate with the experiences of DRE practitioners on the ground. Our research team[1] recently surveyed 23 pico-PV, solar home system (SHS), and minigrid companies currently operating in one or more of the five top performing countries. Together these companies provide energy services (either minigrid connection or lighting products) to over 2.7 million households. Each participant ranked a specific subset of policy instruments according to how important or beneficial the instrument has been or would be to their business. The average ranking of each policy instrument was calculated and compiled into a list of the most important policy instruments in national energy policy, technical regulation, and financial policy (Fig. 5.3). Practitioners on the ground had a clear message:

- *National Energy Policy*: It is important for governments to establish specific DRE targets and to integrate DRE into national electrification planning, thereby demonstrating commitment and setting the stage for market entry.
- *Technical Regulation*: Internationally recognized product quality standards are critical for instilling market confidence for the SHS/pico-PV sector, while streamlined licensing and permitting most significantly reduce technical/operational barriers to entry for minigrid operators.
- *Financial Policy*: The single most effective financial measure that governments can take to catalyze DRE market growth is to remove import duties and tariffs on DRE products.

1. Research conducted jointly by the Renewable and Appropriate Energy Laboratory (RAEL), UC Berkeley, and Power for All, an international education and advocacy campaign for distributed renewable energy. Read more on this research in our published report (Preliminary Monthly, 2017).

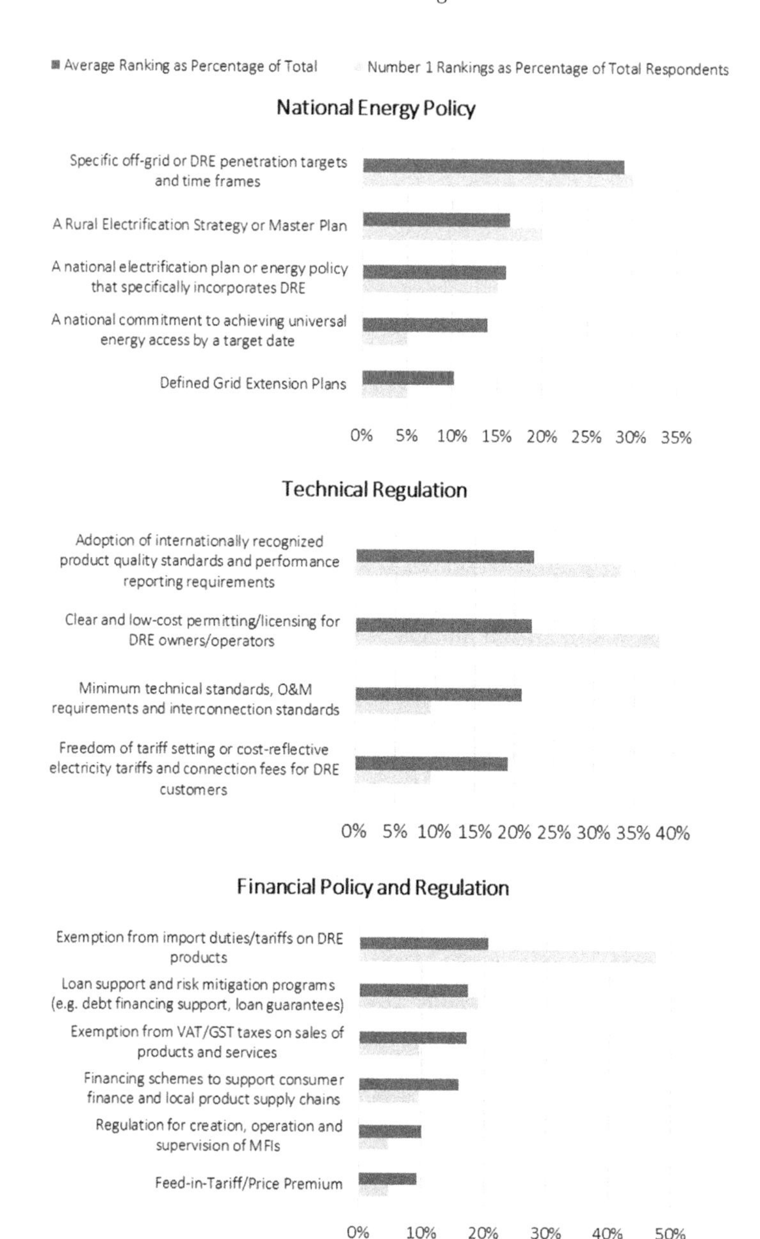

FIGURE 5.3 Average rankings for policy and regulatory instruments ranked by DRE companies operating in case countries.

These two simple exercises clearly highlight five areas for prioritization:

1. Overarching national energy policy that establishes national commitments to electrification
2. Rural electrification plans or programs that integrate DRE as an energy access solution
3. Technical regulation that streamlines licensing procedures for DRE service providers, the first barrier to market entry
4. The adoption of internationally recognized quality standards for DRE products and services
5. Financial policy that reduces or eliminates import duties and tariffs on DRE-related products and that supports the availability of local finance through loans and grants and microfinance

The establishment of energy access targets and incorporation of DRE into planning are identified as the two principal national policy "building blocks," critical for laying the foundation of an enabling environment. The World Bank's Energy Sector Management Assistance Program has recently defined progressive tiers of energy access (the Multi-tier Framework) based on capacity, duration, reliability, quality, and affordability. This helps identify what kinds of intervention are appropriate for moving users to higher tiers, informs investment, and can help countries to set appropriate and realistic energy access and DRE targets and to track that progress more granularly.

It is important to note that policy itself is not a singular solution, and policy design must be matched by implementation. Beyond establishment, policy is often poorly implemented in the country due to limited government capacity; lack of clarity around roles and responsibilities of government ministries, departments, and agencies; and lack of political will from leaders to implement or enforce policy.

In particular, lack of access to finance largely inhibits growth of the DRE sector. Estimates range for the total annual investment necessary, but according to the IEA, $23 million investment in DRE is needed annually to achieve universal energy access goals [17]. Yet the World Bank finds that only a miniscule amount of international financing commitments – 1%, or $200 million a year – now goes to DRE technologies [18]. DRE companies and providers need a range of financing instruments – loans, capital, grants, subsidies and consumer finance. In the next section we describe the case of Sierra Leone, an LEA country that has excelled at policy implementation resulting in the rapid growth of a thriving DRE sector.

FROM THE BOTTOM UP: THE SIERRA LEONE SUCCESS STORY

Sierra Leone is one of the least electrified countries in the world (Fig. 5.4). The WEO estimates that 5 million of its 7 million population are without electricity

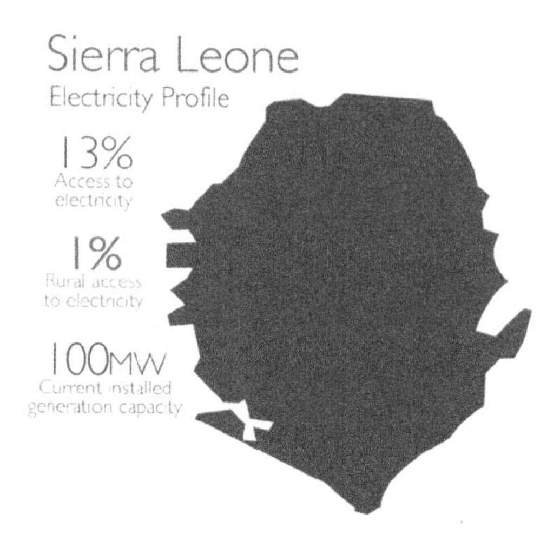

FIGURE 5.4 Sierra Leone's electricity profile.

access. Only 12% and 1% of the urban and rural populations, respectively, is electrified. Particularly after the Ebola outbreak of 2014, Sierra Leone's economy came almost to standstill, with the closure of borders, the cancellation of airline flights, the evacuation of foreign workers, and a collapse of cross-border trade.

Exactly 1 year ago, the government of Sierra Leone embarked on a vision to achieve power for all, signing the first Energy Africa compact with the UK government. This resulted in the launch of the Sierra Leone Energy Revolution, a bold initiative to accelerate access to 250,000 homes by end of 2017 and provide universal electricity access by 2025. The goal, with a focus on household solar solutions, is part of the government's post-Ebola recovery plan.

Commitments were made to

- Supply basic power to all of Sierra Leone's population within 9 years (by 2025)
- Deliver "modern power" to 1 million people by 2020
- Introduce household solar solutions to all 149 chiefdoms in the country
- Eliminate import duties and value-added tax (VAT) on qualified, internationally certified solar product

An Energy Revolution Task Force was created in cooperation between the Ministry of Energy and Power for All, a global campaign to accelerate the end to energy poverty. Established to encourage stakeholder collaboration, enhance communication and to act as a mechanism for implementation of commitments by governments and stakeholders, the Task Force quickly became a platform for market activation: establishing the country's first trade association, the

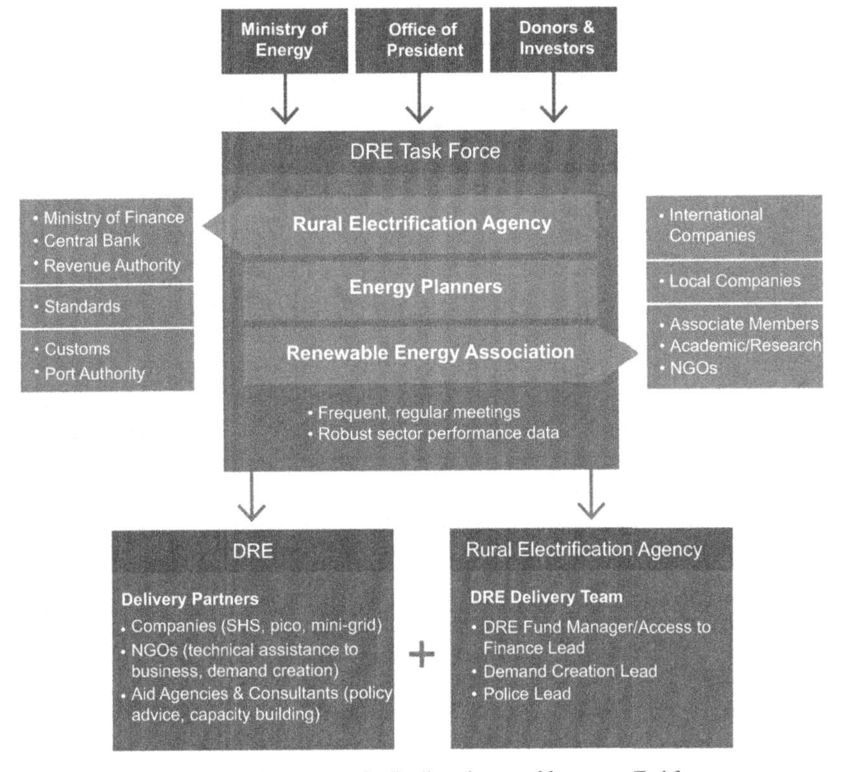

FIGURE 5.5 Sierra Leone's distributed renewable energy Taskforce.

Renewable Energy Association of Sierra Leone (REASL); Africa's first quality-linked VAT/tariff exemption; demand creation and awareness campaigns that led to market growth (including a school campaign by Oxfam IBIS), and development of an micro-finance association to unlock local finance. These market activation mechanisms align directly with those implemented in other high performing DRE markets, as we have seen from our analysis.

One year later Sierra Leone's DRE market had grown over 10 times in size and attracted a dozen key DRE companies like Azuri Technologies, Barefoot Power, Mobile Power, Ignite Power, Greenlight Planet, d.light, and TOTAL (Awango). While the situation in every country is unique, Sierra Leone is a prime market activation case study and offers a successful roadmap for how other countries with the political will can spark their own energy revolution.

CONCLUSIONS

Renewable energy sources continue to experience remarkable growth, both in terms of technological advancement and economic viability. In the context of the United States, developments in the renewable power sector may be most noticeable when considering the large, utility-scale deployments of recent years. Encouraged by historically low installation costs, energy companies are building wind farms and solar PV arrays faster than any other type of electricity generator. Accordingly, the fraction of American electricity derived from clean, renewable sources is growing.

In countries like Sierra Leone, where the benefits of access to advanced energy services are missing for too many people, renewable energy systems are working at a different scale. Minigrids powered by renewables and off-grid solar PV systems that are measured in watts rather than megawatts serve to bring affordable power to populations that have been unelectrified.

The coming years promise to continue the development and deployment of renewable energy, a prospect of great potential benefit to people and planet alike. Once dismissively referred to as "alternative," renewable energy is poised to fuel sustainable communities across the globe.

REFERENCES

Alstone, P., Gershenson, D., Kammen, D.M., April 2015. Decentralized energy systems for clean electricity access. Nature Climate Change 5 (4), 305–314.

2016 Annual U.S. and Global Geothermal Power Production Report. Geothermal Energy Association.

Bloomberg New Energy Finance, Feburary 2016. Off-grid Solar Market Trends Report 2016.

Concentrating Solar Power Projects in the United States. National Renewable Energy Laboratory database.

2015 Cost of Wind Energy Review. National Renewable Energy Laboratory.

Desjardins, S., Gomes, R., Pursani, P., West, C., 2014. Accelerating Access to Energy: Lessons Learned from Efforts to Build Inclusive Energy Markets in Developing Countries. Shell Foundation.

2015 Distributed Wind Market Report. Pacific Northwest National Laboratory.

Electric Power Annual 2015. U.S. Energy Information Administration.

Estimated U.S. Energy Consumption in 2016. Lawrence Livermore National Laboratory and the U.S. Department of Energy.

Estimating Renewable Energy Economic Potential in the United States: Methodology and Initial Results. National Renewable Energy Laboratory.

Franz, M., Peterschmidt, N., Rohrer, M., Kondev, B., 2014. Mini-grid Policy Toolkit: Policy and Business Frameworks for Successful Mini-grid Roll-outs. EU Energy Initiative, Africa EU Renewable Energy Cooperation Programme, Alliance for Rural Electrification, and Renewable Energy Policy Network for the 21st Century.

Geothermal Heat Pumps. Department of Energy. Available: https://energy.gov/energysaver/geothermal-heat-pumps.

Global Tracking Framework, 2017. Progress toward Sustainable Energy (Text/HTML).

Lee, K., et al., Jun. 2016. Electrification for 'under grid' households in rural Kenya. Development Engineering 1, 26–35.

Lighting Global, 2015. Quality Assurance for Off-grid Solar Home Systems: A Comparison of Strategies Employed by Lighting Global and the Bangladesh IDCOL Solar Home System Program.

Living 'Under the Grid' in Nigeria – New Estimates. Center for Global Development. [Online]. Available: https://www.cgdev.org/blog/living-under-grid-nigeria-new-estimates.

Methodology, 2015. Climatescope 2015 [Online]. Available. http://2015.global-climatescope.org/en/methodology/.

Murali, S., et al., 2015. Deployment of Decentralised Renewable Energy Solutions: An Ecosystem Approach. WWF-India and SELCO Foundation.

National Offshore Wind Energy Grid Interconnection Study. ABB, Inc. DOE Award Number EE-0005365.

Fraunhofer ISE press release #26 New World Record for Solar Cell Efficiency at 46% – French-German Cooperation Confirms Competitive Advantage of European Photovoltaic Industry, 2014.

Preliminary Monthly Electric Generator Inventory, March 2017. Energy Information Administration. http://www.powerforall.org/reports/.

Quadrennial Technology Review, 2015. Department of Energy.

Quddus, R., October 26, 2015. After sales service of solar home system (SHS) program under IDCOL in Bangladesh. In: Presented at the 4th International Off-grid Lighting Conference and Exhibition. United Arab Emirates, Dubai.

REN21, 2016. Renewables 2016 Global Status Report. REN21 Secretariat, Paris.

Shedding New Light on the Off-Grid Debate in Power Africa Countries. Center For Global Development. [Online]. Available: https://www.cgdev.org/blog/shedding-new-light-grid-debate-power-africa-countries.

Short-term Energy Outlook, May 2017. Energy Information Administration.

Solar Heat Worldwide, 2013 Solar Heat Worldwide 2013. International Energy Agency.

U.S. Solar Market Insight 2016 Year in Review. Solar Energy Industries Association.

Utility-Scale Solar 2015. Lawrence Berkeley National Laboratory.

Walters, T., Esterly, S., Cox, S., Reber, T., Rai, N., September 2015. Policies to Spur Energy Access: Volume 1 Engaging the Private Sector in Expanding Access to Electricity. National Renewable Energy Laboratory and International Institute for Environment and Development.

WEO – Energy access database. [Online]. Available: http://www.worldenergyoutlook.org/resources/energydevelopment/energyaccessdatabase/.

Wind Technologies Market Report 2015 Lawrence Berkeley National Laboratory.

FURTHER READING

Power for All Decentralized Renewables: From Promise to Progress, March 2017. http://www.powerforall.org/reports/.

Chapter 6

Development Partnership of Renewable Energies Technology and Smart Grid in China

Anjun J. Jin, Wenbo Peng
Huaneng Clean Energy Research Institute, Beijing, China

Chapter Outline

INTRODUCTION

The increase of greenhouse gas emission is creating numerous problems for human health and is also a stable global climate. Growing energy consumption and rising oil prices are also causing increased national security concerns. The scope of challenges in both the energy and climate sectors is far reaching and directly relates to our dependence on traditional carbon-based fossil fuel. This is the heart of our global energy crisis.

There is no shortage of energy flowing into the Earth, since the Sun radiates an enormous amount of power (170,000 TW) onto Earth's surface. Although most of the solar renewable energy is not available to us, acquiring only about 0.01% solar energy is sufficient to meet the world's need today. As

Sustainable Cities and Communities Design Handbook. https://doi.org/10.1016/B978-0-12-813964-6.00006-9

we will discuss soon, the target of using a portfolio of renewable energies is gaining important governmental support and attracting significant private investment. Renewable energy technology is currently developing fast and is becoming economically competitive.

President Barack Obama has championed for renewable energies as well as the smart grid, and he has announced several billion dollars in US government support along with private investment partnership toward this end (Associated Press News Reports). The US President is encouraging the new grid system to be smarter, stronger, and more secure in future. The following sections will discuss how the aforementioned targets can be achieved and what the world has achieved in terms of the renewable energy generation and smart grid power transmission. To date advanced energy technologies have shown us that the development of these technologies could potentially become the linchpin of a new system of modern energy infrastructure. The aforementioned smart grid project would install thousands of new digital transformers and grid sensors in homes and utility substations to enable a grid-smart data system. The sections titled The Solar Electricity Systems and Their Relationship with the Grid and Wind Power show examples of solar electricity and wind power, which are sustainable, abundant, and affordable energies based on excellent uses of natural resources. The section titled Data Response and Power Transmission Lines: Examples of the United States will show that the data response system is to strengthen the grid system. This section along with the one titled The Smart Grid and Market Solution will show the readers that the new system is smarter due to its ability to monitor and control energy consumption comprehensively in real time. The new system is beneficial in terms of demand response, energy efficiency, and compatibility with large supply of renewable energy sources. In the section titled China Rebuilds a Power System and Smart Grid, we will present several cases and explore the key attributes and huge merits of a Chinese-style smart grid.

The resolution to the energy and climate challenges has profound business impacts as well as great societal effect; the effort to meet these challenges is sometimes referred to as the third industrial revolution. Furthermore, most developed nations are facing a major challenge of upgrading the current electricity grid and the energy management infrastructure. The lack of new grid infrastructure forms a bottleneck to the investment in generation of new renewables and in tapping the full power of renewable energy such as the utility-scale solar electricity system.

The strong and rapidly growing consumer demand for clean energy promotes a new resolve to meet global needs with inexpensive electricity from clean energy sources. The advanced energy technology requires a development partnership among the elements of renewable energy, optimized energy efficiency, and smart grid. In the following text, we illustrate the challenges and share our knowledge of employing several renewable energies and optimal grid infrastructure to reduce greenhouse gas emissions by more than 25% by 2020, and by more than 80% by 2050.

THE SOLAR ELECTRICITY SYSTEMS AND THEIR RELATIONSHIP WITH THE GRID

The solar electricity systems are anticipated to be the most likely to become commercially successful without government rebate in a few years. This anticipation is based on the current best knowledge of cost and adoption risks of the solar electricity that offers a superior future technology trajectory. Today as the solar power business grows, we see the cost of solar photovoltaic (PV) panels declining rapidly.

Under Obama's administration in the United States, the President's Solar America Initiative in collaboration with the Department of Energy (DOE) has targeted grid parity wherein the electricity cost based on renewable energy production is the same as the cost of coal-fired traditional power.

More progress is needed in the balance-of-plant aspect of energy production, which is defined as the solar cost per installed system ready for use. Active research to accelerate the progress of cost reduction in this area is underway. In many parts of the world the solar PV system is appropriate for cost-effective distributed generation. All the world's current and expected major electricity load centers are within practical transmission range of excellent solar radiation locations.

As we will show soon in this section, the solar electricity system can utilize the sun's natural energy to generate electricity. The solar electricity generated can be consumed or fed back into the utility grid. This results in less energy that has to be purchased from the utility company, so the consumer's monthly bill decreases (the monthly bill is then just for financing payment). During the solar system warrantee period, the solar PV system is free for electricity generation and free from maintenance. The consumers only pay for the energy used from the grid (and the amortization of the system/installation costs). Solar PV systems connected to grid can be very attractive to reduce the more expensive peak-hour costs.

The term *clean tech* refers to technologies that produce and use energy and other raw materials more efficiently and hence produce significantly less waste or toxicity than prior commercial products. The clean tech energy production industry has developed clean alternative energy utilizing the current benchmark technologies such as solar and wind power. Examples of alternative clean energy sources include (1) wind power, (2) solar PV cell, (3) solar thermal electric power, (4) solar heating, etc. Scientists and engineers continue to search for viable clean energy alternatives to our current traditional power production methods. Even though some renewable energy sources are variable in nature, several renewable energy sources can be integrated into the grid system quite well. For example, studies show the compatibility of sun and wind energy that are complementary and may be quite manageable for integration into a grid system (Abbess, 2009; Clark, 2012; Jin, 2010).

What is a solar electricity system? A solar electricity system is a system for utilizing abundant sun's energy to produce electricity or heat for consumers. For example, the solar PV cell is a physical device that converts light into electricity. Fig. 6.2 is a schematic to illustrate the physical mechanism of a solar PV cell in producing solar electricity. Solar electricity industry is composed of many types of competing technologies with various cost structure, efficiency, and scalability factors that are important to the renewable energy industry sector.

The availability and future prospects are very promising at this time for the following three solar technologies: (1) solar thermal power, (2) solar PV panels, and (3) solar heaters. For example, the utility-scale solar thermal power plants (STPP) have been constructed rapidly in the last two decades. Moreover, the solar PV panels offer scalable power that has been installed today on thousands of rooftops in California.

The solar PV systems, which are made up of individual solar cells, are becoming more affordable and reliable all the time. The solar PV panels are made modular, scalable, and suitable for distributed generation. Moreover, the scalable solar panels can be utilized for utility-scale power plants.

There are several types of devices that may be required to connect solar PV systems to be suitable for individual consumer energy use and/or for supplying power to the electric grid. The most important unit is the inverter. The inverter

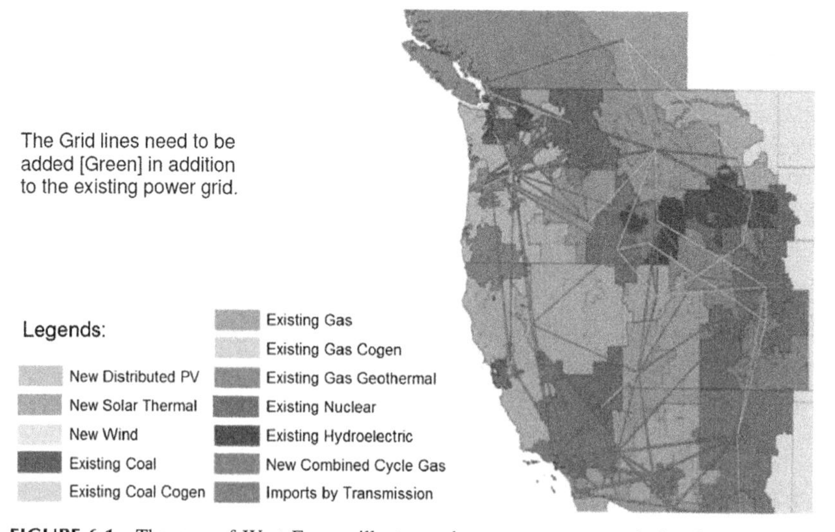

FIGURE 6.1 The map of West Energy illustrates the energy sources and electric transmission grid. The diversity of energy sources and smart grid should enable "utility-grade" power generation that is on par with the current coal-based power plants.

unit is an electronic device that turns direct current (DC) from the solar electricity into an alternating current (AC) that is matched to the incoming main electric utilities standard and that is used by almost all home appliances and electrical devices. The concern for safety also requires the solar electric system to be enabled by circuit breakers for safe maintenance, etc. Circuit breakers are typically connected in both the DC and AC sides of circuitry path.

STPP employs utility-scale steam turbine technology. As shown in Fig. 6.3, STPP collects the solar energy in a large real estate footprint for thermal energy to produce electricity. A circular array of solar light reflectors is used to concentrate the light on a receiver located at the top of a tower. The light is absorbed as heat energy, which heats up the steam gas or air to very high temperatures, which produces pressurized hot gas or air to drive the turbine. STPP has a typical footprint equivalent to the scale of a coal-fired utility power plant.

STPP employs direct sunlight and hence requires its plant site to be in the regions of high solar radiation. The thermal energy storage may be typically

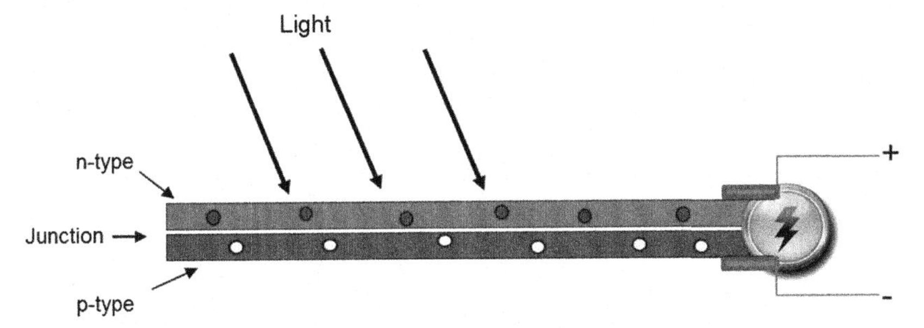

FIGURE 6.2 Schematic of a solar PV cell that converts sunlight to electricity; see text for details.

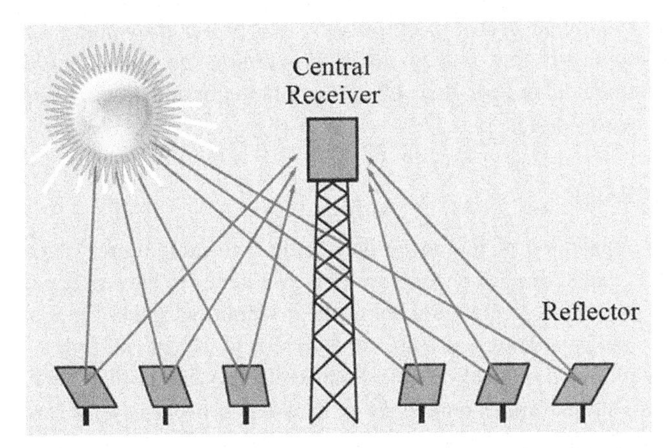

FIGURE 6.3 Solar thermal generator receives highly concentrated sunlight beam, heats up pressurized gas in the receiver till it is very hot, and drives hot gas through a turbine that makes use of the excellent cycle to produce electricity with combined hot gas and turbine operation.

achieved through liquid or solid media to extend the hours of the electric cycle. Fig. 6.3 shows a portion of a typical solar thermal power design.

The United States is the world leader in installed concentrated solar power capacity with 429 MW currently in commission. At present, 3 GW power is operational. In 2010, 7 GW total power was in development. The United States alone anticipates powering 2 million homes by solar thermal power in the year 2020 (Environmental and Energy Study Institute).

All solar power, including solar PV and solar thermal, generated about 0.1% or less of the total US energy supply in 2008, but the installation of solar power is growing quickly. The US DOE's 2009 preliminary forecasts anticipates an annual growth rate in US domestic solar PV generation of 21.3% through 2030 (and some analysts have even higher predictions for the growth rates) (U.S. Energy Information Administration (EIA), 2009). The solar PV technology is scalable. Increasing demand will bring down the cost of PV modules and solar electricity systems when they are in volume production.

Variable but forecastable renewable energies (wind turbines and solar PV power sources) are becoming more reliable in net output when integrated with each other than when used alone. The aforementioned net output is suitable to meet the demand. Risks of security against terrorist attacks or natural disasters can be mitigated by planning geographically dispersed energy sources such as microgrids. One plan is to employ a wide distribution of solar power from many sunny areas, smart grid power systems that will be discussed in the section titled The Smart Grid and Market Solution, backup fuel generators available from natural gases, etc.

Finally, the distributed generation of solar PV electricity can be connected to the grid. Germany is noteworthy for its high-profile feed-in-tariff in promoting solar electricity. Several governments have successfully offered incentive packages to promote renewable energies and energy efficiency in the world. For example, China has significant solar power investment with both a major development plan and an affluent stimulus package detailed in the Chinese National 12th Five-Year plan. The plan is to construct the large-scale renewable energy bases.

WIND POWER

As power is generated by the new utility-scale renewable power plants such as wind power plants, the power transmission grid needs to have capacity to fully deliver the power to consumers without the distribution block. The goals of the US clean energy market can be empowered by addressing the advanced technology platform of a nationwide electric power system. The 2009 ARPA-E grants of the United States have heavily invested in projects with capability to allow intermittent energy sources like wind and solar to provide a steady power flow to the consumers.

We need an interstate power transmission superhighway for the electric power system. Immense solar power farms in the US deserts are facing the transmission challenge for moving through the power grid to the consumers. Congestion of the grid can create significant limitations that can reduce the potential advantage of large renewable power generation to pump power into the electric grid.

What limits the renewable energies (as a commercial bottleneck) is the outdated power transmission grid as mentioned in the last section. For example, the current system cannot accommodate the present and future needs of delivering hundreds of megawatts of wind power to users. In one scenario, the rebuilding of a total of 200,000 miles of power transmission may provoke fights among 500 divided owners and numerous property owners. In another scenario, the large power generation usually requires extra storage system. For today's market, there is no commercially advantageous solution yet. Active development in the storage system area is underway.

The layout of the current power grid should be eventually accessible to a flexible change of total power (e.g., gigawatts) for power interconnection and transmission lines. The current transmission lines cannot increase their transmission capability by hundreds of megawatts needed to meet the challenges. Power pumping shows up as challenges at times even over a distance of a few hundred miles. The commercial pain is the severe congestion today for long-distance power transmission.

We have to achieve clean tech or renewable energy vision. The Kyoto Protocol has set a target that the world needs to reduce greenhouse gas emissions by more than 25% by 2020, and by more than 80% by 2050 (Current Goal). One of the solutions comes from an advanced energy technology, i.e., a wind power generation that it is very cost effective for today's commercial use.

Wind turbine manufacturing is cleaner than the volume production of solar PV cells. Today the United States utilizes barely 1% of the power produced by wind energy. Its goal is to achieve 20% from wind power by the year 2030 (DOE news, 2008). Wind power turbines of the Maple Ridge Wind Farm near Lowville, NY, are capable of producing a total maximum of 320 MW. It has been shutting down at times due to the limitations on pumping capacity of the electric power system. The wind farms too are having power transmission challenges. One cannot easily pump a large amount of power to the grid due to various reasons that we will discuss later. In order for users to utilize the full potential of wind power or other environment-friendly energy, it is imperative for the nation to significantly improve or to rebuild a system of populated and optimized transmission lines.

The wind power is a type of solar-induced energy. Wind is always present on our planet due to uneven heating of its surface by the Sun and due to the so-called Coriolis effect, which relates to the wind being dragged by the constant rotation of Earth on its axis.

The conversion of wind to electrical power is generated by a wind turbine. The modern wind power technology (such as the wind turbine) has been perfected over the last decade. The wind turbine power plant typically generates electricity in much the same way (through electromagnetic induction) as an alternator in a car. As shown in Fig. 6.4A and B, a wind power station is usually positioned such that its rotor always faces the wind. The power engine has a drivetrain system that often includes a gearbox. There is a wealth of information about wind power. Interested readers are referred to the literature (Wealth of Information, 2009).

Wind power depends on three variables: (1) wind velocity, (2) radius of generator, and (3) temperature, which determines the air density. The following is a simplified summary of the aforementioned relationship about the operational state of the wind turbine.

- First, the power increases with the cube of the velocity (e.g., a twofold increase of velocity leads to an eightfold increase of power output).
- Second, the power increases with the square of the radius (e.g., a twofold increase of velocity leads to a fourfold increase of power output).

FIGURE 6.4 (A) Power is generated by a wind turbine. Turbines have increasing capacity rating from left to right in (A). The *dashed line* is 50 m in height. (B) The power is related to the following factors: sweep area (in the rotor radius squared), the wind velocity (v^3), and the temperature variation that affects air density.

- Third, the power increases with decreasing temperature (with about 3.3% of power for the change of every 10°-Celsius in air temperature).

Wind power production not only makes economic sense but also has greater social benefits of clean energy and a sense of personal freedom (by moving toward a zero net energy residence, a type of energy independence). The current transmission lines cannot meet the goals of the advanced energy technologies. With a new grid system, the smart grid can be designed to meet the challenges of and to suit ideally the demands of electricity production, distribution, and utilization.

Wind power makes good sense environmentally and economically. Turbine components are generally either recyclable or inert to the environment. The price of the wind turbine is a critical parameter for the return of investment. The typical payback period for the energy cost is about half a year. Residential wind turbine can be employed in homes or routed to storage such as battery banks. Some farmers have utilized their land for wind farm where the wind power generation does not affect how they farm, produce crops, etc.

DATA RESPONSE AND POWER TRANSMISSION LINES: EXAMPLES OF THE UNITED STATES

Power system management and optimization are really about data management bank, response, and efficiency. When the demand reaches a significantly high level or an energy reduction is needed, the smart demand response should help customers in energy conservation and reduction and thus in enhancing system reliability. The energy security is consistent with our nation's security concern so that the advanced energy technology supports the renewable power standard.

To address the transformation issue of the critical electrical power infrastructure, a major challenge is transmission lines, which the government can adequately support in terms of policy and coordination. The current grid protocol of the power infrastructure has employed a century-old technology, from about the turn of the 20th century (Energy Information Administration). Initially 4000 individual electric utilities owned local grids and operated in isolation. Later, voluntary standards emerged through the electric utility industry to ensure coordination for linked interconnection operations. These voluntary standards were instituted after a major blackout in 1965 that impacted New York, a large portion of the East Coast, and parts of Canada.

Due to the limitation of transmission lines across states, thousands of megawatts of wind projects are stalled or slowed down ,while many solar power deployments are experiencing similar challenges. In the United States, for example, long-distance power transmission has been the major barrier to the success of renewable power standard implementation in certain regions. Moreover, the power grid is considered limiting electricity transmission lines.

For example, California had rolling blackout times in 2001 due to the transmission limitation.

To address the need of a grid transformation, the vision of the energy industry is to employ an internet web model as follows. As shown in Fig. 6.5, the Internet web of a smart grid takes the active system of a nerve network that determines, responds to, and controls the power needed for consumers. The network control system operates under a global scale to dispatch energy, to manage the energy flow protocol, and to distribute control around the system. For example, data response management by the network control system recovers from a power block by circumventing it. This recovery is an attribute of a self-healing power network and has attracted intense research interest.

The information exchange around the Internet web uses the concept of distributed control wherein the web host computer or a designated computer server acts autonomously under a global protocol. Due to the information process capability in the modern Internet web technology, the consumers will benefit in reduced cost by utilizing the Internet to effectively manage the power grid.

The energy efficiency comes from the consumer choosing more efficient energy options over other more costly ones. A smart grid can help utilities to identify losses and to support energy efficiency. The smart grid can manage its effective response to consumers. For energy consumers, power generation owners, buyers, and sellers, the nerve network in the Internet web of a smart grid will be both flexible and economical to extend the services of power purchase transaction. An electricity system would provide supply–demand coordination and would be interconnected in the grid to dispatch power.

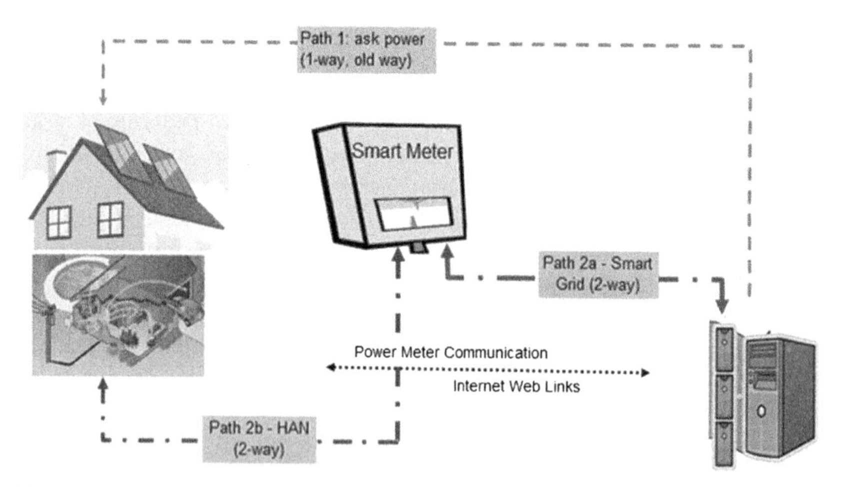

FIGURE 6.5 A smart grid system. A node with a smart meter enables a home automation network (HAN). This control system operates under a global scale to dispatch energy that determines, responds to, and controls the power needed for consumers.

Today, transmission and distribution lines have 500 owners and numerous property owners (Energy Information Administration). The coordination is mostly among three regional interconnections (Western interconnection, Eastern interconnection, and Texas interconnection) and their grid systems are in turn coordinated by the Federal Energy Regulatory Commission.

The diversity of energy sources and the smart grid have designed-in specification that matches the current standard of the existing coal-based power plants. The existing power grid needs to add additional transmission lines, shown in Fig. 6.1 as green lines. Readers are referred to a lecture by Prof. Dan Kammen on October 5, 2009, at UC Berkeley. Additional transmission lines are required for power transmission within each interconnection region and among the three power interconnections. By getting renewable energy sources connected to grid and adding additional transmission lines as required, this will produce sufficient power flowing to consumers. A new power grid can adopt cleaner and more efficient power plants than just the current coal-fired power plant.

THE SMART GRID AND MARKET SOLUTION

In 2009, smart grid companies exhibit significant and fast growing spots alongside the clean tech market need in the United States. The need for smart grid is fundamental to develop the modern energy networks (Major interests, 2010) in the United States, in European Union, in China, and in every grid-connected nation worldwide.

What is a smart grid? The smart grid is a collection of energy control and monitoring devices, software, networking, and communications infrastructure that are installed in homes, businesses, and throughout the electricity distribution grid. This collective system generates a nerve system for the grid and for customers that provides the ability to monitor and control energy consumption comprehensively in real time.

Many tech giants such as Cisco and Google work to bring their products to the smart grid market. Cisco in its May 2009 announcement about its smart grid roadmap stated that the expanding smart grid market is one of its "new market priorities." The advances of these products are most likely to address the challenges of our times in market demand, clean energy need, and greenhouse gas emission reduction.

For example, Cisco expects (Rechargenews) the smart grid market to be bigger than the Internet in reach. The company has identified $15 billion to $20 billion in opportunities globally in the next 5–7 years, Laura Ipsen, Senior Vice President of Cisco's smart grid unit, said at the Reuters Global Climate and Alternative Energy Summit in April, 2012.

However, this market pales when compared with the market opportunity in China. More details will be described in section titled China Rebuilds a Power System and Smart Grid. "A lot of us looking at the China market see $60 billion by 2030 just for China alone," Ipsen said. "A lot of the big companies – the traditional GEs, IBMs, Siemens and others – are over there exploring that market."

For example, the introduction of technology of smart appliances could turn on and off themselves as provided by both the energy management and the smart grid. This technology could help the grid to support a fleet of electric cars. The smart grid would improve the efficiency of transmission power.

Fig. 6.6, which is an expanded version of Fig. 6.5, shows a smart integrated energy system that merges Internet and grid features. This system has a smart grid with the power source(s), the data response, and a load center such as a residential home. Fig. 6.6 illustrates several concepts including data collection, communication, control, and smart grid system. A smart meter collects power usage data for the utilities and consumers, and it has Internet communication capability as mentioned previously.

The real-time data are fed back to a large distribution and transmission power grid. Moreover, the energy storage that assists in load-leveling for transmitting any major power activities comparable to a typical locally rated load center is extremely important. For example, a major power activity could

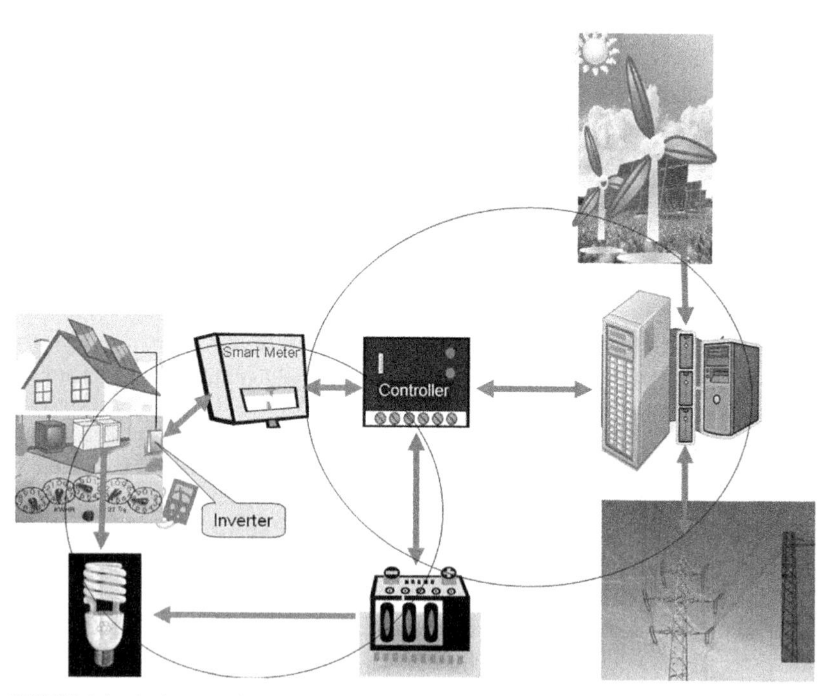

FIGURE 6.6 An integrated energy use system that has a grid to provide power (*upper right circle* shows distribution path), a data response system, and the load center (*lower left circle*) such as the residential home.

be some solar electricity generation, a major charger for an electric car battery, etc. The real-time data are useful for utilities to predict and to hedge power usage.

The smart grid is ideally suited to meet the challenging demands in the production, distribution, and utilization of electricity. The innovation here is to take a century-old power grid infrastructure, turn it upside down and mange it as mentioned at earlier, and connect it to numerous renewable energy sources. The customers are interested in going green today. The cost of fossil-fuel-based energy is rising due to both depleting resources and the cap-and-trade rules on greenhouse gas fuels.

Although solar PV and wind powers may have significant power output, they must be managed with load leveling suited to their output demand profiles. Moreover, much power may be wasted since the power plants such as nuclear power or coal-fired power plants do not shut down when the consumer is asleep. Electricity analysis and management has to be directed toward energy management of existing infrastructure.

As a result, energy efficiency, modeling, and data analysis are needed and can then be fed into the smart grid. The data are for private use, and a computerized electricity management has robust integrity with regard to network security, reliability, and consumer participation. A smart grid may respond to diverse conditions that are indiscriminate to storage, resources, and electricity reliability.

Since over a decade ago, there has been intense interest and investment from both public and private sectors in the field of energy efficiency and carbon-free clean energy production. The smart grid is a bright and fast growing technology sector in recent years. Sustained and deep commitment by regulators, state lawmakers, utilities, and other stakeholders is needed to achieve the cost-effective energy efficiency targets. For example, California utilities are recognized by customers as energy efficiency and demand response experts.

Transformation of the mainstream energy market requires an advanced grid infrastructure with superior energy efficiency and green technology. A smart grid is becoming increasingly important for the wide use of solar, wind, and other renewable energy. This grid reflects excellent criteria for market growth. A smart grid is an investment field in a large market sector and many recognize the opportunity to make a substantial social contribution while providing a good return on the investment. Carefully upgrading the century-old power grids should lead to rebuilding the backbone with a smart grid system.

CHINA REBUILDS A POWER SYSTEM AND SMART GRID

China has sustained over three decades of an "economic boom." The Chinese 12th Five-Year Energy Plan enacted in March 2011 to develop the national renewable energy systems, smart grid, to build its power system has support from the affluent government stimulus package. Rapid development of Chinese smart grid systems has shown and will continue to reap significant social and economic benefits.

Even though this chapter is not intended for a full-set/thorough projection/exploration, we will investigate the merits and attributes of the third-generation grid in China and the great smart grid market success within this section. In general, there is a steady-and-astonishing growth and immense opportunities in China for its smart grid products. The Chinese-style power grid is for achieving integration of power, information, and business flow, and to form a strong and secure grid.

To understand the excitement in Chinese smart grid development, let us review interesting cases in major cities till the end of the year 2011. For example, China has installed smart meters (of over 58 million), planned with smart charging stations and network to serve electric vehicles (with thousands of charging piles completed), and built the world's largest wind, solar energy, and storage systems.

As a case in point on the smart grid, Zhangbei power station (Greenbang) has employed the world's largest hybrid green power station built till the year 2011. It is a demonstration project and has been put into operation on December 25, 2011, in Zhangbei, Hebei province of China. Its grid project includes clean energy sources and energy storage for demonstration.

This demonstration project is constructed by BYD and the State Grid Corporation of China, which is China's largest utilities company. This project is a part of Chinese ambitious smart grid plan. It combines 140 MW of renewable energy generation, efficient energy storage, and a smart power transmission. With an initial investment of $500 million, the project combines 140 MW of renewable energy generation (both wind and solar), 36 MWh of battery power storage, and smart power transmission technologies.

The Zhangbei project is reported to be a great success. Due to its complementation of wind and PV power, the utilization rate of wind turbines has been enhanced by 5%–10%, and its whole renewable energy efficiency has been improved by 5%–10% by using of battery storage system. During its first 100 days safe and stable operation, the power station has already generated over 100 GWh electricity. There may be overflow or excess of energy generation, and the excess energy can be used to feed back to the utility grid once the storage is filled. The Zhangbei project is an excellent story of success, and this project provides a great perspective for renewable energy solution for China and elsewhere.

The other case is about experimental setup of a microgrid (Jin et al., 2012). A smart microgrid PV experimental system of 50 kW has been designed in early this year and is to be set up by China Huaneng Group company in the Future Science and Technology City, Beijing. The company takes up on the microgrid project as a natural innovation for constructing a scalable smart grid in the near future. This project is the first smart microgrid power system and indicates that the company has begun to enter the field of distributed microgrid power generation. Based on microgrid controller, the system has integrated 50 kWPV power, 300 VAh energy storage, grid power, and 30 kW load. Under normal circumstances, the load is completely powered by the PV modules.

When PV power decreases, the controller can deploy battery energy to the load. Under the extreme cases that the DC energy is too small to meet the load, the controller can switch the electricity supply to grid within 8 ms, to ensure stable power supply of the load. Further reviews about Chinese smart grids have shown the following successes.

Merits of the Chinese-Style Smart Grid

The merits of transmission include factors such as safety, reliability, and stability of the smart grid while accommodating production from renewable energy sources. Highly uneven Chinese geographical distribution of electricity production and electricity demand requires China to pay close attention to the power transmission. The backbone of the Chinese smart grid has currently hosted the world's largest wind, solar, and energy storage integrated demonstration project (Popsci).

The key element to achieve smart grid construction, according to the 12th Five-Year development plan of smart grid of Chinese State Grid, has the following goals:

- Generation link: Grid may meet the demand of 60 GW wind generation and 5 GW PV generation in 2015 and 347 GW wind generation and 198 GW PV generation in 2025. The capability of resource-optimized allocation is over 400 GW.
- Transmission link: The next 5 years'' construction is to connect Chinese large-scale energy bases and major load centers, which builds "3 vertical 3 horizontal" backbone EHV grid, which leads to high-level transmission smart grid and transmission line availability factor of 99.6%. "5 vertical 6 horizontal" backbone EHV grid will be built in 2020. Meanwhile, smart grid is fully designed and under construction.
- Transformer link function at high voltages: 6100 smart substations at above 110(66) kV should be completed in 2015, accounting for about 38% of the total substation; 110(66) kV or above smart substations will account for about 65% of the total substations in 2020.

Discussion on Chinese Cases, Investment, and Forecast

This decade set the stage for China's smart grid full-scale construction and improvement. In accordance with the Chinese development plan, from year 2011 to 2015, the smart grid construction investment amount will be over $300 billion and the total investment will reach $600 billion in 2020.

Meanwhile, the national smart grid investment funds will be multiplied tenfold. According to the national smart grid plan, smart grid radiation range is extensive. New industries such as smart city and smart transportation will also be spawned so that the market size is extremely attractive.

Moreover, smart grid construction provides a huge benefit. Specifically, by 2020, the benefits will be as follows:

- The power generation benefits will be around $5.5 billion, saving the system effective capacity investment and reducing power generation costs by RMB 1−1.5 cents/kWh.
- Grid link benefit will be about $3.2 billion, grid loss will reduced by 7 billion kWh, and the maximum peak load will be decreased by 3.8%.
- User benefit will be about $5.1 billion, by offering a variety of services, saving 44.5 billion kWh of electricity.
- The environmental benefits will be about $7 billion, conserving land of about 2000 acres/year, reducing emission of SO_2 of about 1 million tons, and reducing CO_2 emission of approximately 250 million tons.
- Other social benefits will be about $9.2 billion, increasing employment opportunities for 145,000/year, saving the cost of electricity, and promoting balanced regional development.

Historical Review and Attributes of the Third-Generation Grid

Currently, China aims to become the world's largest smart grid user in the power industry. It is imperative to have advantageous elements including a robust and a low-cost smart grid and that accommodate the renewable clean energy.

China's power industry began in 1882 with the birth of the Shanghai Electric Power Company, which produced the first-generation grid of China. Till 1949, Chinese power generation equipment installed capacity was 1.85 GW with a generating capacity of 4.31 billion kWh. The second-generation grids constructed since the 1970s aimed to interconnect the national grid. The Northwest Power Grid 750-kV transmission line started operating in 2005 and China's first 1000-kV UHV transmission lines were built in 2009. Till July 2010, China's transmission lines of 220 kV or above are over 375,000 km in length, which exceeds those of the United States and rank first in the world.

In fact, the total Chinese power installation reached 1 TW by the end of 2011 such that the annual total electricity consumption was 4.7 trillion kWh. The grid-connected new energy power generation capacity reached 51.6 GW, of which there was 45.1 GW of wind power, accounting for 4.27% of the total installed capacity; grid-connected solar PV capacity of 2.1 GW, accounting for 0.2%; biomass-installed power capacity of 4.4 GW, accounting for 0.4%; geothermal power generation capacity of 24 MW; and ocean energy power generation capacity of 6 MW.

The attributes of a Chinese smart grid or third-generation grid are substantially beneficial. The following Here is a brief list of the important attributes (of smart grid):

- Strong: robust and flexible are the basis for the future smart grid.
- Clean/green: the smart grid makes possible the large-scale use of clean energy.
- Transparent: transmission grids, distribution grids, power system status and its network can be monitored for data management.
- Efficient: improve the transmission efficiency, reduce operating costs, and promote the efficient use of energy resources and electricity assets.
- Good interface: compatible with various types of power and user; promote the generation companies and users to actively participate in grid regulation.

Light-Emitting Diode and Energy Efficiency Case Discussion

China offers fast growth business and market opportunities, and it is currently the second largest economy in the world. Chinese investment in the Clean Technology is very positive by its government and has significant efforts in all related areas.

For example, Chinese government's impact on energy efficiency field is noteworthy. Beijing, the Chinese capital city, has pushed strongly on the energy-efficient lighting especially on light-emitting diode technology (LED). LED is a solid-state lighting product. One of Beijing city's goals is to eliminate incandescent lights in a few years. Currently, a consumer can buy 60-W-equivalent LED light bulb for lighting for just a quarter of USD after both discounts and government subsidies. Moreover, LED has attracted huge market in the daily lives for the developed Yangtze Delta cities such as Shanghai, Suzhou, and Wuxi.

In technology front, the latest LED technology delivers highly energy-efficient solution. The LED lighting consumes nearly 80% less energy, lasts much longer, and is environmentally better (without mercury involved in the process) than an incandescent light bulb. According to the Bright Tomorrow Lighting Prize, known as the "L Prize," hosted by the US DOE, there are great products like Philips LED and Cree LED lighting. For example, the Philips' 60 W equivalent LED bulb is coming to stores and can provide 900 Lumens but consumes less than 10 W of energy. It can last 17-years if used 4-h daily. By the way, this Philips product is the winner of the L Prize in August, 2011.

In comparison, an incandescent bulb has less than 2% energy conversion efficiency from electric energy to light energy. The LED technology is has better efficiency and is still maturing. There is a challenge in its thermal management for its lifetime improvement. The limitation of its temperature tolerance in current technology is being addressed through extensive efforts in research and development investment worldwide.

The market potential is huge for LED. Although the current market is still limited, LED provides home lighting and street lighting and lights up city skylines, billboard displays, traffic lights, train and public transits signs and lighting, stage lighting, display lighting in art galleries, automotive headlights, floodlights of buildings, and growth lights for plants, Therefore we can appreciate some Chinese advertisement stating that LED saves you money and beautifies your home.

REFERENCES

Abbess, J., 2009. Wind Energy Variability and Intermittency in the UK. New Reports on Monday. http://www.claverton-energy.com/wind-energy-variability-new-reports.html.

Associated Press News Reports on 10/27/09, Arcadia, FL. The president has set a goal.

Clark, W., 2012. Introduction: the economics of the green industrial revolution. In: Clark, W. (Ed.), The Next Economics: Becoming a Science in Energy, Environment, and Climate Change. Springer-Verlag. Copyright ©.

DOE news – Updated July, 2008. http://www1.eere.energy.gov/windandhydro/pdfs/41869.pdf]. The DOE wind-power news: http://www.energy.gov/news/6253.htm.

Energy Information Administration, http://www.eia.doe.gov/http://tonto.eia.doe.gov/energy_in_brief/power_grid.cfm; Home > Energy in Brief on "What is the electric power grid, and what are some challenges it faces?".

Environmental and Energy Study Institute, http://www.eesi.org/files/csp_factsheet_083109.pdf.

http://www.greenbang.com/china-claims-worlds-largest-battery-storage-station_21041.html.

http://www.popsci.com/science/article/2012-01/china-builds-worlds-largest-battery-36-megawatt-hour-behemoth.

http://www.rechargenews.com/business_area/innovation/article296051.ece.

Jin, A.J., 2010. Transformational relationship of renewable energies and the smart grid. In: Clark, W.W. (Ed.), Sustainable Communities Design Handbook. Elsevier Inc., pp. 217–231. Copyright ©.

Jin, A., Peng, W., et al., 2012. DC-module-based rooftop PV system design and construction. In: Proceedings of the 12th China Photovoltaic Conference, Beijing.

Major interests are dedicated to smart grid: e.g., March 30–31, 2010, Smart Grids Europe 2010. Please refer to the following works as well. a. Refer to Smart grid of European platform in 2006, EUR 22040. http://ec.europa.eu/research/energy/pdf/smartgrids_en.pdf; b. Refer to Smart Grid: Interop. of Energy Tech and Info Tech Operation with the grid: by IEEE, P2030™/ Draft 1.0 Skeletal Outline, 2009®.

The Current Goal from Kyoto is that the world needs to reduce greenhouse gas emission for more than 25 percent by 2020, and for more than 80 percent by 2050. http://www.americanprogress.org/issues/2009/01/pdf/romm_emissions_paper.pdf.

There is a Wealth of Information about wind power, e.g., readers please refer to http://en.wikipedia.org/wiki/Wind_power, and also refer to the Calculation of Wind Power article from the Renewable Energy Website REUK.co.uk. Printed on 5th November 2009 – http://www.reuk.co.uk/Calculation-of-Wind-Power.htm Power = 0.5 x Swept Area x Air Density x Velocity^3, and Air Density = constant * [T,degreeC/(273.15+T,degreeC)]^1.

U.S. Energy Information Administration (EIA), March 2009. Annual Energy Outlook. http://www.eia.doe.gov/oiaf/aeo/index.html.

Chapter 7

Sustainable Towns: The Case of Frederikshavn Aiming at 100% Renewable Energy

Henrik Lund, Poul A. Østergaard
Aalborg University, Aalborg, Denmark

Chapter Outline

INTRODUCTION

This chapter provides an example of how Frederikshavn, within a short period, can be converted into a town supplied 100% by renewable energy. The example is based on a proposal drawn up by a working group in relation to a project called "Energy Camp 2006." The proposal is described in detail in the article "Next City, Frederikshavn - Denmark's renewable energy city."

It should be underlined that this is only a proposal, which may serve as an inspiration for future work. The final project must develop in close dialogue and co-operation with all the actors involved in converting this idea into reality.

Sustainable Cities and Communities Design Handbook. https://doi.org/10.1016/B978-0-12-813964-6.00007-0

DEFINITION OF RENEWABLE ENERGY

Renewable energy is here defined as energy arising from natural resources such as sunlight, wind, rain, waves, tides, and geothermal heat, which are naturally replenished within a span of a few years. Renewable energy includes the technologies that convert natural resources into useful energy services, such as

- Wind, wave, tidal, and hydropower (including micro and river-off hydro)
- Solar power (including photovoltaic), solar thermal, and geothermal power
- Biomass and biofuel technologies (including biogas)
- Renewable fraction of waste (household and industrial waste)

Household and industrial waste are composed of different types of waste. Some fractions are regarded as renewable energy sources, such as potato peel, whereas others are nonrenewable sources, such as plastic products. Only the fraction of waste that is naturally replenished is usually included in the definition. However, in the Energy Town Frederikshavn project, for practical reasons, the entire waste fraction is included as forming part of the renewable energy sources.

When calculating the share of renewable energy sources (RES) in the system, the import and export of power is converted to fuel equivalence, i.e., the fuel needed to produce the power on a power plant with an efficiency of 40%. The same factor has been used when wind power is compared to fuel. Moreover, when calculating the share of RES, the share of wind power has been corrected into the expected production of a normal wind year. In Denmark, wind years vary within the range ±20%.

Definition of Project Area

The project includes the town of Frederikshavn—the three suburbs of Strandby, Elling, and Kilden as well as a limited number of isolated houses. The population of the entire area is approximately 25,000. The delimitation of the project area is in large part established to correspond to the boundary of the local electricity distribution company Frederikshavn Elnet A/S.

The town of Frederikshavn should not be confused with the Municipality of Frederikshavn, which encompasses a larger area extending to the northern tip of Denmark.

The entire area is indicated on the map to the left, where the blue line shows the delimitation. Areas hatched in red are district heating areas. Areas hatched in green are supplied with natural gas. Evidently, even within the contiguously built-up area of central Frederikshavn, there are potentials for expansion of district heating, as some areas are currently supplied by natural gas for heating purposes.

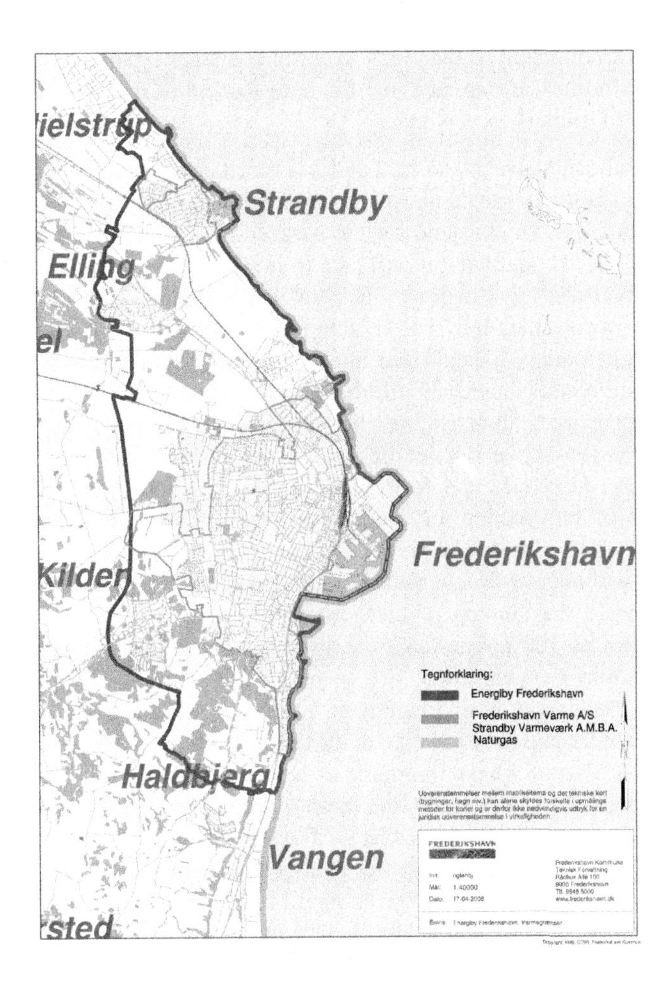

Development Phases

In Frederikshavn, the actual supply in 2007 has a renewable energy share of approximately 20%. Based on this fact, the project works with the following years and development phases:

1. The first step is until 2009 when the United Nations (UN) Climate Summit COP15 was held in Denmark. The objective is to raise the share of renewable energy in Frederikshavn to approximately 40%.
2. Transformation to 100% renewable energy on an annual basis in Frederikshavn by 2015. However, exchange of energy with the surrounding areas is allowed.

3. Further development of the 100% renewable energy system in such a way that possibilities are created for the transformation to 100% renewable energy in Denmark as a whole.

The distinction between phases 2 and 3 is based on the fact that the purpose of the project is not to create an isolated "energy island" with no connections to the surroundings. On the contrary, the purpose is to show what a 100% RES future will look like and what it will take to implement it. Consequently, e.g., vehicles in Frederikshavn fueled by RES in 2015 will have to be able to leave the town. However, they may not be able to refuel in other parts of Denmark until the whole county is converted into a 100% RES system. Also, vehicles from outside Frederikshavn will still have to be able to refuel in the town. A sufficient quantity of "biogasoline" will be produced to cover the transport demand in Frederikshavn, but not all cars will be expected to change to the use of biogasoline. Moreover, cars from Frederikshavn are not expected to be able to refuel with biogasoline in other locations in Denmark. Besides, the exchange of electricity across the project boundary from one hour to another may occur, and biogas from the natural gas network may be used. However, on an annual basis, the amount of fuels for vehicles, electricity, and heat productions based on 100% renewable energy should meet the exact demands of Frederikshavn by the year 2015.

In 2030, the target is to implement a solution in which the amount of biomass resources and the exchange of electricity and fuels will comply with a strategy according to which Denmark as a whole is converted into a 100% renewable energy system. Again, the target is not to entirely avoid exchange. However, it would not be acceptable for Frederikshavn to merely export any imbalances in, e.g., electricity to the areas outside the town, as this would compromise the possibilities of conversion to 100% renewable energy in these areas. Exchange should thus be limited to an appropriate level and should consider the fact that some parts of Denmark may utilize more wind power than others, whereas other parts may utilize more biomass resources. Moreover, Scandinavia may explore the mutual benefits of exchanging, e.g., wind power in Denmark with hydropower in Norway.

THE PRESENT SITUATION: YEAR 2007, APPROXIMATELY 20% RENEWABLE ENERGY

The existing energy supply in Frederikshavn is shown in Fig. 7.1. Most houses and apartments are connected to the public district heating network, but the area also covers a small share of individually heated homes. In addition to this is the energy consumption of industry and transport.

The energy demand consists of the following:

- An electricity demand of 164 GWh/year supplied by the public grid.
- A district heating demand of 190 GWh/year distributed into two separate systems of 175 and 15 GWh. Including grid losses of 52 GWh/year, the annual production in 2007 added up to 242 GWh/year.

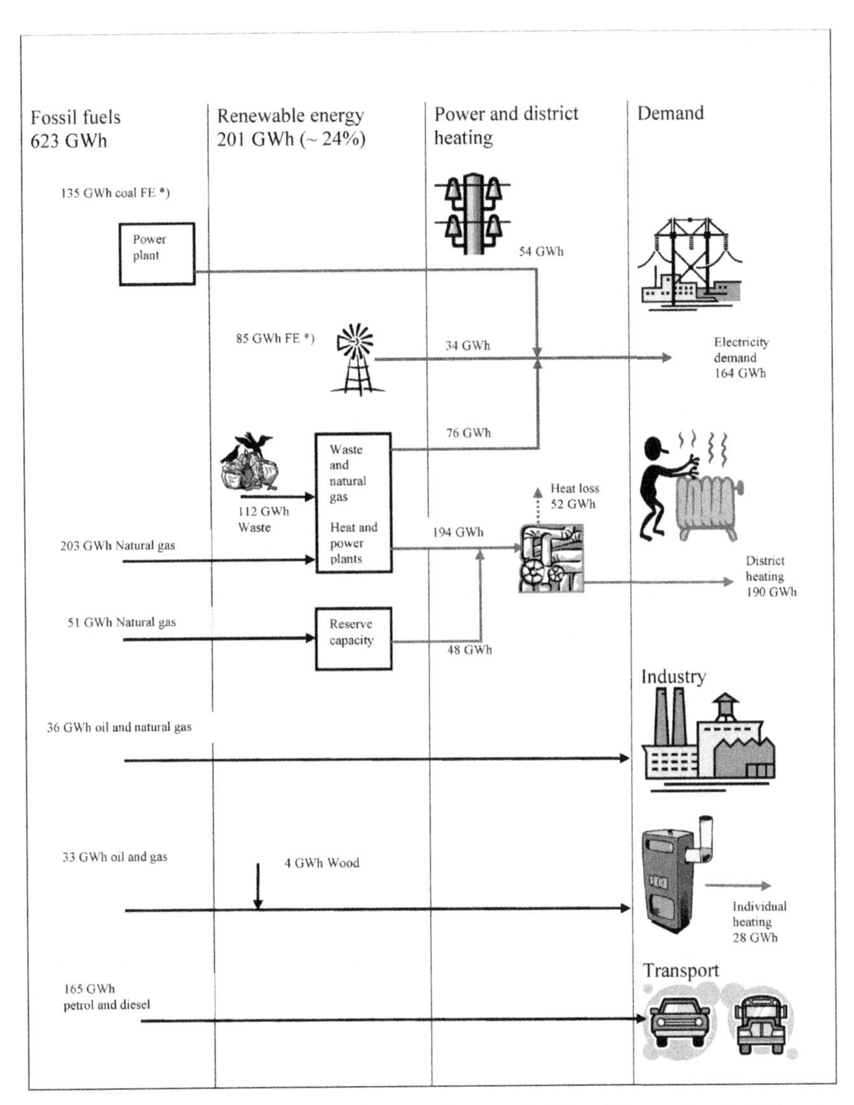

FIGURE 7.1 Frederikshavn 2007 (20% renewable energy). *Electricity as given in fuel equivalents of electricity production of a coal-fired steam turbine with an efficiency of 40%. *FE*, fuel equivalent.

- A transport demand of local transport equal to 165 GWh/year of gasoline and diesel.
- The heating of houses with individual boilers equal to 37 GWh of fuel and estimated 28 GWh of heat.
- Fuel for industry of 36 GWh equal to an estimated room heat demand of 28 GWh and process heat of 3 GWh.

Of the 36 and 37 GWh fuel for individual houses and industry, respectively, an estimated 70% can be converted into district heating.

The district heating in the large system (central Frederikshavn) is produced on combined heat and power (CHP) plants, partly based on waste incineration and partly including peak load boilers fueled by natural gas.

District heating in the small system (the northern suburb Strandby) is produced on a small CHP plant fueled by natural gas, shown in the photograph to the right.

The individual supply is based on oil- and gas-fired burners and a small amount of wood.

Fig. 7.1 shows that a large share of the power demand in Frederikshavn is already met by local wind power and CHP production. In addition to the electricity production of the three mentioned CHP plants, 10.6 MW of near-shore wind power placed at the Frederikshavn harbor (see photograph) covers some of the demand. The rest is imported from the national grid. Here, the latter is assumed to be produced on a coal-fired power station with an efficiency of 40%, equal to the average of Danish condensing-mode power plants.

THE FIRST STEP: FREDERIKSHAVN IN THE YEAR 2009

The year 2009 is chosen as the terminal date of phase 1, because the planned UN Climate Summit in Denmark will provide a good opportunity for promoting the project at an international level. The objective is to raise the share of renewable energy to approximately 40% by implementing the following four projects before the end of 2009:

- 12-MW wind turbines. These wind turbines represent step 1 of a new offshore project of a total expected capacity of 25 MW. The project has been decided, and environmental impact assessments are in progress. The first 12 MW are expected to be implemented during 2009 (see visualization to the right).

- Establishment of 8000 m^2 of thermal solar collectors (see photograph) in combination with an additional 1500 m^3 of water heat storage and an absorption heat pump (see photograph) at the CHP plant of the small district heating supply of Strandby. At present, the project is being implemented. The absorption heat pump will cool the exhaustion gas and increase the total efficiency from the present 94% to 98%, and the solar collectors will generate an annual heat production of approximately 4 GWh.

- Implementation of a facility that upgrades biogas from a local biogas plant outside the town to natural gas quality and transports the gas to a biogas fuel station in Frederikshavn. Furthermore, an investment in 60 bifuel cars will be made. This will supply 7 GWh of biogas. The biogas that will not be used for transport will be used in the CHP plant.

- Establishment of a 1-MW_{th} heat pump at the wastewater treatment plant of the town, which is expected to utilize 2 GWh of electricity annually to extract 4 GWh of heat from the waste water and produce 6 GWh of heat for the district heating supply.

The total budget of this phase is estimated at approximately 200 million DKK, and, as illustrated in Fig. 7.2, it is expected to increase the share of renewable energy to 38%.

FREDERIKSHAVN IN THE YEAR OF 2015

On a continuous basis, the Energy Town Frederikshavn project is in the process of identifying a proper scenario for the implementation of a 100% renewable energy system by the end of the year 2015. Here, a status on the results of such considerations is presented. Each of the proposed projects will be subject to more detailed analyses in the coming period. However, the key components have been identified and are included in the planning and project phases, which are expected to last for several years.

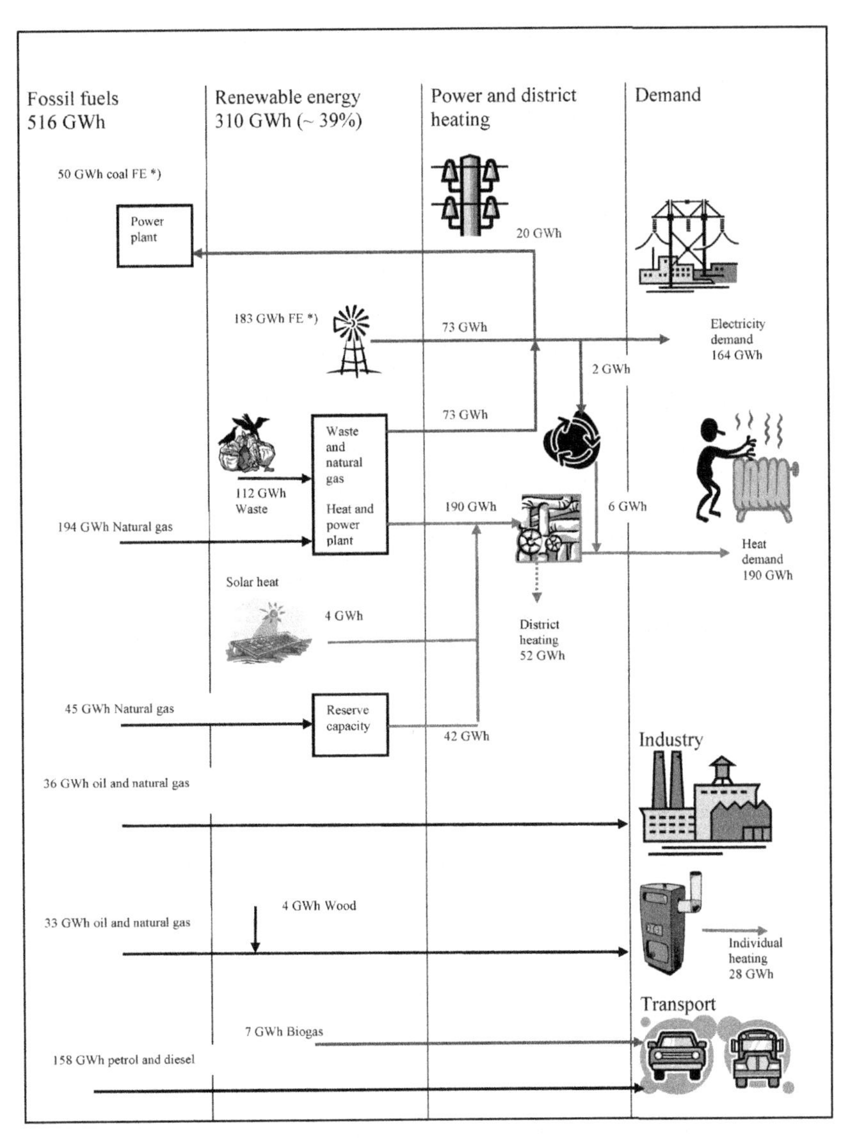

FIGURE 7.2 Frederikshavn 2009 (40% renewable energy). *Electricity as given in fuel equivalents of electricity production of a coal-fired steam turbine with an efficiency of 40%. *FE*, fuel equivalent.

New Waste Incineration Combined Heat And Power Plant

Public waste treatment in Frederikshavn is based on the same principles as in the rest of Denmark, i.e., giving priority to recycling of most of the waste, followed by incineration producing heat and power, and land filling of only very small shares. However, the amounts of waste for incineration now exceed the capacity of the two existing plants in the area (one is beyond the project area in the town of Skagen; the other is shown to the right), and it is planned to build a new waste incineration plant. Consequently, the Energy Town project includes a new waste incineration CHP plant with an expected net electricity efficiency of 23% and a heat efficiency of 64%, and with an incineration capacity of 185 GWh/year, equal to the available amount of local resources.

Expansion of District Heating Grid

The project also includes an expansion of the existing district heating grid, whereby 70% of the heat demand in the presently individually heated houses and industry is replaced by district heating. The demand covered by district heating will thus expand from the present 190 GWh to a total of 236 GWh. The rest of the industry (process heating) will be supplied from biomass boilers, and the remaining individually heated houses will convert to a mixture of solar thermal and electric heat pumps.

Transport

With regard to transport, the project is heading for a solution in which the vehicle fleet consists of bifuel cars (using biogas in combustion engines), electric cars, and plug-in hybrid cars. To implement as much electric driving as possible, it is suggested to implement cars that combine the use of batteries with fuel cell driving based on either methanol or hydrogen. The specific proposal calculated in the following discussion assumes that motorcycles and mopeds (4 GWh) and vans and busses (25 GWh) are converted to biogas, hydrogen, or methanol in the ratio of 1:1. Of the remaining transport demand, 10 GWh is converted into biogas in the ratio of 1:1; 50% of the remaining demand is converted into electric driving (1 kWh of electricity replaces 3 kWh of gasoline due to improved efficiencies); and the rest into fuel cell–based driving, replacing 2 kWh of gasoline by 1 kWh of methanol or hydrogen. In total, 165 GWh of gasoline and diesel are replaced by 10 GWh of biogas, 21 GWh of electricity, and 61 GWh of methanol.

Biogas Plant and Methanol Production

Partly to be able to produce methanol for transportation and partly to replace natural gas for electricity and heat production, the project includes a biogas plant utilizing 34 million tons of manure per year for the production of 225 GWh of biogas. The facility itself consumes 42 GWh of heat to attain the optimal digestion temperature and 7 GWh of electricity.

The biogas can be converted into methanol with an efficiency of 70%. Consequently, the production of 61 GWh of methanol is expected to consume 87 GWh of biogas. However, the production of methanol will provide 17 GWh of heat, which can be utilized for district heating.

Methanol may also be fully or partly produced by electrolysis. Moreover, in the end, the cars may consume hydrogen instead of a certain share of the methanol. In such case, part of the biogas will be replaced by wind power instead.

Geothermal and Heat Pumps

The town of Frederikshavn is located on potential geothermal resources, which may be included in the project. The resources can supply hot water at a temperature of approximately 40°C. However, the temperature can be increased to district heating level by the use of an absorption heat pump, which can be supplied with steam from the waste incineration CHP plant. It has been calculated that an input of steam of 13.3 MW in combination with a geothermal input of 8.7 MW can produce 22 MW of district heating. The steam input will decrease the electricity production from the CHP plant by only 1.3 MW and the heat production by 11.9 MW. Marginally, the absorption heat pump has a coefficient of performance (COP) of more than 7.

The geothermal plant in Thisted, Denmark, produces 15.4 GWh of heat per year. (Thisted Varmeforsyning).

Additional compression heat pumps may be applied to utilize the exhaust gases from the CHP plants and the boilers supplemented by other sources, such as waste water, as already included in the plans for 2009. A potential of 10 MW_{th} output is included by use of a heat pump with a COP of 3.

Combined Heat and Power Plants and Boilers

The project includes a biogas CHP plant of 15 MW and efficiencies of 40% electricity and 55% heat. The rest of the heat production will be supplied from a biomass boiler burning straw with an efficiency of 80%.

Wind Power

Finally, the project includes enough wind turbines to cover the rest of the electricity supply, i.e., a total of around 40 MW, of which already more than half will be implemented by the year 2009.

ENERGY SYSTEM ANALYSIS

By the use of the EnergyPLAN model, a series of detailed energy system analyses of the expected year 2015 system have been conducted to identify the hourly balances of heat supply and exchange of electricity.

The EnergyPLAN model is a deterministic input/output model. General inputs are demands, renewable energy sources, energy station capacities, costs, and a number of optional different regulation strategies emphasizing import/export and excess electricity production. Outputs are energy balances and resulting annual productions, fuel consumption, import/exports of electricity, and total costs including income from the exchange of electricity. See the figure on the following page.

The model can be used to calculate the consequences of operating a given energy system in such way that it meets the set of energy demands of a given year. Different operation strategies can be analyzed. Basically, the model distinguishes between technical regulation, i.e., identifying the least fuel-consuming solution, and market economic regulation, i.e., identifying the consequences of operating each station on an electricity market with the aim of optimizing the business economic profit. In both situations, most technologies can be actively involved in the regulation. And in both situations, the total costs of the systems can be calculated.

The model includes a large number of traditional technologies, such as power stations, CHPs, and boilers, as well as energy conversion and technologies used in renewable energy systems, such as heat pumps, electrolyzers, and heat, electricity, and hydrogen storage technologies including compressed air energy storage. The model can also include a number of alternative vehicles, for instance, sophisticated technologies such as V2G (vehicle to grid), in which vehicles supply electricity to the electric grid. Moreover, the model includes various renewable energy sources, such as solar thermal and photovoltaic, wind, wave, and hydropower.

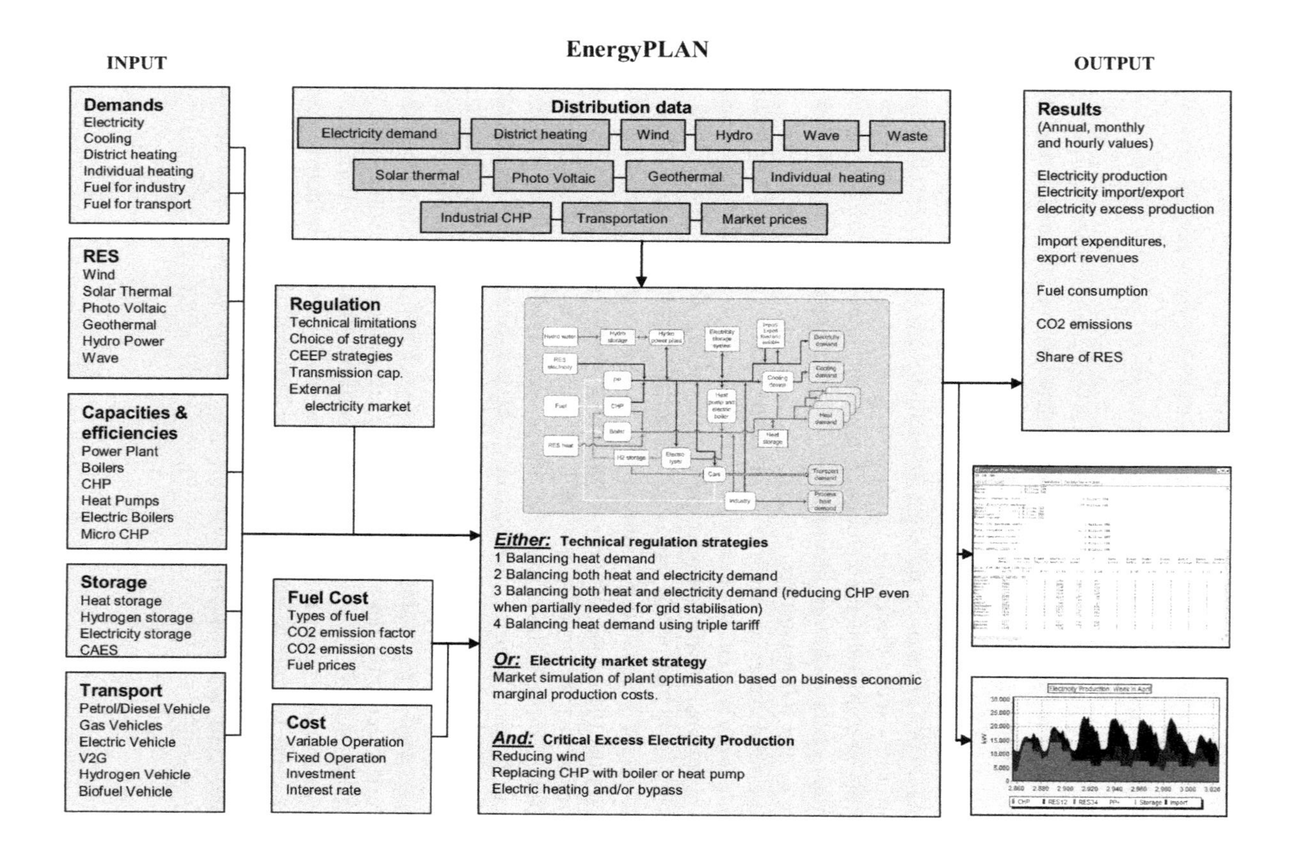

EnergyPLAN
INPUT
OUTPUT
Demands
Electricity
Cooling
District heating
Individual heating
Fuel for industry
Fuel for transport
RES
Wind
Solar Thermal
Photo Voltaic
Geothermal
Hydro Power
Wave
Capacities & efficiencies
Power Plant
Boilers
CHP
Heat Pumps
Electric Boilers
Micro CHP
Storage
Heat storage
Hydrogen storage
Electricity storage
CAES
Transport
Petrol/Diesel Vehicle
Gas Vehicles
Electric Vehicle
V2G
Hydrogen Vehicle
Biofuel Vehicle
Distribution data
Electricity demand
District heating
Wind
Hydro
Wave
Waste
Solar thermal
Photo Voltaic
Geothermal
Individual heating
Industrial CHP
Transportation
Market prices
Regulation
Technical limitations
Choice of strategy
CEEP strategies
Transmission cap.
External electricity market
Fuel Cost
Types of fuel
CO2 emission factor
CO2 emission costs
Fuel prices
Cost
Variable Operation
Fixed Operation
Investment
Interest rate
Either: Technical regulation strategies
1 Balancing heat demand
2 Balancing both heat and electricity demand
3 Balancing both heat and electricity demand (reducing CHP even when partially needed for grid stabilisation)
4 Balancing heat demand using triple tariff
Or: Electricity market strategy
Market simulation of plant optimisation based on business economic marginal production costs.
And: Critical Excess Electricity Production
Reducing wind
Replacing CHP with boiler or heat pump
Electric heating and/or bypass
Results
(Annual, monthly and hourly values)
Electricity production
Electricity import/export
electricity excess production
Import expenditures, export revenues
Fuel consumption
CO2 emissions
Share of RES

Key data in the analyses are hourly distributions of demands and fluctuating renewable energy sources, as shown in the diagrams. Fluctuations in electricity demand are based on actual measurements of the demand of Frederikshavn in the year 2006, and the district heating demand has been based on a typical Danish distribution adjusted by monthly values of the district heating demand of Frederikshavn in 2007. Wind power is based on the actual production of the existing wind turbines in 2006. However, the productions have been corrected for downtime and adjusted to the expected annual production of an average wind year. The case of solar thermal uses a typical Danish production of solar thermal power supplying district heating systems.

The results of the energy system analyses reveal that if the use of waste for incineration is increased from the present 112 to 185 GWh in a plant with efficiencies and district heating demand corresponding to the present level, the summer heat production will exceed the demand by 10 GWh. Such excess production will be even higher if heat from methanol production is included. However, by building a new plant with higher electric output, expanding the district heating network, and adding the heat consumption of the biogas plant, excess heat production is avoided.

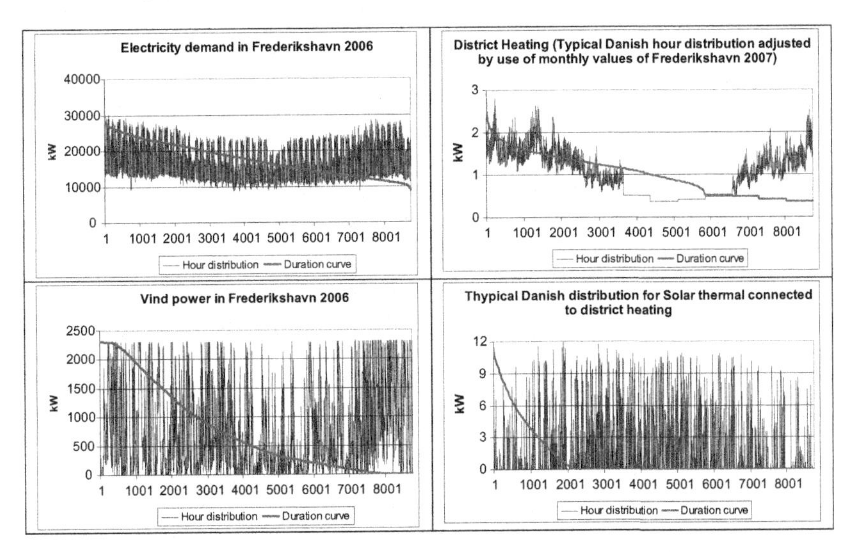

Moreover, the analyses show that, on an annual basis, all energy demands in the Energy Town Frederikshavn can be met by 100% renewable energy by using 185 GWh of waste, 225 GWh of biogas, 48 GWh of straw, 5 GWh of solar thermal power, 48 GWh of geothermal heat, and 130 GWh of wind power (equal to a fuel equivalence of 325 GWh). However, the production of electricity is not able to meet the demands during all hours. The analyses indicate that the system needs an exchange of approximately 25 GWh of imported electricity. However, on an annual basis, such import is compensated by a similar export during other hours.

The system configuration is shown in the diagram in Fig. 7.3.

PHASE 3: FREDERIKSHAVN IN THE YEAR 2030—100% RENEWABLE ENERGY AND LESS BIOMASS

Phase 3 involves the reduction of biomass including waste to a level corresponding to Frederikshavn's share of the national resources. A potential of 165–400 PJ of domestic biomass in Denmark has been identified, depending on the scale of energy crops. Residual resources (straw, biogas, wood chips, and waste) account for 165 PJ/year. Based on the size of the population, Frederikshavn's share ranges from 220 to 500 GWh/year. The expected system of year 2015 utilizes around 450 GWh. This is in the upper end of the range and will then have to be adjusted accordingly. However, the use of biomass is in the right order of magnitude with regard to the long-term objective of the project.

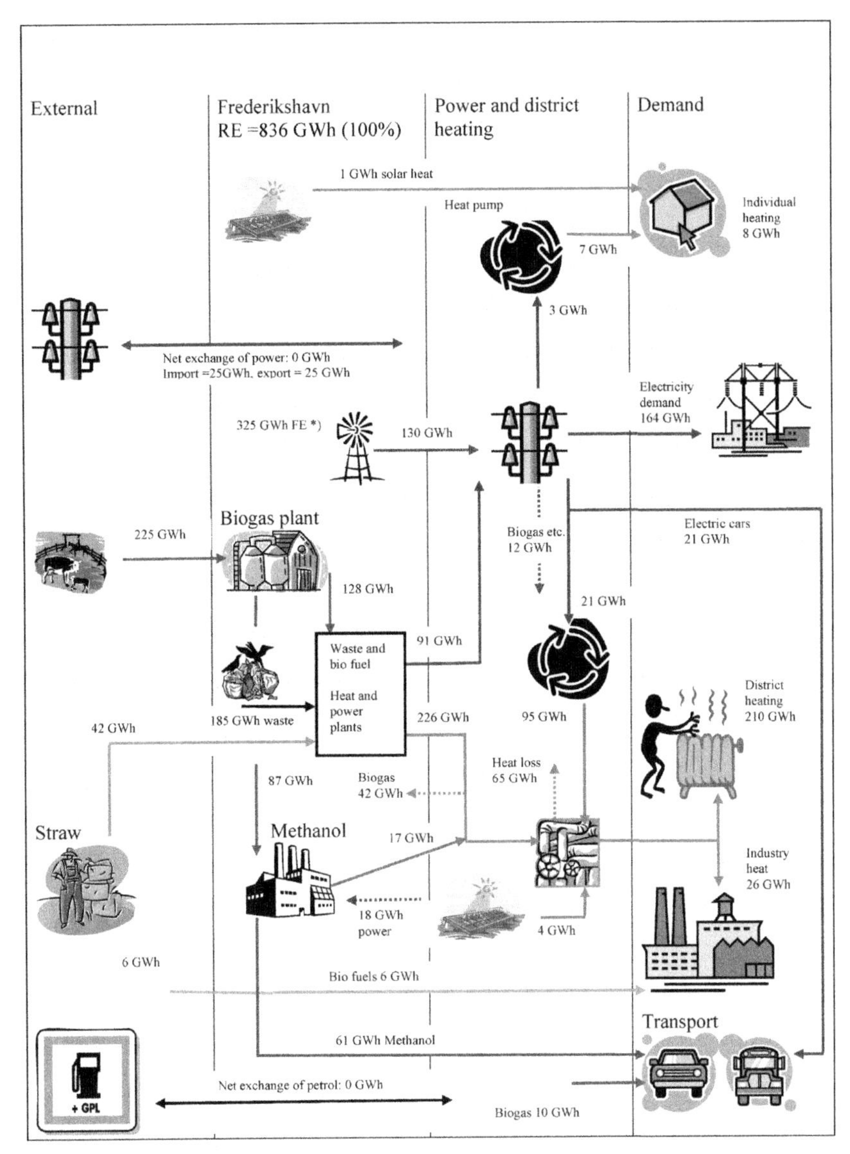

FIGURE 7.3 Frederikshavn 2015 (100% renewable energy). *Electricity as given in fuel equivalents of electricity production of a coal-fired steam turbine with an efficiency of 40%. *RE*, renewable energy.

Phase 3 will, among other things, include better insulation of homes, power savings and an increased efficiency in the industry, as well as further transition to electric vehicles in transport. The changes in phase 3 must be coordinated with conditions and activities in the rest of the country. The changes are not made more specific, and no attempt has been made to assess the need for investments.

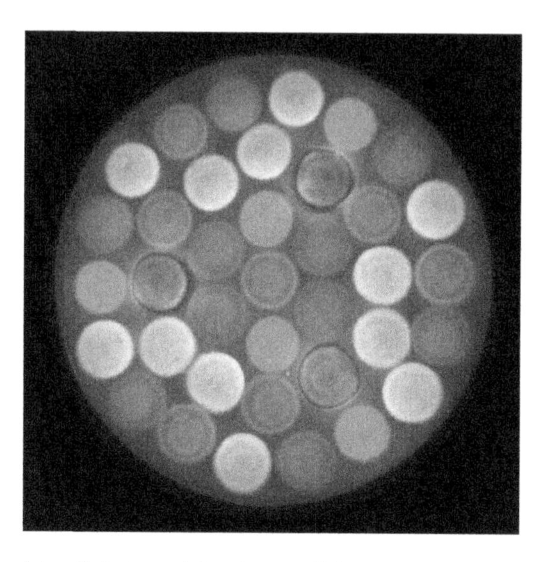

A 2-W light-emitting diode lamp giving the same light output as a 25-W filament light bulb.

Chapter 8

Life Cycle Analysis Versus Cost Benefit of Renewable Energy: Solar Systems Photovoltaics in Public Private Partnerships

Tom Pastore
Sanli Pastore & Hill, Los Angeles, CA, United States

Chapter Outline

INTRODUCTION

This is an update of my prior paper written 10 years ago. It presents the types, use, and reasoning for applying financial analyses in determining the feasibility of implementing renewable energy systems. The renewable energy sources discussed herein focus primarily on solar photovoltaic ("PV") systems. However, these analyses can apply to other renewable energy sources, e.g., wind or tidal or hybrid systems such as hydrogen storage and solar units.

Over the past decade, there have been large-scale vast changes in the landscape of noncombustible clean energy applications. Witness the emergence of Tesla electric cars as reckoning force in the otherwise staid and entrenched automobile industry. Renewable energy applications offer a broad array of applications for personal residences, commercial buildings, and government infrastructures. Incentivized by a combination of social and

Sustainable Cities and Communities Design Handbook. https://doi.org/10.1016/B978-0-12-813964-6.00008-2

financial factors, homes, businesses, and cities are installing solar PV systems across the globe.

Governments have been a key player in renewable energy. Solar arrays appear on municipal buildings and parking lots, in transit way meters, and across landfills. For government entities, the implementation of PV systems and related financial feasibility analyses are complicated. There are location and size constraints and nonprofit status may limit the use of financial incentives. This chapter presents a financial feasibility analysis of a city PV system installation.

Energy Challenges are Enormous

Multiple factors must be taken into consideration when forming an energy plan. In addition to issues regarding energy security and reliability, economic growth could affect the conditions of an energy contract. Furthermore, natural disasters could have an adverse impact on energy supply, and the environmental impact of an energy system may affect future conditions under which energy is produced. An agile energy plan incorporating renewable/green energy can address these considerations and factors.

SolarCity—Solar PV Array Installations on City Facilities

SolarCity (the "City"), located in Southern California, is considering the installation of solar PV arrays on several City facilities. The goal is to reduce greenhouse gas ("GHG") emissions while also lowering electricity costs. As part of its analyses, the City wants to assess the economic and technical feasibility of installing solar PV arrays on its City Hall, the initial project. The City plans to install solar PV arrays on the rooftop of its City Hall (the "City Hall Solar Project") to reduce its impact on existing structures and the building's exterior appearance.

Due Diligence Procedures

Proper due diligence requires various analyses of the proposed PV systems. These analyses include:

1. Engineering analyses of design proposals, installation sites, and ongoing maintenance.
2. Financial analyses of PV systems' implementation costs, financing costs, federal and state incentives, operating costs, and maintenance costs.
3. Legal analyses of proposed contracts between the City, the PV systems developer/owner, and the electric utility.
4. Project management analyses from the perspective of the City.

All of the above analyses should be conducted to determine the feasibility of renewable energy central plants. This chapter focuses in detail on the financial analysis procedures involved with a life cycle analysis of a PV system.

LIFE CYCLE ANALYSIS OF A PV SYSTEM FROM A FINANCIAL PERSPECTIVE

The life cycle analysis must encompass all cash flows during the life of a PV system. Considerations important to this analysis include:

1. The need for any upfront capital investment.
2. Applications for all available incentives.
3. In the case with a nonprofit municipality, consideration of the structure of a power purchase agreement ("PPA") with a third party who will design, build, and maintain the system for the life of the PPA.
4. Current energy costs and a future escalation factor as the baseline for savings.
5. Plans for financing the PV system.

Furthermore, some characteristics of a PV system that affect financial analyses include the maintenance costs of a PV system, which are typically minimal, the fact that PV panels are usually guaranteed for at least 20 years, and that a PV system could operate for as many as 25–40 years. The following is a cost and incentives analysis of the City's proposed 150-kW PV system for the City Hall Solar Project (Table 8.1):

As noted earlier, the costs of *PV Panels* and *Structures and Installation* compose a majority of the total cost.

Available Incentives

Monetary incentives, which reduce the cost of PV systems, are available from both state and federal programs. There are two primary federal incentives. First, there is an income tax deduction of up to 30% of the equipment costs for commercial solar systems, referred to as the investment tax credit ("ITC"). This ITC was set to expire by the end of 2016; however, it has been extended as follows: ITC remains at 30% through the end of 2019; it then declines to 26% in 2020, 22% in 2021, and 10% from 2022 onward. Second, there are depreciation expense deductions under the internal revenue service Modified Accelerated Cost Recovery System ("MACRS"). Under MACRS, 5-year depreciation-based deductions allow recovery of up to 26% of system costs on a present value basis. The economic life of such PV systems usually ranges from 25 to 40 years, so this incentive allows for a relatively rapid recovery of investments compared with the expected economic life of the property installed.

TABLE 8.1 Cost and Incentives Analysis

Hard and Soft Cost of Panel Installation				
				% of Total Cost
Cost of photovoltaic panels	$293,250	$ per watt:	$1.96	46
Cost of ancillary electrical equipment	$57,375	$ per watt:	$0.38	9
Cost of structures and installation	$184,875	$ per watt:	$1.23	29
Cost of engineering	$12,750	$ per watt:	$0.09	2
Cost of construction management	$25,500	$ per watt:	$0.17	4
Cost of general site work	$6375	$ per watt:	$0.04	1
Other costs (commissioning, performance guarantee, bonds, fees, and permits, OH&P)	$57,375	$ per watt:	$0.38	9
Total Hard + Soft Cost	**$637,500**	**$ per watt:**	**$4.25**	**100**
Incentives				
Federal income tax benefit	a			
Present value of accelerated depreciation recognition	a			
Total Incentives	**a**			
Effective Cost of Project Installation				
Total Hard + Soft Cost−Incentives	**$637,500**	**$ per watt**	**$4.25**	**100%**

OH&P, overhead and profit.
[a]*As discussed in the next section, the main incentives are Federal income tax benefits and the accelerated depreciation of solar equipment, which cannot be used by a nonprofit agency such as a city. However, a city can take advantage of these incentives by partnering with a for-profit, third-party developer.*

Many states, local governments, and utilities offer solar PV incentives that assist in increasing the financial feasibility of installing PV systems. These incentives consist of upfront cash payments, property and sales tax reductions or credits, and performance-based incentives. It is beyond the scope of this article to delve into the myriad incentives available on a state-by-state basis. In addition, incentives can change considerably over time. It is important for the financial analyst to keep abreast of changes in existing and formation of new incentives. Information on all state and federal incentive programs around the country is available at the Database of State Incentives for Renewables and Efficiency at http://www.dsireusa.org/.

Power Purchase Agreement Business Model

Under this model, the City would engage in a public–private partnership and allow a solar project developer to design, build, own, maintain, and operate a renewable energy facility on municipal property. This City would enter into a PPA with the developer to purchase all or part of the electricity generated by the system. A PPA is a contractual agreement whereby the developer/owner sells electricity to the City at fixed price per kilowatt hour over a long contractual term, typically 15–20 years. Such an agreement allows the City, which is a nonprofit entity that cannot fully utilize all available incentives, to still indirectly benefit from them by entering into a contract with a third party that utilizes all available incentives and sells the produced electricity to the nonprofit entity at a rate defined by the PPA.

Overall, the financial advantages of the PPA, i.e., the savings passed on to the City by the developer/owner taking advantage of federal tax credits, usually make it the most attractive business model for a municipality. Other advantages to the City include avoiding installation and equipment costs as well as ongoing operations and maintenance expenses, and locked-in fixed electricity price over the term of PPA contract.

Some disadvantages of a PPA include high legal costs involving complicated contract negotiations and multiparty negotiations between the City, developer/owner, and electric utility for net metering, grid tie-in, and offtake.

Calculating Utility Rates

Determination of the financial feasibility of a renewable energy system includes a comparison of its annual costs, typically through the PPA, with the current and expected future annual utility costs. For example, Southern California Edison utility rates involve charges for energy use, by customer, and by demand. Energy charges involve delivery service and generation charges based on time of use ("TOU"). Customer charges and facilities-related and power factor adjustment demand charges are not TOU charges. Time-related demand charges involve TOU during summer (12 a.m. on the first Sunday in June through 12 a.m. of the first Sunday in October) and winter (the remainder of the year).

TOU rates are based on three time periods: on peak, mid peak, and off peak, with maximum demand rates established for each of the time periods based on the maximum average kilowatt input recorded during any 15-min interval during the month. On-peak hours are noon through 6 p.m. on summer weekdays, except holidays. Mid-peak hours are 8 a.m. to noon and 6–11 p.m. on summer weekdays, except holidays, and off-peak hours account for all remaining hours.

Monthly energy rates are determined based on the above-mentioned factors.

Financial Analyses

Table 8.2 presents a cost comparison of the City Hall's annual utility costs with the City's proposed 150-kW PV system over an estimated 30-year economic life, financed over 20 years. The City will enter into a PPA with a developer to purchase electricity generated by the PV system.

There are multiple methods of financial analysis to consider for a PV system. The first is the payback period, the length of time required to recover an initial investment through cash flows generated by the investment. The payback period is important when considering the City's financial ability to implement a PV system.

The internal rate of return ("IRR") is the discount rate that makes the project have a zero net present value ("NPV"). Under this method, the project should be implemented if the IRR is greater than its hurdle rate, i.e., required rate of return. The NPV is the sum of the present values of the annual cash flows minus the present value of the investments. The discount rate accounts for the time value of money and uncertainties present in the NPV. This method is important as it shows the net value of the PV system from year to year. The project should be implemented if the NPV is positive.

To explain the strengths and weaknesses of the methods described earlier, Charles T. Horngren writes (Horngren et al., 2003):

> *One big advantage of the NPV method is that it expresses computations in dollars, not in percent. Therefore, we can sum NPVs of individual projects to calculate NPV of a combination of projects. In contrast, IRRs of individual projects cannot be added or averaged to represent IRR of a combination of projects…Two weaknesses of the payback method are that (1) it fails to incorporate the time value of money and (2) it does not consider a project's cash flows after the payback period…Another problem with the payback method is that choosing too short a cutoff period for project acceptance may promote the selection of only short-lived projects. [If they use only the pay-back method] An organization will tend to reject long-run, positive-NPV projects.*

The three methods described previously often do not yield the same result. Table 8.3 displays a financial analysis, based on the above-mentioned methods, of the proposed solar PV array installation:

Under this example, the payback period is approximately 10 years. However, the project's NPV does not become positive until year 11, and the IRR also does not exceed its required rate of return, i.e., 5% in the aforementioned scenario, until year 11. By the 30th year, the system is projected to have experienced $374,846 total paid back and have an NPV of $111,505.

TABLE 8.2 Cost Comparison

Year	Solar Electricity Produced (kWh)	Utility Electricity Cost ($/kWh)	PPA Solar Electricity Sale Price ($/kWh)	Total Utility Electricity Cost ($)	PPA Electricity Cost ($)	Difference in Cost Between Utility and Solar Electric Production ($)	Fixed Capacity Charge ($)	Net Benefit ($)
	A	B	C	D = A × B	E = A × C	F = D–E	G	H=F–G
1	160,000	0.120	0.075	19,200	12,000	7200	8000	(800)
2	159,200	0.124	0.077	19,677	12,298	7379	8000	(621)
3	158,404	0.127	0.080	20,166	12,604	7562	8000	(438)
4	157,612	0.131	0.082	20,667	12,917	7750	8000	(250)
5	156,824	0.135	0.084	21,181	13,238	7943	8000	(57)
6	156,040	0.139	0.087	21,707	13,567	8140	8000	140
7	155,260	0.143	0.090	22,247	13,904	8342	8000	342
8	154,483	0.148	0.092	22,799	14,250	8550	8000	550
9	153,711	0.152	0.095	23,366	14,604	8762	8000	762
10	152,942	0.157	0.098	23,947	14,967	8980	8000	980
11	152,177	0.161	0.101	24,542	15,339	9203	8000	1203
12	151,417	0.166	0.104	25,152	15,720	9432	8000	1432
13	150,660	0.171	0.107	25,777	16,110	9666	8000	1666
14	149,906	0.176	0.110	26,417	16,511	9906	8000	1906
15	149,157	0.182	0.113	27,074	16,921	10,153	8000	2153
16	148,411	0.187	0.117	27,746	17,341	10,405	8000	2405

Continued

TABLE 8.2 Cost Comparison—cont'd

Year	Solar Electricity Produced (kWh)	Utility Electricity Cost ($/kWh)	PPA Solar Electricity Sale Price ($/kWh)	Total Utility Electricity Cost ($)	PPA Electricity Cost ($)	Difference in Cost Between Utility and Solar Electric Production ($)	Fixed Capacity Charge ($)	Net Benefit ($)
	A	B	C	D = A × B	E = A × C	F = D−E	G	H=F−G
17	147,669	0.193	0.120	28,436	17,772	10,663	8000	2663
18	146,930	0.198	0.124	29,142	18,214	10,928	8000	2928
19	146,196	0.204	0.128	29,867	18,667	11,200	8000	3200
20	145,465	0.210	0.132	30,609	19,131	11,478	8000	3478
21	144,737	0.217	0.000	31,369	0	31,369	0	31,369
22	144,014	0.223	0.000	32,149	0	32,149	0	32,149
23	143,294	0.230	0.000	32,948	0	32,948	0	32,948
24	142,577	0.237	0.000	33,767	0	33,767	0	33,767
25	141,864	0.244	0.000	34,606	0	34,606	0	34,606
26	141,155	0.251	0.000	35,466	0	35,466	0	35,466
27	140,449	0.259	0.000	36,347	0	36,347	0	36,347
28	139,747	0.267	0.000	37,250	0	37,250	0	37,250
29	139,048	0.275	0.000	38,176	0	38,176	0	38,176
30	138,353	0.283	0.000	39,125	0	39,125	0	39,125
Total	**4,467,701**	**0.190**		**840,919**	**306,073**	**534,846**	**160,000**	**374,846**

PPA, power purchase agreement.

TABLE 8.3 Financial Analysis

Year	Net Benefit (i.e., Savings)		
	Pay Back Period ($)	Internal Rate of Return (%)	Net Present Value ($)
1	(800)		(762)
2	(1421)		(1325)
3	(1859)		(1703)
4	(2109)		(1909)
5	(2166)		(1954)
6	(2026)		(1849)
7	(1683)		(1606)
8	(1133)		(1234)
9	(371)		(742)
10	609	3.82	(141)
11	1,812	8.74	563
12	3,244	12.33	1,360
13	4,910	15.01	2,244
14	6,816	17.04	3,207
15	8,969	18.60	4,242
16	11,374	19.81	5,344
17	14,037	20.77	6,506
18	16,966	21.53	7,723
19	20,166	22.14	8,989
20	23,644	22.63	10,300
21	55,013	25.25	21,560
22	87,162	26.71	32,550
23	120,110	27.64	43,277
24	153,877	28.28	53,747
25	188,482	28.73	63,966
26	223,948	29.06	73,940
27	260,295	29.31	83,676

Continued

TABLE 8.3 Financial Analysis—cont'd

	Net Benefit (i.e., Savings)		
Year	Pay Back Period ($)	Internal Rate of Return (%)	Net Present Value ($)
28	297,545	29.49	93,178
29	335,721	29.63	102,453
30	374,846	29.73	111,505

Consideration of Externalities

Beyond consideration of payback and NPV analyses, there are both qualitative and quantitative externalities involved with PV systems that must be considered in a complete financial analysis. Quantitative benefits to the entity operating the PV system include possible employee health care savings as a result of a cleaner environment.

Qualitative externalities include reduction of pollution and GHG emissions, and reduced dependency on utility providers as well as control over energy-price volatility for the campuses by eliminating that cost entirely. In addition, PV systems can provide power during traditional power outages, whether due to natural disasters or any other reason. Finally, there is virtually no maintenance cost associated with PV systems, with long-term reliability of 25–40 years. Table 8.4 presents a hypothetical analysis of externality factors to provide a quantitative concept of the effect of externalities on NPV.

TABLE 8.4 Externality Factors

Financial Analysis	
Net present value (NPV) at year 30	$111,505
Add	
NPV of health care costs savings[a]	$30,000
NPV of savings with externalities	**$141,505**

[a]*For illustrative purposes only.*

CONCLUSION

A proper financial analysis is one part of evaluating the feasibility of an energy plan such as the example of the City herein. As outlined in this chapter, it should take into account all factors in the life cycle of the PV system, including the PPA, investment costs, available incentives, energy costs, and externalities involved with the implementation of the PV system. Proper use of financial analyses to determine the financial viability of a PV system provides a critical portion of the proper due diligence in analyzing a PV system proposal.

REFERENCE

Horngren, C.T., Datar, S.M., Foster, G., 2003. Cost Accounting: A Managerial Emphasis, eleventh ed. Pearson Education, Inc., Upper Saddle River, NJ, pp. 720–725.

Chapter 9

Public Buildings and Institutions: Solar Power and Energy Conservation as Solutions

Legal and Financial Mechanisms for Sustainable Buildings

Douglas N. Yeoman

Parker and Covert LLP, Sacromento, CA, United States

Chapter Outline

ALTERNATIVE ENERGY PUBLIC POLICY

The California Legislature in 1974 passed Public Resources Code section 25007, which established a state policy "to employ a range of measures to reduce wasteful, uneconomical, and unnecessary uses of energy, thereby reducing the rate of growth of energy consumption, prudently conserve energy resources, and assure statewide environmental, public safety, and land use goals." With continued increases in energy usage and energy costs, the Legislature revised the statewide policy in 1981, adding that it was further the policy of the state to "promote all feasible means of energy and water conservation and all feasible uses of alternative energy and water supply

Sustainable Cities and Communities Design Handbook. https://doi.org/10.1016/B978-0-12-813964-6.00009-4

sources [including, but not limited to solar technologies]." (Public Resources Code section 25008).

With the establishment of the California Renewables Portfolio Standard in 2002, series of Executive Orders and Legislation have followed, the most recent being Governor Brown's signing into law Senate Bill 350 in October 2015, requiring retail sellers and publicly owned utilities to procure "half of the state's electricity from renewable sources by 2030." When signing SB 350, Governor Brown stated that "California has taken groundbreaking steps to increase the efficiency of our cars, buildings and appliances and provide ever more renewable energy."

According to the California Energy Commission, as of October 31, 2016, almost 9400 MW of distributed generation capacity was operating or installed in California, with an additional 900 MW pending. The 12,000-MW goal for distributed generation by 2020 set by Governor Brown in his 2011 term is expected to be met or exceeded.

LEGAL MECHANISMS FACILITATING DEVELOPMENT OF ALTERNATIVE ENERGY SOURCES

California Clean Energy Jobs Act

In the November 6, 2012, General Election, the voters approved Proposition 39, known as the California Clean Energy Jobs Act, which has encouraged and provided significant funds to local educational agencies (LEAs) (including county offices of education, school districts, charter schools, and state special schools) and community college districts for the purpose of funding eligible projects that create jobs in California improving energy efficiency and expanding clean energy generation. For a period of five fiscal years through 2017–18, Proposition 39 requires annual appropriations by the Legislature from the General Fund into the Clean Energy Job Creation Fund. For fiscal years 2013–14 through 2016–17, a total of $1.4 billion has been allocated to LEAs and another $165,437,000 has been made available to community college districts for eligible energy efficiency projects.

Energy Management Agreement by Community College Districts

Education Code sections 81660 through 81662 authorize a community college district to enter into an energy management agreement for energy management systems (i.e., solar energy or solar and energy management systems) with the *lowest responsible bidder*, considering the net cost or savings to the district, less the projected energy savings to be realized from the energy management system. The maximum term of such an agreement shall be the estimated useful life of the energy management system, or 15 years, whichever is less.

Energy Service Contract and Facility Ground Lease by Public Agencies

To implement the public policy set forth in Public Resources Code section 25008, the California Legislature in 1986 adopted and added Chapter 3.2 of the Government Code (sections 4217.10 through 4217.18), which authorizes public agencies to enter into energy service contracts for the development of energy conservation, cogeneration, and alternate energy supply sources, *without competitive bidding* (referred to hereinafter as "4217 Contracts"). School districts, community college districts, counties, cities, districts, joint powers authorities, or other political subdivisions are included in the definition of "public agency." (Gov. Code Section 4217.11, subd. (j).)

The use of 4217 Contracts has been utilized more than any other contracting mechanism by LEAs and community college district to implement both energy conservation and energy generation projects because of its flexibility and straight forward compliance requirements.

Although a direct energy service contract and related facility ground lease may be entered into without competitive bidding if Proposition 39 funds are not intended to be used, a limitation to using this contracting method is the requirement that the contract only involve alternate energy equipment, including but not limited to solar, maintenance, load management techniques, or other conservation measures that result in the *reduction of energy use or makes for a more efficient use of* energy. (Gov. Code, Section 4217.11, subds. (a) and (c).) For example, an energy service contract may not include the installation of air conditioning where no form of air conditioning existed previously, as the addition of the air conditioning would result in an increase, not a reduction of energy use. In this case, the air conditioning component of the project would require competitive bidding, assuming the estimated cost would exceed the bidding amount threshold of the particular public agency.

As a condition to entering into an energy service contract and any necessarily related facility ground lease, the governing board must determine that entering into such agreements are in the best interests of the public agency. Except for State agency heads who can make the findings described later without holding a public hearing, the governing boards of all other public agencies must make the "best interests" determination at a regularly scheduled public hearing in which public notice has been given at least 2 weeks in advance. To support this determination, the board must find: (1) that the anticipated cost to the agency for thermal or electrical energy or for the "energy conservation facility" under the contract will be less than the anticipated marginal cost to the agency of thermal, electrical, or other energy that would have been consumed by the district in the absence of those purchases; and (2) that the difference, if any, between the fair rental value for the real property subject to the facility ground lease and the agreed rent, is anticipated to be offset by below-market energy purchase or other benefits provided under the energy service contract. (Gov. Code, Section 4217.12.) The term "energy conservation facility" is defined at Government Code section 4217.11(e) to mean: "alternate energy equipment, cogeneration equipment, or conservation measures located in public buildings or on land owned by public agencies."

Government Code section 4217.13 also authorizes a public agency to enter into a facility financing contract and a facility ground lease on terms determined by the board to be in the best interest of the agency if the determination is made at a regularly scheduled public hearing and if the governing body finds that funds for the repayment of the financing or the cost of design, construction and operation of the energy conservation facility, or both, are projected to be available from revenues generated from the sale of electricity or thermal energy from the facility or from funding that otherwise would have been used for the purchase of electrical, thermal, or other energy required by the agency in the absence of the energy conservation facility or both. As with energy service contracts, State agency heads may make these findings without holding a public hearing.

Public agencies typically support the findings required earlier for entering into an energy service contract and facility financing contract upon a preliminary survey of the agency's existing energy equipment and usage conducted by the prospective contractor. Government Code section 4217.15 authorizes a public agency to base its findings on projections for electrical and thermal energy rates from the following sources: (1) the public utility that provides thermal or electrical energy to the public agency; (2) the State Utilities Commission; (3) the State Energy Resources Conservation and Development Commission; or (4) the projections used by the Department of General Services for evaluating the feasibility of energy conservation facilities at state facilities located within the same public utility service area as the public agency.

Under this legislative scheme, public agencies may, "notwithstanding any other provision of law," enter into contracts for the sale of electricity, electrical generating capacity, or thermal energy produced by the energy conservation facility at rates and on such terms as may be approved by the governing board. (Gov. Code, Section 4217.14.)

Although no competitive selection process is required, a public agency may wish to solicit proposals to ensure it is receiving the greatest available energy savings. Section 4217.16 of the Government Code provides for the option of seeking proposals in stating:

> *Prior to awarding or entering into an agreement or lease, the public agency may request proposals from qualified persons. After evaluating the proposals, the public agency may award the contract on the basis of the experience of the contractor, the type of technology employed by the contractor, the cost to the local agency, and any other relevant considerations. The public agency may utilize the pool of qualified energy service companies established pursuant to Section 388 of the Public Utilities Code and the procedures contained in that section in awarding the contract.*

Public Utilities Code section 388 referenced earlier, which is applicable to State agencies, authorizes agencies to "enter into an energy savings contract with a qualified energy service company for the purchase or exchange of thermal or electrical energy or water, or to acquire energy efficiency or water conservation services, or both energy efficiency and water conservation services for a term not exceeding 35 years, at rates and upon those terms approved by the agency."

Paragraph (b) of Public Utilities Code section 388 provides the option for State agencies and local agencies intending to enter into an energy savings contract or a contract for an energy retrofit project to establish a pool of qualified energy service companies, which are updated at least every 2 years based on qualification, experience, pricing, or other pertinent factors. The paragraph further provides that energy service contracts for individual projects may be awarded through a competitive selection process to individuals or firms identified in the pool. As used in section 388, an "energy retrofit project" does not include the erection or installation of a power generation system, a power purchase agreement, or a project utilizing a site license or site lease.

Government Code section 4217.18 concludes Chapter 3.2 on energy conservation contracts by emphasizing the intended flexibility of the aforementioned sections by stating:

The provisions of this chapter shall be construed to provide the greatest possible flexibility to public agencies in structuring agreements entered into hereunder so that economic benefits may be maximized and financing and projects may be minimized. To this end, public agencies and the entities with whom they contract under this chapter should have great latitude in characterizing components of energy conservation facilities as personal or real property and in granting security interests in leasehold interests and components of the alternate energy facilities to project lenders.

As previously noted, should an LEA or community college district intend to implement an energy conservation or generation project utilizing Proposition 39 funds, the public agency may not use a sole source process to award funds. Proposition 39 requires that a competitive selection process be used, which may utilize "best value criteria" (Public Resources Code Section 26235(c)).

To participate in the Proposition 39 program, an LEA or community college district must follow an eight-step process as outlined in the latest version of the Program Implementation Guidelines adopted by the California Energy Commission ("CEC") from initially providing the CEC access to the utility usage and billing data at the school site level to the submittal and approval by the CEC of the agency's Energy Expenditure Plan, including but not limited to meeting the current savings to investment ratio of 1.01, and subsequent energy project tracking and reporting requirements upon completion of the project.

Utilizing the 4217 Contract statutory scheme may be advantageous where a public agency desires to implement an energy conservation project involving conservation measures where the cost of design, construction, and operation is projected to be recovered from energy savings over the life expectancy of the conservation measures. Conversely, if (1) new energy conservation measures are being considered, (2) the cost of design, construction, and operation of energy conservation measures is not projected to be recovered over the life expectancy of the energy conservation measures, or (c) the public agency either does not have the funds or does not desire to finance the new energy conservation measures, the agency may want to consider entering into a Power Purchase Agreement as discussed in the following section.

Power Purchase Agreement by Governmental Agency

To assist local governmental agencies in infrastructure financing of energy or power production projects, Assembly Bill 2660 was passed in 1996, which added Government Code sections 5956 through 5956.10 (referred to herein as the "Power Purchase Provisions"). The Power Purchase Provisions grant the authority to a city, county, school district, community college district, public district, county board of education, joint powers authority, transportation commission or authority, or any other public or municipal corporation (collectively, "governmental agency"), "to utilize private investment capital to study, plan, design, construct, develop, finance, maintain, rebuild, improve, repair, or operate, or any combination thereof, fee-producing infrastructure facilities." (Gov. Code Section 5956.1.) The term "fee-producing infrastructure project" is defined as the "operation of the infrastructure project or facility ... paid for by the persons or entities benefited by or utilizing the project or facility." (Gov. Code Section 5956.3(c).) Any combination of private infrastructure financing, federal or local funds may be utilized under this statutory scheme. (Gov. Code Section 5956.9.) State Agencies are specifically prohibited from utilizing the Power Purchase Provisions. (Gov. Code Section 5956.10.)

The Power Purchase Provisions provide that an agency may solicit proposals as part of a competitive negotiation process when selecting a contractor for the studying, planning, design, developing, financing, construction, maintenance, rebuilding, improvement, repair, or operation, or any combination thereof, for fee-producing infrastructure projects. *Neither competitive bidding nor compliance with any other provision of the Public Contract Code or Government Code relating to public procurements is required.* Notwithstanding, should the governmental agency intend to use Proposition 39 funds, a competitive selection process similar to that discussed earlier for 4217 Contracts must be utilized. Projects may be proposed by a private entity and selected by the governmental agency in its discretion, subject to the following selection criteria being considered: (1) demonstrated competence and qualifications of the private entity; and (2) the proposed facility must be operated at fair and reasonable prices to the user of the infrastructure facility services. The competitive negotiation process must specifically prohibit practices that may result in unlawful activity, including, but not limited to, rebates, kickbacks, or other unlawful consideration, and any prohibited conflict of interest involving the employees of the governmental agency in violation of Government Code section 87100, which states:

> *No public official at any level of state or local government shall make, participate in making or in any way attempt to use his official position to influence a governmental decision in which he knows or has reason to know he has a financial interest.*

The Power Purchase Agreement to be entered into with the private entity is required to contain provisions to ensure the following:

1. Provide whether the facilities will be owned by the agency or contractor during the term of the Agreement. If the facilities are leased to the contractor, the Agreement must provide for a complete reversion of ownership in the facility at the expiration of the term (may not exceed 35 years), *without charge* to the governmental agency.
2. Compliance with the California Environmental Quality Act ("CEQA") commencing at Public Resources Code section 21000 before the commencement of project development. Although cogeneration projects at existing facilities may be categorically exempt from CEQA if the conditions set forth at Title 14, California Code of Regulations, section 15329 are satisfied, typically, a negative declaration or mitigated negative declaration will be required to demonstrate that the facility will not have a significant adverse impact on the environment.
3. Performance bonds as security to ensure completion of the construction of the facility and contractual provisions that are necessary to protect the revenue streams of the project. Insurance provisions (example under item 11 below), hold harmless and indemnity clauses also provide such protection.
4. Adequate financial resources of the private entity to design, build, and operate the facility, after the date of the agreement.
5. Authority for the governmental agency to impose user fees for use of the facility in an amount sufficient to protect the revenue streams necessary for projects or facilities. The user fee revenue must be dedicated exclusively to payment of the private entity's direct and indirect capital outlay costs for the project, direct and indirect costs associated with operations, direct and indirect user fee collection costs, direct costs of administration of the facility, reimbursement for the direct and indirect costs of maintenance, and a negotiated reasonable return on investment to the private entity.

Before taking action to impose or increase a user fee, the governmental agency must conduct at least one public hearing on the proposed fee. Notice of the public hearing(s) must be given (1) by mail not less than 14 days before the meeting to any interested party who has requested notice of the meeting, and (2) by publication not less than 10 days before the meeting in a newspaper of general circulation in the jurisdiction of the governmental agency. All data in support of the proposed user fee must also be available for public inspection at least 10 days preceding the meeting. All costs incurred by the governmental agency in providing the required notice and holding the public hearing(s) may be recovered from the fees to be charged.

Action to impose or increase a user fee must be taken by ordinance or resolution by the governing board of the governmental agency. The established fee may not exceed the estimated amount required to provide the service for which the fee is charged and a reasonable rate of return on investment.

6. Require that any revenues in excess of the actual cost and a reasonable rate of return on investment, be applied by the governmental agency to either reduce any indebtedness incurred by the private entity with respect to the project, be paid into a reserve account to offset future operation costs, be paid into the appropriate government account, be used to reduce the user fee or service charge creating the excess, or a combination of these sources.
7. Require the private entity to maintain the facility in good operating condition at all times, including the time the facility reverts to the governmental agency.
8. Preparation by the private entity of an annual audited report accounting for the income received and expenses to operate the facility.
9. Provision for a buyout of the private entity by the governmental agency in the event of termination or default before the expiration of the term of the Power Purchase Agreement.
10. Provision for appropriate indemnity promises between the governmental agency and private entity.
11. Provision requiring the private entity to maintain insurance with those coverages and in those amounts that the governmental agency deems appropriate. A sample insurance provision is as follows:
    A. *Contractor's Insurance*. The Contractor shall provide and maintain insurance, at the Contractor's own cost and expense, against all claims or losses that may arise from or in connection with the performance of services by the Contractor. The obligation to maintain insurance shall not in any way affect the indemnity provided in or by Section __. District's acceptance of Contractor's insurance hereunder shall not in any way act as a limitation on the extent of Contractor's liability.

B. Coverages, Subcontractor, Subconsultant Insurance.

(1) Contractor shall maintain and shall require that every Subcontractor and Subconsultant, of any Tier, performing or providing any portion of the Work obtain and maintain, for the duration of its performance of the Work and for the full duration of all guarantee or warranty periods set forth in the Contract Documents (and such longer periods as required below for completed operations coverage), the insurance coverage outlined in (a) through (d) below, and all such other insurance as required by Applicable Laws; provided, however, that Subcontractors not providing professional services shall not be required to provide Professional Liability coverage and except where District has given its written approval to waive said limits for a specific Subcontractor.

(a) Commercial General Liability and Property Insurance, on an "occurrence" form covering occurrences (including, but not limited to those listed below) arising out of or related to operations, whether such operations be by the Contractor, a Subcontractor or Subconsultant or by anyone directly or indirectly employed by any of them, or by anyone for whose acts any of them may be liable, involving damage or loss, of any kind: (1) because of bodily injury, sickness or death of any person other than the Contractor's, Subcontractor's or Subconsultant's employees; (2) sustained (a) by a person as a result of an offense directly or indirectly related to employment of such person, or (b) by another person; (3) other than to the Work itself, because of injury to or destruction of tangible property including loss of use resulting therefrom; (4) because of bodily injury, death of a person or property damages arising out of ownership, maintenance or use of a motor vehicle; (5) contractual liability insurance; and (6) completed operations, with limits as follows:

- $2,000,000 per occurrence for Bodily Injury and Property Damage.
- $2,000,000 General Aggregate - other than Products/Completed Operations.
- $1,000,000 Products/Completed Operations Aggregate for the duration of a period of not shorter than 1 year after Final Completion and Acceptance of the Project.
- $1,000,000 Personal & Advertising Injury.
- Full replacement value for Fire Damage.
- And including, without limitation, special hazards coverage for:
- $1,000,000 Material hoists
- $1,000,000 Explosion, collapse & underground

(b) Auto Liability insurance, on an occurrence form, for owned, hired and non-owned vehicles with limits of $1,000,000 per occurrence

(c) Professional Liability insurance (only to be provided by Subconsultants or Subcontractors performing professional services), written on a "claims-made" form, with limits of:
- $1,000,000 per claim
- $1,000,000 aggregate

(d) Excess Liability insurance, on an "occurrence" form, in excess of coverages provided for Commercial General Liability, Auto Liability, Professional Liability and Employer's Liability, with limits as follows:
- $1,000,000 each occurrence (or, in the case of coverage in excess of Professional Liability, each claim).
- $1,000,000 aggregate

(2) *Evidence of Insurance*. Upon request of District, Contractor shall promptly deliver to District Certificates of Insurance evidencing that the Subcontractors and Subconsultants have obtained and maintained policies of insurance in conformity with the requirements of this Section 29. Failure or refusal of Contractor to do so may be deemed by District to be a material default by Contractor of the Contract.

(3) *Builder's Risk "All-Risk" Insurance*. Builder's Risk "All Risk" Insurance will be purchased by District, which shall include primary coverage protecting the insured's interest in materials, supplies, equipment, fixtures, structures, and real property to be incorporated into and forming a part of the Project and with policy limits protects up to the Estimated Maximum Value of the Project for any one loss or occurrence and with deductibles of between $5000–$25,000 per occurrence. Said Builder's Risk policy shall be endorsed to add Contractor and its Subcontractors of the first Tier and Subconsultants of the first Tier as additional named insureds, as their interests may appear, and to waive the carrier's right of recovery under subrogation against Contractor and all Subcontractors and Subconsultants whose interest are insured under such policy. If a claim results from any construction activity of Contractor or a Subcontractor or Subconsultant, then Contractor or the Subcontractor or Subconsultant having care, custody and control of the damaged property shall pay the deductible amount. Any loss or damage covered by the Builder's Risk Policy shall be adjusted by and payable to District, or its designee, for the benefit of all Parties as their interest may appear. District shall not be responsible for loss or damage to and shall not obtain and/or maintain in force insurance on temporary structures, construction equipment, tools or personal effects, owned, rented to, or in the care, custody and control of Contractor or any Subcontractor or Subconsultant. In the event of loss or damage is caused by the acts or omissions of Contractor or its Subcontractors or Subconsultants that is not covered by the Builders Risk policy, the cost of the repair and/or replacement of such loss or damage shall be at Contractor's own expense. District, Contractor and all Subcontractors and Subconsultants each and all waive rights of subrogation against each other to the extent that said Builder's Risk policy covers property damage arising out of the perils of fire or other casualty also covered by Contractor's or a Subcontractor's or Subconsultant's insurance policy.

(4) *Policy Requirements and Endorsements.* Except as otherwise stated in this Paragraph 4, each policy of insurance required to be provided by Subcontractors and Subconsultants shall comply with the following:

(a) The commercial general liability insurance policy shall contain a waiver of subrogation rights against District, members of the Board of Trustees, District's Consultants, and each of their respective agents, employees, and volunteers, and the State Allocation Board

(b) The insurance policies provided for Commercial General Liability, Auto Liability, as well as any Excess Liability coverage in excess thereof shall be endorsed to include, individually and collectively, the District, members of the Board of Trustees, District's Consultants, and each of their respective agents, employees, and volunteers, and the State Allocation Board, as additional insureds.

(c) The insurance policies shall provide that the insurance is primary coverage with respect to District and all other additional insureds, shall not be considered contributory insurance with any insurance policies of the District or any other additional insureds, and all insurance coverages provided by District and any other additional insureds shall be considered excess to the coverages provided by the Subcontractor or Subconsultant.

12. In the event of a dispute, both parties shall be entitled to all available legal or equitable remedies.

13. Payment bonds issued by an admitted surety insurer in an amount not less than 100% of the contract amount to secure the payment of claims of laborers, mechanics, and materials suppliers employed on the work under the contract.

14. Require that the plans and specifications for the project be constructed in compliance with all applicable governmental design standards and shall utilize private sector design and construction firms to design and construct the infrastructure facilities.

15. Comply with all applicable laws relating to public property and public works projects, including the payment of prevailing wages.

Although not required by statute, it is suggested that governmental agencies consider including a guarantee provision in the Power Purchase Agreement whereby the energy provider will guarantee a minimum energy output, which if not met, will result in a monetary penalty on behalf of the power provider, such as requiring the energy provider to pay the difference between what the governmental agency is required to pay the utility company for the power shortage and what the agency would have been required to pay the energy provider had the guaranteed energy output been delivered. An example of such a provision is set forth below.

> *Guarantee. Power Provider shall provide a Cumulative Output Guarantee from the Generating Facility commencing on the date of Commercial Operation and continuing until the twentieth (20th) anniversary of the Commercial Operation Date or achievement of the twentieth year cumulative output guarantee of _______________kWh, whichever comes first. The guarantee is defined to be 90% of the expected annual production from the Generating Facility to be measured in kWh.*
>
> *In order to control for variations in weather, the actual output will be compared to the Cumulative Output Guarantee on a cumulative basis on the third (3rd), sixth (6th) ninth (9th), twelfth (12th), fifteenth (15th) and twentieth (20th) year during the cumulative output Guarantee Term. Actual production shall accrue to the cumulative balance each year and be compared on the anniversary dates noted above of the Commercial Operation Date to the aggregate cumulative output guarantee for the years in that measurement period as indicated in the table below. In the event that the Guaranteed Energy Output is not achieved as described above during the term of this Agreement (the "Guaranteed Energy Output Shortage"), and Purchaser is required to purchase replacement kWhac from Southern California Edison, then Power Provider shall refund the difference between the amount Purchaser pays Southern California Edison for the replacement power and the annual rate as specified in Exhibit __. The Southern California Edison replacement power price is defined as the blended average annual TOU-8 tariff for that portion of kWhac representing the Guaranteed Energy Output Shortage. This guarantee shall immediately terminate if the Generating Facility title is transferred to Purchaser.*
>
> *Example of hypothetical shortfall payment calculation. In year 3, The governmental agency consumes 7 million kWh of electricity and pays Southern California Edison one million fifty thousand dollars for its total annual energy use under the TOU-8 rate. Therefore the blended average annual TOU-8 rate is equal to $0.15 per kWh ($1,050,000 / 7,000,000 kW h = $0.15 per kWh). The*

cumulative output guarantee in year 3 is 4,193,486 kW h. The actual delivered cumulative output is 4,100,000 kW h. The shortfall is therefore 93,486 kW h. The PPA rate is $0.14333 per kWh in year 3. The shortfall payment paid by Power Provider to Purchaser is $624 (93,486 kW h x [$0.15 - $0.1433] = $624).

Lease of Photovoltaic System

The Los Angeles Department of Water and Power ("DWP") has instituted a Solar Photovoltaic Incentive Program (the "Incentive Program") consistent with the California Solar Initiative set forth in Senate Bill One (SB 1, Murray), which was approved during the 2005–06 Legislative Term with the goal of installing 3000 MW of net-metered solar energy systems by December 31, 2016. On December 6, 2016, DWP agreed to continue the Incentive Program and implement a new program, which effective January 3, 2017, reduced incentives of Government/Nonprofit organizations from $1.15 per watt to $0.95.

Public Utilities Code section 2854(b) requires that on or before January 1, 2008, a local publicly owned electric utility must offer monetary incentives for the installation of solar energy systems of at least $2.80 per installed watt, or for the electricity produced by the solar energy system, measured in kilowatt hours, as determined by the governing board of a local publicly owned electric utility, for photovoltaic (PV) solar energy systems. The incentive level is scheduled to decline each year thereafter at a rate of no less than an average of 7% per year.

For a local publicly owned electric utility to institute a solar energy program, Public Utilities Code section 2854(d) requires the program to be consistent with all the following:

(1) That a solar energy system receiving monetary incentives comply with the eligibility criteria, design, installation, and electrical output standards or incentives established by the State Energy Resources Conservation and Development Commission pursuant to Section 25782 of the Public Resources Code.

(2) That solar energy systems receiving monetary incentives are intended primarily to offset part or all of the consumer's own electricity demand.

(3) That all components in the solar energy system are new and unused, and have not previously been placed in service in any other location or for any other application.

(4) That the solar energy system has a warranty of not less than 10 years to protect against defects and undue degradation of electrical generation output.

(5) That the solar energy system be located on the same premises of the end-use consumer where the consumer's own electricity demand is located.

(6) That the solar energy system be connected to the electric utility's electrical distribution system within the state.

(7) That the solar energy system has meters or other devices in place to monitor and measure the system's performance and the quantity of electricity generated by the system.
(8) That the solar energy system be installed in conformance with the manufacturer's specifications and in compliance with all applicable electrical and building code standards."

In implementing the DWP Incentive Program consistent with the abovementioned criteria, DWP customers have been given an alternative to purchasing and owning the PV system. The customer may lease the system from a third party, provided that the following conditions are met:

1. The lease is guaranteed for at least 10 years (to cover the anticipated period of energy production that the incentive is based on.
2. The PV system is operational and operated at the expected generation capacity for a 10-year term.
3. The lease provides for customer ownership by the end of the 10-year term.
4. The lease payments may not be based on energy production from the equipment, which could be interpreted as the sale of electricity.
5. The incentive payment will be paid directly to the customer and is not assignable to a third party, but the customer may request that DWP forward the customer's incentive payment directly to the customer's contractor or the manufacturer, installer, or owner of the solar PV system.

The Incentive Program requires that the lease agreement for the equipment be provided to DWP for review should the lease be modified or amended.

TREATMENT OF ENVIRONMENTAL INCENTIVES

When a governmental agency is considering use of one of the aforementioned legal mechanisms for an alternative energy program, it is important that the governmental agency control, if possible, as many of the environmental attributes, environmental incentives, and reporting rights as possible. As used herein, the terms "environmental attributes," "environmental incentives," and "Reporting Rights" are defined as follows:

> "Environmental Attributes" means the characteristics of electric power generation at the Generating Facility (the electric power generation equipment, controls, meters, etc. connected to the energy delivery point as a fixture on the site) that have intrinsic value, separate and apart from the Energy Output (total quantity of all actual net energy generated), arising from the perceived environmental benefits of the Generating Facility of the Energy Output, including but not limited to all environmental and other attributes that differentiate the Generating Facility or the Energy Output from energy generated by fossil-fuel based generation units, fuels or resources, characteristics of the Generating Facility that may result in the

avoidance of environmental impacts on air, soil or water, such as the absence of emission of any oxides of nitrogen, sulphur or carbon or of mercury, or other gas or chemical, soot, particulate matter or other substances attributable to the Generating Facility or the compliance of the Generating Facility or the Energy Output with the law, rules and standards of the United Nations Framework Convention on Climate Change (the "UNFCCC") or the Kyoto Protocol to the UNFCCC or crediting "early action" with a view thereto, or laws or regulations involving or administered by the Clean Air Markets Division of the Environmental Protection Agency or successor administrator or any state or federal entity given jurisdiction over a program involving transferability of Environmental Attributes and Reporting Rights.

"Environmental Incentives" means all rights, credits (including tax credits), rebates, benefits, reductions, offsets, and allowances and entitlements of any kind, howsoever entitled or named (including carbon credits and allowances), whether arising under federal, state or local law, international treaty, trade association membership or the like arising from the Environmental Attributes of the Generating Facility or the Energy Output or otherwise from the development or installation of the Generating Facility or the production, sale, purchase, consumption or use of the Energy Output. Without limiting the forgoing, "Environmental Incentives" includes green tags, renewable energy credits, tradable renewable certificates, portfolio energy credits, the right to apply for (and entitlement to receive) incentives under the Self-Generation Incentive Program, the Emerging Renewables Program, the California Solar Initiative, or other incentive programs offered by the State of California and the right to claim federal income tax credits under Sections 45 and/or 48 of the Internal Revenue Code.

"Reporting Rights" means the right of private Power Provider to report to any federal, state, or local agency, authority or other party, including without limitation under Section 1605(b) of the Energy Policy Act of 1992 and provisions of the Energy Policy Act of 2005, or under any present or future domestic, international or foreign emissions trading program, that Power Provider owns the Environmental Attributes and the Environmental Incentives associated with the Energy Output.

An example of how the Environmental Attributes, Incentives, and Reporting Rights may be treated in an agreement is as follows:

"**(a)** *Delegation of Attributes to Power Provider.* Notwithstanding the Generating Facility's presence as a fixture on the Site, Power Provider shall own, and may assign or sell in its sole discretion, all right, title and interest associated with or resulting from the development and installation of the Generating Facility or the production, sale, purchase or use of the Energy Output including, without limitation:

(i) all Environmental Incentives except for Solar Renewable Energy Credits associated with the Generating Facility; and

(ii) the Reporting Rights and the exclusive rights to claim that: (A) the Energy Output was generated by the Generating Facility; (B) Power Provider is responsible for the delivery of the Energy Output to the Energy Delivery Point; (C) Power Provider is responsible for the reductions in emissions of pollution and greenhouse gases resulting from the generation of the Energy Output and the delivery thereof to the Energy Delivery Point; and (D) Power Provider is entitled to all credits, certificates, registrations, etc., evidencing or representing any of the foregoing.

(b) *Delegation of Attributes to Purchaser (governmental agency).* Purchaser shall own, and may assign or sell in its sole discretion, all right, title and interest associated with or resulting from the following:

(i) all Environmental Attributes and Solar Renewable Energy Credits associated with the Generating Facility; and

(ii) the Reporting Rights and the exclusive rights to claim that Purchaser is entitled to all Solar Renewable Energy Credits evidencing or representing any of the foregoing."

Based upon the public policy espoused by the California Legislature over the past 40 years, a number of legal mechanisms have been authorized to provide and encourage the development and installation of alternative energy sources. With the ever growing awareness and publicity regarding the continuing erosion of our environment, the support and advancement of alternative energy source technologies can be expected to continue for the foreseeable future.

Chapter 10

Life Cycle Analysis: The Economic Analysis of Demand-Side Programs and Projects in California

Woodrow W. Clark, II[1], Arnie Sowell[2], Don Schultz[3]

[1]*Clark Strategic Partners, Beverly Hills, CA, United States;* [2]*California State Assembly Office, Sacramento, CA, United States;* [3]*California Rate Payer Commission, Sacramento, CA, United States*

Chapter Outline

Sustainable Cities and Communities Design Handbook. https://doi.org/10.1016/B978-0-12-813964-6.00010-0

THE BASIC METHODOLOGY

Efficiency, conservation, and load management (C&LM) programs have been promoted since the 1970s by the California Public Utilities Commission (CPUC) and the California Energy Commission (CEC) as alternatives to -power plant construction and gas supply options. C&LM programs have been implemented in California by the major utilities through the use of ratepayer money and by the CEC pursuant to the CEC legislative mandate to establish energy efficiency standards for new buildings and appliances. The result is that California has been ranked consistently as one of the most energy-efficient states.

While cost effectiveness procedures for the CEC standards are outlined in the Public Resources Code, there was no such official guideline for utility-sponsored programs. With the publication of the Standard Practice for Cost-Benefit Analysis of Conservation and Load Management Programs in Feb. 1983, this void was substantially filled. With the informal "adoption" 1 year later of an appendix that identified cost-effectiveness procedures for an "All Ratepayers" Test, C&LM program cost-effectiveness consisted of the application of a series of tests representing a variety of perspectives—participants, nonparticipants, all ratepayers, society, and the utility. The Standard Practice Manual (SPM) was revised again in 1987–88. The primary changes (relative to the 1983 version) were:

1. renaming the "Non-Participant Test" to the "Ratepayer Impact Test";
2. renaming the "All-Ratepayer Test" to the "Total Resource Cost (TRC) Test";
3. treating the "Societal Test" as a variant of the "TRC Test";
4. expanding the explanation of "demand-side" activities that should be subjected to standard procedures of benefit–cost analysis.

Further changes to the manual captured in the 2002 (Clark et al., 2002) version were prompted by the cumulative effects of changes in the electric and natural gas industries, and a variety of changes in California statute related to these changes. As part of the major electrical industry restructuring legislation of 1996 (AB1890), for example, a public goods charge was established that ensured minimum funding levels for "cost-effective conservation and energy efficiency" for the 1998–2002 period, and then (in 2000) this was extended until the year 2011. Additional legislation in 2000 (AB1002) established a natural gas surcharge for similar purposes. Later in that year, the Energy

Security and Reliability Act of 2000 (AB970) directed the CPUC to establish, by the Spring of 2001, a distribution charge to provide revenues for a self-generation program and a directive to consider changes to cost-effectiveness methods to better account for reliability concerns.

In the Spring of 2001, a new state agency—the Consumer Power and Conservation Financing Authority—was created. This agency was expected to provide additional revenues—in the form of state revenue bonds—that could supplement the amount and type of public financial resources to finance energy efficiency and self-generation activities. By 2003, the agency closed due to lack of demand for funds.

The modifications to the SPM[1] reflect these more recent developments in several ways. First, the "Utility Cost Test" has been renamed the "Program Administrator Test" to include the assessment of programs managed by other agencies. Second, a definition of self-generation as a type of "demand-side" activity is included. Third, the description of the various potential elements of "externalities" in the Societal version of the TRC Test is expanded. Finally, the Limitations section outlines the scope of the manual and elaborates upon the processes traditionally instituted by the implementing agencies to adopt values for these externalities and to adopt the policy rules that accompany the manual.

Demand-Side Management Categories and Program Definitions

One important aspect of establishing standardized procedures for cost-effectiveness evaluations is the development and use of consistent definitions of categories, programs, and program elements.

This chapter uses the general program categories that distinguish between different types of demand-side management (DSM) programs—conservation, load management, fuel substitution, load building, and self-generation. Conservation programs reduce electricity and/or natural gas consumption during all or significant portions of the year. "Conservation" in this context includes all "energy efficiency improvements." An energy efficiency improvement can be defined as reduced energy use for a comparable level of service, resulting from the installation of an energy efficiency measure or the adoption of an energy efficiency practice. The level of service may be expressed in such ways as the volume of a refrigerator, temperature levels, production output of a manufacturing facility, or lighting level per square foot. Load management programs may either reduce electricity peak demand or shift the demand from on-peak to nonpeak periods.

1. Over 50 state analyses and experts participated in the Standard Practices Manual revision. They are listed in the Appendix. The revision of the SPM was truly a collaborative effort that lasted over 8 months but resulted in significant accounting and analytical changes for projects and programs in California.

Fuel substitution and load building programs share the common feature of increasing annual consumption of either electricity or natural gas relative to what would have happened in the absence of the program. This effect is accomplished in significantly different ways, by inducing the choice of one fuel over another (fuel substitution) or by increasing the sales of electricity, gas, or electricity and gas (load building). Self-generation refers to distributed generation (DG) installed on the customer's side of the electric utility meter, which serves some or all of the customer's electric load, that otherwise would have been provided by the central electric grid.

In some cases, self-generation products are applied in a combined heat and power manner, in which case the heat produced by the self-generation product is used on site to provide some or all of the customer's thermal needs. Self-generation technologies include, but are not limited to, photovoltaics, wind turbines, fuel cells, microturbines, small gas-fired turbines, and gas-fired internal combustion engines.

Fuel substitution and load building programs were relatively new to DSM in California in the late 1980s, born out of the convergence of several factors that translated into average rates that substantially exceeded marginal costs. Proposals by utilities to implement programs that increase sales had prompted the need for additional procedures for estimating program cost effectiveness. These procedures may be applicable in a new context. AB 970 amended the Public Utilities Code and provided the motivation to develop a cost-effectiveness method that can be used on a common basis to evaluate all programs that will remove electric load from the centralized grid, including energy efficiency, load control/demand-responsiveness programs, and self-generation. Hence self-generation was also added to the list of DSM programs for cost-effectiveness evaluation. In some cases self-generation programs installed with incremental load are also included since the definition of self-generation is not necessarily confined to projects that reduce electric load on the grid. For example, suppose an industrial customer installs a new facility with a peak consumption of 1.5 MW, with an integrated on-site 1.0-MW gas-fired DG unit. The combined impact of the new facility is *load building* since the new facility can draw up to 0.5 MW from the grid, even when the DG unit is running. The proper characterization of each type of DSM program is essential to ensure the proper treatment of inputs and the appropriate interpretation of cost-effectiveness results.

Categorizing programs is important because in many cases the same specific device can and should be evaluated in more than one category. For example, the promotion of an electric heat pump can and should be treated as part of a conservation program if the device is installed in lieu of a less efficient electric resistance heater. If the incentive induces the installation of an electric heat pump instead of gas space heating, however, the program needs to be considered and evaluated as a fuel substitution program. Similarly, natural gas–fired self-generation, as well as self-generation units using other nonrenewable fossil fuels, must be treated as fuel substitution. In common with other types of fuel substitution, any cost of gas transmission and distribution, and environmental externalities, must be accounted for. In addition, cost-effectiveness analyses of self-generation should account for utility interconnection costs. Similarly, a thermal energy storage device should be treated as a load management program when the predominant effect is to shift load. If the acceptance of a utility incentive by the customer to install the energy storage device is a decisive aspect of the customer's decision to remain an electric utility customer (i.e., to reject or defer the option of installing a gas-fired cogeneration system), then the predominant effect of the thermal energy storage device has been to substitute electricity service for the natural gas service that would have occurred in the absence of the program.

In addition to Fuel Substitution and Load Building Programs, recent utility program proposals have included reference to "load retention," "sales retention," "market retention," or "customer retention" programs. In most cases, the effect of such programs is identical to either a Fuel Substitution or a Load Building program—sales of one fuel are increased relative to sales without the program. A case may be made, however, for defining a separate category of program called "load retention." One unambiguous example of a load retention program is the situation where a program keeps a customer from relocating to another utility service area. However, computationally the equations and guidelines included in the manual to accommodate Fuel Substitution and Load Building programs can also handle this special situation.

Basic Methods

The chapter identifies the cost and benefit components and cost-effectiveness calculation procedures from four major perspectives: Participant, Ratepayer Impact Measure (RIM), Program Administrator Cost (PAC), and TRC. A fifth perspective, the Societal, is treated as a variation on the TRC Test. The results of each perspective can be expressed in a variety of ways, but in all cases it is necessary to calculate the net present value (NPV) of program impacts over the lifecycle of those impacts.

Table 10.1 summarizes the cost-effectiveness tests addressed in the manual. For each of the perspectives, the table shows the appropriate means of expressing test results. The primary unit of measurement refers to the way of expressing test results that is considered by the staff of the two commissions as the most useful for summarizing and comparing DSM program's cost effectiveness. Secondary indicators of cost effectiveness represent *supplemental* means of expressing test results that are likely to be of particular value for certain types of proceedings, reports, or programs.

This chapter does not specify how the cost effectiveness test results are to be displayed or the level at which cost effectiveness is to be calculated (e.g., groups of programs, individual programs, and program elements for all or

TABLE 10.1 Cost-Effectiveness Tests

Primary	Secondary
Participant	
NPV (all participants)	Discounted payback (years)
	BCR
	NPV (average participant)
Ratepayer Impact Measure	
Lifecycle revenue impact per unit of energy (kWh or therm) or demand customer (kW)	Lifecycle revenue impact per unit
	Annual revenue impact (by year, per kWh, kW, therm, or customer)
	First-year revenue impact (per kWh, kW, therm, or customer)
NPV	BCR
Total Resource Cost	
NPV	BCR
	Levelized cost (cents or dollars per unit of energy or demand)
	Societal (NPV, BCR)
Program Administrator Cost	
NPV	BCR
	Levelized cost (cents or dollars per unit of energy or demand)

BCR, benefit–cost ratio; *NPV*, net present value.

some programs). It is reasonable to expect different levels and types of results for different regulatory proceedings or for different phases of the process used to establish proposed program funding levels. For example, for summary tables in general rate case proceedings at the CPUC, the most appropriate tests may be the RIM lifecycle revenue impact, TRC, and PAC Test results for programs or groups of programs. The analysis and review of program proposals for the same proceeding may include Participant Test results and various additional indicators of cost effectiveness from all tests for each individual program element. In the case of cost-effectiveness evaluations conducted in the context of integrated long-term resource planning activities, such detailed examination of multiple indications of costs and benefits may be impractical.

Rather than identify the precise requirements for reporting cost-effectiveness results for all types of proceedings or reports, the approach taken in the manual is to:

- specify the components of benefits and costs for each of the major tests
- identify the equations to be used to express the results in acceptable ways
- indicate the relative value of the different units of measurement by designating primary and secondary test results for each test.

It should be noted that for some types of DSM programs, meaningful cost-effectiveness analyses cannot be performed using the tests in the SPM. The following guidelines are offered to clarify the appropriated "match" of different types of programs and tests:

1. For generalized information programs (e.g., when customers are provided generic information on means of reducing utility bills without the benefit of on-site evaluations or customer billing data), cost-effectiveness tests are not expected because of the extreme difficulty in establishing meaningful estimates of load impacts.
2. For any program where more than one fuel is affected, the preferred unit of measurement for the RIM Test is the lifecycle revenue impacts per customer, with gas and electric components reported separately for each fuel type and for combined fuels.
3. For load building programs, only the RIM Tests are expected to be applied. The TRC and PAC Tests are intended to identify cost effectiveness relative to other resource options. It is inappropriate to consider increased load as an alternative to other supply options.
4. Levelized costs may be appropriate as a supplementary indicator of cost per unit for electric C&LM programs relative to generation options and gas conservation programs relative to gas supply options, but the levelized cost test is not applicable to fuel substitution programs (since they combine gas and electric effects) or load building programs (which increase sales).

Delineation of the various means of expressing test results in Table 10.1 is not meant to discourage the continued development of additional variations for expressing cost effectiveness. Of particular interest is the development of indicators of program cost effectiveness that can be used to assess the appropriateness of program scope (i.e., level of funding) for General Rate Case proceedings. Additional tests, if constructed from the net present worth in conformance with the equations designated in the manual, could prove useful as a means of developing methodologies that will address issues such as the optimal timing and scope of DSM programs in the context of overall resource planning.

Balancing the Tests

The tests set forth in the manual are not intended to be used individually or in isolation. The results of tests that measure efficiency, such as the TRC Test, the Societal Test, and the PAC Test, must be compared not only with each other but also with the RIM Test. This multiperspective approach will require program administrators and state agencies to consider trade-offs between the various tests. Issues related to the precise weighting of each test relative to other tests and to developing formulas for the definitive balancing of perspectives are outside the scope of the manual. The manual, however, does provide a brief description of the strengths and weaknesses of each test (Sections Participant Test, The Ratepayer Impact Measure Test, Total Resource Cost Test, and Program Administrator Cost Test) to assist users in qualitatively weighing test results.

Limitations: Externality Values and Policy Rules

The list of externalities identified in section Total Resource Cost Test, in the discussion on the Societal version of the TRC Test is broad, illustrative and by no means exhaustive. Traditionally, implementing agencies have independently determined the details such as the components of the externalities, the externality values, and the policy rules that specify the contexts in which the externalities and the tests are used.

Externality Values

The values for the externalities have not been provided in the manual. There are separate studies and methodologies to arrive at these values. There are also separate processes instituted by implementing agencies before such values can be adopted formally.

Policy Rules

The appropriate choice of inputs and input components vary by program area and project. For instance, low-income programs are evaluated using a broader set of nonenergy benefits, which have not been provided in detail in the manual. Implementing agencies traditionally have had the discretion to use or not use these inputs and/or benefits on a project- or program-specific basis. The policy rules that specify the contexts in which it is appropriate to use the externalities, their components, and tests mentioned in the manual are an integral part of any cost-effectiveness evaluation. These policy rules are not a part of the manual.

To summarize, the manual provides the methodology and the cost–benefit calculations only. The implementing agencies (such as the CPUC and the CEC) have traditionally utilized open public processes to incorporate the diverse views of stakeholders before adopting externality values and policy rules, which are an integral part of the cost-effectiveness evaluation.

PARTICIPANT TEST

Definition

The Participant Test is the measure of the quantifiable benefits and costs to the customer due to participation in a program. Since many customers do not base their decision to participate in a program entirely on quantifiable variables, this test cannot be a complete measure of the benefits and costs of a program to a customer.

Benefits and Costs

The benefits of participation in a demand-side program include a reduction in the customer's utility bill(s), any incentive paid by the utility or other third parties, and any federal, state, or local tax credit received. The reductions to the utility bill(s) should be calculated using the actual retail rates that would have been charged for the energy service provided (electric demand or energy or gas). Savings estimates should be based on gross savings, as opposed to net energy savings (refer to point 1 in Notes).

In the case of fuel substitution programs, benefits to the participant also include the avoided capital and operating costs of the equipment/appliance not chosen. For load building programs, participant benefits include an increase in productivity and/or service, which is presumably equal to or greater than the productivity/service without participating. The inclusion of these benefits is not required for this test, but if they are included then the Societal Test should also be performed.

The costs to a customer of program participation are all out-of-pocket expenses incurred as a result of participating in a program, plus any increases in the customer's utility bill(s). The out-of-pocket expenses include the cost of any equipment or materials purchased, including sales tax and installation; any ongoing operation and maintenance costs; any removal costs (less salvage value); and the value of the customer's time in arranging for the installation of the measure, if significant.

How the Results Can Be Expressed

The results of this test can be expressed in four ways: through a net present value per average participant (NPV_{avp}), a NPV for the total program, a benefit−cost ratio (BCR), or discounted payback. The primary means of expressing test results is the NPV for the total program; discounted payback, BCR, and per participant NPV_{avp} are secondary tests.

The discounted payback is the number of years it takes until the cumulative discounted benefits equal or exceed the cumulative discounted costs. The shorter the discounted payback, the more attractive or beneficial the program is to the participants. Although "payback period" is often defined as undiscounted in the textbooks, a discounted payback period is used here to approximate more closely the consumer's perception of future benefits and costs (refer to point 2 in Notes).

NPV_P gives the net dollar benefit of the program to an average participant or to all participants discounted over some specified time period. An NPV above 0 indicates that the program is beneficial to the participants under this test.

BCR_P is the ratio of the total benefits of a program to the total costs discounted over some specified time period. It gives a measure of the rough rate of return for the program to the participants and is also an indication of risk. A BCR above 1 indicates a beneficial program.

Strengths of the Participant Test

The Participant Test gives a good "first cut" of the benefit or desirability of the program to customers. This information is especially useful for voluntary programs as an indication of potential participation rates.

For programs that involve a utility incentive, the Participant Test can be used for program design considerations such as the minimum incentive level, whether incentives are really needed to induce participation, and whether changes in incentive levels will induce the desired amount of participation.

These test results can be useful for program penetration analyses and developing program participation goals, which will minimize adverse ratepayer impacts and maximize benefits.

For fuel substitution programs, the Participant Test can be used to determine whether program participation (i.e., choosing one fuel over another) will be in the best interest of the customer in the long run. The primary means of establishing such assurances is the NPV, which looks at the costs and benefits of the fuel choice over the life of the equipment.

Weaknesses of the Participant Test

None of the Participant Test results (discounted payback, NPV, or BCR) accurately capture the complexities and diversity of customer decision-making processes for DSM investments. Until or unless more is known about customer attitudes and behavior, interpretations of Participant Test results continue to require considerable judgment. Participant Test results play only a supportive role in any assessment of C&LM programs as alternatives to supply projects.

Formulas

The following are the formulas for discounted payback, NPV_p, and BCR_p for the Participant Test.

$$NPV_p = B_p - C_p$$

$$NPV_{avp} = (B_p - C_p)/P$$

$$BCR_p = B_p/C_p$$

$$DPp = \text{Min j such that } B_j \geq C_j$$

where

NPV_p = NPV to all participants
NPV_{avp} = NPV to the average participant
BCR_p = BCR to participants
DP_p = Discounted payback in years
B_p = NPV of benefit to participants
C_p = NPV of costs to participants
B_j = Cumulative benefits to participants in year j
C_j = Cumulative costs to participants in year j
P = Number of program participants
J = First year in which cumulative benefits are cumulative costs
d = Interest rate (discount)

The Benefit (B_p) and Cost (C_p) terms are further defined as follows:

$$B_p = \sum_{t=1}^{N} \frac{BR_t + TC_t + INC_t}{(1+d)^{t-1}} + \sum_{t=1}^{N} \frac{AB_{at} + PAC_{at}}{(1+d)^{t-1}}$$

$$C = \sum_{t=1}^{N} \frac{PC_t + BI_t}{(1+d)^{t-1}}$$

where

BR_t = Bill reductions in year t
BI_t = Bill increases in year t
TC_t = Tax credits in year t
INC_t = Incentives paid to the participant by the sponsoring utility in year t (refer to point 3 in Notes)
PC_t = Participant costs in year t to include:

- initial capital costs, including sales tax (refer to point 4 of Notes)
- ongoing operation and maintenance costs including fuel cost
- removal costs, less salvage value
- value of the customer's time in arranging for installation, if significant

PAC_{at} = Participant-avoided costs in year t for alternative fuel devices (costs of devices not chosen)
Ab_{at} = Avoided bill from alternative fuel in year t

The first summation in the B_p equation should be used for C&LM programs. For fuel substitution programs, both the first and second summations should be used for B_p.

Note that in most cases, the customer bill impact terms (BR_t, BI_t, and AB_{at}) are further determined by costing period to reflect load impacts and/or rate schedules, which vary substantially by the time of day and season. The formulas for these variables are as follows:

$$BR_t = \sum_{i=1}^{I}(\Delta EG_{it} \times AC{:}\,E_{it} \times K_{it}) + \sum_{i=1}^{I}(\Delta DG_{it} \times AC{:}\,D_{it} \times K_{it}) + OBR_t$$

$AB_{at} = BR_t$ formula, but with rates and costing periods appropriate for the alternate fuel utility

$$BI_t = \sum_{i=1}^{I}(\Delta EG_{it} \times AC{:}\,E_{it} \times (K_{it} - 1)) + \sum_{i=1}^{I}(\Delta DG_{it} \times AC{:}\,D_{it} \times (K_{it} - 1)) + OBI_t$$

where

ΔEG_{it} = Reduction in gross energy use in costing period i in year t
ΔDG_{it} = Reduction in gross billing demand in costing period i in year t
$AC{:}E_{it}$ = Rate charged for energy in costing period i in year t
$AC{:}D_{it}$ = Rate charged for demand in costing period i in year t
$K_{it} = 1$ when ΔEG_{it} or ΔDG_{it} is positive (a reduction) in costing period i in year t, and 0 otherwise
OBR_t = Other bill reductions or avoided bill payments (e.g., customer charges, standby rates)
OBI_t = Other bill increases (i.e., customer charges, standby rates)
I = Number of periods of participant's participation.

In load management programs such as TOU rates and air-conditioning cycling, there are often no direct customer hardware costs. However, attempts should be made to quantify indirect costs that customers may incur that enable them to take advantage of TOU rates and similar programs.

If no customer hardware costs are expected or estimates of indirect costs and value of service are unavailable, it may not be possible to calculate the BCR and discounted payback period.

THE RATEPAYER IMPACT MEASURE TEST (REFER TO POINT 5 OF NOTES)

Definition

The RIM Test measures the effect on customer bills or rates of changes in utility revenues and operating costs caused by the program. Rates will go down if the change in revenues from the program is greater than the change in utility costs. Conversely, rates or bills will go up if revenues collected after program implementation are less than the total costs incurred by the utility in implementing the program. This test indicates the direction and magnitude of the expected change in customer bills or rate levels.

Benefits and Costs

The benefits calculated in the RIM Test are the savings from avoided supply costs. These avoided costs include the reduction in transmission, distribution, generation, and capacity costs for periods when load has been reduced and the increase in revenues for any periods in which load has been increased. The avoided supply costs are a reduction in total costs or revenue requirements and are included for both fuels for a fuel substitution program. The increase in revenues are also included for both fuels for fuel substitution programs. Both the reductions in supply costs and the revenue increases should be calculated using net energy savings.

The costs for this test are the program costs incurred by the utility, and/or other entities incurring costs and creating or administering the program, the incentives paid to the participant, decreased revenues for any period in which load has been decreased, and increased supply costs for any period when load has been increased. The utility program costs include initial and annual costs, such as the cost of equipment, operation and maintenance, installation, program administration, and customer dropout and removal of equipment (less salvage value). The decreases in revenues and the increases in the supply costs should be calculated for both fuels for fuel substitution programs using net savings.

How the Results Can Be Expressed

The results of this test can be presented in several forms: the lifecycle revenue impact (cents or dollars) per kWh, kW, therm, or customer; annual or first-year revenue impacts (cents or dollars per kWh, kW, therms, or customer); BCR; and NPV. The primary units of measurement are the lifecycle revenue impact, expressed as the change in rates (cents per kWh for electric energy, dollars per kW for electric capacity, cents per therm for natural gas), and the NPV. Secondary test results are the lifecycle revenue impact per customer, first-year and annual revenue impacts, and the BCR. The lifecycle revenue impact (LRI_{RIM}) values for programs affecting electricity and gas should be calculated for each fuel individually (cents per kWh or dollars per kW and cents per therm) and on a combined gas and electric basis (cents per customer).

The LRI is the one-time change in rates or the bill change over the life of the program needed to bring total revenues in line with revenue requirements over the life of the program. The rate increase or decrease is expected to be put into effect in the first year of the program. Any successive rate changes such as for cost escalation are made from there. The first-year revenue impact is the change in rates in the first year of the program or the bill change needed to get total revenues to match revenue requirements only for that year. The annual revenue impact (ARI) is the series of differences between revenues and revenue requirements in each year of the program. This series shows the cumulative rate change or bill change in a year needed to match revenues to revenue requirements. Thus the ARI_{RIM} for year 6 per kWh is the estimate of the difference between present rates and the rate that would be in effect in year 6 due to the program. For results expressed as lifecycle, annual, or first-year revenue impacts, negative results indicate favorable effects on the bills of ratepayers or reductions in rates, whereas positive test result values indicate adverse bill impacts or rate increases.

NPV_{RIM} gives the discounted dollar net benefit of the program from the perspective of rate levels or bills over some specified time period. An NPV above 0 indicates that the program will benefit (lower) rates and bills.

The benefit–cost ratio (BCR_{RIM}) is the ratio of the total benefits of a program to the total costs discounted over some specified time period. A BCR above 1 indicates that the program will lower the rates and bills.

Strengths of the Ratepayer Impact Measure Test

In contrast to most supply options, DSM programs cause a direct shift in revenues. Under many conditions, revenues lost from DSM programs have to be made up by ratepayers. The RIM Test is the only test that reflects this revenue shift along with the other costs and benefits associated with the program.

An additional strength of the RIM Test is that the test can be used for all DSM programs (conservation, load management, fuel substitution, and load building). This makes the RIM Test particularly useful for comparing impacts among DSM options.

Some of the units of measurement for the RIM Test are of greater value than others, depending upon the purpose or type of evaluation. The lifecycle revenue impact per customer is the most useful unit of measurement when comparing the merits of programs with highly variable scopes (e.g., funding levels) and when analyzing a wide range of programs that include both electric and natural gas impacts. BCR can also be very useful for program design evaluations to identify the most attractive programs or program elements.

If comparisons are being made between a program or group of conservation/load management programs and a specific resource project, lifecycle cost per unit of energy and annual and first-year net costs per unit of energy are the most useful way to express test results. Of course, this requires developing lifecycle, annual, and first-year revenue impact estimates for the supply-side project.

Weaknesses of the Ratepayer Impact Measure Test

Results of the RIM Test are probably less certain than those of other tests because the test is sensitive to the differences between long-term projections of marginal costs and long-term projections of rates, two cost streams that are difficult to quantify with certainty.

RIM Test results are also sensitive to assumptions regarding the financing of program costs. Sensitivity analyses and interactive analyses that capture feedback effects between system changes, rate design options, and alternative means of financing generation and nongeneration options can help overcome these limitations. However, these types of analyses may be difficult to implement.

Additional caution must be exercised in using the RIM Test to evaluate a fuel substitution program with multiple end use efficiency options. For example, under conditions in which the marginal costs are less than the average costs, a program that promotes an inefficient appliance may give a more favorable test result than a program that promotes an efficient appliance. Although the results of the RIM Test accurately reflect rate impacts, the implications for long-term conservation efforts need to be considered.

Formulas: the formulas for LRI_{RIM}, NPV_{RIM}, BCR_{RIM}, the first-year revenue impacts, and annual revenue impacts are as follows:

$$LRI_{RIM} = (C_{RIM} - B_{RIM})/E$$

$$FRI_{RIM} = (C_{RIM} - B_{RIM})/E \quad \text{for } t = 1$$

$$\begin{aligned} ARI_{RIMt} &= F_{RIM} \quad \text{for } t = 1 \\ &= (C_{RIMt} - B_{RIMt})/E_t \quad \text{for } t = 2, \ldots N \end{aligned}$$

$$NPV_{RIM} = B_{RIM} - C_{RIM}$$

$$BCR_{RIM} = B_{RIM}/C_{RIM}$$

where

LRI_{RIM} = Lifecycle revenue impact of the program per unit of energy (kWh or therm) or demand (kW) (the one-time change in rates) or per customer (the change in customer bills over the life of the program). (Note: an appropriate choice of kWh, therm, kW, and customer should be made)
FRI_{RIM} = First-year revenue impact of the program per unit of energy, demand, or per customer
ARI_{RIM} = Stream of cumulative annual revenue impacts of the program per unit of energy, demand, or per customer. (Note: the terms in the ARI formula are not discounted; thus they are the nominal cumulative revenue impacts. Discounted cumulative revenue impacts may be calculated and submitted if they are indicated as such. Note also that the sum of the

discounted stream of cumulative revenue impacts *does not* equal the LRI_{RIM})
NPV_{RIM} = NPV levels
BCR_{RIM} = BCR for rate levels
B_{RIM} = Benefits to rate levels or customer bills
C_{RIM} = Costs to rate levels or customer bills
E = Discounted stream of system energy sales (kWh or therms) or demand sales (kW) or first-year customers. (See Appendix C for a description of the derivation and use of this term in the LRI_{RIM} test.)

The B_{RIM} and C_{RIM} terms are further defined as follows:

$$B_{RIM} = \sum_{t=1}^{N} \frac{UAC_t + RG_t}{(1+d)^{t-1}} + \sum_{t=1}^{N} \frac{UAC_{at}}{(1+d)^{t-1}}$$

$$C_{RIM} = \sum_{t=1}^{N} \frac{UIC_t + RL_t + PRC_t + INC_t}{(1+d)^{t-1}} + \sum_{t=1}^{N} \frac{RL_{at}}{(1+d)^{t-1}}$$

$$E = \sum_{t=1}^{N} \frac{E_t}{(1+d)^{t-1}}$$

where

UAC_t = Utility avoided supply costs in year t
UIC_t = Utility increased supply costs in year t
RG_t = Revenue gain from increased sales in year t
RL_t = Revenue loss from reduced sales in year t
PRC_t = Program administrator program costs in year t
E_t = System sales in kWh, kW, or therms in year t or first-year customers
UAC_{at} = Utility-avoided supply costs for the alternative fuel in year t
RL_{at} = Revenue loss from avoided bill payments for alternate fuel in year t (i.e., device not chosen in a fuel substitution program).

For fuel substitution programs, the first term in the B_{RIM} and C_{RIM} equations represents the sponsoring utility (electric or gas) and the second term represents the alternative utility. The RIM Test should be calculated separately for electric, gas, and combined electric and gas.

The utility-avoided cost terms (UAC_t, UIC_t, and UAC_{at}) are further determined by costing period to reflect time-variant costs of supply:

$$UAC_t = \sum_{i=1}^{I} (\Delta EN_{it} \times MC: E_{it} \times (K_{it})) + \sum_{i=1}^{I} (\Delta DN_{it} \times MC: D_{it} \times K_{it})$$

UAC_{at} = UAC_t formula, but with marginal costs and costing periods appropriate for the alternate fuel utility.

$$UIC_t = \sum_{i=1}^{I}(\Delta EN_{it} \times MC: E_{it} \times (K_{it} - 1)) + \sum_{i=1}^{I}(\Delta DN_{it} \times MC: D_{it} \times (K_{it} - 1))$$

where:

(only terms not previously defined are included here.)
ΔEN_{it} = Reduction in net energy use in costing period i in year t
ΔDN_{it} = Reduction in net demand in costing period i in year t
$MC{:}E_{it}$ = Marginal cost of energy in costing period i in year t
$MC{:}D_{it}$ = Marginal cost of demand in costing period i in year t

The revenue impact terms (RG_t, RL_t, and RL_{at}) are parallel to the bill impact terms in the Participant Test. The terms are calculated in exactly the same way with the exception that the net impacts are used rather than gross impacts. If a net-to-gross ratio is used to differentiate gross savings from net savings, the revenue terms and the participant's bill terms will be related as follows:

$$RG_t = BI_t \times (\text{net-to-gross ratio})$$

$$RL_t = BR_t \times (\text{net-to-gross ratio})$$

$$RL_{at} = AB_{at} \times (\text{net-to-gross ratio})$$

TOTAL RESOURCE COST TEST (REFER TO POINT 6 OF NOTES)

Definition

The TRC Test measures the net costs of a DSM program as a resource option based on the total costs of the program, including both the participants' and the utility's costs.

The test is applicable to conservation, load management, and fuel substitution programs. For fuel substitution programs, the test measures the net effect of the impacts from the fuel not chosen versus the impacts from the fuel that is chosen as a result of the program. TRC Test results for fuel substitution programs should be viewed as a measure of the economic efficiency implications of the total energy supply system (gas and electric).

A variant of the TRC Test is the Societal Test. The Societal Test differs from the TRC Test in that it includes the effects of externalities (e.g., environmental, national security), excludes tax credit benefits, and uses a different (societal) discount rate.

Benefits and costs: this test represents the combination of the effects of a program on both the customers participating and those not participating in a program. In a sense, it is the summation of the benefit and cost terms in the

Participant and the RIM Tests, where the revenue (bill) change and the incentive terms intuitively cancel (except for the differences in net and gross savings).

The benefits calculated in the TRC Test are the avoided supply costs—the reduction in transmission, distribution, generation, and capacity costs valued at marginal cost—for the periods when there is a load reduction. The avoided supply costs should be calculated using net program savings, savings net of changes in energy use that would have happened in the absence of the program. For fuel substitution programs, benefits include the avoided device costs and avoided supply costs for the energy-using equipment not chosen by the program participant.

The *costs* in this test are the program costs paid by both the utility and the participants plus the increase in supply costs for the periods in which load is increased. Thus *all* equipment costs, installation, operation and maintenance, cost of removal (less salvage value), and administration costs, no matter who pays for them, are included in this test. Any tax credits are considered a reduction to costs in this test. For fuel substitution programs, the costs also include the increase in supply costs for the utility providing the fuel that is chosen as a result of the program.

How the Results Can Be Expressed

The results of the TRC Test can be expressed in several forms: as NPV, BCR, or as a levelized cost. The NPV is the primary unit of measurement for this test. Secondary means of expressing TRC Test results are BCR and levelized costs. The Societal Test—expressed in terms of NPV, BCR, or levelized costs—is also considered a secondary means of expressing results. Levelized costs as a unit of measurement are inapplicable for fuel substitution programs, since these programs represent the net change of alternative fuels, which are measured in different physical units (e.g., kWh or therms). Levelized costs are also not applicable for load building programs.

NPV_{TRC} is the discounted value of the net benefits to this test over a specified period of time. It is a measure of the change in the TRCs due to the program. A NPV above 0 indicates that the program is a less expensive resource than the supply option upon which the marginal costs are based.

BCR_{TRC} is the ratio of the discounted total benefits of the program to the discounted total costs over some specified time period. It gives an indication of the rate of return of this program to the utility and its ratepayers. A BCR above 1 indicates that the program is beneficial to the utility and its ratepayers on a TRC basis.

The levelized cost is a measure of the total costs of the program in a form that is sometimes used to estimate costs of utility-owned supply additions. It presents the total costs of the program to the utility and its ratepayers on a per kW, per kWh, or per therm basis levelized over the life of the program.

The Societal Test is structurally similar to the TRC Test. It goes beyond the TRC Test in that it attempts to quantify the change in the TRCs to society as a whole rather than to only the service territory (the utility and its ratepayers). In taking society's perspective, the Societal Test utilizes essentially the same input variables as the TRC Test, but they are defined with a broader societal point of view. More specifically, the Societal Test differs from the TRC Test in at least one of five ways. First, the Societal Test may use higher marginal costs than the TRC Test if a utility faces marginal costs that are lower than other utilities in the state or than its out-of-state suppliers. Marginal costs used in the Societal Test would reflect the cost to *society* of the more expensive alternative resources. Second, tax credits are treated as a transfer payment in the Societal Test and thus are left out. Third, in the case of capital expenditures, interest payments are considered a transfer payment since society actually expends the resources in the first year. Therefore capital costs enter the calculations in the year in which they occur. Fourth, a societal discount rate should be used (refer to point 7 in Notes). Finally, marginal costs used in the Societal Test would also contain externality costs of power generation not captured by the market system. An illustrative and by no means exhaustive list of "externalities and their components" is given below (refer to the Limitations section for elaboration). These values are also referred to as "adders" designed to capture or internalize such externalities. The list of potential adders would include, for example:

1. The benefit of avoided environmental damage: the CPUC policy specifies two "adders" to internalize environmental externalities, one for electricity use and one for natural gas use. Both are statewide average values. These adders are intended to help distinguish between cost-effective and non-cost-effective energy efficiency programs. They apply to an average supply mix and would not be useful in distinguishing among competing supply options. The CPUC electricity environmental adder is intended to account for the environmental damage from air pollutant emissions from power plants. The CPUC-adopted adder is intended to cover the human and material damage from sulfur oxides, nitrogen oxides, volatile organic compounds (sometimes called reactive organic gases), particulate matter at or below 10 micron diameter, and carbon. The adder for natural gas is intended to account for air pollutant emissions from the direct combustion of the gas. In the CPUC policy guidance, the adders are included in the tabulation of the benefits of energy efficiency programs. They represent reduced environmental damage from displaced electricity generation and avoided gas combustion. The environmental damage is the result of the net change in pollutant emissions in the air basins, or regions, in which there is an impact. This change is the result of direct changes in power plant or natural gas combustion emission resulting from the efficiency measures,

and changes in emissions from other sources, that result from those direct changes in emissions.

2. The benefit of avoided transmission and distribution costs is that the energy efficiency measures that reduce the growth in peak demand would decrease the required rate of expansion to the transmission and distribution network, eliminating costs of constructing and maintaining new or upgraded lines.
3. The benefit of avoided generation costs is that the energy efficiency measures reduce consumption and hence avoid the need for generation. This would include avoided energy costs, capacity costs, and a time and date line.
4. The benefit of increased system reliability: the reductions in demand and peak loads from customers opting for self-generation provide reliability benefits to the distribution system in the forms of:
   a. avoided costs of supply disruptions
   b. benefits to the economy of damage and control costs avoided by customers and industries who need greater than 99.9 level of reliable electricity service from the central grid since these industries depend on the electronics delivered from electrical systems
   c. marginally decreased System Operator's costs to maintain a percentage reserve of electricity supply above the instantaneous demand
   d. benefits to customers and the public of avoiding blackouts.
5. Nonenergy benefits: Nonenergy benefits might include a range of program-specific benefits such as saved water in energy-efficient washing machines or self-generation units and reduced waste streams from an energy-efficient industrial process.
6. Nonenergy benefits for low-income programs: Low–income programs are social programs that have a separate list of benefits included in what is known as the "low-income public purpose test." This test and the specific benefits associated with it are outside the scope of the manual.
7. Benefits of fuel diversity include considerations of the risks of supply disruption, the effects of price volatility, and the avoided costs of risk exposure and risk management.

Strengths of the Total Resource Cost Test

The primary strength of the TRC Test is its scope. The test includes total costs (participant plus program administrator) and also has the potential for capturing the total benefits (avoided supply costs plus, in the case of the societal test variation, externalities). To this extent supply-side project evaluations also include total costs of generation and/or transmission; the TRC Test provides a useful basis for comparing demand- and supply-side options.

Since this test treats incentives paid to participants and revenue shifts as transfer payments (from all ratepayers to participants through increased revenue requirements), the test results are unaffected by the uncertainties of projected average rates, thus reducing the uncertainty of the test results. Average rates and assumptions associated with how other options are financed (analogous to the issue of incentives for DSM programs) are also excluded from most supply-side cost determinations, again making the TRC Test useful for comparing demand- and supply-side options.

Weakness of the Total Resource Cost Test

The treatment of revenue shifts and incentive payments as transfer payments—identified previously as a strength—can also be considered a weakness of the TRC Test. While it is true that most supply-side cost analyses do not include such financial issues, it can be argued that DSM programs *should* include these effects since, in contrast to most supply options, DSM programs do result in lost revenues.

In addition, the costs of the DSM "resource" in the TRC Test are based on the total costs of the program, including costs incurred by the participant. Supply-side resource options are typically based only on the costs incurred by the power suppliers.

Finally, the TRC Test cannot be applied meaningfully to load building programs, thereby limiting the ability to use this test to compare the full range of DSM options.

Formulas

The formulas for NPV_{TRC}, BCR_{TRC}, and levelized costs are as follows:

$$NPV_{TRC} = B_{TRC} - C_{TRC}$$

$$BCR_{TRC} = B_{TRC}/C_{TRC}$$

$$LC_{TRC} = LCRC/IMP$$

where

NPV_{TRC} = NPV of total costs of the resource
BCR_{TRC} = BCR of total costs of the resource
LC_{TRC} = Levelized cost per unit of the total cost of the resource (cents per kWh for conservation programs; dollars per kW for load management programs)
B_{TRC} = Benefits of the program
C_{TRC} = Costs of the program
LCRC = TRCs used for levelizing
IMP = Total discounted load impacts of the program
PCN = Net participant costs.

The B_{TRC}, C_{TRC}, LCRC, and IMP terms are further defined as follows:

$$B_{TRC} = \sum_{t=1}^{N} \frac{UAC_t + TC_t}{(1+d)^{t-1}} + \sum_{t=1}^{N} \frac{UAC_{at} + PAC_{at}}{(1+d)^{t-1}}$$

$$C_{TRC} = \sum_{t=1}^{N} \frac{PRC_t + PCN_t + UIC_t}{(1+d)^{t-1}}$$

$$LCRC = \sum_{t=1}^{N} \frac{PRC_t + PCN_t - TC_t}{(1+d)^{t-1}}$$

$$IMP = \sum_{t=1}^{N} \frac{\left[\left(\sum_{i=1}^{I} \Delta EN_{it}\right) \text{ or } (\Delta DN_{it} \text{ where } i = \text{peak period})\right]}{(1+d)^{t-1}}$$

(All terms have been defined in previous sections).

The first summation in the B_{TRC} equation should be used for C&LM programs. For fuel substitution programs, both the first and second summations should be used.

PROGRAM ADMINISTRATOR COST TEST

Definition

The PAC Test measures the net costs of a DSM program as a resource option based on the costs incurred by the program administrator (including incentive costs) and excluding any net costs incurred by the participant. The benefits are similar to the TRC benefits. Costs are defined more narrowly.

Benefits and Costs

The benefits for the PAC Test are the avoided supply costs of energy and demand—the reduction in transmission, distribution, generation, and capacity valued at marginal costs—for the periods when there is a load reduction. The avoided supply costs should be calculated using net program savings, savings net of changes in energy use that would have happened in the absence of the program. For fuel substitution programs, benefits include the avoided supply costs for the energy-using equipment not chosen by the program participant only in the case of a combination utility where the utility provides both fuels.

The costs for the PAC Test are the program costs incurred by the administrator, the incentives paid to the customers, and the increased supply costs for the periods in which the load is increased. Administrator program costs include initial and annual costs, such as the cost of utility equipment, operation and maintenance, installation, program administration, and customer dropout and removal of equipment (less salvage value). For fuel substitution programs, costs include the increased supply costs for the energy-using equipment chosen by the program participant only in the case of a combination utility, as mentioned earlier.

In this test, revenue shifts are viewed as a transfer payment between participants and all ratepayers. Although a shift in revenue affects rates, it does not affect revenue requirements, which are defined as the difference between the net marginal energy and capacity costs avoided and program costs. Thus, if $NPV_{pa} > 0$ and $NPV_{RIM} < 0$, the administrator's overall total costs will decrease, although rates may increase because the sales base over which revenue requirements are spread has decreased.

How the Results Can Be Expressed

The results of this test can be expressed as NPV, BCR, or levelized costs. The NPV is the primary test, and the BCR and levelized cost are the secondary tests.

NPV_{pa} is the benefit of the program minus the administrator's costs, discounted over some specified period of time. An NPV above 0 indicates that this demand-side program would decrease the costs to the administrator and the utility.

BCR_{pa} is the ratio of the total discounted benefits of a program to the total discounted costs for a specified time period. A BCR above 1 indicates that the program would benefit the combined administrator and utility's total cost situation.

The levelized cost is a measure of the costs of the program to the administrator in a form that is sometimes used to estimate the costs of utility-owned supply additions. It presents the costs of the program to the administrator and the utility on a per kW, per kWh, or per therm basis levelized over the life of the program.

Strengths of the Program Administrator Cost Test

As with the TRC Test, the PAC Test treats revenue shifts as transfer payments, meaning that the test results are not complicated by the uncertainties associated with long-term rate projections and associated rate design assumptions. In contrast to the TRC Test, the Program Administrator Test includes only the portion of the participant's equipment costs that is paid for by the administrator in the form of an incentive. Therefore for purposes of comparison, costs in the PAC Test are defined similarly to those supply-side projects that also do not include direct customer costs.

Weaknesses of the Program Administrator Cost Test

By defining device costs exclusively in terms of costs incurred by the administrator, the PAC Test results reflect only a portion of the full costs of the resource.

The PAC Test shares two limitations noted previously for the TRC Test:

1. by treating revenue shifts as transfer payments, the rate impacts are not captured
2. the test cannot be used to evaluate load building programs.

Formulas

The formulas for NPV, BCR, and levelized cost are as follows:

$$NPV_{pa} = B_{pa} - C_{pa}$$

$$BCR_{pa} = B_{pa}/C_{pa}$$

$$LC_{pa} = LC_{pa}/IMP$$

where

NPV_{pa} = NPV of PACs
BCR_{pa} = BCR of PACs
LC_{pa} = Levelized cost per unit of PAC of the resource
B_{pa} = Benefits of the program
C_{pa} = Costs of the program
LC_{pc} = Total PACs used for levelizing

$$B_{pa} = \sum_{t=1}^{N} \frac{UAC_t}{(1+d)^{t-1}} + \sum_{t+1}^{N} \frac{UAC_{at}}{(1+d)^{t-1}}$$

$$C_{pa} = \sum_{t=1}^{N} \frac{PRC_t + INC_t + UIC_t}{(1+d)^{t-1}}$$

$$LCpc = \sum_{t=1}^{N} \frac{PRC_t + INC_t}{(1+d)^{t-1}}$$

(All variables are defined in previous sections.)

The first summation in the B_{pa} equation should be used for C&LM. For fuel substitution programs, both the first and second summations should be used.

NOTES

1. Gross energy savings are considered to be the savings in energy and demand seen by the participant at the meter. These are the appropriate program impacts to calculate bill reductions for the Participant Test. Net savings are assumed to be the savings that are attributable to the program. That is, net savings are gross savings minus those changes in energy use and demand that would have happened even in the absence of the program. For fuel substitution and load building programs, gross-to-net considerations account for the impacts that would have occurred in the absence of the program.
2. It should be noted that if a demand-side program is beneficial to its participants ($NPV_p \geq 0$ and $BCR_p \geq 1.0$) using a particular discount rate, the program has an internal rate of return of at least the value of the discount rate.
3. Some difference of opinion exists as to what should be called an incentive. The term can be interpreted broadly to include almost anything. Direct rebates, interest payment subsidies, and even energy audits can be called incentives. Operationally, it is necessary to restrict the term to include only dollar benefits such as rebates or rate incentives (monthly bill credits). Information and services such as audits are not considered incentives for the purposes of these tests. If the incentive is to offset a specific participant cost, as in a rebate-type incentive, the full customer cost (before the rebate) must be included in the PC_t term.
4. If money is borrowed by the customer to cover this cost, it may not be necessary to calculate the annual mortgage and discount this amount if the present worth of the mortgage payments equals the initial cost. This occurs when the discount rate used is equal to the interest rate of the mortgage. If the two rates differ (e.g., a loan offered by the utility), then the stream of mortgage payments should be discounted by the discount rate chosen.
5. The RIM Test has previously been described under what was called the "Non-Participant Test." The Non-Participant Test has also been called the "Impact on Rate Levels Test."
6. This test was previously called the All Ratepayers Test.
7. Many economists have pointed out that use of a market discount rate in social cost—benefit analysis undervalues the interests of future generations. Yet if a market discount rate is not used, comparisons with alternative investments are difficult to make.

Appendix A

Inputs to equations and documentation.

A comprehensive review of procedures and sources for developing inputs is beyond the scope of the manual. It would also be inappropriate to attempt a complete standardization of the techniques and procedures for developing inputs for such parameters as load impacts, marginal costs, or average rates. Nevertheless, a series of guidelines can help to establish acceptable procedures and improve the chances of obtaining reasonable levels of consistent and meaningful cost-effectiveness results. The following "rules" should be viewed as appropriate guidelines for developing the primary inputs for the cost-effectiveness equations contained in the manual:

1. In the past, marginal costs for electricity were based on production cost model simulations that clearly identify key assumptions and characteristics of the existing generation system as well as the timing and nature of any generation additions and/or power purchase agreements in the future. With a deregulated market for wholesale electricity, marginal costs for electric generation energy should be based on forecast market prices, which are derived from recent transactions in California energy markets. Such transactions could include spot market purchases as well as longer term bilateral contracts, and the marginal costs should be estimated based on components for energy as well as demand and/or capacity costs, as is typical for these contracts.
2. In the case of submittals in conjunction with a utility rate proceeding, average rates used in demand-side management (DSM) program cost-effectiveness evaluations should be based on proposed rates. Otherwise, average rates should be based on current rate schedules. Evaluations based on alternative rate designs are encouraged.
3. Time-differentiated inputs for electric marginal energy and capacity costs, average energy rates, demand charges, and electric load impacts should be used for
   a. load management programs
   b. any conservation program that involves a financial incentive to the customer
   c. any fuel substitution or load building program.

   Costing periods used should include, at a minimum, summer and winter, on- and off-peak; further disaggregation is encouraged.
4. When program participation includes customers with different rate schedules, the average rate inputs should represent an average weighted by the estimated mix of participation or impacts. For General Rate Case proceedings it is likely that each major rate class within each program will be considered as program elements requiring separate cost-effectiveness analyses for each measure and each rate class within each program.
5. Program administration cost estimates used in program cost-effectiveness analyses should exclude costs associated with the measurement and evaluation of program impacts unless the costs are a necessary component to administer the program.

6. For DSM programs or program elements that reduce electricity and natural gas consumption, costs and benefits from both fuels should be included.
7. The development and treatment of load impact estimates should distinguish between gross (i.e., impacts expected from the installation of a particular device, measure, appliance) and net (impacts adjusted to account for what would have happened anyway, and therefore not attributable to the program). Load impacts for the Participant Test should be based on gross, whereas for all other tests the use of net is appropriate. Gross and net program impact considerations should be applied to all types of DSM programs, although in some instances there may be no difference between the gross and net.
8. The use of sensitivity analysis, i.e., the calculation of cost-effectiveness test results using alternative input assumptions, is encouraged, particularly for the following programs: new programs, programs for which authorization to substantially change direction is being sought (e.g., termination, significant expansion), and major programs that show marginal cost-effectiveness and/or particular sensitivity to highly uncertain input(s).

The use of many of these guidelines is illustrated with examples of program cost effectiveness contained in Appendix B.

Purpose

These worksheets, developed by the California Interagency Green Accounting Working Group, are designed to calculate the various cost-benefit tests as prescribed in the ***Standard Practice Manual: Economic Analysis of Demand-side Programs and Projects*** (October, 2001).

Currently in place on the sheets are examples of the application of the cost effectiveness tests to various self-generation and energy efficiency programs. Future versions of this workbook may include sample calculations for load management programs.

Using the Spreadsheet

The input values in the worksheets can be modified for those who wish to use these worksheets to evaluate the cost-effectiveness of an actual energy efficiency or self-generation program or project. All values that should be modified are in blue. Changes in the input values on these worksheets to conduct analyses of actual programs will produce cost-benefit results that conform with the *SPM*.

All other values in black or red; any changes to the formulas or values in the black cells may produce cost-effectiveness results that do not conform with the *SPM*

Avoided Costs

The avoided cost values used in the analysis in this spreadsheet—the primary parameter for establishing the benefits or reduced purchases of electricity from the central grid–are based on long term forecasts developed in the year 2000, and are currently used to estimate the lifecycle costs and benefits of energy efficiency and self generation programs under the regulatory oversight of the California Public Utilities Commission. When these avoided cost forecasts are updated, the updated forecasts will be incorporated into this spreadsheet by replacing the values shown in the Avoided Cost worksheet.

Appendix B

Summary of equations and glossary of symbols.

Basic Equations

Participant Test

$$NPV_P = B_P - C_P$$

$$NPV_{avp} = (B_P - C_P)/P$$

$$BCR_{avp} = B_P/C_P$$

$$DP_P = \text{min j such that } B_j > C_j$$

Ratepayer Impact Measure Test

$$LRI_{RIM} = (C_{RIM} - B_{RIM})/E$$

$$FRI_{RIM} = (C_{RIM} - B_{RIM})/E \quad \text{for } t = 1$$

$$ARI_{RIMt} = F_{RIM} \quad \text{for } t = 1$$
$$= (C_{RIMt} - B_{RIMt})/E_t \quad \text{for } t = 2, \ldots N$$

$$NPV_{RIM} = B_{RIM} - C_{RIM}$$

$$BCR_{RIM} = B_{RIM}/C_{RIM}$$

Total Resource Cost Test

$$NPV_{TRC} = B_{TRC} - C_{TRC}$$

$$BCR_{TRC} = B_{TRC}/C_{TRC}$$

$$LC_{TRC} = LCRC/IMP$$

Program Administrator Cost Test

$$NPV_{pa} = B_{pa} - C_{pa}$$

$$BCR_{pa} = B_{pa}/C_{pa}$$

$$LC_{pa} = LCpa/IMP$$

Benefits and Costs

Participant Test

$$Bp = \sum_{t=1}^{N} \frac{BR_t + TC_t + INC_t}{(1+d)^{t-1}} + \sum_{t=1}^{N} \frac{AB_{at} + PAC_{at}}{(1+d)^{t-1}}$$

$$Cp = \sum_{t=1}^{N} \frac{PC_t + BI_t}{(1+d)^{t-1}}$$

Ratepayer Impact Measure Test

$$B_{RIM} = \sum_{t=1}^{N} \frac{UAC_t + RG_t}{(1+d)^{t-1}} + \sum_{t=1}^{N} \frac{UAC_{at}}{(1+d)^{t-1}}$$

$$C_{RIM} = \sum_{t=1}^{N} \frac{UIC_t + RL_t + PRC_t + INC_t}{(1+d)^{t-1}} + \sum_{t=1}^{N} \frac{RL_{at}}{(1+d)^{t-1}}$$

$$E = \sum_{t=1}^{N} \frac{E_t}{(1+d)^{t-1}}$$

Total Resource Cost Test

$$B_{TRC} = \sum_{t=1}^{N} \frac{UAC_t + TC_t}{(1+d)^{t-1}} + \sum_{t=1}^{N} \frac{UAC_{at} + PAC_t}{(1+d)^{t-1}}$$

$$C_{TRC} = \sum_{t=1}^{N} \frac{PRC_t + PCN_t + UIC_t}{(1+d)^{t-1}}$$

$$L_{TRC} = \sum_{t=1}^{N} \frac{PRC_t + PCN_t - TC_t}{(1+d)^{t-1}}$$

$$IMP = \sum_{t=1}^{N} \frac{\left[\left(\sum_{i=1}^{N} \Delta EN_{it}\right) \text{ or } (\Delta DN_{it} \quad \text{where I} = \text{peak period})\right]}{(1+d)^{t-1}}$$

Program Administrator Cost Test

$$B_{pa} = \sum_{t=1}^{N} \frac{UAC_t}{(1+d)^{t-1}} + \sum_{t=1}^{N} \frac{UAC_{at}}{(1+d)^{t-1}}$$

$$C_{pa} = \sum_{t=1}^{N} \frac{PRC_t + INC_t + UIC_t}{(1+d)^{t-1}}$$

$$LCPA = \sum_{t=1}^{N} \frac{PRC_t + INC_t}{(1+d)^{t-1}}$$

GLOSSARY OF SYMBOLS

AB_{at}	Avoided bill reductions on bill from alternate fuel in year t
$AC{:}D_{it}$	Rate charged for demand in costing period i in year t
$AC{:}E_{it}$	Rate charged for energy in costing period i in year t
ARI_{RIM}	Stream of cumulative annual revenue impacts of the program per unit of energy, demand, or per customer. Note that the terms in the ARI formula are not discounted, thus they are the nominal cumulative revenue impacts. Discounted cumulative revenue impacts may be calculated and submitted if they are indicated as such. Note also that the sum of the discounted stream of cumulative revenue impacts does not equal the LRI_{RIM}*
BCR_p	Benefit–cost ratio to participants
BCR_{pa}	Benefit–cost ratio of program administrator and utility costs
BCR_{RIM}	Benefit–cost ratio for rate levels
BCR_{TRC}	Benefit–cost ratio of total costs of the resource
BI_t	Bill increases in year t
B_j	Cumulative benefits to participants in year j
B_p	Benefit to participants
B_{pa}	Benefits of the program
B_{RIM}	Benefits to rate levels or customer bills
BR_t	Bill reductions in year t
B_{TRC}	Benefits of the program

C_j	Cumulative costs to participants in year i
C_p	Costs to participants
C_{pa}	Costs of the program
C_{RIM}	Costs to rate levels or customer bills
C_{TRC}	Costs of the program
D	Discount rate
ΔDG_{it}	Reduction in gross billing demand in costing period i in year t
ΔDN_{it}	Reduction in net demand in costing period i in year t
DP_p	Discounted payback in years
E	Discounted stream of system energy sales - (kWh or therms) or demand sales (kW) or first-year customers
ΔEG_{it}	Reduction in gross energy use in costing period i in year t
ΔEN_{it}	Reduction in net energy use in costing period i in year t
E_t	System sales in kWh, kW, or therms in year t or first-year customers
FRI_{RIM}	First-year revenue impact of the program per unit of energy, demand, or per customer
IMP	Total discounted load impacts of the program
INC_t	Incentives paid to the participant by the sponsoring utility in year t First year in which cumulative benefits are ≥ cumulative costs
K_{it}	1 when ΔEG_{it} or ΔDG_{it} is positive (a reduction) in costing period i in year t, and 0 otherwise
LC_{pa}	Total program administrator costs used for levelizing
LC_{pa}	Levelized cost per unit of program administrator cost of the resource
LCRC ·	Total resource costs used for levelizing
LC_{TRC}	Levelized cost per unit of the total cost of the resource
LRI_{RIM}	Lifecycle revenue impact of the program per unit of energy (kWh or therm) or demand (kW)—the one-time change in rates—or per customer—the change in customer bills over the life of the program.
$MC{:}D_{it}$	Marginal cost of demand in costing period i in year t
$MC{:}E_{it}$	Marginal cost of energy in costing period i in year t
NPV_{avp}	Net present value to the average participant
NPV_P	Net present value to all participants
NPV_{pa}	Net present value of program administrator costs
NPV_{RIM}	Net present value levels
NPV_{TRC}	Net present value of total costs of the resource
OBI_t	Other bill increases (i.e., customer charges, standby rates).
OBR_t	Other bill reductions or avoided bill payments (e.g., customer charges, standby rates).
P	Number of program participants
PAC_{at}	Participant-avoided costs in year t for alternate fuel devices
PA_t	Program administrator costs in year t
PC_t	Participant costs in year t to include: • Initial capital costs, including sales tax • Ongoing operation and maintenance costs • Removal costs, less salvage value • Value of the customer's time in arranging for installation, if significant
PCN	Net participant costs
PRC_t	Program administrator program costs in year t
RG_t	Revenue gain from increased sales in year t
RL_{at}	Revenue loss from avoided bill payments for alternate fuel in year t (i.e., device not chosen in a fuel substitution program)

RL_t	Revenue loss from reduced sales in year t
TC_t	Tax credits in year t
UAC_{at}	Utility-avoided supply costs for the alternate fuel in year t
UAC_t	Utility-avoided supply costs in year t
UIC_t	Utility-increased supply costs in year t

Appendix C

Derivation of RIM Lifecycle Revenue Impact Formula

Most of the formulas in the manual are either self-explanatory or are explained in the text. This appendix provides additional explanation for a few specific areas where the algebra was considered to be too cumbersome to include in the text.

Rate Impact Measure

The Ratepayer Impact Measure lifecycle revenue impact Test (LRI_{RIM}) is assumed to be the one-time increase or decrease in rates that will reequate the present valued stream of revenues and stream of revenue requirements over the life of the program.

Rates are designed to equate long-term revenues with long-term costs or revenue requirements. The implementation of a demand-side program can disrupt this equality by changing one of the assumptions upon which it is based: the sales forecast. Demand-side programs by definition change sales. This expected difference between the long-term revenues and revenue requirements is calculated in the NPV_{RIM}. The amount by which the present valued revenues are below the present valued revenue requirements equals $-NPV_{RIM}$.

The LRI_{RIM} is the change in rates that creates a change in the revenue stream that, when present valued, equals the -NPV_{RIM}*. If the utility raises (or lowers) its rates in the base year by the amount of the LRI_{RIM}, revenues over the term of the program will again equal revenue requirements. (The other assumed changes in rates, implied in the escalation of the rate values, are considered to remain in effect.)

Thus the formula for the LRI_{RIM} is derived from the following equality where the present value change in revenues due to the rate increase or decrease is set equal to the $-NPV_{RIM}$ or the revenue change caused by the program.

$$-NPV_{RIM} = \sum_{t=1}^{N} \frac{LRI_{RIM} \times E_t}{(1+d)^{t-1}}$$

Since the LRI_{RIM} term does not have a time subscript, it can be removed from the summation, and the formula is then:

$$-NPV_{RIM} = LRI_{RIM} \times \sum_{t=1}^{N} \frac{E_t}{(1+d)^{t-1}}$$

Rearranging terms, we then get:

$$LRI_{RIM} = -NPV_{RIM} \Big/ \sum_{t=1}^{N} \frac{E_t}{(1+d)^{t-1}}$$

Thus,

$$E = \sum_{t=1}^{N} \frac{E_t}{(1+d)^{t-1}}$$

REFERENCE

Clark II, W.W., Sowell, A., Schultz, D., 2002. Standard practice manual: the economic analysis of demand-side programs and projects in California. International Journal of Revenue Management X.

Chapter 11

The Next Economics: Civic–Social Capitalism

Woodrow W. Clark, II

Clark Strategic Partners, Beverly Hills, CA, United States

Chapter Outline

INTRODUCTION

The energy crisis in California was a challenge for all its citizens. The "design flaws" or "restructuring," as some economists now label it, were not the only problems. As new energy systems are envisioned and constructed to respond to the crisis, policy makers must reformulate the basic premises that led to the crisis in the first instance. This means reevaluating the political–economic foundations that led to deregulation. These assumptions must be recast to provide a new direction that will provide cheap, reliable, and environment-friendly electricity without relying on price competition as the main economic tool.

The extent to which the old basic premises need to be replaced is clear not only from the failure in California but also from the more widespread problems with deregulation or privatization in other states and nations. Growing evidence appears to indicate serious problems with energy sectors worldwide

Sustainable Cities and Communities Design Handbook. https://doi.org/10.1016/B978-0-12-813964-6.00011-2

where unrestrained competition has been tried (Kapner and McDonough, 2002), and this is not just a minor adjustment but a huge mistake. The energy crises during the summer of 2003 in northern United States and southern Canada along with those in Europe point dramatically to something being wrong with the deregulation and privatization economic models.

The current energy crisis created a challenge that provides the opportunity to look at energy economics in a new and different manner. Neoclassical or conventional economic theory looks at energy from the perspective of the market, whereas energy economics needs to be examined from the perspective of the society in general. The object of an energy system or sector should not be to maximize corporate profits, but to assure that civic interests are protected for all citizens and best developed for future generations.

In the current predominant deregulation model, the pursuit of profit is assumed to lead to public good. As evidence mounts, the pursuit of the public good when it leads to profit only is a disaster for not only the company but also the general public. This "public good" argument is parallel to Hawkin and Lovin's concept of "natural capitalism" rather than to the neoclassical economic theory. It is consistent with the findings that socially and environmentally responsible firms are profitable, sometimes more profitable than average (Angelides, 2003a).

However, this alternative economic framework is only beginning to be articulated. Although it is not possible to present a complete new theory of civic capitalism here (Clark and Lund, 2001), this chapter will outline the basic economic elements of a new approach to electricity structure that supports the development of an "agile energy system." "Civic markets" define the role of government and regulatory oversight that is embedded in public–private partnerships. This is not a socialist or communist model (Clark and Li, 2004). The public good is not just maximized by central planning and control or by the elimination of private ownership.

Cooperation between the public–private sectors in the form of partnerships, collaborations, rule making, setting codes and standards, and implementing programs is the new civic market model. This approach to economics and politics is an alternative to the theory that competitive market forces would increase the public good of any nation state. By letting private monopolies control the supply (or demand) of any infrastructure sector like energy, government opens the door for mistakes like the one that happened in the California energy crisis between 2000 and 2002.

The worldwide energy crisis has reinforced the basic tenet that all governments must adhere to higher standards for the public good. Leaving energy, water, environment, or waste, among other infrastructure sectors, to the "market" or "competitive forces" of supply and demand was wrong in the first instance. The predictable results were that private monopolies gained legal control of energy supply and generation. These market forces only replaced the publicly regulated monopolies that had supplied California with power for a century of economic growth.

This chapter argues for a new set of tools by which the economics of the power sector can be reformulated to create new solutions and opportunities for the future of all citizens. The old neoclassical competitive model that gave deregulation to California, most of the United States, and now the world, needs to be replaced with a new energy/environmental economic model that builds on networks, flexibility, and innovation. Such a new economic model is well rooted in civic markets.

A New Framework for Understanding Energy and Economics Within the Context of Civic Society

The concept of "civic markets" is put forth in this chapter to highlight the differences that need to be addressed in managing a complex industry such as electricity. However, civic markets also apply to other infrastructure sectors like water, waste, transportation, and education where market forces can be relied upon either technically or financially to be honest. In addition, civic markets are likely to be in the new economy and concentrated in industries that are expanding rather than contracting.

This is most clearly seen not only in monopolist industries such as energy but also in industries involving other public infrastructure such as airlines and airports and information and telecommunications, industries with high environmental impacts such as the natural resource industries, as well as service industries such as health and welfare. Even industries that depend on a steady stream of innovation from university and government research laboratories such as pharmaceuticals, life sciences, and biotechnology are moving rapidly toward civic markets or partnerships between public and private sectors. The framework for the new economics is rapidly evolving and is reflected in a growing body of thought in politics as well as business and economics (Clark and Lund, 2001).

According to conventional neoclassical economics, companies should operate with little or no government interference. Ideally companies have no regulations and taxes, etc., but contribute to societal needs on their own. Adam Smith's (rev. 1934) concept of the "invisible hand" and more recently the Bush administration's (2001) application of it in outlining its "energy plan," are good examples of this neoclassical economic perspective: government should not be involved in energy business activities, especially regulations. In any industry, as in any country, it is argued that there is a "balance" between supply and demand that keeps prices low due to competition among the companies for customers. It is the supply–demand balance that is the basis for all energy economics and the rational for deregulation in California as well as similar conventional economic justifications elsewhere in the United States and worldwide.

The energy system points out the limitations to the conventional economic models and gives priority to new concepts. Many of the contrasts between the neoclassical and civic market models are matters of degree and centrality; the civic issues are "externalities" in the current models used by modern economists rather than being at the core. Civic market functions must take prominence in framing competition and market economics. The main differences are as follows:

- Neoclassical economic models are based on concepts of independent firms competing to gain advantage over other firms because of efficiencies, product, technology, and price, and thus meeting the public interest because they better produce what the public wants at the lowest cost. In contrast today, we better understand that firms are in networks in which innovation, efficiencies, and price are the result of the interfirm sharing and cooperation rather than simply competition.
- Neoclassical models assume private sector involvement, whereas the new system is based on an increasing number of public–private partnerships and shared responsibility between the public and private sectors. Shared ownership and management control are at the root of the programs that blend the public good with private initiative.
- Neoclassical models are based on premises that markets and technological systems are largely self-regulating and that government's role is limited to protecting against market power and unfair competition by enforcing laws preventing price gouging, protecting patents, enforcing contracts, and prohibiting malicious misrepresentations or corruption, etc. In contrast, we now see an expanded role for government that goes well beyond rules to creating the context for public good in expanding markets, promoting employment, and protecting the environment.
- Neoclassical models left innovation and technological change to the marketplace, whereas the new model relies on government leadership to introduce and stabilize markets for innovations that serve the public good but may not be in the short-term private interest of market leaders.
- Neoclassical models make minimal distinction between industries where it is easy for companies to enter or leave, compared with companies in grid or network industries where control of the grid constitutes a public obligation to serve and a natural monopoly. In fact, barriers to entry in a number of industries are growing because of increased interdependency and specialized materials, information, and markets that limit participation in the industry to those already involved.

The transformation away from the neoclassical and now conventional economic model that was the basic philosophical and theoretical bases, along with a bipartisan political agenda, and hence responsible for the deregulation framework that led to the California energy crisis and to changes in the electricity system structures in other nations must be discussed in some detail.

It is important to understand that neither all existing economic philosophies and theories nor are the accomplishments of neoclassical economics in solving other industrial and business problems dismissed. Nonetheless, a full discussion is necessary.

Neoclassical economics and its conventional contemporary proponents that are derived from a particular economic philosophy are not appropriate for the energy and many other infrastructure sectors. Furthermore, there are other economic philosophical paradigms that lead to very different economic principles and rules (Clark and Fast, 2008). In short, the explanation of economic issues surrounding the electricity industry require new tools, frameworks models based upon a different social science philosophical paradigm. It is this paradigm, called "interactionism," discussed by Clark and Fast (2008), which is framed by the civic market theory.

Interactionism is, in short, the theory that because people (actors) interact in specific situations (everyday behavior such as business), companies and their behaviors are better understood. The decisions of business actors does not depend on numbers, figures, and statistics alone. Instead, business people form strategies and plans, such as deregulation public policy, knowing that they can maneuver the newly formed markets. A key component in understanding business in the interactionism paradigm is to also know, influence, or control the role of government. Much has been written on this subject, but the "invisible hand" of government needs to be influenced to do as business wants it to do.

Economists want to be scientific and therefore ignore this influence over government. Instead, they tend to think that the use of statistics and numbers place them above the interaction between people. Economists see themselves akin to the hard and natural sciences. There is almost a sense in economics that if the field is not scientific (e.g., statistical or numbers oriented) then it is not professional. For most economists, however, the perspective and view or definition of what science is and does bears little proof in reality (Blumer, 1969).

Science is not a simple matter of statistics or numbers (Perkins, 1996). Although some field or qualitative studies have been conducted, especially on productivity, economists remain steadfast in their belief that fieldwork is the main research area for sociologists and journalists. Yet, the need to explore the "productivity paradox" as Nobel Laureate, Robert Solow, called it in 1987, promoted statistical research in the 1990s only to crash-land with the explosive truths behind the "productivity miracle" of that decade by the turn of the next century.

Clearly statistics did not tell the "truth" about productivity in the 1990s. The popular journal, *The Economist* (see *The Economist* issues 1998–2002), often tries to "sugarcoat" or marginalize the accounting scandals of CEOs, major American corporations, corporate governance, and bankruptcies in

2002, as simply downward revisions of company financials, when in fact these crises represent only the beginning of corporate illegal misbehavior. The issue of validation and verification of economic data is simply not statistics and numbers (quantitative) versus fieldwork and observation (qualitative) data to prove points or hypothesis but a combination of both (Casson, 1996).

Implementation of energy economics today has been traditionally done (prior to deregulation, privatization, or liberalization) through a variety of "mechanisms" by energy experts. HarvardWatch (2002) looked behind the scenes of public policy and discovered, however, questionable direct links between the objective experts at some universities and the energy private sector. The "links" between scholars and experts and the companies violate the credibility of economics as being either objective or scientific. Far more important are the "networks" of people who develop and implement government policies that impact the public through the private sector. As will be described in the following sections, government policies do not just mean regulations, tax, and incentive programs. They should also include, as California has championed, economic accounting for projects/programs (Schultz, 2001) and the creation of market demand.

At this point, it is important to make note of how California government found itself in the middle of redefining energy economics. The energy crisis can never be fully explained (CEC, 2002), but one basic economic issue is clear: the state government had to play an active role in solving the crisis. For California, this meant that a number of measures and legal steps had to be taken from long-term energy supply contracts, to emergency funds for conservation and efficiency programs, to incentives such as buy-down and rebate programs, to expedited siting of new power plants.

As discussed earlier in some detail, other economists such as Borenstein et al. (2002), Woo and Sachs (2001), and Nobel Laureates (2001) all agreed that the energy crisis could be averted and changed if the government simply took off all the price caps on energy. What is ignored traditionally by economists is a focus on the firm itself (Teece, 1998). Energy economics, however, only discusses the companies as end users of energy such that energy flow, and hence costs, should be controlled by the consumer's awareness of "real-time prices" (Borenstein et al., 2002).

Elsewhere (Clark, 2004a,b) argues that "qualitative economics" is a new area of economics, within the interactionism economic paradigm. The purpose of qualitative economics is to understand how companies work. Much of the field is concerned with case studies as well as corporate descriptions of operations and people. However, the most significant concern of qualitative economics is to gather data to understand what the numbers mean. For companies when they add, as Enron allegedly did, 2 plus 2 and got 5, the meaning of those numbers is critical. The issue is that economics must understand how businesses work and can only do that with deeper definitions, meanings, and backgrounds of organizations, people, and their interactions.

The goal of economics must be to take quantitative and qualitative data and derive rules. Economics needs to expose universal rules based and tested in reality on a combination of statistics and interactions. From rules, laws can be articulated (Perkins, 1996). Scientists in fields such as linguistics (Chomsky, 1968) and developmental cognition (Cicourel, 1974) have long investigated science in terms of developing universal rules and laws. Just like the natural sciences, natural sciences use observation, description, and hypothesis testing (Chomsky, 1980). The science argument is spelled out in other works (Clark and Fast, 2008) and some aspects of the qualitative economic theories are presented in the following discussion.

The Advantages of Cooperation Over Competition

The first economic principle supporting an agile energy system is that cooperation among companies, and among consumers, helps increase civic good and societal purpose. Clearly an important side benefit, however, is sustainable development. Or to put it another way, the growth of environmental and clean energy-friendly businesses is needed as an overall public policy. As California State Treasurer put it in a speech delivered to the United Nations in November 2003, California must "mobilized financial capital in new and innovative ways, consistent with the highest fiduciary standards, to meet our 'double bottom line' goals of achieving positive financial returns and fostering sustainable growth and sound environmental practices" (Angelides, 2003a,b, p. 1).

One of the lessons of the personal computer technological revolution is that by having an open architecture and allowing other firms to follow standards and achieve standardization, the IBM model succeeded over other firms that were trying to go alone. Indeed, the notion of firm cooperation is not new or unique to alternative economics, but it is clear that it plays a central role in thinking about how to introduce a new technology and energy system.

Competition will not, and should not, go away in California. Governor Schwarzenegger appears to be headed into a middle pathway as well, following much of the "civic market" approach to economics. Today, California has an electrical generation system, and a minimally competitive retail market. Due to both legal and practical reasons, there is little reason to consider reverting to the market structures of the vertically integrated energy system dominated by three monopolies that the state had before deregulation. The companies that purchased the bulk of the state's generation capacity after deregulation now, since the energy crisis, have medium-term contracts of up to 10 years to sell power to the state's utilities at fixed rates. This is a significant shift from the spot market mechanisms.

The California Power Exchange, which is now bankrupt, was created by deregulation in 1996. It still holds (2004) over $2 billion in assets as the courts try to decide on how to disperse the funds. Nonetheless, the dozen or so exchange staff remain on payroll. After some significant conflicts, due to the California energy crisis, the Federal Energy Regulatory Commission (FERC) and the California Independent System Operator are now cooperating.

Although the midterm contracts were negotiated while the state was in its energy crisis, and by all estimates are overpriced, significant reductions were later negotiated. In addition, most of the new power plants that are being proposed or built are seeking and usually obtaining long-term contracts on more reasonable terms. The large number of fixed contracts has reduced price competition among electricity generators on the day-ahead and spot markets to a very small proportion of power that is needed. Looking ahead, the state now has a modified market structure for buying and selling power, which the FERC is expected to approve in 2004. That market structure will continue to operate. As the existing long-term contracts end, this market may become more robust for supplying the core demand of the utilities. However, given the past performance of the market mechanisms, there will be a need for alternative plans.

On the retail side, some limited competition may emerge over the next few years. Now, the existing utilities serve virtually all customers in their geographical areas, with the exception of a few large consumers who had direct access contracts before the energy crisis. However, pressure remains on the California Public Utility Commission (CPUC) to allow other large users similar contracts. Modified direct access and limited exit fees have allowed the continued development for on-site generation especially with new solar/ photovoltaic (PV), microturbine, and fuel cell technologies.

In addition, some interest groups and municipalities are intensely interested in aggregation of customers to obtain power from their own contracts. The California Stationary Fuel Cell Collaborative (CSFCC) and the California Power Authority developed such aggregated master purchase contacts for new advanced on-site technologies. The premise of the CPUC and most legislators is that those customers who want to bypass their utilities and the existing system must continue to pay their share of the energy debt and overpriced contracts that are a residual of the energy crisis. If these "exit fees" are added, the advantages of leaving the traditional utility become limited except in some exceptional cases. So compromises have been worked out to mitigate the negative impact on business developed for retail customers. A significant side benefit is the economic development of new technologies and companies providing the technologies, services, and operations.

The issue, then, is not whether price competition will be eliminated in California—some competition is here to stay and may beneficially increase. The issue, instead, is to identify what will replace competition as the organizing premise for the evolving energy system in California. What should be the organizing premise for other states and countries looking to restructure their power system, given the California experiences?

From an economic perspective, the strategic driving force of civic capitalism is to increase cooperation not competition.

The premise is that since full and legitimate price competition in electricity is not possible, and indeed its pursuit is catastrophic, the alternative is to structure multiple options for cooperative solutions to the energy problem of how to deliver the best electrical services to the population with the lowest possible cost. In a particularly interesting article, Henrik Lund contrasted the conflict model of energy policy to a "democratic" model that involved broad public participation in decisions to reduce fossil fuel dependence and innovate with renewable and conservation alternatives in Denmark. This country is the world leader in innovative solutions to the energy crisis without focusing solely on getting firms to compete on short-term prices.

The vitality of cooperation and collaboration among firms is highly valued in studies of economic success, for example, Saxenian's study (1994) of the conditions leading to the ascendance of Silicon Valley over Boston Route 128 in microelectronic technology. She showed that the networking and interaction among competitors in Silicon Valley gave these high-tech companies a competitive edge. One high-tech entrepreneur from Silicon Valley, for example, made a fortune in the doy.com/information technology industry, but then turned his attention to sustainable environmental companies. He has formed networks and raised capital for such firms.

In addition, firms in many industries have collaborated to find solutions to major technical problems that affect the whole industry. In electricity, the Energy Producers Research Institute (EPRI) model is an example of how the utilities have contributed the funds necessary to do fundamental research on power plant design, maintenance, operations, modeling, and many other factors. Deregulation and competition have led to serious cuts (over 33% since the mid-1990s) in funding and programs at EPRI because the new market players have no incentive to look to long-term collaborations and financial participation in programs to research, develop, and deploy new advanced technologies.

Finally, the potential of cooperation is seen in building regional networks of firms that can take advantage of specialized infrastructure and supplier industries, becoming a type of regional node. This type of industrial concentration that gives individual firms competitive advantage has been called an industrial cluster by Bradshaw, King, and Wahlstrom. The cluster concept has been used in other less specific ways by Porter and others, but what is important is that research is showing that industries concentrate in certain areas and that this gives them an advantage. This economic advantage is one that increases the economic value of each of the firms in the region.

The California State Assembly took the lead in understanding and seeing the need for regional public policy. Under Speaker Robert Hertzberg and then continuing under Herb Wesson, the state issued a report recognizing the realities of the its diversity in both cultural and physical make-up. Guided by a series of regional convened forums throughout 2002, the Public Policy Institute gathered opinions and perspectives from citizens and leaders throughout the Golden State. A constant theme was the value and concern for protecting the environment. Because of this, the state legislature took the initiative in passing laws to protect the environment as well as plan for the future growth of California.

A good example of the regional approach undertaken by the Legislature is the California Environmental Goals Policy Report passed as the Wiggins Bill (AB #857) in September 2002 that requires the Governor to form a statewide environmental plan that must go to the Legislature for approval. This statewide plan is the first such on in over 25 years but does not dictate to the local communities and cities as to what to do. It basically establishes public policy and provides some direction as to the state's plans for growth. The first report was issues at the end of Governor Davis' term in office before the recall took effect (EGPR, 2003).

The Economic Advantages of Civic Capitalism

The second theme that underlies civic capitalism is the importance of public–private partnerships. Building on the first theme of cooperation, the public is a favored cooperator with private firms. From an economic perspective, the public role increases in complex industries and economies because of the special role of public participants in facilitating business activities that are beneficial to business and that pursue the public good. The partnerships in energy have gotten more complex over time and the beneficial partnerships are markedly different from the early regulatory relationship. Regulators primarily tried to keep the monopoly utilities from raising rates too high, whereas the new partnerships try to use public resources to help better meet civic interests. In the past the public role with utilities was the power to say "no"; however, the emerging role is for the public to say "yes."

What is the role of government in business? Much of the economic literature has been focused on the dichotomy between free markets and tight regulation as in the historical electrical industry. The emerging era of public–private partnerships is neither. The justification for public involvement in the power industry is twofold. First, the transmission and distribution monopoly and technological nature of electricity networks mean that the public has an interest in overseeing the private suppliers of such an essential part of modern life. This point has been made consistently in previous chapters. Second, the public has many social and environmental interests that intersect with the provision of electricity, such as environmental protection, public safety, equity, economic development, and long-term reliability. Simply put, given the extensive public agenda, it is more effective to try to reach these goals through partnerships than rule making. This is not unique to the electricity industry, although it stands out in very clear relief.

In Denmark, for example, the free market, has historically involved a partnership between government and business. If shared societal goals (free universal education, national health care, jobs, strong social services, and high standard of living) are to be achieved, then business and government must work together toward common economic goals. The "partnership" between government is not always smooth or cooperative, but it remains dedicated to the shared values for the common good.

Government is deeply involved in many industries in more than a regulatory role. For example, government provides over $16 billion annually to the US Department of Energy and its over dozen "national" laboratories. Two of these scientific laboratories receive over $1 billion annually in research funds: Los Alamos National Laboratory in New Mexico and Lawrence Livermore National Laboratory in California. Both of these laboratories as well as Lawrence Berkeley National Laboratory are operated by the University of California System, which receives over $25 million annually as a management fee. The amount of research funds flowing through these and other laboratories clearly influences both public policy and business strategies in the United States and worldwide.

In passed 2017 national energy bill included assistance for coal and nuclear power as well as expanded incentives for oil and gas, as well as most parts of the electricity industry. The electronics industry credits high-price defense contracts with giving them the capacity to develop and market early transistors and integrated circuits when there would have been no private markets for these products given their costs. In addition, the US agricultural incentives have become hotly contested by Europe and Asian countries claiming unfair competition in trade. Also, the Bush administration's favoring of government support for industry is seen in the prescription drug bill recently passed. In short, the myth of industry operating without government support and control is hopelessly inadequate.

The local and regional level is also a critical resource for public–private partnerships. The role of local governments is often forgotten, but together they have extensive planning and program activities because their residents and constituencies need and want it. Thus local-level governmental entities, such as government and counties, are one focal point for renewable energy generation and hence noncentral grid energy systems. In 2000, the voters of California passed Initiative #38, which allowed local governments or districts to use finance measures such as bond measures.

One of the most successful has been the Community College Districts, the largest college system in the world with 1.3 million students on 108 campuses. By the spring of 2002, six districts followed the lead of the Los Angeles Community College District (LACCD) and its Board of Directors who passed a bond for $1.3 billion. At least half of the bond measure funds are being used for "sustainable" (green) buildings in LACCD under international green building standards. In other words, the public colleges are leading the way to renewable energy in their facilities.

Part of the evidence for the political and economic success rests in the fact that the Board of Directors for the California Community College District appointed the Chancellor for the LACCD (Mark Drummond) to head the entire state system in 2004, the largest higher education systems in the United States with over 108 campuses and over 1.2 million students. The advancement of "green" college buildings throughout the state will certainly be far more rapid and cost effective. The political and economic repercussions to this are significant. Local communities are the market drivers for renewable and clean energy systems.

In California, the mass purchasing of sustainable systems has reduced the price and expedited the implementation of these systems. The state and the agencies and programs that it funds are a huge market, and the mobilization of this market is large enough to change the economics for the production of many items. The state mandated some level of internal consumption of recycled paper, green building, and renewable energy, for example, which creates enough of a market to help establish these industries at a level in which they have economies of scale and become cost competitive in open and unsubsidized markets. Many examples exist of this financing practice such as the state purchase of police and other vehicles, which are zero emission, to the Energy Star program for energy-saving appliances and efficient equipment. The Department of General Services has led the state in this effort and included new technologies, such as fuel cells and solar devices (CSFCC, 2002).

Government partnerships rather than regulatory power help create more successful programs to meet civic goals. Using moral persuasion and the legitimacy of the state, governments can lead by demonstrating their willingness to invest in what they are telling private companies to do. California has the public policies and mechanisms in place for local and regional clean distributed energy systems. The State Government Code already provides for Community Energy Authorities (No. 5200) for local and regional energy systems (not municipal utilities). With local governments, the private sector, educational and research institutions, as well as nonprofits all working together as partners, clean energy is viable along a business model for recovering costs and providing for innovation and change.

Promotion of dispute resolution and conflict mediation as a way to resolve differences between private firms and state agencies or programs. The competitive model is based on a win–lose model, whereas economic growth and public interests are increased with win–win responses to differences of opinion on what direction the power system should take.

The public role in partnerships is often and most importantly the collection of information and data on power demand, technological change, environmental resources and pollution, and national and global trends. As discussed in earlier chapters, one of the major contributions of the California Energy Commission when it was formed was to help the utilities come up with a standard methodology forecast of power demand, which led to the reduction in the expected number of nuclear power plants needed. The fact that these plants were not built protected the state from even worse disasters of stranded assets and high costs.

Resource mapping and technological feasibility studies are also an essential function that government can provide. Again, the Energy Commission had as one of its initial mandates the collection and dissemination of information on the location and extent of renewable resources such as geothermal hot spots, wind resources, untapped hydro capacity, solar capacity etc.; these resource inventories became essential for firms looking to invest in these technologies in the state. These studies were done in partnership with the industry, pooling their private information and setting up methodologies for collecting additional data. Because the data came from a partnership involving the firms intending to use the data, it had a higher level of credibility than if it had been developed to force business partners to do something.

Provision of additional resources and technical assistance must be done to implement best practices, including green building and site design, product development, and installation of energy efficiency systems. The public role in partnerships can also include developing and providing training, conferences, best practice reports, and consultations with staff experts that will eventually lead to more successful private developments meeting state standards and agendas.

Finally, the partnership can encourage California-based philanthropies and the commercial media to work with the public sector on public education and participation and inform readers and viewers on energy issues. The partnerships in the state that encouraged conservation that eventually broke the price spiral of the energy crisis show that broad partnerships can work very effectively.

The implementation of new policies and programs cannot be done until basic public finance issues are addressed and resolved (Clark and Sowell, 2002). The state investment in conservation, new environmentally advantageous technologies, and public energy infrastructure is not an expenditure but a down payment that will generate considerable return. State Treasurer Phil Angelides had implement two major policies that leverage the vast $300 billion State CalPERS retirement fund. Two policies were launched since 1999 (1) sustainable development as "Smart Investment" with a $12 billion investment and (2) The Double Bottom Line, which is investing in California's "Emerging Markets" at $8 billion and another $1 billion to developing countries.

In addition, the public sector has many financial tools and mechanisms to stimulate development while providing environmental protection and safeguards. For example, the California Infrastructure and Economic Development

Bank makes funds available at lower rates and to projects that do not necessarily qualify for bank financing. The Energy Commission has identified a number of financing tools available to assist private firms obtain the loans necessary to start their businesses. In particular, a number of new tools are available for regional and community-distributed generation (DG) capacity, purchase of energy savings equipment, retrofits, etc.

For the consumer market in partnership with utilities and homebuilders, the USA Federal Housing Authority has established a new Energy Efficient Mortgage that has been rolled out by the California Housing Finance Agency. Publicity about it and its benefits to the state can increase its use and the energy savings associated with housing that goes well beyond minimum standards.

Finally, it is important to note that the public role is not likely to include much in the way of funding in the future. The state budget is over $25 billion short, and the new governor has taken deep cuts in virtually all existing programs, and has had to borrow large amounts of money from bond markets. This means that even state-supported bonds, which have lower interest rates, will not be available for other uses because there are limits to the amount of money the state can raise without becoming fiscally unstable. That point may near.

The Civic Markets at Work

The third issue involved in the new economics of energy concerns the interface between electricity firms and the state. Under regulation and deregulation the issue was to control the market power of the large utilities to assure that they did not overcharge; the new energy economics aims to assure that the market is broadened to include the broader concept of public good including environmental protection, economic growth, and long-term stability. Especially after the energy crisis the commodity markets for power are narrowly based on short-term prices for power. The new economics recognizes that the important market for the public good is broad and long term. This difference between narrow and short-term markets versus broad and long-term markets is not easily managed by making adjustments to how electricity is sold. It is not something easily woven into a numbers of Power pool. The new comprehensive market reflects the diversity of markets that constitute an agile power system.

With globalization and increasing scale of corporate structures, the conditions for and effect of competition needs reassessment. Today more multinational companies have significant shares of the global and local markets and they shape demand rather than respond to it. Although large parts of the economy are strongly competitive, many trends are establishing the consolidation of firms, interlocking ownership, and corporate–government linkages contrast with innovativeness and price competition characteristic of small business markets.

The new civic economic model must take into account the growing power of large firms and the inability of competitive markets to work effectively without oversight. It is the role of oversight that restrains these large firms from being uncompetitive and to exercise market power. In James Scott's words, "Today, global capitalism is perhaps the most powerful force for homogenization, whereas the state may in some instances be the defender of local difference and variety." Thus the civic market is not built on the premise that a competitive market must be created and maintained; instead it is built on the premise that such a competitive market is impossible to guarantee and that the public good must be served and assured by active public partnerships between empowered state agencies and innovative and socially responsible companies.

A number of economists have argued (see various issues in *The Economist* in February–March, 2002) that the collapse of Enron proves how neoclassical economics works. As the argument goes, if a company cannot perform in the market, it fails. Other companies come in and take its place. The free market economy moves on.

The problem with this argument is that it ignores what allegedly Enron did in the first place—set up and influence the deregulation of the energy market, then manipulate it, and finally, use or profit from investors so that in the end the collapse of the company was based on inflated shareholder value and unsecured creditors who lost their funds. Employees lost their retirement with the collapse in the value of the stock, thousands of stockholders outside the company as well as mutual funds and retirement portfolios suffered. Thus the consequence was not just a failed company; it was a cascade of pain for people all across the nation, and it was the leading contributor to the $40 billion electricity charge to the citizens of California.

However, there is a growing environmental economics tradition in which strategies for valuing the economy are made. For example, in business, making a profit may not be the only motive for the firm itself, the individuals working within it, and the shareholders themselves. Ritzau indicates in his study of Danish researchers that success is not defined totally by monetary reward. The survey results from 63% of about 350 researchers indicate that there are other rewards (what we call deep structures) for inventing and patenting new ideas and innovations.

As Ritzau puts it "for the majority of the scientist economic benefits are not the primary motivating factor for them to patent their research. Much more so is it the prospect of being able to contribute to the development of society and at the same time secure their own research projects. An economic benefit is considered nice and indeed motivating, but seems not to be the primary incentive of patenting."

Kuada indicates much the same in his work with African entrepreneurs and companies. Ritzau summarized the universal concept even more precisely with a focus upon "Scientists who have filed patent applications" notes that these successful scientists, by any quantitative measure or scale, have placed "emphasize even less the chances of any personal economic benefit. They focus on personal satisfaction of realizing their own ideas and ensuring funds for their own research."

The most common supply-side economic mechanism is "integrated resource management" in which the supply of energy into a system is based on what is known as "base load" or firm supply of power. Base load is usually dependable fossil fuel sources such as coal, oil, natural gas, and hydroelectric or nuclear. However, in the past few decades, geothermal has been considered base load. Renewable energy such as wind, solar, and even biomass are considered nonbase load or intermittent. Hence the challenge to have renewable energy calculated and treated as base load power supply.

The issue of energy management is critical for understanding how the economics of energy works. In short, the integrated resource management approach places the burden on the power system for delivering energy. Hence a "firm" or base load must be established that has reliable power from traditional fossil energy such as coal, oil, or natural gas. Nuclear and hydroelectric power are also considered firm load as the fuel can be constant and uninterrupted. The problem is that each of these fuel sources (hydroelectric usually has social and environmental side effects) pollute the atmosphere and have waste disposal problems. Thus the use of current "technologies" for firm load are unacceptable, given the increase in greenhouse gases and atmospheric particulates.

The problem in the energy sector, therefore, is the definition and operation of the market itself. When government does not regulate and directly oversee the energy sector, the private sector can manipulate and game the system as it is constructed. What is also not often stated or acknowledged is that the private sector, unlike the public sector, need *not* invest or even explore renewable technologies since they can not foresee immediate profits. As Schumpeter noted many decades ago, innovation for companies will only occur when they see it in their own best interests.

Stimulating Innovation for Sustainable Environment and Energy Systems

The fourth change from the neoclassical model is the observation that in marginally competitive situations (mature or established industries) innovation is limited because it would render otherwise profitable investments obsolete. Lovins, among others, has demonstrated the "factor of four" whereby new technologies get twice the productivity from half the input of resources. These technologies are sensible and environment-friendly innovations that not only are cheaper to use but also are cheaper to install in the first place. One can ask the same question about most conservation strategies and many cogeneration (combined heat power) projects, as well as other renewable projects.

Why is this type of sensible innovation so difficult to introduce to mass markets? Simply, according to the logic of the alternative strategies, it is because existing industry does not look for opportunities to cut costs and increase output, and when it sees opportunities it perceives the change too difficult. Costs and profits do not drive this behavior, and aversion to change does.

For example, the US automakers failed to invest in change until faced with catastrophic drops in sales as foreign cars flooded the market with fuel-efficient, safe, and attractive vehicles. The American energy companies still have exploited only a tiny fraction of potential cogeneration, which could be its cheapest generation source as shown in California under Public Utility Rate Purchase Agreement. Conservation programs that saved utilities billions of dollars were proposed and mandated by regulators and not the industries that benefited from them (along with huge consumer benefits). In short, the classical economic model is not adequate to establish advanced programs that develop and implement cost-effective and profitable innovations in dominant industrial systems.

Networks and Social Relations as Opposed to Only Numbers

Entry to the grid or the networks for large consumers, like industries, shopping plazas, and commercial users, is critical. Most companies want to have reliable and inexpensive power. Technically, having dispersed generation or DG is not much more difficult to manage than dispersed consumption, but for a long time the grid has been managed in the simplest way possible by a centralized utility able to control a small number of generation plants. Generation can be tailored to meet demand on a moment's notice, and generation levels can be shifted to assure that transmission capacities do not get exceeded. Dispersed production poses new challenges since a single utility does not own all the power generation plants making input to the system and there are many more of them.

The key issue is that local, distributed power generation is potentially far more cost effective and reliable than central grid supplied energy. The importance of local energy generation means that control and oversight can be far more effective and "democratic" in that citizens have control in their communities including on-site generation and local power systems. Networks are critical in this DG process. Grabher correctly argues that networks, and not just social ones, have become of increasing interest to researchers. Hakansson and Johanson point out that "there is an important difference between these social networks and the industrial networks of interest... Social networks are dominated by actors and their social exchange relations."

Hakansson and Johnson argue that "activities and resources in interaction are the more significant factors" in networks. Network theories can fit into the basic social constructionist framework and work within a subjectivist perspective for an understanding of everyday business life. Nevertheless, many scholars in the field find themselves rapidly moving into the objectivist paradigm because it offers structures that provide predefined and convenient explanations of the business activities. Thus in 1994, Hakansson and Snehota argue:

> *We are convinced that adopting the relationship perspective and the network approach has rather far-reaching theoretical as well as managerial implications. It seems to open up a quite new and different theoretical world compared to the traditional way of conceptualizing companies within markets. It offers new perspectives on some broad traditional problems of business management and* yields some novel and perhaps unexpected normative implications for business management.

Emphasis ours, Hakansson and Johanson (1993, pp. 1–4)

Therefore, "Relationships between companies are a complex knitting of episodes and interactions. The various episodes and processes that form business relationships are often initiated and triggered by circumstances beyond the control of people in companies. They are however never completely random, they form patterns." In order for the authors to understand a network, they revert to "structural characteristics," such as "continuity," "complexity," "symmetry," and "informality."

Enron appears to prove to be a good negative example of networks or the building of business relationships (also known as cronies and the "old boy" network) to influence and control markets. In the American energy sector, Enron indeed used "some novel" management skills and tools that have shown how it manipulated the energy markets (HarvardWatch, 2002).

Various court cases are proving how Enron executives misled regulators, cheated their customers and clients, pocketed unreasonable personal gains, and influenced public energy policy through social and business networks. Much of that influence was paid for by Enron corporation and its executives in donations and grants to scholars and politicians who deregulated without anticipating the negative consequences.

While all the evidence is not in as of late 2002, it is clear that these networks that Enron built were to influence political decision makers. This use of networks and personal relationships was supported by money, which influenced entire sectors of the economy and therefore markets. The behavior of Enron and other firms is not unusual, but in this case, the end result was not an open or free market but initially their control and domination of an entire sector. The plan almost worked.

The key to networks is "trust." This can only be achieved after people work together on a common problem and find that others are able to keep secrets, share valuable information, and exchange new ideas. The interactionist economic theory (Clark and Fast, 2008) best describes, understands, and explains how networks are created and operate. In themselves, networks are neither theory or scientific. Instead, the understanding of networks allows both the scholar and the business person to pursue shared goals. If networks are institutionalized as formal permanent structures, they will implode from their own administrative weight. The very notion or idea of a network is something that exists at a moment or situation in time to accomplish some task(s). People know one another and form the network to solve the problem at hand.

The basic issue for most people in firms is how to make the company survive and grow during any particular point in time. They must not only be free to move in the marketplace but also be secretive enough to protect its privacy. More importantly, firms must have concern for others and their environment. The public good must be protected because it is in the best interests of the firm to maximize its profits for shareholders and executives alike. The protection of the public good is essential in various infrastructures and sectors (Clark and Lund, 2001). In recent months, the protection of the transportation sector must be embedded in the government. The energy crisis in California has clearly shown that the energy sector is a public trust. The same could be argued for environment, water, as well as telecommunications.

In a highly dispersed system grid managers do not control much of the system and must have control over ancillary service capacity to respond to changes in generation quantity and location as they occur rather than have control over all the generators. With new computer and monitoring technology, this is more feasible and technologically efficient, although given the crisis the utilities have been reluctant to engage in any experimentation or investment in system change. The goal for the new civic markets is to help bypass the control roadblocks of the old system and to facilitate the transition to an agile system in the interest of the public good, even if the changes are not necessarily beneficial to existing institutions in the short run.

A significant component at the microeconomic level is the use of "networks" or relationship between people and organizations. These personal connections, partnerships, and relationships between technical staff and separately between business executives are collaborations in which often intense exchanges of information are commonplace. Some networks form in many different ways, but primarily link businesses with compatible strengths (and in some cases weaknesses) to achieve common goals. Other networks form between government and private industry. Networks can form on horizontal and vertical plains, depending on the nature of the interactions between the actors, organizations, and situations involved.

ECONOMIC COLLABORATIONS AND PARTNERSHIPS IN ACTION

The CSFCC is an example of civic markets at work initially but turned into an organization that wanted funding from a source—the natural gas industry at the turn of the 21st century wanted to get new technologies using their fossil fuel for power. The CSFCC participants include all the major Japanese and American car manufacturers, as well as firms interested in hydrogen economy who were (and still are) natural gas companies. The collaborative works on a variety of technical and marketing issues needed to get the technology introduced. One of the most significant trends in many industries is the collaborative programs undertaken by industry associations. To move forward, California Governor Schwarzenegger issued an Executive Order in early 2004 for California to have a Hydrogen Highway. However, although the past Governor Davis' staff had already written the Executive Order, he also saw the need for the state as well as the private sector to supply funding. The new Governor in 2004 who issued the Executive Order did not. Instead over 100 companies worked for 3 years trying to develop a plan but failed to do that and implement one.

It took a new Governor (Jerry Brown) to provide some state funds matched by private sector funds to create and implement a California Hydrogen Highway refueling system. By 2016 and then into 2017, this Highway has begun as there are four active car companies leasing Hydrogen Fuel Cell Cars (HFC) in California. The demand for hydrogen refueling stations has grown as over 2000 HFC are leased and more are coming. California now leads the world in this new technology, with Japan, Germany, Norway, and China not far behind.

Another example is the National Association of Home Builders, in the fiercely competitive homebuilding industry, which sponsors many programs to increase research on new materials, techniques, and markets for members. The biotechnology industry also has collaborative industry programs. In sum, even in the sectors of industry that are most heralded as being models of the free market, most have deep cooperative and sharing strategies for building capacity.

The most interesting new addition to the economics of power is the growth of networks of firms and consumers that strive together for advantage. The pervasive networks of firms are not only the generators who are connected to the grid but also many firms in diverse industries that are networked in buyer–supplier networks. For example, equipment manufacturers work with installers, parts manufacturers, computer venders, software designers, and hundreds of other related companies, and all benefit when a new generator is ordered. Understanding the existence and nature of networks in the new economy is one of the most pressing challenges for civic economics. In many countries the energy system includes equipment manufacturers as well as suppliers of transmission towers and transformers.

Cooperation between suppliers and the manufacturers of generators will also pay off. Although this form of cooperation has long been active and beneficial, supplying equipment for changing and expanding markets is very competitive. For example, in the case of pollution control equipment, multifuel plants that can use renewable resources is another example of how different parts of the industry can cooperate with utilities and grid suppliers.

One of the mistaken assumptions about the regulated California electricity system is that it was not competitive. However, in the United States the technology and equipment side of generation is not regulated. In several countries the utilities own the equipment manufacturers, and this has led to serious problems. In the current situation the manufacturers and the utilities can collaborate to produce more efficient and less polluting plans.

Renewable energy generator companies can also cooperate. For example, wind generators have in the past worked together to build and secure the transmission capacity to bring their power to market. More importantly, they cooperated with each other and the utilities to resolve problems around intermittent production. Mention has been made already of the LACCD with not only its funding for green buildings but also meeting the silver LEED standards. More bond financing for other college districts appears to be spreading the LACCD initiative not only statewide but also nationally.

In July 2003, for example, the University of California System Board of Regents passed its own green building initiative for all 10 campuses. A similar resolution will be considered by the California State University System with its 26 campuses.

Governor Schwarzenegger appeared to stand by the green of the state buildings as well to mark his support. During the last 6 months of the Governor Davis administration in 2003, a number of "green building" initiatives were under way based on both Governor Davis' Executive Order on Public Buildings in 2001 and the Consumer Agency Road Map Report (2001) reflecting the need to have renewable technologies for on-site generation in public buildings. From these governmental perspectives grew a number of private sector investments and programs (Ziman, 2003; Kenidi, 2003).

Most significantly in public sector has been the State's Green Driving Working Group Team under the Consumer Agency whose goal has been to develop demand and specifications for new "green" vehicles in the state fleet. This team continues to meet and move aggressively in this sector. Moreover, the Team has explored and put into place a "sustainable historical building" initiative that provides over $128 million in associated bond funds along with state specifications and coordination with the State Architect's Office as well as the professional architect associations.

Similarly, the Governor Davis Administration's Office of Planning and Research lead an effort to quantify the costs for installing and maintaining renewable and "green" technologies for on-site power generation. The result of an Interagency Working Group (over 50 active participants) was the revision of the 1987 California Standard Practices Manual (SPM) to a version in 2001 that uses life cycle analysis and externalities to get the correct costs for renewable energy in buildings. The SPM is becoming widely used by state and local governments in costing out their installation of energy-saving technologies (Clark and Sowell, 2002). Governor Schwarzenegger is expected to continue these programs and efforts.

Nonetheless, two critical issues remain and appear to be solvable. First is the need to aggregate the demand for green building technologies and services. The State community College District did just that under its Foundation, which has a program to provide a competitive central purchasing contract using LEED standards, criteria, and codes. This initiative implements the attempts of the certified public accountants and CSFCC to do the same on a voluntary basis. With a "competitively aggregated contracting" mechanism substantial amounts of funds can be saved and controlled.

The second and perhaps even more significant issue is the standards, codes, and rules themselves. The California Fuel Cell Partnership took the lead in implementing a civic market mechanism in 1999. In answer to the need to get industry and government to identify and solve mutual issues surrounding the California "clean" air acts, especially the zero emission laws enacted in the early 1990s, the Partnership holds semiannual meetings in Sacramento, which requires paid memberships not only among the auto industry but also with fuel suppliers (oil and gas), research laboratories, universities, and technology companies. As noted earlier it collapsed and changed in the second decade of the 21st century.

Finally, one of the most successful collaborative efforts was advanced by the Governor Davis administration and appears to be duplicated and expanded under Governor Schwarzenegger: setting renewable energy goals. Davis set in 2002 a Renewable Energy Portfolio Standard of 20% by 2017. Schwarzenegger announced in 2003 a shortened time period for 20% renewables by 2010. Moreover, Schwarzenegger, is far more aggressive in pushing for a "hydrogen economy" in the state sooner than later. However, it failed as he was *not* willing to "invest" state funds into it (Clark, 2004a,b).

CONCLUSION: MAXIMIZING THE PUBLIC GOOD IN ENERGY ECONOMICS

Civic markets have many attributes that are different from neoclassical theories, although none are topics unfamiliar to those working in energy economics. What is different is the emphasis that is given to besides the fact that it is an emerging focus.

At the root is the fact that the objective is not to maximize sales and profits but to maximize the public good.

The public good is not assumed from economic theory to follow from firms competing to lower prices through higher productivity, but lower prices and a higher value are assumed to follow from maximizing the public good. Although evidence on this topic is increasing, we acknowledge that it is not definitively proven. However, we simply know that the old economics did not work.

California is the global laboratory for the agile energy system and the first "sustainable nation state" that is building its energy system on a new set of economic assumptions after being blacked out by deregulation built on the old assumptions. To fully develop the agile energy system, Californians must continue to put civic concerns over private profits, and build an energy infrastructure based on what is good for the public. As a "bellwether" nation state, California has the opportunity to lead the world in the new energy system.

REFERENCES

Angelides, P., November 21, 2003a. A New Era of Environmental Investment and Responsibility. Institutional Investor Summit on Climate Risk. UN Headquarters, NY, NY, pp. 1–4. unpublished.

Angelides, P., December 2003b. Acceptance Speech for "Lifetime Award for the Environment". California League of Conservation Voters, Tenth Annual Meeting, Los Angeles, CA.

Blumer, H., 1969. Symbolic Interaction - Perspective and Method. Prentice-Hall, Englewood Cliffs, NJ, p. 1986.

Borenstein, S., Bushnell, J., Wolak, F., December 2002. Measuring market inefficiencies in California's deregulated wholesale electricity market. American Economic Review 92.

California Energy Commission, February 2002. Report on Summer 01. Sacramento, CA.

California Stationary Fuel Cell Collaborative (CSFCC), 2002.

Casson, M., 1996. Economics and anthropology – reluctant partners. Human Relations 49 (9), 1151–1180.

Chomsky, N., 1968. Language and Mind. University of California Press, Berkeley.

Chomsky, N., 1980. Rules and Representations. Columbia University Press, NY.

Cicourel, A.V., 1974. Cognitive Sociology: Language and Meaning in Social Interaction. Free Press, New York. http://www.sociologyguide.com/ethnomethodology/aaron-cicourel-cognitive.php.

Clark, W.W., January 2004a. The California Hydrogen Economy. Data and Forum Summary released on CD.

Clark, W.W., 2004b. Qualitative Economics. unpublished paper.

Clark, W.W., Fast, M., 2008. Qualitative Economics. Coxmoor Press. Updated second ed. Elsevier Press, 2018.

Clark, W.W., Li, X., 2004. Social capitalism: transfer of technology for developing nations. International Journal of Technology Transfer and Commercialization 3 (1). Inderscience, London, UK.

Clark II, W.W., Lund, H., December 2001. Civic markets in the California energy crisis. International Journal of Global Energy Issues 16 (4), 328–344. Inderscience, UK.

Clark II, W.W., Sowell, A., November 2002. Standard practices manual: life cycle analysis for project/program finance. International Journal of Revenue Management. London, UK: Inderscience Press.

Environmental Goals Policy Report, October 2003. Governor's Office of Planning and Research, Available through the Local Government Commission. Sacramento, CA. http://www.lgc.org.

Governor Schwarzenegger Executive Order, February 2004. California's Hydrogen Highway. Sacramento, CA.

HarvardWatch, May 2002. De-regulation and Enron. Cambridge, MA. http://www.harvardwatch.org/.

Kapner, K., McDonough, R., March 2002. Doing your homework on individual equity futures. Futures Magazine.

Kenidi, R., June 2003. Personal Communications.

Laureates, N., 2001. The Sveriges Riksbank Prize in Economic Sciences. www.nobelprize.org/nobel_prizes/economic-sciences/laureates/2001/press.html.

Perkins, L.J., 1996. What is physics and why is it a "science"?. In: Lecture Presented at Graduate Physics Seminar, UC Berkeley. University of California.

Saxenian, A., 1994. Regional Advantage: Culture and Competition in Silicon Valley and Route 128. Harvard University Press, Cambridge, MA.

Schultz, D., 2001. Standard Practices Manual (SPM).

Teece, D.J., Spring 1998. Capturing value from knowledge assets: the new economy, markets for know-how, and the intangible assets. California Management Review 40 (3), 55–79.

Woo, W.T., Sachs, J.D., 2001. Understanding Chi economic performance. The Journal of Policy Reform 4 (1), 1–50.

Ziman, R., December 2003. Acceptance Speech for "Sustainable Business Award". California League of Conservation Voters, Tenth Annual Meeting, Los Angeles, CA.

FURTHER READING

Aubrecht, G., 1994. Energy. In: Supplies of Energy and Related Materials Form the Basis of the World Economy, second ed., p. xiii Englewood Cliffs, NJ.

Banks, F., October 2003. Economic theory and update on electricity deregulation failure in Sweden. In: Unpublished Paper Presented at Arne Ryde Memorial Conference on Nordic Electric Markets. Lund University, Sweden.

Blinder, A.S., Baumol, W., 1979. Economics: Principles and Policy, first ed. Harcourt Brace (Dryden Press). thirteen Edition in 2015.

Bollman, N., 2002. Regionalism: Public Policy Perspective for the State of California. Public Policy Institute, San Francisco, CA.

California Air Resources Board (CARB), September 2001. Secondary Benefits of Zero Emission Vehicle Rules. Sacramento, CA.

California Sustainable Building Task Force, February 2002. Building Better Buildings: A Blueprint for Sustainable State Facilities (Blueprint). Governor in Executive Order D-16–00. http://www.ciwmb.ca.gov/GreenBuilding/TaskForce.

CalPirg, June 2002. Renewables Work: Job Growth from Renewable Energy Development in California. Sacramento, CA.

CARB, 2000. Secondary Benefits.

Chomsky, N., 1957. Syntactic Structures. Mouton & Co., The Hague.

Chomsky, N., 1975. Reflections on Language. Pantheon Books, NY.

Clark II, W.W., 2002a. Greening technology. International Journal of Environmental Innovation Management. Inderscience, London, UK.

Clark II, W.W., 2002b. Entrepreneurship in the commercialization of environmentally sound technologies: the American experience in developing nations. In: Kuada, J. (Ed.), Culture and Technological Transformation in the South: Transfer or Local Development. Samfundslitteratur Press, Copenhagen, Denmark.

Clark II, W.W., 2003. California's Next Economy. Governor's Office Planning & Research.

Clark II, W., Jensen, J.D., 1997. Economic Models: The Role of Government in Business Development for the Reconversion of the American Economy.

Clark II, W.W., Paolucci, E., July 1997. An international model for technology commercialization. Journal of Technology Transfer. Washington, D.C.

Clark II, W.W., Paulocci, E., 2001. Commercial development of environmental technologies for the automotive industry: toward a new model of technology innovation. International Journal of Environmental Technology and Management 1 (4), 363–383.

Clark, W.W., Kune, A., Feinberg, T., Kaplan, A., October 2003. Sustainable Environmentally Sound Technologies (EST); the role of public policy. Unpublished. California Labor and Workforce Development Agency.

Coase, R.H., November 1937. The nature of the firm. Economica 4, 386–405.

Coase, R.H., 1988. The Firm, the Market, and the Law, Chicago. University of Chicago Press.

Coase, R.H., 1993. The nature of the firm: influence. In: Williamson, O.E., Sidney, G.W. (Eds.), The Nature of the Firm: Origins, Evolution, and Development. Oxford University Press.

Davis, G., January 31, 2002. Letter to FERC Chairman Wood.

Demirag, I., Clark II, W.W., Bline, D., 1998. Financial markets, corporate governance and management of research and development: reflections on US managers' perspectives. In: Demirag, I. (Ed.), Corporate Governance, Accountability and Pressures to Perform: An International Study. Oxford University Press, Oxford, UK.

Fitzgerald, G., Bolinger, M., Wiser, R., September 2003. Green Buildings: The Expanding Role of State Clean Energy Funds. Lawrence Berkeley National Laboratory, US Department of Energy, pp. 1–11.

Hawken, P., Lovins, A., Hunter Lovins, L., 2000. Natural Capitalism: The Next Industrial Revolution. Little, Brown and Company, New York, NY. www.environmentandsociety.org/.../hawken-paul-hunter-lovins-and-amory-lovins-na.

Isherwood, W., Ray Smith, J., Aceves, S., Berry, G., Clark, W.W., Johnson, R., Das, D., Goering, D., Seifert, R., Fall 2000. January 1998. Published in Energy Policy. Remote Power Systems with Advanced Storage Technologies for Alaskan Village, vol. 24. University of Calif., Lawrence Livermore National Laboratory, UCRL-ID-129289, pp. 1005–1020.

Jawetz, P., et al., 2003. Renewable Energy Technology Diffusion. Final Report, pp. 1–32. San Francisco, CA.

Kammen, D.J., September 2006. The rise of renewable energy. Clean Power, Special Issue Scientific American 295 (3), 84–93. www.sciam.com.

Marshal, A., Guillebaud, C.W., 1961. Principles of Economics, vol. 2. Macmillan for the Royal Economic Society. https://www.questia.com/library/economics-and-business/economists/alfred-marshall.

Porter, M.E., 1980. Competitive Strategy: Techniques for Analyzing Industries and Competitors. Free Press, New York.

Porter, M.E., 1990. The Competitive Advantage of Nations. Free Press, NY, NY.

Solar Catalyst Group, Clean Edge Inc., 2003. Solar Opportunity Assessment Report, pp. 1–67. Berkeley, CA.

Solow, R.M., 1970. Growth Theory – An Exposition. Oxford University Press, ISBN 978-0195012958 second edition 2006.

Sorensen, G., 1983. Democracy and Democratization. Westview Press.
Southern California Edison, 2003a. Selecting the Right Technology for Your Business: A Small Business Guide. Economic and Development, Rosemead, CA. www.sce.com/ebd.
Southern California Edison, 2003b. Financial Resources Manual: A Guide to Capital Access and Financial Information. Economic and Development, Rosemead, CA. www.sce.com/ebd.
University of California, 2016. Energy and Environment. School of Law, Berkeley. https://www.law.berkeley.edu/research/clee/research/climate/climate-change-and-busines.
Vesper, K.H. (Ed.), 1990. New Venture Strategies. Prentice Hall, Englewood Cliffs, NY.
Williamson, O.E., 1996. The Mechanisms of Governance. Oxford University Press.
Williamson, O.E., Sidney, G.W., 1993. The Nature of the Firm: Origins, Evolution, and Development. Oxford University Press.
Wiser, R., Bollinger, 2003. M. Lawrence Berkeley National Laboratory, Berkeley, CA.

Chapter 12

Urban Circular Economy: The New Frontier for European Cities' Sustainable Development

Danilo Bonato[1], Raimondo Orsini[2]
[1]*Remedia Consortium, Milano, Italy;* [2]*Sustainable Development Foundation, United Kingdom*

Chapter Outline

THE URBAN CIRCULAR ECONOMY IN EUROPE

Circular economy is our best alternative to the traditional make, use, and dispose linear economy model. Implementing circular economy's models, we do our best to keep resources in use for as long as possible, extract the maximum value from them while in use, and then recover and regenerate products and materials at the end of each service life. Circular economy is about our everyday life, it concerns our role as citizens and consumers, and it will probably shape our future.

Circular economy is about people, and people live in the cities; in 2050, 75% of the global population will reside in cities. Cities are the engine of the

Sustainable Cities and Communities Design Handbook. https://doi.org/10.1016/B978-0-12-813964-6.00012-4

economic growth, as about 85% of the global gross domestic product is generated by cities. Such rapid growth puts an enormous pressure on urban resources, carrying capacities, and quality of life. The reason for the future success of the circular economy in European cities stands in its higher ability to face the challenges and risks brought about by the traditional model: economic losses and structural waste, price risks, supply risks, and natural systems degradation.

Cities are places where complex and interdependent challenges related to resource depletion, climate change impacts, environmental degradation, pollution, health issues, and social exclusion must be faced.

Typical paradoxes of the linear economy in European cities can be described with the following examples:

- A typical European car is parked 92% of the time;
- Almost10%–15% of building material is wasted during construction;
- About 50% of most city land is dedicated to streets and roads, and other vehicle-related infrastructure.

In Europe, the role of cities as key actors and incubators for innovative solutions that tackle these challenges has been acknowledged in a new "Urban Agenda," and it is recognized that a systemic and cross-sectorial *urban ecosystem* approach is needed.

Luckily enough, in Europe we are experiencing an overwhelming response by major European cities to several Horizon 2020 initiatives, the mainstream innovation structural program of the European Union. Through cross-cutting activities in the *Smart and Sustainable Cities* focus area, Europe will continue to invest in actions to enhance the innovation capacity of cities to act as hubs of innovation in designing and implementing their transition pathways toward resilient, sustainable, low-carbon, resource-efficient, and inclusive cities with a reduced environmental footprint.

Complementary research and innovation programs were also activated to strengthen the cooperation between Europe and other major economies, such as the United States, China, and India, on sustainable urbanization and to support the overall innovation capacity of the cities, assisting them in designing and implementing transition pathways toward environmental, social, economic, institutional, and cultural sustainability and resilience, within a systemic circular economy framework.

A key role will be played by the so-called *nature-based solutions*, which under the current EU research and innovation policy framework are defined as: "living solutions inspired and supported by nature that simultaneously provide environmental, social and economic benefits and help to build resilience." The challenge is to bring more nature and natural features and processes into cities, landscapes, and seascapes, through locally adapted, resource-efficient, and systemic interventions.

Cities are successfully embracing the circular economy and are looking on how to capture and repurpose key resources. Cities represent a fundamental catalytic power to drive resource efficiency and circular economy forward, and make them among the greatest beneficiaries of such a transition. Clever European city leaders are systematically reengineering the current urban systems, to explore new ways of value creation and optimization, while ensuring long-term prosperity, resource sufficiency, economic viability, and well-being in urban centers.

As urban populations continue to rise, smart leaders must find ways to cope with acute demands for resources and space, and the mainstream strategy to do this is the *urban circular economy.*

Applying the urban circular economy principles, planners and policy makers are rethinking the way our current urban systems operate but for this they need to learn from previous mistakes and implement these lessons to enable long-term resilience, resource optimization, economic prosperity, and human well-being.

From roads made out of waste plastic to bricks made from old construction waste, many cities are developing projects that involve local administrations, businesses, and citizens in a completely new way of cooperating and sharing value.

In this context, built-up urban areas create perfect conditions for the development of *peer-to-peer business models*. In Europe, for instance, where almost 90% of households own a car, car sharing has been shown to help reduce traffic without taking away people's freedom to drive. It also helps cut the cost of maintaining and parking cars. The question is how to go beyond individual examples of circular innovation to build a citywide circular system. In this respect, business will have a major role in developing circular cities, where peer-to-peer lending is becoming an exciting reality.

THE BIG CITIES EXAMPLES: AMSTERDAM, LONDON, PARIS, AND MILAN ON THE LEADING EDGE OF THE URBAN CIRCULAR ECONOMY

Big cities can really make the difference, showing how to implement the required changes. The city of Amsterdam in the Netherlands clearly understood that urban circular economy can help to preserve and enhance natural capital, optimize resource usage, achieve resilience through diversity, and design out waste. Along with its systems change approach, urban circular economy has the potential to provide the necessary framework for building a resilient and prosperous community, which in turn provides a supportive operating environment for businesses.

With a dedicated research center established to explore how circular economy principles can be leveraged for closing the resource loops on a municipal level and by that enhancing the local economy, Amsterdam ranks among the most prominent front-runners with everything related to urban circularity. The developed solutions are being translated into pilot projects, including the living laboratory for the circular city in Buiksloterham, a testing ground for the optimization of material flows and prevention of CO_2 emissions and the investigation of what can be achieved through the introduction of more effective regulations for new-build homes. Neighboring Haarlemmermeer, home to the Netherlands' Schiphol Airport and the surrounding trade park, is another case where ambitious resource management is underway. Projects include Park 20/20, a full-service, cradle-to-cradle working environment, and the Schiphol Trade Park, which aims to become a center for regenerative activities.

London was one of the first large European cities to launch a roadmap for Circular economy transformation. London's Waste and Recycling Board estimates that the transition to a more effective economy could be worth $10 billion annually to the city's economy. The Greater London Authority seeks to achieve this through looping resource flows, while also creating over 12,000 net new jobs through the reuse, remanufacturing, and maintenance industries. The governing bodies of London, Amsterdam, and Copenhagen announced a collaboration focused on designing plastic packaging out of their respective waste streams. The launch came shortly after the release of a report by the Ellen MacArthur Foundation, which estimated that plastic packaging waste amounts up to $120 billion annually. International initiatives like these demonstrate that policy makers, at least in Europe, have identified significant economic opportunities in circular models. For the London metropolitan area, embedding circular economy principles into the urban environment can open up new opportunities for symbiosis between cities and businesses, potentially paving the way for greater resilience and prosperity for the growing urban population around the world.

Paris is also actively pioneering the incorporation of circularity within its urban ecosystem. The French capital launched the city's white paper on circular economy, which provides an overview of the main challenges in the field of resources, economy, environment, and society. The white paper presents a comprehensive strategy on how to address these through 65 circular economy–based initiatives, covering areas such as education and public awareness building, public procurement, as well as fiscal and regulatory measures. The Greater Paris Metropolis (GPM), with a population of nearly 7 million, exercises four major authorities: metropolitan area planning; local housing policy; economic, social, and cultural development and planning; and air and environmental protection. The GPM is also responsible for harmonizing electricity, gas, as well as heating and cooling distribution networks. Thanks to its planning and operational authorities, the GPM is able to promote the circular economy on a wide scale. For instance, the development of resource stores and recycling centers contributes to the creation of regional ecosystems, which promote human beings and their potential and enable the emergence of a new economy that respects the environment, through local hires and community channels. As far as energy is concerned, In the Marne-la-Vallée region, the heat produced by data centers is recovered for the Paris Val d'Europe business park network: water at a temperature of 48°C is channeled directly from the exchanger outlet to the municipal heating pipes of the business park. Ready to use, it is particularly virtuous from an economic and environmental point of view. The city of Paris is asking the private sector to invest itself to accelerate the momentum surrounding urban regeneration and the adaptation to new economic, social, and environmental challenges. Each new project has to demonstrate its contribution to a sustainable and intelligent city through its design, its technical specificities, its planning, and its place in the immediate and metropolitan environment. This is expressed via the use of new circular economy technologies, as well as the ability to adapt and accompany, and even generate, new lifestyles.

Milan, the Italian economic capital, in spite of being part of a country with the highest rate of private cars per inhabitants in Europe (700 cars per 1000 people), has become in just a few years a European benchmark of sustainable transport and shared mobility, a laboratory of new economical and social models. How? Very simply: the city leaders have decided to shift from a mobility system based on private vehicles to a newer one, focused on accessibility, investing, and sustainable planning.

In a few years the number of shared mobility users and the number of nonprivate vehicles and operators has increased steadily, leading to citizens' higher appreciation of more comfortable, practical, cheaper, and cleaner services.

The enabling conditions to reach this result have been the abandoning of subsidies to private vehicles, improvement of public transport services, new funding to bike and pedestrian infrastructure, and, in parallel, supporting the new model of sharing mobility systems.

The main result of this innovative city policies has been in 2016 a reduction of private vehicles use, reaching only 30% share of private vehicles trip in Milan, against 65% in Rome, and achieving the following:

- A car sharing explosion: 4% of vehicles from car sharing schemes, 10,000 car sharing rentals per day in the city;
- The bike sharing success: 5000 shared bicycles and 270 stations operating in the city;
- A car pooling growth: first city in Italy with urban car pooling use;
- A low-emission zone: 136 km^2 in which only clean vehicles may enter without paying toll.

Milan has shown that mobility cannot be treated separately from urban planning. A major reason for Milan's success is the idea of investing on the concept of urban regeneration and maintain a "compact city," the circular model for building and architecture, and trying to limit urban sprawling. Milan is ranked high among European cities with regard to the three indicators of a compact city: LCPI (Largest Class Patch Index), ED (Edge Density), and ID (Dispersion Index), and it is the Italian city with the highest investments on urban projects of green architecture.

WHERE TO FIND THE RESOURCES TO INVEST IN URBAN CIRCULAR ECONOMY?

It is true that Europe is facing a difficult economic period: most EU members are trying to lower the public debt rates and find technical and legislative barriers in increasing public investments. There is a renovated and serious attention on reducing public spending, and new limits have been set by EU legislation to member states.

On the other hand, it is generally accepted that urban investments and radical changes toward a greener and circular economy in cities depend mainly on public funding.

How to Overcome This Contradiction?

If we consider the case of shared mobility business model as example, we can realize that public investment is not the only way of making circular economy real in European cities. The shift to a new model may also include a shift of the financial paradigm, giving a stronger role to the "public–private partnerships" (PPP), a historical reality in the United States that has big potentialities for a rapid growth in Europe also.

A new car sharing scheme, which can take away from the city streets thousands of parked private cars, does not need any subsidy or fund from the city government. It creates a sustainable business in itself. Furthermore, it can take fresh money to the city, through the payment of parking spaces, access to limited traffic zones, etc. Also, the new bike sharing systems do not need direct funding either. They are provided generally by advertising companies, who pay the investments and the costs for the service, in return of the possibility of having free access to advertising spaces in the city.

One of the main advantages of circular economy is that private companies and association of companies can become the main driver of success in urban regeneration, not only through designing and selling more sustainable services and products but also by creating new systems of financial cooperation with city administrators and city leaders.

As a consequence, many municipalities are starting new models of PPP in Europe. Let us also take a look at the field of renewable energies and energy efficiency: new partnerships are taking a great variety of forms such as Energy Service Companies (ESCos) and Multiutility Service Companies (MUSCos), in many cases formed by public and private partners who team up to finance, build, and manage energy and utility services in urban areas.

With a very interesting project of urban regeneration, using an estimated budget of £1.5 billion, London's Southwark Council has redeveloped 70 acres of city property in its Elephant and Castle and Aylesbury neighborhoods. By creating a decentralized energy system that will bring heat, hot water, communications, and infrastructure to 9700 residential units and 38,000 m^2 of commercial space, the council aims to create a mixed-use site that is at the "forefront of sustainability." The regeneration is about more than new homes, community facilities, and improved open spaces. The partnership is committed to ensuring that local people experience the social and economic benefits of regeneration, such as employment, education and training, and improvements in health and well-being. Construction of the new homes has started in 2016, and the entire regeneration project is expected to finish in 2032. To finance plans for the massive overhaul of the area's energy system, the council contracted an MUSCo to "plan, design, contract, finance and operate the plan and infrastructure required to deliver low carbon energy." Composed of three partners, including Thameswey Energy Ltd, the London ESCo Veolia Water Outsourcing Ltd. & BskyB, as well as the private company Dalkia & Three Valleys Water, and partially supported by the Clinton Climate Initiative, the MUSCo is expected to absorb 100% of the commercial risk to the council while providing a long-term vehicle for private investment in the area.

The lesson we can take from examples like these is that great benefits to European circular economy in cities can come from PPP, particularly

- Creating vehicles for long-term investment;
- Reducing public expenditure and raising additional capital;
- Encouraging local and bottom-up participation;
- Enhancing technological innovation and incorporate expertise;
- Promoting knowledge sharing;
- Handing over commercial risk.

An interesting exercise of PPP is the interest in city regeneration and investments by big private companies, as quality of life and circular economy business models are now in the top of the big groups' list of priorities. In this way, a private actor can support cities in their choices of circular economy models, offering indicators first, then advising on possible solutions, and lastly offering financial partnership and business cooperation.

Siemens, one of the leading actors in European urban regeneration, engaged in several levels of research on cities to start its involvement in PPP. A "Green City Index" was first developed by Siemens in collaboration with the Economist Intelligence Unit. The Index analyzes more than 130 cities for best practices, ranks their environmental performance over several infrastructure areas, and shares the results with the public to contribute to the debate on sustainable cities.

As a second step, Siemens built in London "The Crystal," a big exhibition on the future of cities and a state-of-the-art meeting place for city decision makers and influencers, urban planners, architects, and public infrastructure operators, being also a model for sustainable buildings. It generates its own energy through solar power and heat pumps. A highly efficient building management system optimally distributes electrical power, as well as heating and cooling. Rainwater is harvested for use, and all waste water is recycled.

Then it went through concrete business actions of partnership with cities. Berlin had annual energy costs of €17.2 million for its 185 public buildings. Siemens implemented an energy management system for heat generation and distribution, and for water control, monitoring, and maintenance, helping the city fund this large investment through sales of receivables, a framework agreement with attractive terms and conditions, and accounting and tax clarification. In all, Berlin invested €28.5 million in this project, with contract terms valid for 9–12 years. Berlin now saves €5.3 million per year in energy costs and has reduced its CO_2 emissions by 25% per year. This project served as a model for how PPP can help municipalities achieve their economic and environmental goals.

THE ROLE OF SMALL CITIES: THE ITALIAN EXPERIENCE

Does urban circular economy only work for big cities? We believe that you do not need to be big as Paris, London, or any other large capital to benefit from what circular models can do for your people and the environment in which they live.

Many midsize cities in Italy, having from a few thousand up to 300,000 inhabitants, are really enjoying the benefit of urban circular economy approaches. Urban centers such as Belluno, Bolzano, Ferrara, Parma, and Treviso have a rich and historical cultural heritage, and their city centers are a protected United Nations Educational, Scientific and Cultural Organization site. They are all nice examples of urban circular economy, with a strong focus on *quality tourism*. The main investments are on the sustainable mobility, where ecovehicles and cycle routes represent a key resource. The use of renewable energy is widely applied, and there are systems for monitoring air pollution and waste prevention. Local communities are deeply involved in awareness-raising initiatives, actions, and programs concerning recycling and reuse, exploitation of green spaces, energy efficiency, and food waste prevention.

However, Italy can also provide good examples of how circular economy and sustainable policies can improve the quality of life even in smaller towns. Let us take the municipality of Varese Ligure located in the Liguria region of northwest Italy, a very small town with less than 5000 habitants. Transformed from a degrading municipality in the 1980s to a thriving location today, Varese Ligure is fully powered by renewable energy. In the 1980s the municipality was in trouble, due to lack of jobs, no industry, decaying properties, and the absence of essential services. However, the local administrators did not give up, as they focused on turning the municipality's weaknesses into strengths. The valley had clean air and unspoiled land: why not become a sustainable tourist destination by introducing renewable energy and implementing organic farming? The citizens were given an offer—if the public authorities funded an upgrade of the roads, sewers, aqueducts, and street lighting, would the citizens be willing to repair and renovate their ancient stone houses?

This first step was successful, and although some of the funding for renovations came from the European Union, the majority of it came from the citizens themselves. Like most urban planning, the plan relied on citizen buy-in, which occurred once the citizens accepted the vision between themselves. The Environmental Education Center was created to educate the local young generation about organic agriculture, renewable energy, and sustainability. Organic farming practices were taught to the local farmers. And this involved encouraging the farmers to stop using chemical fertilizers. As the farmers learnt that organic products could be sold at a higher price, and they could get help with EU grants for organic farms, they became interested in certifying the farms as organic. People returned to the municipality to be involved in organic farming, beekeeping, and cheese-making. Today there are 108 organic farms including organic porcini mushrooms, chestnuts, cheeses, honey, fruit, vegetables, meat, and dairy products. Good restaurants opened. Artists and craftspeople arrived. The Vara valley is now known as the "Organic Valley," and in 1999 it became Europe's first valley to be certified for environmental management under ISO 14001. To supply the municipality with renewable electricity, four wind turbines were installed on a ridge 1100 m above sea level where the average annual wind speed is 7.2 m per second. These turbines generate 8 million kWh of electricity per year, three times the amount needed by the municipality, which is fed into the local grid. The electricity produced from the wind turbines reduces carbon emissions by 8000 tons per annum. The wastewater treatment plant has a 4-kW photovoltaic system, and the municipality pool is heated by solar power as well. The shift to renewable energy has added 140 jobs, stabilized the population, and added an additional $500,000 in annual tax revenues for the municipality.

Although small and medium municipalities may seem not so relevant in terms of the number of people involved in the circular transformation, they can still provide, even on a limited scale, very interesting examples on how to address a circular economy approach in large urban areas. We can clearly see that with a combination of vision, strategy, technology, capital investment, and most importantly, political and social will, these transformations are possible.

REFERENCES AND SOURCES

- Closing the loop: an EU action plan for the circular economy, European Commission – 2015.
- Urban biocycles, Ellen MacArthur Foundation – 2017.

- The circular economy: a wealth of flows, Ellen MacArthur Foundation – 2015.
- Understanding smart cities: an integrative framework, Hafedh Chourabi et al. – 2012.
- The EU moves towards the circular economy, W. Clark – D. Bonato – Huffington Post 2016.
- The Sharing mobility in Italy, Sustainable Development Foundation – 2016.
- The city of the future manifesto, Sustainable Development Foundation – 2017.
- Smart cities implementation plan, European Innovation Partnership Smart cities and communities – 2015.
- The new plastics economy: catalyzing actions, Ellen MacArthur Foundation – 2016.
- Smart London Plan, City of London – 2016.
- White paper on the circular economy of Greater Paris, City of Paris – 2015.
- Developing a roadmap for the first circular city: Amsterdam, Circle Economy - 2016.
- Website www.smart-cities.eu.
- Website www.sustainablecities.eu.

CHART 1: CIRCULAR ECONOMY AND ITS IMPACTS ON CITIZENS' DAY-TO-DAY LIFE

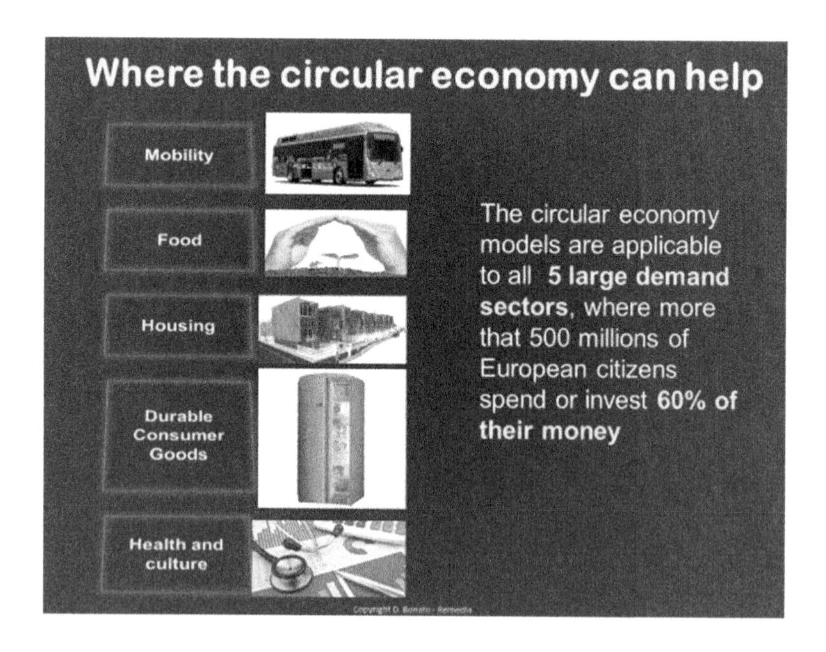

CHART 2: THE EUROPEAN UNION CIRCULAR ECONOMY ACTION PLAN AND THE ROLE OF THE CITIES

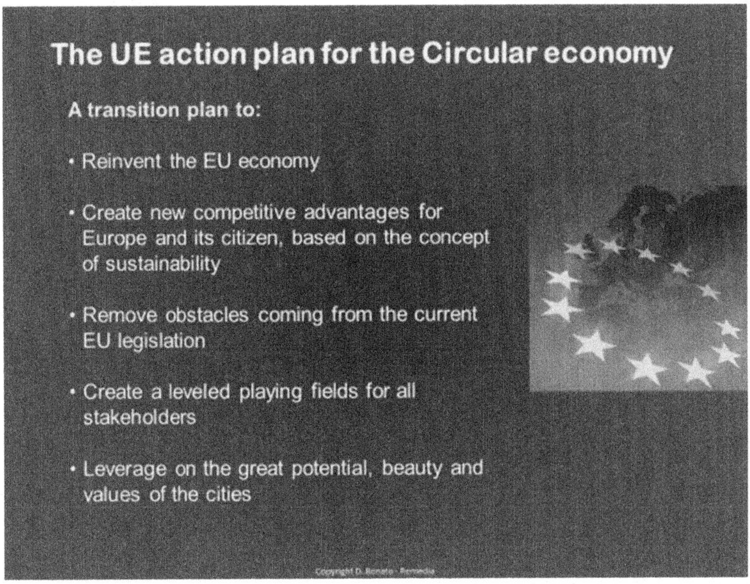

The UE action plan for the Circular economy

A transition plan to:

- Reinvent the EU economy
- Create new competitive advantages for Europe and its citizen, based on the concept of sustainability
- Remove obstacles coming from the current EU legislation
- Create a leveled playing fields for all stakeholders
- Leverage on the great potential, beauty and values of the cities

Chapter 13

Big Heart Intelligence in Healthy Workplaces and Sustainable Communities*

Julian Gresser
Alliances for Discovery, Santa Barbara, CA, United States

Chapter Outline

In the early days of the industrial revolution Charles Dickens captured perhaps better than any author the brutal, uncompromising, and fearful face of the unsustainable workplace:

> *It was a town of red brick, or of brick that would have been red if the smoke and ashes had allowed it; but, as matters stood it was a town of unnatural red and black like the painted face of a savage. It was a town of machinery and tall chimneys, out of which interminable serpents of smoke trailed themselves for ever and ever, and never got uncoiled. It had a black canal in it, and a river that ran purple with ill-smelling dye, and vast piles of buildings full of windows where there was a rattling and a trembling all day long, and where the piston of the steam-engine worked monotonously up and down, like the head of an elephant in a state of melancholy madness.*
>
> Dickens (1854)

* To see the illustrated online version of this chapter please see http://alliancesfordiscovery.net/bhi/bhi-in-sustainability/.

Sustainable Cities and Communities Design Handbook. https://doi.org/10.1016/B978-0-12-813964-6.00013-6

LICENSED FOR PRINT FROM ALAMI

He described the drudgery of the despairing souls who lived and worked in such places. The language is poetic and in the dialect of a worker, a family man of that time and place, but its poignant echo reverberates across two centuries to the feelings of those who are burned out today.

> *'Sir, I canna, wi' my little learning an' my common way, tell the genelman what will better aw this - though some working men o' this town could, above my powers - but I can tell him what I know will never do 't. The strong hand will never do 't. Vict'ry and triumph will never do 't. Agreeing fur to mak one side unnat'rally awlus and for ever right, and toother side unnat'rally awlus and for ever wrong, will never, never do 't. Nor yet lettin alone will never do 't. Let thousands upon thousands alone, aw leading the like lives and aw faw'en into the like muddle, and they will be as one, and yo will be as anoother, wi' a black unpassable world betwixt yo, just as long or short a time as sich-like misery can last. Not drawin nigh to fok, wi' kindness and patience an' cheery ways, that so draws nigh to one another in their monny troubles, and so cherishes one another in their distresses wi' what they need themseln - like, I humbly believe, as no people the genelman ha seen in aw his travels can beat - will never do 't till th'*

Sun turns t' ice. Most o' aw, rating 'em as so much Power, and reg'latin 'em as if they was figures in a soom, or machines: wi'out loves and likens, wi'out memories and inclinations, wi'out souls to weary and souls to hope - when aw goes quiet, draggin on wi' 'em as if they'd nowt o' th' kind, and when aw goes onquiet, reproachin 'em for their want o' sitch humanly feelins in their dealins wi' yo - this will never do 't, sir, till God's work is onmade.'

Dickens (1854)

This is what the workspace looked like then.

HOW FAR HAVE WE COME?

We have, of course, made considerable progress. In the United States and many other industrialized countries, Leadership in Energy and Environmental Design (LEED) has become an international standard of environmental consciousness backed with certification by the US Green Building Council and its counterparts abroad. A new field of "Buildingonomics" is rapidly extending the boundaries of LEED certification (Naturalleader-a). On its website, The Delos Group is advocating a new integrative "WELL Building Standard and offers its own employees a generous range of options reflecting the Standard: gym membership reimbursement, complimentary bike share programs, wearable technology devices and company-wide competitions, nutritious snacks, subsidized fitness races, organized volunteer events; guided meditation, curated wellness discounts, and educational programming and materials to increase employee awareness of personal well-being."

Complementing the development of an International WELL Building Standard is a powerful movement driven by Google, Facebook, and other social media giants toward "smart" green buildings, focused particularly on renewable energy, energy efficiency technologies, and best design, architectural, and construction practices (Usgbc). We would seem today to be a far cry from the hellholes described by Charles Dickens.

And yet burnout continues to be an unresolved problem, costing the civil society $300 billion and worsening by the year. It is estimated that 90% of burnout derives from the workplace (Wikipedia-A). To what extent can smart green buildings address this problem, and if many such projects currently do not, what might we be missing? Is it possible that by focusing exclusively on cognitive impairment, we may be overlooking something even more fundamental that Dickens so eloquently addressed—the Heart? Indeed, the relentless grind toward efficiency (perhaps even green efficiency) that crushes the soul and breaks the heart is precisely the malady Dickens sought in his life to correct. We must be cautious in our enthusiasm in elevating a new form of efficiency to replace the old, but failing to address the deeper human yearnings of the heart. This is the new frontier of "Big Heart Intelligence (BHI)".

BIG HEART INTELLIGENCE BASICS

BHI refers to a state of awareness where the visionary, ethical, spiritual, and other capacities of the Heart flourish in harmony with the power of the brain and mind. In cultivating BHI its explorers report that they experience enhanced vitality, empowerment, relaxation, joy, perception, compassion, balance, and flow. BHI is easily cultivated in a few hours of practice, and its beneficial results can be validated, replicated, and measured. A simple and enjoyable introduction is *Laughing Heart–A Field Guide to Exuberant Vitality for All Ages—10 Essential Moves* (Alliancesfordiscovery-A).

BHI is not simply about glorious design. For small hearts and narrow minds and burnout can rule in the most transcendent spaces. BHI seeks to nourish a different kind of continuing transaction among the hearts and minds of everyone who is engaged in these environments. The best green designs, of course, do not simply render spaces healthy by making them free of chemical toxins. They also seek to inspire and elevate occupants of the minds and spirits. BHI extends these principles, conceiving spaces as noted later as organic, living, and intelligent environments, and evolving with their users. Just as the information technology industry is increasingly aware of the importance of integrating hardware and software, BHI encourages an integration of form and flow in architecture and design with the goal of creating spaces that nourish vitality, joy, and happiness for everyone. Here are two examples of how BHI can be practically introduced at an early stage in the planning of such projects.

EXAMPLE 1: BIG HEART INTELLIGENCE IN THE DESIGN OF HOSPITALS AND HEALTH CARE FACILITIES

Most hospitals in the United States and other industrialized countries at least from a patient's perspective are sterile, cold, and fearful places, where joy, kindness, and cheer appear only seldom, and when they do, they appear as transitory rays of sunshine. Most hospitals are businesses for profit measured by bed count turnover and other hard metrics. They are designed as venues to treat serious illnesses or injuries, not to accelerate healing and transformation. They mirror accurately the values, economics, and infrastructure of contemporary medicine.

It is not surprising that burnout is rampant among physicians, nurses, caregivers, and supporting staff in most modern hospitals and health care facilities, and that burnout often manifests among caregivers in soaring rates of cardiovascular disease, suicide, and other most serious maladies (Ncbi). Burnout of caregivers necessarily translates into deteriorating quality of care of patients, especially the elderly and the poor, who are also most vulnerable to nosocomial infections. At the same time it is estimated that caregivers confer annually over $500 billion in services to the civil society, a significant amount of which is uncompensated. Without their contributions, it is fair to say that modern society would collapse (States).

In recent years a global movement has developed to transform hospitals and health care facilities into healing and hope. One example is Health Care Without Harm —an international coalition of visionary green hospitals (Noharm). In such hospitals, BHI will find receptive and fertile soil.

Explorers Wheel: Inspiration, Conception, Design

At the earliest stage of a project the Explorers Wheel offers a systematic way to connect wide domains of knowledge and experience (Alliancesfor discovery-B). It can help architects and designers solve problems for clients in a new and powerful way that we call BHI-inspired "intertidal" thinking. Here is how practically an architect or designer might use the Explorers Wheel.

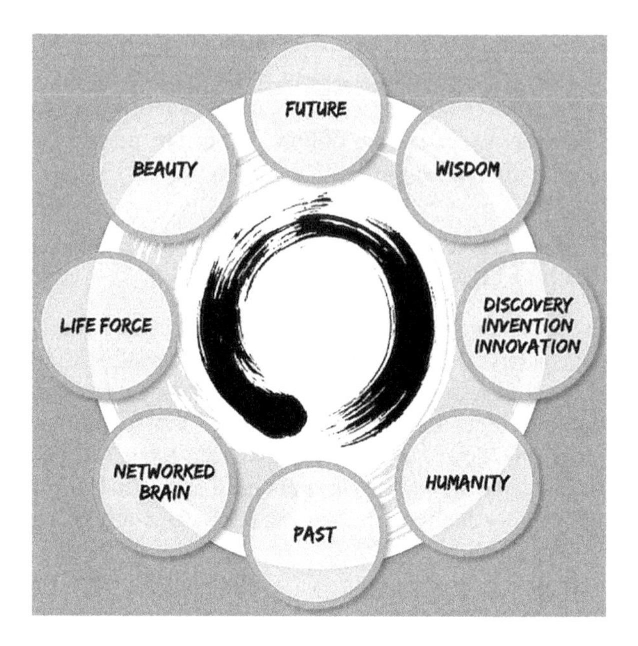

- *Enso*: Seeding the Process/Beginning Your Journey—to start the Explorers Wheel spinning all you need do is write your Core Discovery Challenge inside the *enso* at the bulls eye of the Explorers Wheel. In Japanese culture, an *enso* is rendered by masters with a single stroke of the brush. It is considered a passage way that connects the present to alternative (deeper) levels of consciousness and reality. The goal is to express an essential design challenge, defined as manifesting BHI in an optimal way to address a client's aspirations, requirements, and budget (Alliancesfordiscovery-C).
- *Let Laughing Heart Be Your Guide*: Now simply let your Laughing Heart guide you to which "moon" seems most relevant. You may feel a slight "tug" or an inner voice inviting you to come to a particular moon. You can easily capture its special resonance for you. You do not need to bear down or force the process. It is unnecessary to think your way through. The key is to listen to your Laughing Heart and let the process inspire you.
- *Finding New Connections*: The next step is to draw lines connecting other moons to your questions and to each other. In this way you can discover the "intertidal" linkages among the moons.
- *Collaborative Exploration and Discovery:* The Explorers Wheel will gain momentum, the more you "play" with colleagues, clients, family, and friends, and even perfect strangers! Actually, we are all connected in many fascinating ways we scarcely understand. Discovering this shared consciousness can be a source of enormous vitalizing power. The Explorers Wheel and Laughing Heart offer an alternative to our ordinary consensus-based versions of reality.

Oxygen for Caregivers

A great many nurses report that the hospitals they work for require them to leave their hearts at home. As it is increasingly recognized that love and immunity are closely linked (Scientificamerican), it is hardly surprising that nurses are among the first to suffer "compassion fatigue," "empathetic hyperarousal," and burnout (Private communication from Simon Fox).

Oxygen for Caregivers is one of the several innovative programs developed by Adventures in Caring Foundation (AICF) that addresses the specific challenges of burnout among health care professionals and volunteers. It has identified five precipitating factors in the psychology of burnout: (1) no control over circumstances, (2) cannot predict what will happen next, (3) facing it alone, (4) no escape, and (5) no hope of it getting better. AICF's remedy is three-pronged: (1) enhancing control by reestablishing work/life balance through a program of self-care, (2) developing foresight by training to warning signs, and (3) establishing connections through narrative and social support (Adventuresincaring).

The BHI state of heightened vitality, empowerment, perception, balance, and flow is a precise antidote to this set of conditions. When you feel truly powerful and alive, you do not care very much about controlling circumstances because you are happy to surf the wave. Similarly, without struggling with what will happen next, you are focused on what is happening now. Since you have a palpable sense of being deeply connected to the rolling world, you rarely feel alone and even if you do, you know that such feelings are impermanent and that every day brings a noble chance. There is a famous *koan* by the Chinese Zen master Yúnmén that aptly describes this state of being, "The whole world is medicine" (Wikipedia-B).

How might these principles be further enhanced by designers and architects? One interesting approach, reflective of BHI principles, might be to begin conceiving a building less as a static physical structure, the end of a story ("frozen music" is how Goethe described architecture), but rather as an invitation for new engagement, exploration, and discovery.

Big Heart Intelligence Process

If we begin to look at a building or other physical space as a dynamic process, there is a striking parallel with negotiation. Drawing upon the second definition in the Webster's dictionary, a building becomes the initiation of a continuous process of exploring and meeting of challenges as in "negotiating" a river or a mountain. BHI has a special contribution to make here because the combination of the Heart and Mind is a far more reliable Global Positioning Satellite than a system that operates solely on the brain or the mind. Moreover, a cardinal tenet of BHI is to translate the sense of vitality, empowerment, and gratefulness with continuous acts of "paying forward"; in other words, it is the act of passing on benefits received to others without asking anything in return. From an energetic perspective this prevents the *qi* or life force from stagnating and it will generate powerful eddies of generosity that compound the total vitality of the system. This is called realizing "BHI or Laughing Heart Advantage" (Alliancesfordiscovery-D).

So conceived the entire life cycle of an architectural enterprise—client engagement, preliminary design, mock-up, modeling, prototyping, blueprinting, construction, financing, community outreach, marketing—becomes a coherent, seamlessly integrated process, with BHI as its DNA. In such environments people flourish because *qi* and love are balanced and flow easily.

EXAMPLE 2: RETIREMENT COMMUNITIES LINKED TO UNIVERSITIES AND COLLEGE TOWNS

Retirement can be viewed either as the beginning of a downward spiral or an inflection point leading to the next stage life's exciting adventure. For the great majority of people this is an existential choice.

BHI was originally designed based on the personal experience of the author to assist other voyagers who have passed the landmark age of 55 years. Here is an excerpt from the *Laughing Heart Guide* specifically addressing the challenges of our later years (Alliancesfordiscovery-E).

> *Declining powers*—even as our physical powers weaken, we can harvest new reservoirs of health, energy, strength, and vitality.
> *Vulnerability*—when our heart opens, untapped inner resources become available. With the wisdom of our later years, we realize life has many colors and forms. By flowing with life, we are less vulnerable.
> *Relevance*—Laughing Heart enables us to find new and useful applications for our special gifts and experience.
> *Meaning*—By paying forward and bringing joy into the world, we discover new meaning in ordinary things.
> *Closing pathways*—Some paths close with aging but others open. The wide world beckons for exploration and discovery.
> *Listlessness and passivity*—Laughing Heart can hold and sustain us when we feel listless and blocked by life.
> *Disengagement*—When we are renewed, we have the energy to reengage.
> *Loss of loved ones, of being cared for and caring*—A Laughing Heart enables us to better bear the grief and sorrows of life and can nourish us in our darkest moments.
> *The end of the story*—Although there may be no definitive answers to our common dilemma, we do have a fierce and courageous option: to leave this world with a great roaring Laughing Heart.

It is not surprising that these same sentiments are shared by millions of retirees around the world, who constitute one of the fastest growing sectors,

The Longevity Economy (a study by the Aarp). It is also logical that retirees in search of new opportunities for lifelong learning would look to join communities of older explorers that are closely linked to universities (Bankrate). The question, again, is what is the unique added value that BHI can bring to the design and implementation of these projects?

The answer is easy. Vitality is the essence of BHI and also the transformative force in retirement. BHI is also about deeply connecting humans to one another and also to Nature. We are only just now exploring how to cleanse and rebalance the *qi* and to recharge ourselves by reestablishing these profound energetic connections with Nature (Alliancesfordiscovery-F). Moreover, the Explorers Wheel process is itself likely neuroenhancing and neuroregenerative, especially when combined with practices that reconnect us to Nature. An axiom of neuroscience is neurons that "fire together, wire together." The process of exploring in this way is therefore likely to help retired persons, especially those who are active in these communities, stave off the natural deterioration of neurological capacities.

However, there is an important additional element. Retired persons have a wealth of knowledge, expertise, and life experience that many are happy to share when the opportunity presents itself with others in the society. This corpus of insights and practical experience is a treasure waiting for retirement communities to harvest. Empowerment also has a magical dimension: the more you give away, the more powerful you become. We call this "Creating Your Own Luck". This "Move" (#9) lends a playful and entertaining dimension to what is actually a very deep and interesting process (Alliancesfordiscovery-G).

A natural next frontier is smart technologies and buildings. Might it be possible to endow "smart" buildings with a Heart?

SMART TECHNOLOGIES WITH A HEART

BHI offers the possibility for a continuously refreshed experience that can be expressed, differentiated, and evolved throughout the life span of a building or community initiative.

- *Integrating a Smart Explorers Wheel with Computer-Aided Design (CAD)*: It is relatively easy to integrate an Explorers Wheel with existing CAD systems and advanced search capability. This will create a powerful means to iterate rapidly from initial inspiration, search, rendition, feedback, assessment, and implementation.
- *BHI Platform*: The BHI Platform including a suite of apps will become increasingly intelligent, interactive, and personalized. This will enable any project, especially those designed for retirees, to transform into a forum for continuous lifelong learning.

- *Augmented Reality*: BHI can be easily integrated with advanced augmented reality technologies such as Microsoft's Hololens. Unlike conventional virtual reality technologies augmented reality expands the potential of what actually is by continuously providing viewers with a sense of "what if." As the potential for enhancing BHI exists in every project, augmented reality is a perfect match, because it can provide users with a palpable multisensorial experience of what a BHI-inspired future workplace or retirement community will look and feel like.
- *BHI Buildings and Sustainable Communities as Continuously Evolving Living Environments*: Smart technologies combined with BHI have the capacity today to enable 21st century buildings and even older structures to evolve physically along with their occupants. With an array of 21st century technologies just now coming into the market, the design, materials, electronics, and every other aspect of a structure can be organic, flexible, adaptable, modular, reprogrammable, recyclable, and reconfigurable, in ways that optimally reflect the enhanced BHI experience intended by its imagineers, namely, clients, architects, planners, and users.

EVOLVING BUILDINGS WITHIN AN EMERGING NEW PARADIGM IN GLOBAL HEALTH

Individual buildings and communities today, just like people, have a choice whether to evolve in isolation or to affiliate with any number of emerging Global Collaborative Innovation Networks (COINs). One premier forum is the World Health Innovation Summit (WHIS) based in the United Kingdom (www.worldhealthinnovationsummit.com). WHIS is ideally positioned to accelerate experimentation, testing, and differentiation of BHI design and architectural principles in these 21st century workplaces and communities. BHI provides not only a *common language* for easy communication of best solutions rapidly across the WHIS network but also a *common goal*. Although still a hypothesis, it seems highly plausible that the cascading effects in a BHI-embodied COIN will accelerate and multiply; their penetration will be deeper, wider, and richer; and the threshold inflection point will occur earlier, especially in a COIN such as WHIS that itself embodies and nourishes the creative energies of *qi* and love.

We are still early into the 21st century, and the warning signs are everywhere of imminent nuclear, biologic, and economic catastrophes. And yet, at the same time, we are poised for some of the greatest scientific and technological advances the world has ever witnessed (Michelson, 2017). Hence as we look backward over the centuries to these troubled times how happy we will be to recall the marvels we achieved by artfully designing our future, harvesting the prodigious power of our brains and also guided by the timeless wisdom of the Heart.

REFERENCES

http://www.aarp.org/content/dam/aarp/home-and-family/personal-technology/2013-10/Longevity-Economy-Generating-New-Growth-AARP.pdf.

http://www.adventuresincaring.org/resources/.

www.alliancesfordiscovery.org.

http://alliancesfordiscovery.org/guide/laughing-heart/move-6-the-explorers-wheel-connecting-laughing-heart-with-everything/.

http://alliancesfordiscovery.org/how-to-use-the-explorers-wheel-in-a-nutshell/.

http://alliancesfordiscovery.org/guide/laughing-heart/laughing-heart-advantage/.

http://alliancesfordiscovery.org/guide/laughing-heart/laughing-heart-and-your-later-years/.

See fascinating interview with Regina and Cecil Esquivel Obregon and accompanying materials. http://alliancesfordiscovery.org/guide/laughing-heart/conversation-with-regina-esquil-obregon-and-cecil-esquil-obregon/.

http://alliancesfordiscovery.org/guide/laughing-heart/move-9-creating-your-own-luck/.

http://www.bankrate.com/finance/retirement/retirement-communities-university-ties-2.aspx; http://www.campuscontinuum.com/resources.htm; http://www.bankrate.com/finance/retirement/retirement-communities-university-ties-1.aspx; http://www.bestguide-retirementcommunities.com/Collegelinkedretirementcommunities.html.

Dickens, C., 1854. Hard Times.

Michelson, J., May 10, 2017. Prepare for a new supercycle of innovation. WSJ A19.

http://naturalleader.com/thecogfxstudy/study-2/buildingomics/.

http://naturalleader.com/thecogfxstudy/study-2/higher-cognitive-function/.

https://www.ncbi.nlm.nih.gov/pmc/articles/PMC4554995/.

https://noharm-global.org/issues/global/global-green-and-healthy-hospitals.

Private communication from Simon Fox, Director of Adventures in Caring. https://www.adventuresincaring.org/.

https://www.scientificamerican.com/article/how-happiness-boosts-the-immune-system/; See also Bernie Siegel, Love, Medicine, and Miracles https://www.amazon.com/Love-Medicine-Miracles-Self-Healing-Exceptional/dp/0060919833.

http://states.aarp.org/family-caregivers-provide-522-billion-in-uncompensated-care-per-year/.

http://www.usgbc.org/articles/2016-building-energy-summit-creating-smart-green-buildings; https://urbanland.uli.org/planning-design/ulx-10-smart-green-buildings.

https://en.wikipedia.org/wiki/Occupational_burnout; http://pulmccm.org/main/2016/review-articles/almost-50-intensivists-feel-severe-burnout-report-says/.

https://en.wikipedia.org/wiki/K%C5%8Dan; http://zenosaurus.blogspot.com/2014/07/sickness-medicine-revised.html.

Chapter 14

The European Union: Nordic Countries and Germany

Tor Zipkin
Aalborg University, Denmark

Chapter Outline

GERMANY

Germany has a variety of goals to reduce the amount of its greenhouse gas (GHG) emissions in the future by transforming its energy system and usage, dubbed the Energiewende. Goals include phasing out of nuclear energy, reducing GHG by 90% from 1990 levels by 2050, and supplying 60% of renewable energy by 2050. The country has utilized a variety of methods large and small, from expanding the electrical grid to installing heat pumps, with creative and innovative solutions continuously developed. Cities such as Freiburg and Bottrop have furthermore proved themselves as examples how to go green through reducing energy demand and utilizing renewables.

Bottrop

Bottrop, Germany, is a city of approximately 117,00 people located in the "Ruhr industrial area" or "Ruhr region," Germany's largest urban agglomeration comprising multiple cities and a population of over 8 million people in total. Historically Bottrop was defined by a strong coal industry plus other

Sustainable Cities and Communities Design Handbook. https://doi.org/10.1016/B978-0-12-813964-6.00014-8

FIGURE 14.1 The Bottrop city pilot area for sustainable redevelopment (InnovationCity Ruhr, 2017).

industrial processes and enterprises. In the year 2010 Initiativkreis Ruhr GmbH, put on a competition throughout the Ruhr region to select a city that would focus on sustainability and serve as an example for all the other cities in the region how to reduce GHG emissions. Innovation City Ruhr is a private–public partnership project associated with approximately 70 businesses coming from energy, services, trade, logistics, and consulting, one such member being Siemens. The city of Bottrop was selected because of its broad base of stakeholders that could involve themselves in the project such as those from the local government and businesses. The goal of the city is to reduce its CO_2 emissions by 50% of 2010 by 2020 through a variety of sustainable measures. A company was set up, called Innovation City Management GmbH (ICM), to manage the project. The company is made up of urban planning experts, energy professionals as well as communications experts and consists of 30 people. A majority of the company is owned and supported by Initiativkreis Ruhr GmbH, with the city of Bottrop, a local energy company, an industry and public sector consultancy, and real estate company, each being 10% shareholders. The projects are limited to the inner city of Bottrop plus the surrounding districts, comprising a total of 67,000 people and representing the variety of the Ruhr region as a whole (InnovationCity Ruhr, 2017) (Fig. 14.1).

A Blueprint for Success

A master plan titled "Climate friendly urban conversion" of approximately 1300 pages was created by collecting a wide range of information from citizens and experts regarding almost 350 projects throughout Bottrop, about 200 of such having been completed by that time. The plan was supposed to serve as a blueprint for the future of the city and outlines the major areas of project focus throughout the city. It was presented to the city government, which decided it was to serve as a general guideline for future development and to assist in further development of the projects outlined within the plan. This was significant, for it attached governmental approval of and commitment to the plan.

The creation of the plan had heavily involved the citizens of Bottrop and gave multiple opportunities for said citizens to involve themselves in the planning process. For a period of almost 7 months, residents had the ability to contribute their ideas regarding future development online, which generated the collection of 100 ideas. Furthermore, five residents' workshops were held where a total of 300 ideas and suggestions were collected. These suggestions and ideas were considered in the creation of the master plan.

The blueprint categorizes projects into five categories: "Living" projects focusing on retrofitting of residential areas, "Working" projects focusing on retrofitting of companies, "Energy" projects focusing on renewable and regenerative energy, "Mobility" projects focusing on electric and sustainable mobility, and "City" projects focusing on green urban development.

The focus of living projects is to make energy conscious refurbishments throughout the Bottrop residential sector. The project area encompasses around 12,500 residential buildings, 10,000 being privately owned. Refurbishments are varied depending on the building, many being focused on reducing heating demand. Significant focus was put into creating three energy plus buildings that generate more energy than they use. This was done for a single-family home, multifamily house, and commercial building and they were meant to serve as technical examples for others to develop and refurbish buildings with sustainability in mind.

The single-family home from the 1960s, coined the innogy Future House, was refurbished completely with solar panels, improved insulation, a heat pump heating system, and LED lighting. The building now meets passive house standards and in total the heating and power consumption was reduced by 99%.

The multifamily home was conducted by the company Vivawest and refurbished also with a heat pump, and timed control for electrical appliances, heating, and lighting, BluEvolution 92 energy efficient windows, and various increases in wall-based thermal insulation. Furthermore, 90 solar panels were installed with a total of 24.30 kWp providing an annual 1B200 kW h on the building's roof, and photovoltaic (PV) wall panels were installed, each panel producing around 75 kWh (zukunftshaus, 2017) (Fig. 14.2).

The commercial building, Covestro Zukunftshaus, became the world's first commercial building renovation to achieve an energy plus standard. The building uses 100% renewable energy and produces over 7500 kWh than it

FIGURE 14.2 The Vivawest multifamily house after its sustainable refurbishment (InnovationCity Ruhr, 2017).

uses a year. This was achieved through a combination of energy efficient technologies, such as LED lighting including presence control and daylight control systems per office, glass fiber lighting technology using sunlight without additional energy supply, ceiling heating system (heating and cooling using activation of concrete ceiling core), district heating pumps and heating control devices, decentralized ventilation with 90% minimum heat recovery, energy efficient lifts with 75% energy recovery (Build up EU, 2016) plus renewable energy installations including a geothermal heat pump, 108 rooftop solar installations, and a 300W wind plant.

The working projects focused on the refurbishment of nonresidential buildings, such as the Covestro Zukunftshaus, of which there were about 2000, including buildings used for commerce, industry, services, recreation, and the public. Projects include energy-efficient refurbishments, adding batteries to existing solar systems (redox flow battery storage system), investigation of coupling industrial areas' energy systems with residential energy systems, for example, coordinating and storage of energy between the two, developing a green gas station powered via solar cells to reduce its energy demand by 50%, and installing solar power to meet the energy demands of industrial processes. The Bottrop company metal processing company Technobox installed 70 kW of solar power, producing 60,000 kWh more than the company's electricity demand (InnovationCity Ruhr, 2017).

Mobility-focused projects reduce the amount of CO_2 that is emitted due to transportation. These included expanding the range and number of electric vehicle charging stations, increasing the amount of electrical vehicles available for rental, transportation management studies meant to increase the efficiency of transportation, efficient distribution of goods and city logistics to reduce traffic, and a goal to reduce transport-based CO_2 emissions by 30%.

Energy-based projects deem to optimize energy production and do not focus on energy efficiency. Energy projects are based on low CO_2 emissions, high efficiency generation, decentralized power, and intelligent supply demand coordination. Projects include:

- Simulating load variable tariffs: this project is meant to determine how well people respond to different electricity prices throughout the day. Participants are told of the next day's electricity prices so they can run appliances accordingly. Twenty-four electricity consumers were involved and utilized special electricity measuring equipment.
- Expanding the district heating network.
- In partnership with the consulting company E.ON, a study on dual demand side management, an innovative energy storage method at the city district level utilizing the thermal storage capacity of buildings, was completed on Bottrop.
- New gas heat pump pilot projects in residential single-family homes and medium- to large-scale buildings.
- Installation and operation of 100 micro cogeneration power plants in existing buildings meant to represent the normal building stock of Germany to serve as example of how such technology can be utilized. These decentralized power plants are monitored to present their successes and continually be optimized and adjusted based on the building they are located in.

City projects relate to urban planning, use of open spaces, and water management. Projects include the planning of an energy and technology park at the site of old coal and oil plants in the Welheimer Mark quarter of the city, research projects on energy efficient urban development in partnership with research universities, the greening of roofs and facades with plants, and the usage of LED street lamps throughout the city.

The success of Bottrop can be seen in large part because of the planning and collaboration techniques taken, such as the use of a central planning authority, the ICM, and its involvement and ease of access for local citizens. Every 2 weeks, representatives from ICM, the municipality, and the private sector meet to review projects and proposals, and to discuss new ideas and overall progress. These meetings are furthermore provided with input from collaborators on the state, business, and academic levels, all of whom meet quarterly to help further Bottrop's goals and provide advice and support for Bottrop's energy transition. They are as follows:

- An interministerial governmental working group was created with representatives from state-level ministries such as the State Chancellery, the Ministry of Economy and Transport, the Ministry of Environment, and the Ministry of Innovation

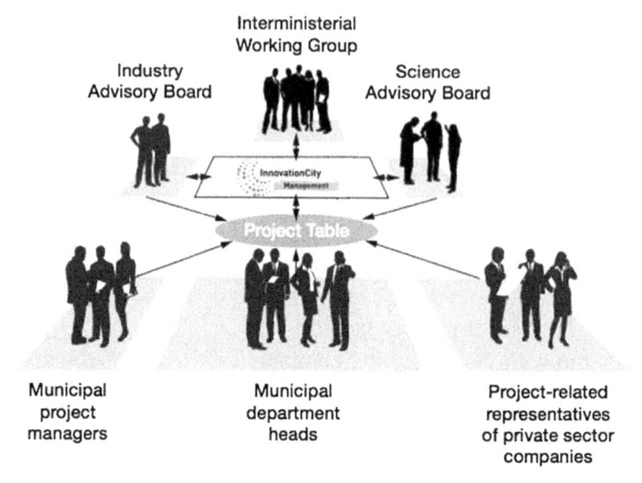

FIGURE 14.3 The different advisory boards and management representatives that has allowed for Bottrop's success (ICLEI, 2016).

- An industry advisory board (as mentioned earlier) consisting of companies in partnership with ICM with approximately 70 representatives
- A science advisory board chaired by the Wuppertal Institute for Environment, Energy, and Climate consisting of 25 members from various research centers (Fig. 14.3)

Involvement with local citizens has also contributed to the success of Bottrop. Through ICM residents and businesses are offered energy efficiency consulting services by the Centre of Information and Advice (Zentrum für Information und Beratung – ZIB) through individual building analysis. This has in turn led to almost 8% of all buildings in the target area being refurbished, compared with the national 1%. District management committees were created for each of Bottrop's seven districts within the pilot area. These committees are meant to ensure the public continues its part in Bottrop's transition and are included in the decision-making process. These committees that are made up of those within the communities they serve ensure that the Bottrop projects are not implemented on a purely top-down level and provide input as to how projects will affect their areas on the technical, social, and economic levels (ICLEI, 2014; InnovationCity Ruhr, 2017).

DENMARK

Denmark is a country with over 70 inhabited islands ranging in population from below 100 people to over 2 million (Zealand), with the goal to generate 50% of total electricity production via wind power by 2020 and to stop using fossil fuels by 2050. Owing to their close proximity with good wind resources and small population size, some islands in Denmark have aggressively reduced their carbon footprint and can serve as blueprints for other islands of similar stature.

Ærø

Ærø is located in the Baltic Sea south of the island of Fyn, is approximately 88 km^2, has a population of around 6300, and generates 55% of its energy from wind, solar, and biomass. The island has had a history of interest in renewable energy, starting in 1981 with the establishment of the Ærø Energy and Environment Office comprising 200 local residents with the goal of bringing renewable energy to the island. By 1985, 11 × 55 kW wind turbines were erected on the island, all financed and owned by a cooperative of 128 local shareholders (Aeroe Energy and Environment Office). In the early 2000s, plans to replace the old wind turbines with two wind parks of 3 × 2 MW (12 MW total) turbines were pursued, continuing the theme of local investment in local electricity production. The wind turbine investment cost was divided into shares, of which people living on the island or owning a house there were guaranteed up to 20 of, the rest being available on a first come first serve basis. For the first wind park, two organizations were created to take advantage of different tax regulations, one comprising six people owning one turbine, the other 550 people owning the other two. The second wind park was financed in a similar way, yet only secured 200 investors, with individual investors owning more shares. Banks provided loans for investment in turbine shares, with the shares many times serving as collateral. Payback time for investment in the shares was around 8 years. The wind turbines are 100% community owned through 650 people, approximately 10% of the island's population, and cover 130% of the island's electricity usage (Zipkin et al., 2015).

The island also utilizes the sun both for electricity and heat. Although 500 kW of rooftop PV capacity is installed with an annual production of 400 MWh, it is the island's use of solar heating that is impressive. The island heavily uses solar collectors to supply heat via the island's three district heating networks. The island's largest district heating network supplies heat to 1460 people and is Europe's largest solar district heating plant with 33,000 m^2 of solar collectors. When not using solar heat, the district heating systems utilize biomass resources to power combined heat and power (CHP) generators, resulting in 100% renewable district heat systems. District heating supplies 65% of the island's heating supply (Sunstore 4, 2013) (Figs. 14.4 and 14.5).

Samsø

Samsø is an island off the Danish Jutland peninsula 114 km^2 and with a 2017 population of around 3700. Starting in 1998, Samsø began planning a 10-year

FIGURE 14.4 One section of the solar collecting district heating plant on Ærø (Denmark.dk).

FIGURE 14.5 Water storage for the MARSTAL DH plant (Sunstore 4, 2013).

project to become a renewable energy island with all its electricity needs coming from sustainable sources. Between 1990 and 2000, 10 onshore wind turbines with a total capacity of 11 MW were erected, and starting in 2002, construction of 10 offshore wind turbines at 2.3 MW each was begun (these turbines were meant to compensate for the continued usage of fossil fuel–burning cars on the island and oil-based heat production and the CO_2 emissions from such). Furthermore, 60% of the island's heat demand was to be supplied via district heating and 40% via individual boilers. Those using individual boilers were encouraged and supplied information regarding biomass boilers, solar collectors, and heat pumps, with the help of local tradesmen, resulting in half of those not connected to the district heating network converting. About 70% of the heat supply on the island comes from sustainable sources. Community meetings, held once a month, were commonplace during the planning process, where information about the energy transition was shared and discussed, such as financial costs of the project and turbine visualizations (Energy Academy, 2011) (Fig. 14.6).

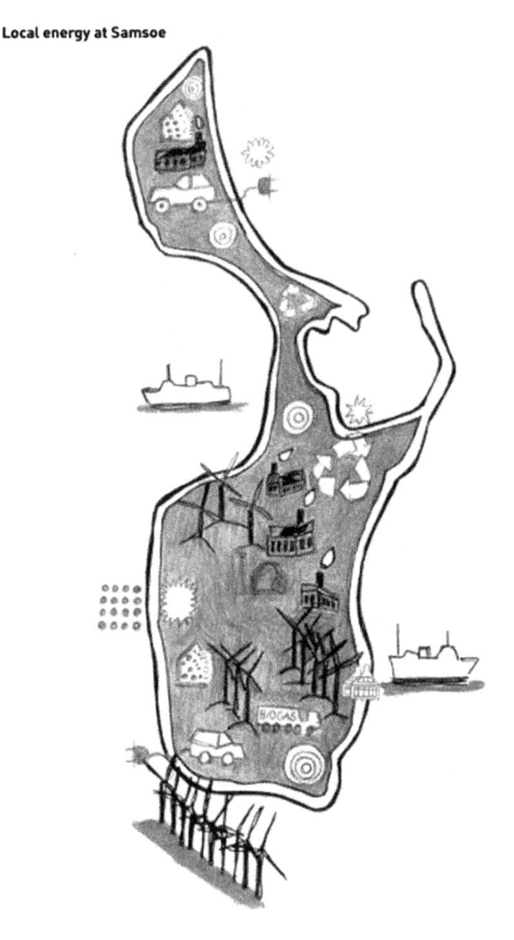

FIGURE 14.6 Depiction of the renewable energy island Samsø (Energy Institute).

As in Ærø, the ownership structure of the Samsø projects heavily includes the local population and government. The municipality owns 5 of the 10 offshore wind turbines; profits from these wind turbines are to be invested in future energy projects as per Danish regulations. As for the other five, private groups made up mainly of local farmers own three, and the other two are owned by cooperatives. One cooperative is locally managed and it consists of 450 people who own shares, and the other is national (people from throughout Denmark can buy shares), consists of 1100 people who own shares, and is managed by a professional investment foundation (Energy Academy, 2011; Zipkin et al., 2015). Local farmers own nine of the onshore turbines independently and the other two are owned by local cooperatives. The two local cooperatives for onshore turbines consist of 5400 shares that were offered to

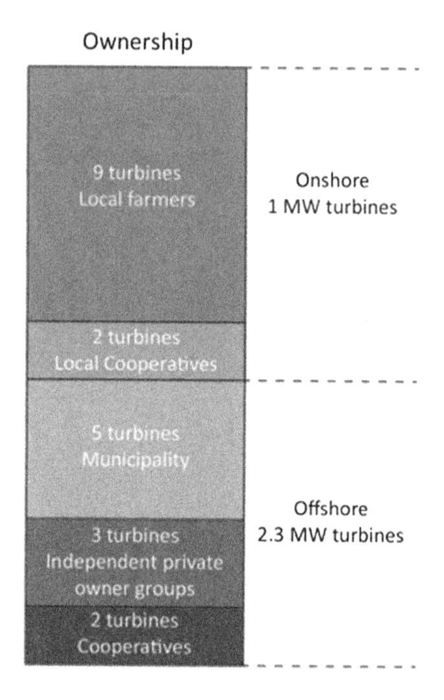

FIGURE 14.7 Ownership structure for the Smasø wind turbines (Zipkin et al., 2015).

the public over the course of 6 months. Residents had the opportunity to purchase shares in packages of 1 share, 10 shares, or 30 shares, again with local banks providing loans with the shares serving as collateral (Fig. 14.7).

Bornholm

Bornholm is the easternmost Danish island positioned closer to Sweden than to Denmark. It is 588 km^2 and has a population of 39,664, giving it a significantly larger energy demand than the islands mentioned earlier. The island is currently connected to Sweden via a 60-kV 70-MW undersea power cable, with a peak load electricity demand around 55 MW. The island hopes to become reliant on 100% renewable energy, with 36-MW wind turbines already supplying between 30% and 40% of its yearly electricity demand and at many times throughout the year wind power supplying more than the electricity demand (Madsen et al., 2012; EcoGrid EU, 2015) (Fig. 14.8).

Bornholm's singular connection to a larger grid system, desire to further rely on intermittent renewable energy in the future makes, and representation of Denmark on a smaller scale, made it a good choice to study and develop an advanced smart grid, formally named the EcoGrid EU. EcoGrid EU is Europe's largest smart grid project and is funded by the European Union. As part of the project, 1900 houses and 18 industrial/commercial electricity users

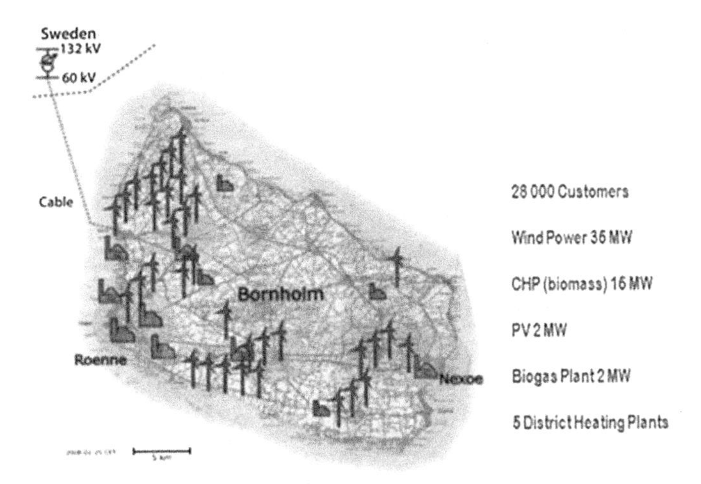

FIGURE 14.8 Bornholm energy system (EcoGrid EU, 2015).

on the island participated in flexible electricity demand response to the real-time electricity price signals using demand response appliances and smart controllers. Participants were divided into four groups, and every 5 min electricity prices were updated and the amount of electricity available was determined. Depending on the group, this information was accessed by one's own initiative, shared via a notification to one's cell phone, or based on the electricity price electric heating systems and heat pumps were automatically turned on or off. This is used to study how participants can stabilize the fluctuating nature of wind power to ensure a stable grid system while saving money on their electricity bills (EcoGrid EU, 2015). The objective of the EcoGrid was to develop and demonstrate real-time market solutions that can be used into the future as energy systems continue to fluctuate because of high penetrations of renewable energy as a way to balance grid fluctuations (Fig. 14.9).

Currently the technology used on Bornholm applied only to house-heating systems; however, in the future it will include home appliances and electric vehicles. A test house was furthermore created, a normal residential house equipped with the EcoGrid Technology, to teach those participating in the project how the EcoGrid project could affect their energy usage and to demonstrate how the system operates, and over 600 houses took part in group training sessions.

SWEDEN

Sweden is a country heavily dedicated toward sustainability, enacting a variety of measures and practices with the environment in mind, such as the heavy use

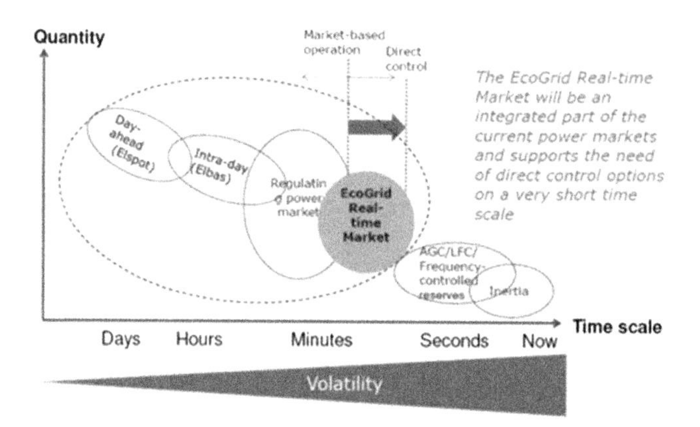

FIGURE 14.9 Electricity regulating by the EcoGrid Real Time Market (EcoGrid EU, 2015).

of waste to energy, and throughout the country seeing the development of both large and small projects devoted to the reduction of GHG emissions.

Växjö

Växjö is a Swedish city with a population of around 65,000 that has been committed to sustainability since 1996, when it decided on the goal to eliminate all fossil fuel usage by 2030, because of overwhelming local pollution at that time. In 2011 the city updated its goals for the year 2015:

- reduce the final energy usage per capita by 15%
- reduce CO_2 emissions by 55% compared with 1993 levels
- reduce electricity usage 20% per capita compared with 1993 levels
- reduce the municipality owned locations' energy use by 17% compared with 2002/2003 levels
- reduce city transport CO_2 emissions by 30% compared with 1999

The city furthermore signed the Covenant of Mayors, an agreement to go beyond the EU 20-20-20 sustainability goals and targets (20% share of renewable energy, 20% reduction of CO_2 emissions compared with 1990, 20% greater energy efficiency). Växjö therefore has committed to reduce CO_2 emissions by 65% per capita from 1993 levels by 2020 (Kommun, 2015) (Fig. 14.10).

In 2010 Växjö had a renewable energy production share of 53%, due in large part to the usage of CHP plants utilizing regional forestry waste as a biomass fuel to supply heat via the city's district heating network, with electricity from CHP usage accounting for a third of the city's electricity consumption. The Sandvik CHP Station is a 105-MW CHP plant and the city's largest provider of energy, with an energy utilization factor of 92%. It runs mainly on a variety of biomass fuels collected as waste products from the

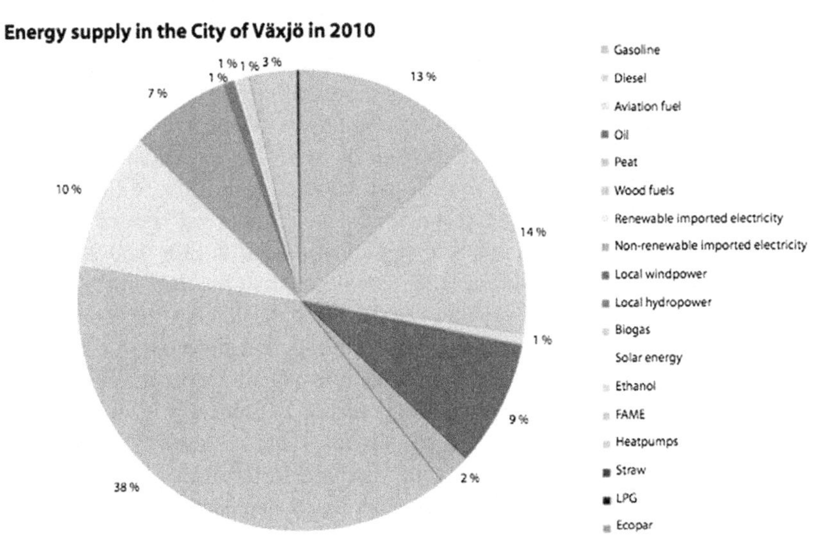

FIGURE 14.10 Energy supply in Växjö, the largest section representing biomass (Kommun, 2015).

regional forest industry such as forest residue, sawdust, woodchips, bark, and peat (Ramboll; Force Technology).

The city has a variety of future goals to continue its reduction of CO_2 emissions and increase its usage of renewable energies.

- Upgrading and increasing the capacity of existing hydro power plants within the municipality
- Expand the district heating network
- Pursue the investment in large-scale wind turbines
- Reduce the use of peat as fuel because of its contribution of GHG emissions and phase it out by 2020
- Energy efficiency standards for new municipal buildings
- Increasing the access to renewable fuels for transportation and electric vehicles.

(Kommun, 2015)

Hammarby Sjöstad (Hammarby Lake City)

Hammarby Sjöstad began as an old industrial area in Stockholm that beginning in the 1990s and throughout the 2000s was continuously transformed and developed into what is now a model of sustainable urban development that has served as an inspiration worldwide. With the original goal to develop the area in preparation for an (eventually failed) Olympic bid, environmental performance was to be "twice as good" as modern developments at the time and energy was to come from renewable and local sources, with the desire to decrease energy flowing into and out of the district as much as possible, i.e., "close the loop." The Stockholm municipality connected the Stockholm energy, water, and waste infrastructure companies to work together to help develop this sustainable district, instead of working individually in their respective fields, considered the status quo for similar projects in large cities. In 1997 an outline for the future sustainable development was agreed upon by these companies and the Stockholm municipality and is now known as the Hammarby Model; this outlines an "integrated infrastructural system" connecting energy and material flows and usages throughout the local infrastructure in the hopes of "closing the loop" while providing energy, water, and waste and sewage services for residential housing and offices. The model called for the usage of technology already in use in Stockholm, such as CHP and district heating, and newer technologies such as a local wastewater treatment plant the waste of which could be used as fertilizer or converted to biogas and PV cells and solar collectors, stressing interactions between technologies to "close the loop." (The potential of a closed infrastructural system of Hammarby Sjöstad in Stockholm, Sweden.) Although this model was not technically perfect, it proved to be essential for the project success and has been examined and used in other sustainable urban development projects, such as the Caofeidian Ecocity development in China (Iveroth et al., 2013, 2012) (Fig. 14.11).

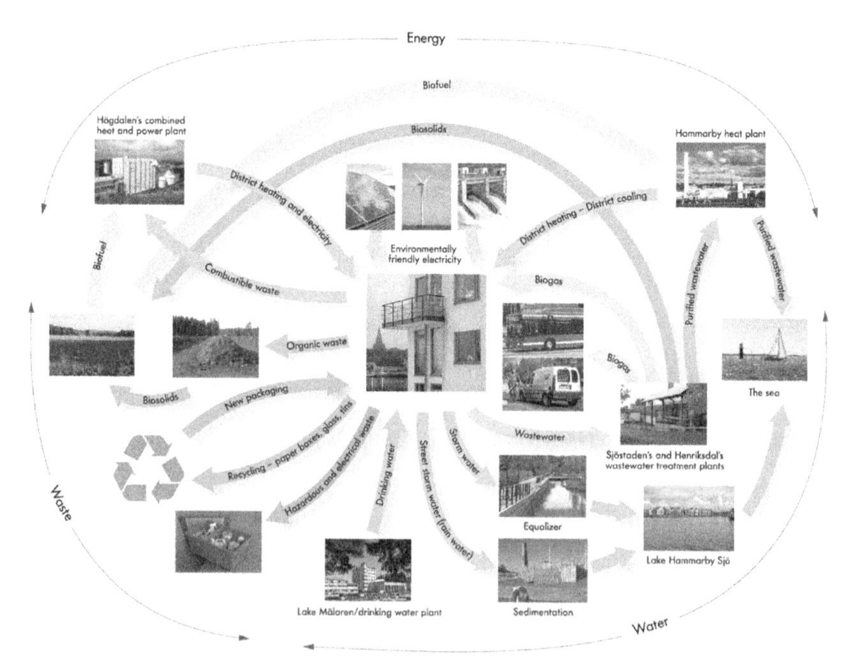

FIGURE 14.11 The Hammarby model (Hammarby Sjostad).

As of 2013, 26,000 people live in the district in the 11,000 apartments located there; construction completion is projected for 2017 with the ability to accommodate up to 35,000 residents. The district implemented a variety of transportation measures to reduce GHG emission, including a light rail system that accounts for one-third of all travel within the district (of which every residence is within 300 m of a stop), a free ferry service, a carpooling service utilizing electric vehicles (18% of households having signed up by 2010), and an extensive bike and walking path network. Based on a 2007 survey, the goal for 80% of transportation in the city to be based on public transportation was nearly reached, with 79% of all transportation being nonprivate (Jernberg et al., 2015). Hundred percent of the houses in the district are heated through a district heating network supplied by the nearby waste incinerating Högdalen CHP plant and heat pumps using treated wastewater at the Hammarby thermal power plant. Most of the heat and electricity used comes from the burning of the waste generated in the district at the CHP plant. Furthermore, the sludge from the district's wastewater treatment process is converted into enough biogas to supply the district with its gas demand. Although there were no green building goals, the average energy usage for the districts buildings is 118 kWh/m^2, lower than Stockholm's average. One building in particular that garners attention is the GlashusEtt, which serves as an environmental information center displaying sustainable infrastructure and building techniques resulting in an energy consumption 50% less than similar glass buildings. Such features include smart monitoring system, PV cells, hydrogen fuel cell, and heat pumps (Jernberg et al., 2015) (Fig. 14.12).

The district has been viewed as a success, yet criticism has been voiced that most of the waste generated comes from outside the district, for example, food waste comes from food throughout the world, counterintuitive to the "close the loop" idea. Also noted is that despite as originally envisioned, PV, solar collectors, and wind turbines on a large scale were never developed, with only a few houses having solar collectors or solar PV cells (Iveroth et al., 2013).

FIGURE 14.12 Glasshuset (Stockholm Stad).

Malmö

Malmö is Sweden's third largest city with a population over 300,000 and the goal to use 100% renewable energy by 2030. Like Stockholm it has heavily focused on sustainable urbanization and in the year 2016 won the European Union's Sustainable Urban Mobility Award. The city has multiple sustainable construction projects, having started since 1998, committed to reducing the amount of GHG emitted.

- Sustainable Hilda is one such area, a housing cooperative targeting 767 apartments and 2400 residents with the goal to become energy independent by 2020. The 2014 goals include achieving a 50% reduction of CO_2 emissions, reducing energy and water consumption 40%, generating 10,000 kWh of renewable energy, and the participation of 70% of target residents. Renovations such as façade insulation retrofitting and low-flow faucets have been utilized to increase efficiency, as well as the installation of solar panels and rain collection systems (Malmo Stad).
- The Hyllie area is a planned sustainable district to encompass 10,000 homes and have power by 100% renewable or recycled resources by 2020. The city of Malmö, water and sewage municipal authority VA SYD, and consulting company E.ON partnered in 2011 to create the most climate friendly district in the region. The area is to integrate electricity, heating, and cooling, and utilize a smart grid system to optimize energy resources, measure and influence energy usage, and use storage capacity when energy supply and demand are not balanced. Five local developers have received grants from the European Union for the BuildSmart project, meant to demonstrate residential and commercial buildings with an energy usage less than 60 kWh/m^2, less than half of Sweden's average building energy usage. Regarding transportation, buses are to be powered by fossil-free fuels, such as waste by-product biogas, and the effects of electric vehicles on the smart grid are to be studied in greater detail by E.ON (Malmo Stad).
- The Western Harbor (Västra Hamnen) is another brownfield redevelopment project like Hammarby Sjöstad in Stockholm. The development of the area began in 2001 with the Bo01 project and has since continued. Developers of the Bo01 were required to follow a green space factor guideline to ensure water permeability and a green point guideline following 35 point options such as reusing gray water in courtyards, having green roofs, and scoring above a certain number. Energy in this original Bo01 development came from 100% renewable sources, such as a 2-MW wind turbine in close by Norra Hamnen, PV cells, and heat pumps, with district heating utilizing geothermal energy and 1200 m^2 of solar collectors (in 2016 there were 3000 m^2) supplying the rest of the heating and cooling demands. Buildings were given a target energy usage, and although they did not reach this

FIGURE 14.13 Western Harbor Malmo gov source (Malmö City Planning Office, 2015).

target energy, many buildings achieved very low energy usage (http://www.collegepublishing.us/jgb/samples/JGB_V8N3_a02_Austin.pdf). Development in the Western Harbor has continued with a focus on sustainability, including the construction of 200 passive grade houses in 2012. By 2030 it is planned to inhabit 20,000 people, with a population of almost 7000 in 2013 (Malmö City Planning Office, 2015) (Fig. 14.13).

Rotterdam

Rotterdam is the second largest city in the Netherlands with a population of 633,471. As it is located on the coast of the Netherlands, Rotterdam is vulnerable to rising sea levels; however, it is the location of a large port, which in turn results in the region emitting 16% of the country's GHG emissions while contributing to only 8.5% of the country's GDP. About 88% of the city's GHG emissions come from the industry and energy-generating facilities in the port area. In 2007 the city established the Rotterdam Climate Initiative (RCI), bringing together the government, business, and public to achieve a more sustainable city with the three goals of reducing GHG emission to half of 1990 levels, designing a city that can mitigate the effects of future climate change, and achieving sustainable economic development all by 2025. Goals include:

- Provide 20% of energy by 2020 that comes from renewable sources, and by 2025 use 3 million tons of sustainably produced biomass and have 350 MW of wind turbines installed to supply energy.
- Increase the use of public transport by 40% and bicycle use by 30% by 2025.
- Encourage planting of trees, creation of green roofs, and practice of urban farming.
- Development of carbon capture storage beginning with the Rotterdam Capture and Storage Demonstration Project (ROAD) storing carbon into the North Sea (City of Rotterdam, 2017).

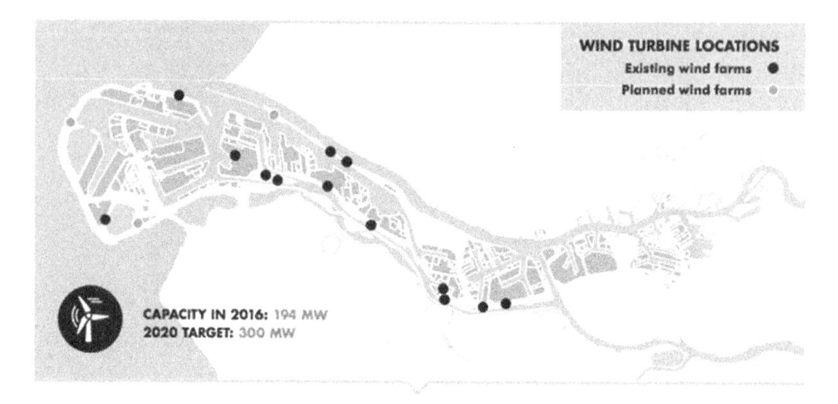

FIGURE 14.14 Wind power at the Rotterdam Port (Port of Rotterdam, 2016).

As the Port of Rotterdam is the most polluting aspect of the city, focus has been given to reducing its overall energy usage and GHG emissions. The port area was in 2016 the location of almost 200 MW of installed wind turbine capacity. Infrastructure investments are furthermore continuously made to utilize the waste heat generated by the port area industrial process for district heating and other industrial processes. A new port, Maasvlakte 2, was also created as a demonstration port, with strict sustainability requirements for companies that wish to operate there, with fully electric container terminals running on wind power (Port of Rotterdam) (Fig. 14.14).

The RCI adopted the Rotterdam Energy Approach and Planning (REAP) approach to help achieve its sustainability goals in 2009. REAP is a three-step energy initiative emphasizing energy efficiency, waste energy flows, and the use of renewable energy. Eight interactive workshops and 20 meetings were held at the beginning of its adoption by the city for politicians, civil servants, economists, the regional European Parliament, and local energy companies to explain and discuss the development plan. Examples of REAP in practice include the Hart van Zuid retrofitting project, Stadshaven redevelopment project, and also increasing of the district heating network with plans to increase connection to 155,000 homes by 2035 (Lenhart et al., 2015).

DISCUSSION

When looking at the European energy transition at a community level, a couple of methods are constantly seen, perhaps the most constant being the use of district heating. The centralized production of heat, many times via CHP power plants, centralizes the combustion fuel allowing for greater emission control versus the use of individual boilers, and allows for a greater variety of fuels to be used that have lesser carbon footprints than coal or oil. This is seen in Sweden, where biomass waste from the regional forestry industry is used, or waste is burned rather than landfilled. District heating networks also have been proved to help mitigate the fluctuating electricity supply that is associated with energy systems with large amounts of renewable energy, especially in countries such as Denmark, where district heating combined with heat storage supplies heat to a majority of Danes.

Another important theme in successful European energy transitions is the inclusion of local populations. Including the local citizens of an area in planning processes, information campaigns, and especially investments, as seen in Bottrop and throughout the Danish islands, can increase support and even be a driving force in the redevelopment of communities. Yet perhaps just as essential is the centralized actor, normally in the form of government, that sets goals and brings the multiple actors together that are needed to create and implement the blueprints for the future. In most of the cases mentioned, an overseeing organization set ambitious goals, brought stakeholders together, and managed the multiple projects all of which were needed to reduce the emissions of GHGs. In conclusion, what should be taken away is that a successful transition toward a renewable and efficient future is both a top-down and bottom-up process, one that needs support from both sides for maximum success.

Europe is considered to lead the way in the transition toward a sustainable future; the cases given here are at the forefront and serve as an inspiration throughout the world. Lessening energy demand through energy efficiency in buildings and industry like in Germany and the Netherlands, utilizing local resources for power like in Sweden and Denmark, and including local populations in the planning and execution of projects have all been proved as best practices and should be remembered and looked toward as examples when government, business, and citizens are creating a better tomorrow.

REFERENCES

Build up EU, June 30, 2016. Covestro Zukunftshaus Bottrop. Retrieved from Build Up The European Portal For Energy Efficiency In Buildings. http://www.buildup.eu/en/practices/cases/covestro-zukunftshaus-bottrop-2.

City of Rotterdam, 2017. Investing in Sustainable Growth Rotterdam Programme on Sustainability and Climate Change, pp. 2010–2014 (City of Rotterdam).

Denmark.dk., n.d. Retrieved from http://denmark.dk/en/~/media/Denmark/Images/Green%20living/Sustainable%20projects/Sunmark1MarstalDistrictHeatingPlant580.jpg?w=580&h=360&as=1.

EcoGrid EU, 2015. The Bornholm Test Site. Retrieved from EcoGrid EU. http://www.eu-ecogrid.net/ecogrid-eu/the-bornholm-test-site.

Energy Academy, 2011. RE-ISLAND. Retrieved from energiakademiet. https://energiakademiet.dk/en/vedvarende-energi-o/.

Environment Office., n.d.. Ærø – a Renewable Energy Island. Retrieved from Aeroe Energy and Environment Office: http://aeroe-emk.dk/eng/aeroe_energy_island.html.

Energy Institute., n.d. Retrieved from http://arkiv.energiinstituttet.dk/504/1/Folder%20DK_v12_FINAL.pdf.

Force Technology., n.d. VÄXJÖ ENERGI AB, SANDVIK 2, SWEDEN. Retrieved from https://bio-chp.force.dk/downloads/chp-plants-key-figures/vaxjo-energi-ab-sandvik-2-sweden/.

ICLEI, 2014. Bottrop, Germany InnovationCity Ruhr – model City Bottrop: revitalizing an industrial region through low-carbon redevelopment and active public-private partnerships. ICLEI.

ICLEI, 2016. Innovative City Business Collaboration. ICLEI.

Iveroth, S.P., Johansson, S., Brandt, N., August 8, 2013. The potential of the infrastructural system of Hammarby Sjöstad in Stockholm, Sweden. Energy Policy 59.

Iverotha, S.P., Vernayb, A.-L., Mulderb, K.F., Brandt, N., 2012. Implications of systems integration at the urban level: the case of Hammarby Sjöstad, Stockholm + Stockholm municipality from that source. Journal of Cleaner Production.

Jernberg, J., Hedenskog, S., Huang, C., 2015. Hammarby Sjöstad an Urban Development Case Study of Hammarby Sjöstad in Sweden, Stockholm. China Development Bank Capital).

Kommun, V., 2015. Energy Plan for the City of Växjö (Växjö Kommun).

Lenhart, J., van Vliet, B., Mol, A.P., November 2015. New roles for local authorities in a time of climate change: the Rotterdam Energy Approach and Planning as a case of urban symbiosis. Journal of Cleaner Production 107.

Madsen, H., et al., 2012. Evaluation of Energy Storage System to Support Danish Island of Bornholm Power Grid. Technical University of Denmark, Center for Electric Technology.

Malmö City Planning Office, 2015. Västra Hamnen Current Urban Development. Retrieved from. http://malmo.se/vastrahamnen: http://malmo.se/download/18.76b7688614bb5ccea09157af/1491304414891/Current+urban+development+in+Western+Harbour+%282015%29.pdf.

Malmo Stad., n.d.-a. Climate-smart Hyllie – Testing the Sustainable Solutions of the Future. Retrieved from Malmo Stad: http://malmo.se/download/18.760b3241144f4d60d3b69cd/1491305302024/Hyllie+klimatkontrakt_broschyr_EN_2013.pdf.

Malmo Stad., n.d.-b. SUSTAINABLE HILDA. Retrieved from Malmo Stad: http://malmo.se/download/18.72a9d0fc1492d5b743fc837b/1491301239246/Sustainable+Hilda.pdf.

Port of Rotterdam., n.d. Sustainabilty. Retrieved from Port of Rotterdam: https://www.portofrotterdam.com/en/the-port/sustainability.

Port of Rotterdam, 2016. Port of Rotterdam Authority Highlights of the 2016 Annual Report (Port of Rotterdam).

Ramboll., n.d. Empowering Växjö to become fossil-free. Retrieved from Romboll: http://www.ramboll.com/projects/re/veab-sandvik-power-station-unit-3?alias=vaxjo.

InnovationCity Ruhr, 2017. Retrieved from InnovationCity Ruhr. http://www.icruhr.de/.

Sjostad., n.d. The Hammarby Model. Retrieved from http://www.hammarbysjostad.eu/2-hammarby-sjostad-the-full-story-the-hammarby-model/.

Stockholm Stad., n.d. Retrieved from http://www.stockholm.se/OmStockholm/Stadens-klimat-och-miljoarbete/Manadens-klimatsmart-Stockholm/GlashusEtt–Internationell-kandis-i-Sjostaden/.

Sunstore 4, 2013. Why a R&D Project? Retrieved from SUNSTORE 4 100% renewable district heating. http://sunstore4.eu/understand/why-a-r1d-project/.

Zipkin, T., et al., 2015. Appendix 13-a case of community involvement in wind turbine planning. In: The Green Industrial Revolution. Butterworth Heinemann.

zukunftshaus, 2017. Retrieved from: http://www.zukunftshaus.org/.

Chapter 15

Mauritius Island Nation: 100% Renewable Energy System by 2050

A. Khoodaruth[1], V. Oree[1], M.K. Elahee[1], Woodrow W. Clark, II[2]
[1]*University of Mauritius, Reduit, Mauritius;* [2]*Clark Strategic Partners, Beverly Hills, CA, United States*

Chapter Outline

INTRODUCTION

Energy is widely acknowledged as being one of the most critical and pervasive issues that will challenge decision makers globally during this century. On the one hand, the importance of energy in fueling sustained economic growth and development has prompted governments to ensure that their relentlessly rising energy demands are met adequately at all times. On the other hand, the 82% share of fossil fuels in the global energy mix raises concerns about detrimental environmental impacts and energy security (IEA, 2013). New power systems and integration of several energy resources are becoming more and more common throughout the world (Clark and Cooke, 2011). Harnessing renewable energy (RE) sources can contribute to decarbonizing the energy system and upholding long-term energy security worldwide. Since the turn of the century,

Sustainable Cities and Communities Design Handbook. https://doi.org/10.1016/B978-0-12-813964-6.00015-X

the share of RE in the energy mix of many developed countries has increased substantially, and this trend is expected to continue in the future. Nevertheless, the various technical and economic challenges posed by a high level of RE integration have led many experts to dismiss the development of a 100% RE mix as a utopian ideal. However, recent studies point out that the design of carbon-neutral energy systems is technologically achievable (Jacobson et al., 2014; Elliston et al., 2013). The prospects for a completely "green" system improve each year as favorable market economics and continuing technological innovation contribute to make RE technologies less expensive, more efficient, and reliable. A host of developed countries, including Denmark (Lund and Mathiesen, 2009), Portugal (Krajačić et al., 2011), Ireland (Connolly et al., 2011), and New Zealand (Mason et al., 2010), have already developed elaborate roadmaps that detail the optimized combinations of RE resources and ancillary capabilities to meet their energy demands exclusively through clean energy sources in the long term.

Developing countries have a key role to play in the concerted effort toward the global transition to greater reliance on RE resources. Rising energy demand is characteristic of most developing nations as their expenditures on energy-consuming services that provide enhanced productivity, leisure, and comfort escalate. The International Energy Outlook 2013 report projects that growth in world energy consumption will predominantly emanate from non–Organisation for Economic Co-operation and Development (OECD) countries in the future (IEO, 2013). Thus energy use in non-OECD countries is expected to grow by 2.2% annually and their share of global energy consumption is projected to rise from 54% in 2010 to 65% in 2040 (IEO, 2013). With this in mind, developing nations also account for 95 of the 138 countries with RE support policies in 2014, up from 15 in 2005 (REN21, 2014). In addition, five developing countries (Uruguay, Mauritius, Costa Rica, South Africa, and Nicaragua) led the investments per gross domestic product (GDP) on RE technologies in 2013, spending between 0.8% and 1.6% of their GDP in this area (REN21, 2014). These initiatives underscore the fact that developing and emerging countries appear to be poised to make the switch to 100% RE given the necessary financial, political, and technological support. Owing to their unique characteristics and vulnerabilities, small island developing states (SIDS) denote a distinctive case for sustainable development among developing economies. The absence of exploitable natural resources in SIDS implies a massive reliance on imported fossil fuels for their energy requirements causing their small economies to be extremely vulnerable to fuel price fluctuations. This vulnerability is mainly due to their remote geographic location and economic structure (Levantis, 2008). Energy prices in SIDS are much higher than in continental countries because of high fuel transportation costs and constraints on exploiting scale economies through bulk purchasing. The economies of SIDS also rely on fuel-intensive activities such as tourism. As a result, a significant share of the national budget is devoted to oil imports. In general, fuel imports represent up to 20% of the annual import costs of SIDS, and between 5% and 20% of their GDP (Walker-Leigh, 2012). The vulnerability of economies of SIDS to fossil fuel price increases was clearly illustrated in Mauritius when the 2008 oil crisis increased the costs by 28% to reach USD 921.2 million when compared with USD 721.3 million in the previous year (MOENDU, 2010). The electricity utilities in SIDS are typically state-owned and have overall control over the distribution and transmission of electrical power. In addition, SIDS have high exposure to multiple and extreme vagaries of climate change, adding to the imperative of moving toward a sustainable future. These factors indicate that small islands constitute excellent test beds for decarbonization of the energy system. A related contemporary issue confronting power system planners worldwide is the merging of the central grid with on-site or distributed power. This new approach to energy generation can help provide power while reducing greenhouse gas (GHG) emissions and protecting the environment. Today's more "agile" systems (Clark and Bradshaw, 2004) interconnect power systems with renewable on-site power generation that may not be controlled by the public utility, as in the case of Mauritius and elsewhere. Agile systems can underpin the new energy paradigm for SIDS.

We assess the energy situation in Mauritius, a small island state, and present the main building blocks of a new energy paradigm aiming at achieving a 100% RE target by the year 2050. The present energy mix is critically analyzed to make recommendations for a 100% renewable system on the island by 2050. Although the Long-Term Energy Strategy (LTES) for the period 2009–25 devised by the Government of Mauritius sets pathways for a sustainable future, it does not evaluate scenarios that would achieve the objectives set. In addition, it limits the scope of the RE targets to 35% and the planning horizon up to 2025 and considers electrical power only. This chapter proposes to extend the analysis to 2050 by looking beyond electricity only to assess the energy system comprehensively in terms of alternative resources for primary uses, including energy-intensive transportation and cooling. A longer planning horizon is considered here because the new energy paradigm requires new energy infrastructure investments with lengthy lead times. Moreover, novel energy technologies can be slow to mature and fully penetrate the market. For these reasons, most international efforts to curb GHG emissions and to reach ambitious RE integration goals, such as the 21st Conference of Parties agreement and the EU decarbonization roadmap, use the year 2050 as their target. Furthermore, as mentioned in Introduction, many studies have been performed in different countries that are pledging their efforts toward 100% RE. Most of these studies have specified 2050 as a target based on an understanding that a full turnover of the electricity sector by 2050 is need to avoid an increase in the global average temperature beyond 2°C (Verbruggen and Lauber, 2009).

The rest of the chapter is structured as follows. The section Current Energy Status of Mauritius provides an account of the present energy mix of Mauritius and mentions the main challenges faced by the energy sector. The section The Government's Vision of Renewable Energies up to 2025 critically analyzes the energy portfolio proposed by the government through 2025. The section titled Options for a 100% Renewable Energy System by 2050 elaborates on realistic options that Mauritius could exploit to achieve the 100% RE target by 2050. The chapter concludes by highlighting the prerequirements for the successful implementation of this ambitious energy transition.

CURRENT ENERGY STATUS OF MAURITIUS

Primary Energy Requirements

Mauritius is located in the southwest of the Indian Ocean, off the eastern coast of Madagascar. With a population of 1.2 million for a total surface area of 1865 km^2, Mauritius has one of the highest population densities in the world (SM, 2011). The country depends highly on imported fossil fuels for its primary energy requirements, with 84% consisting mostly of coal and petroleum products, as shown in Fig. 15.1 (SM, 2016a). The primary energy requirements have grown steadily during the period 2006–2015, at a yearly rate of around 5%. Per capita energy requirements for Mauritius are high when compared with those of most developing nations in Africa but still low when compared with those of developed countries (World Bank, 2016).

In developing countries like Mauritius, the growth in energy consumption is largely due to the pace of economic development, resulting in associated financial burden together with the adverse environmental implications of consuming fossil fuels. Energy statistics reveal that the transportation sector is

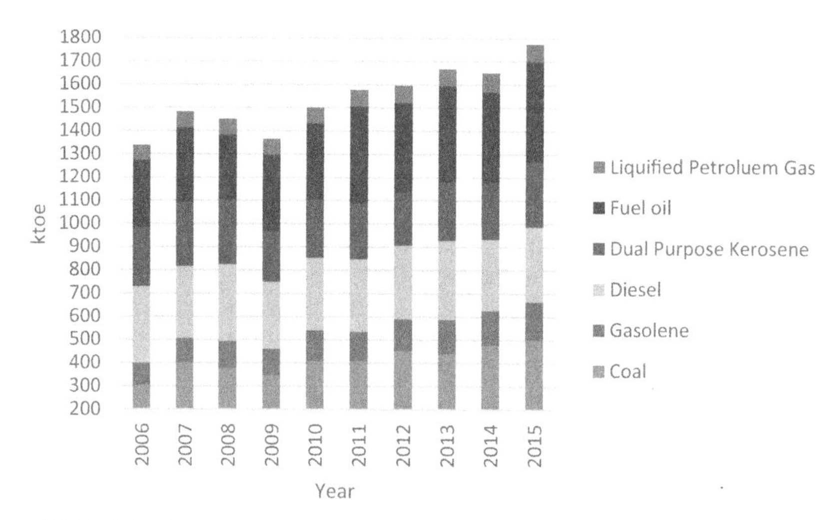

FIGURE 15.1 Importation of fossil fuels during the period 2006–15 (SM, 2015a, 2016a).

by far the highest final energy consumer, accounting for 50.7% of the total final energy consumption of the country, followed by manufacturing, households, commercial/distributive trade, and agriculture, which represented 23.7%, 14.2%, 10.5%, and 0.5% of the energy consumed, respectively (SM, 2016a). The higher standards of living ushered by economic growth of the country have brought a considerable increase in the number of vehicles. The number of registered vehicles has increased significantly by 52% to reach 486,124 during the period 2006–15, as illustrated in Fig. 15.2 (SM, 2015a,b, 2016a,b), whereas the corresponding growth in petroleum imports for transportation was only 20%. The importation of more efficient vehicles and the availability of better quality fuels since 2006 have contributed greatly in mitigating the surge in transport-related fuel imports. Between 15,000 and 20,000 vehicles are added annually on the roads, thereby worsening traffic problems despite the construction of new motorways. One study estimated that traffic congestion cost the Mauritian economy about USD 33 million worth of surplus petroleum products annually (MEF, 2007). Nevertheless, being the biggest end user of energy, the oil-dominated transport sector has considerable energy security and GHG emission implications. To curtail the energy use by transportation, the national government has several projects in development. These include enlargement of motorways to cater for a special bus lane and introduction of Light Rail Transport to provide an efficient and quality mass public transport system (MPILT, 2004; Hansard, 2014). Although the initial investments will be substantial, the various outcomes in terms of reduced imports of fossil fuels, GHG emissions, and traffic congestion along with increased traffic safety and health are expected to be beneficial for the country in the long term.

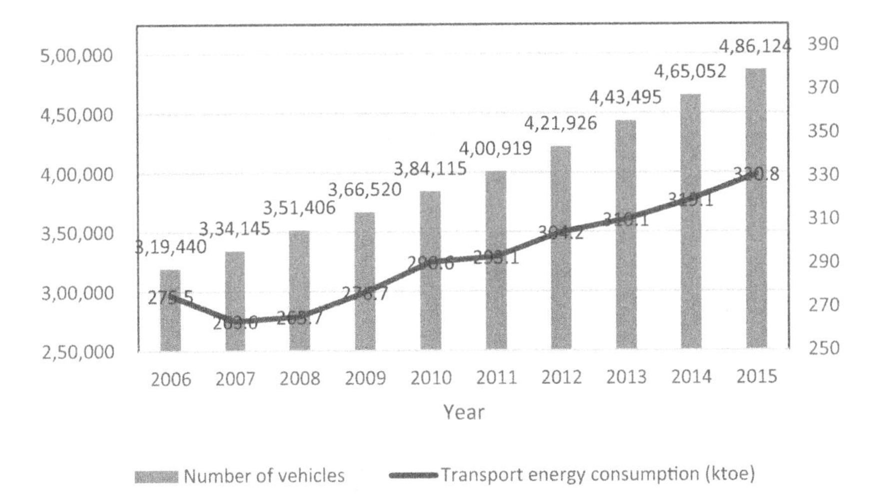

FIGURE 15.2 Stock of registered vehicles in Mauritius and final energy consumption by transportation sector for the period 2006–15 (SM, 2015a,b, 2016a,b).

Fig. 15.3 illustrates the final energy consumption of all major consumers. Unlike transportation, the manufacturing sector faced a net decrease of 19% in energy consumption during the period 2006–15. This drop can mainly be attributed to the closure of many factories in the local textile industry due to severe competition from other countries in which labor was cheaper. By comparison, households and commercial/distributive trade experienced a relatively linear annual growth in energy consumption of about 2% and 5%, respectively, during the same period. Finally, the energy consumption of the agricultural sector has been constant over the years at around 4.5 ktoe (1000 tons of oil equivalent).

Electricity Generation

The total electricity generation in Mauritius has increased by about 27% over the last 10 years, from 2320 GWh in 2006 to around 2956 GWh in 2015 (SM, 2015a, 2016a). Electricity is primarily generated from fossil fuels, as depicted in Fig. 15.4. This rise in usage generally follows the worldwide trend in electricity production, which experienced a 38% growth during the same period (Enerdata, 2016). Meanwhile, electricity generation in Mauritius from RE resources increased to 22.7% (a growth of 0.5%) due to increase in the use of hydropower despite a small decrease in bagasse-based generation attributable to a drop in sugarcane production. Fig. 15.4 shows that the 77% share of fossil fuels in electricity generation is less than its 85% stake in primary energy requirements due mainly to the extensive use of bagasse in the production of electricity.

The national Central Electricity Board (CEB) is solely responsible for the transmission, distribution, and supply of electricity in Mauritius. The CEB

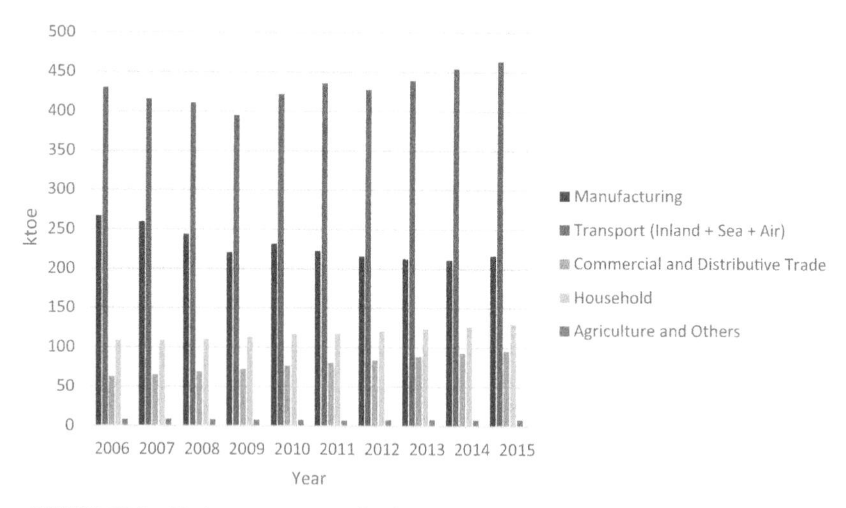

FIGURE 15.3 Final energy consumption by sector from 2006 to 2015 (SM, 2015a, 2016a).

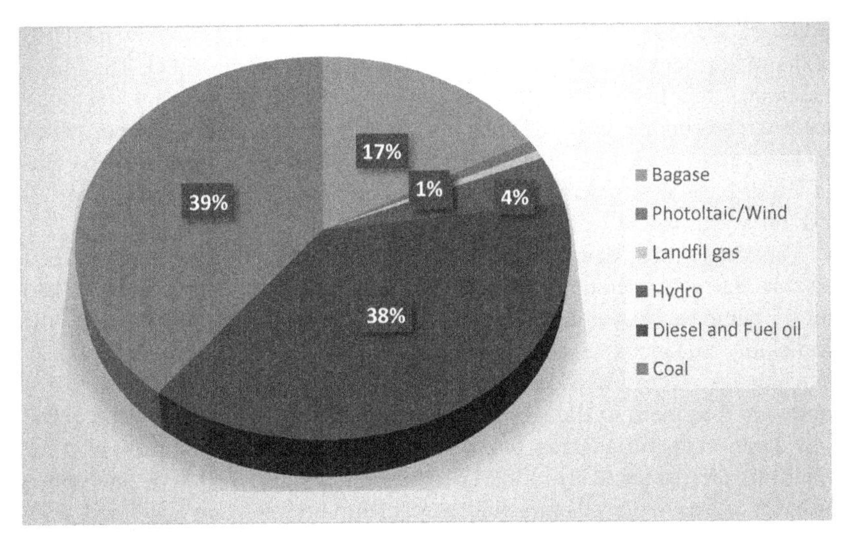

FIGURE 15.4 Electricity generation by source of energy for 2015 (SM, 2016a).

also produced about 41% of the total electricity consumed in 2015 (SM, 2016a). For this purpose, it operates thermal power plants running on fuel oil and kerosene that provide 37% of the total electricity generated and 11 hydroelectric power stations that generate 4% of the total electricity (SM, 2016a). Until the start of this century, the sugarcane industry was one of the main pillars of the Mauritian economy, as the country enjoyed trade privileges with the European Union that guaranteed big sugar quotas at favorable prices. Since the end of the preferential rates and ensuing reduction of sugar prices starting in 2006, about 2000 ha of land were converted annually into residential/commercial plots, which guarantee better returns.

Consequently, the sugar industry has reengineered itself to support the biorefinery concept. Production of bioethanol for the transportation sector and biofuel for the electricity sector have since been promoted. Ethanol plants, refinery facilities, and combination bagasse and coal power plants have been implemented at sugar mills and operate all year round. The owners of these power plants are commonly referred to as Independent Power Producers (IPPs). The contribution from IPPs started in 1995 with the power purchasing agreements (PPAs) meant to encourage use of bagasse, a by-product of sugarcane. PPAs also allowed the use of coal, even outside the intercrop season. In fact, the weak financial incentive to use bagasse instead of coal is associated with increased coal use by IPPs. In 2015, for example, the IPPs used 198 ktoe of bagasse as opposed to 424 ktoe of coal (SM, 2016a). In the wake of more strict environmental regulations, the PPAs will be renegotiated soon, and it is expected that they will be amended to favor the use of biomass, including bagasse. New PPAs with solar photovoltaic (PV) and wind energy producers are also anticipated. Since 2007, the IPPs have overtaken the CEB as the major generator of electricity in Mauritius (Elahee, 2011). They have maintained this trend since then and generated 59.2% of the total electricity needs of the country in 2013. They operate five thermal plants, four of which are annexed to sugar mills. One of the IPPs has proposed to import wood chips on a trial basis as a greener alternative to coal. The fifth plant runs solely on coal on a year-round basis.

The location of Mauritius just above the Tropic of Capricorn means that average daily temperatures around 30°C are common during the hot and humid summer season from November to April. During this period, the availability of adequate energy sources to cater to the average and peak demands for electricity caused by heavy air-conditioning loads (CEB, 2013) represents the main challenge for power system planners in Mauritius. The load duration curve indicates that the base load, semiload, and peak load for Mauritius are, respectively, 170 MW, between 170 and 370 MW, and above 370 MW, respectively. Contractual agreements between the CEB and IPPs presently require CEB to provide for peak and semipeak loads, whereas the typically cheaper base load electricity is provided by the IPPs. This scenario is not ideal for the CEB as it neither fully exploits its generating capacity nor benefits financially. Nevertheless, the urgency to meet the ever-increasing demand of electricity led the government to offer financial incentives to the IPPs to produce electricity. Concurrently, it represented a unique opportunity to exploit a green resource in the form of bagasse for electricity generation. The IPPs that owned the sugar factories were thus motivated to invest heavily in either new steam generators or upgrade to higher pressure and higher temperature boilers and use condensing extraction steam turbines.

The effective capacities of existing power plants in Mauritius, the types of load they serve, and their ownerships are given in Table 15.1. Although the total effective capacity of CEB is higher than that of the IPPs, the latter export all of their surplus generated electricity to the grid, which serves the base load as per the PPAs. The CEB supplies the demand mostly during peak periods, implying a low utilization factor. Kerosene-propelled gas turbines are operated only during peak times or in case of unexpected breakdowns from other power plants as their unit cost of generated electricity is higher. In contrast, the lower cost of coal has seen it become the preferable option for serving base loads. Thus coal has gradually replaced heavy fuel oil (HFO) in the overall thermal electricity production with its share rising from 42% in 2006 to 50% in 2015 (SM, 2015a, 2016a).

Fig. 15.5 illustrates that the maximum peak demand has been increasing on an average by 3% annually during the period 2006–15, corresponding to an effective yearly growth of about 9 MW (SM, 2015a, 2016a). Meanwhile, authorities have ensured reliable electrical supply by upgrading the total effective capacity to keep pace with the peak demand (SM, 2015a, 2016a). In 2015, the recorded peak load was 459.9 MW, whereas while the combined

TABLE 15.1 Details of Existing Power Plants in Mauritius

Sector	Type	Total Effective Capacity (MW)	Fuel	Load	Estimated Cost of Electricity ($/kWh)
CEB: government-owned	Diesel engine	312.90	Heavy fuel oil	Semibase	0.20–0.22
	Gas turbine	75.00	Kerosene	Peak	0.60
	Hydro	53.80	Water	Peak	0.03
IPPs: owned by private sector	Steam turbine	237.50	Bagasse/Coal	Base	0.09–0.13
	Gas turbine	2.00	Landfill gas	Base	Not available
	PV	16.42	Solar	Base	0.22

CEB, Central Electricity Board; *IPPs*, Independent Power Producers; *PV*, photovoltaic.
Adapted from NEC, October 2013. National Energy Commission (NEC), Republic of Mauritius. Making the Right Choice for a Sustainable Energy Future: The Emergence of a "Green Economy". Available at: http://publicutilities.govmu.org/English//DOCUMENTS/NEC%20FINAL%20REPORT.PDF.

FIGURE 15.5 Effective capacity and peak power demand for the period 2006–15 (SM, 2015a, 2016a).

effective generating capacity of the CEB and IPPs was 714.4 MW (SM, 2015a, 2016a). Although a reserve margin of more than 50% of the peak load may indicate a reliable power system, there are other aspects that need to be considered. Several power plants are either nearing their retirement or have already exceeded their normal operational lifetimes but have not been phased out as their replacements are not ready. In addition, not all IPP plants can be operated simultaneously and hydroelectric plants cannot be relied upon during summer when peak loads occur because of droughts. As a result, an effective reserve margin of 43 MW only is available to address any unforeseen generator failure (ADB, 2014a). Even a large margin will be inadequate if a major power plant fails unexpectedly. In the next few years, the projected increase in peak demand combined with the expected decommissioning of some existing generators implies that the CEB must urgently plan for generation expansion to mitigate potential shortages in electrical supply.

Finally, end use data show that the total electricity consumption is relatively equally distributed among the domestic, commercial, and industrial sectors, as depicted in Fig. 15.6. This distribution contrasts noticeably with the disparate distribution of primary energy use among these three end users detailed in Primary Energy Requirements section. The manufacturing sector accounts for a bigger portion of the primary energy consumption as in addition to electricity, it consumes large amounts of fuel for its plant and machinery, such as boilers, furnaces, and heat pumps. Households require more primary

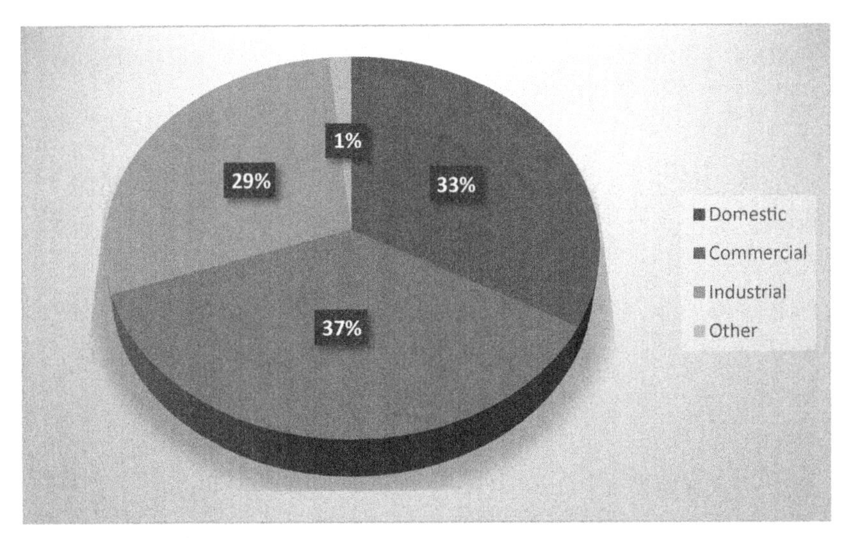

FIGURE 15.6 Percentage share of electricity sold by tariff group in 2015 (SM, 2016a).

energy than the commercial sector due to the heavy use of liquefied petroleum gas (LPG) for cooking and water heating. A domestic census carried out in 2011 indicated that 97.6% households use LPG for cooking while 61%, 13%, and 12% of total households are equipped with gas water heater, solar water heater (SWH), and electric water heater, respectively, for bathing (SM, 2011).

THE GOVERNMENT'S VISION OF RENEWABLE ENERGIES UP TO 2025

Based on this profile, it is obvious that a business-as-usual scenario is economically unsustainable in the long term as it exposes the national economy to high risks due to the volatility of the oil market.[1] For economic and environmental reasons, it is imperative for Mauritius to reduce its dependency on fossil fuels and expand the use of renewable resources (Clark and Bradshaw, 2004). The fact that the island is endowed with abundant and varied sources of RE, particularly solar and wind, suggests that this green plan for the power sector is feasible. Recently, Mauritius has started to leverage its resources in earnest through the development of solar and wind farms. Ostensible progress in this direction is evidenced by the increase of 1% in electricity generated from RE sources during the period 2010–15, to reach about 681 GWh, as illustrated in Table 15.2.

Driven by key objectives like improved energy security, economic resilience, and environmental protection, Mauritius must rely on substantial commitments

1. Although oil prices have been falling for several years, they might rise again, therefore creating a risky situation for countries that are heavily reliant on fossil fuels for energy production.

TABLE 15.2 Evolution of the Share of RE in the Electricity Mix During the Past 5 years (SM, 2015a, 2016a)

	2010		2015	
	GWh	%	GWh	%
Bagasse	474.1	17.8	509.8	17.0
Landfill		0.0	20.4	0.7
Photovoltaic	2.5	0.1	25.9	0.9
Wind		0.0	2.7	0.1
Hydro	100.7	3.8	121.9	4.1
Total RE	577.3	21.7	680.7	22.7
Total electricity generated	2656.6	100	2995.6	100.0

RE, renewable energy.

by public policy makers. In response to these considerations, the government devised the LTES in 2009 for the period 2010–25 (MREPU, 2009). The highlight of the report was to improve the self-sufficiency of the country in electricity supply from 20% in 2010 to 35% in 2025. To this end, progressive targets in terms of percentage of total electricity generation by various RE technologies over the period 2010–25 were set, as shown in Table 15.3. The highest forecasted growth until 2025 is envisaged from wind energy, which is expected to contribute 8% of the total electricity generated by the end of the report horizon, although its present input is negligible.

The forecast share of coal in the electricity mix will fall only marginally and will still represent the main energy source for electricity in 2025, with a 40% involvement, down from 43% in 2010. The main reason is that a new coal plant is planned based on the fluidized bed combustion technology. Accordingly, Mauritius will rely on coal for electrification of the country for some decades despite associated environmental problems. Conversely, the targets for solar PV and wind energy appear low in view of the great potential to harness these natural resources. Mauritius is located in subtropical areas below the equator with an average of 7.7 h of bright sunshine daily (SM, 2015c) and an average wind speed of 6.5 m/s (MMS, 2015). In contrast, Germany, one of the world leaders in the implementation of RE technologies, has on an average only 3.1 sunshine hours per day (NREL, 2016) and an average wind speed of 4.5 m/s (EEA, 2009). Unlike the plans for many other countries, the government's LTES does not set any target for the reduction in GHG emissions.

The government has initiated the enactment of appropriate legislation to accompany the transformation of Mauritius into a sustainable island nation. A

TABLE 15.3 Targets of Electricity Generation for the Period 2010–25 (MREPU, 2009)

Fuel Sources		Percentage of Total Electricity Generation (%)			
		2010	2015	2020	2025
Renewable	Bagasse	16	13	14	17
	Hydro	4	3	3	2
	Waste to energy	0	5	4	4
	Wind	0	2	6	8
	Photovoltaic	0	1	1	2
	Geothermal	0	0	0	2
	Subtotal	20	24	28	35
Nonrenewable	Fuel oil	37	31	28	25
	Coal	43	45	44	40
	Subtotal	80	76	72	65
	Total	100	100	100	100

carbon tax was introduced in 2011 for the transportation sector. The law was amended to enable charging of a carbon dioxide (CO_2) levy for vehicles with a CO_2 emission rate higher than a set threshold. The amendment also offered a rebate on excise duty payable if the emission rate is lower than the threshold. The latter was set at 158 g/km in 2013 and was revised to 150 g/km in 2014. In the European Union, which enforces very stringent emissions standards for cars and lighter commercial vehicles, the current threshold of 125 g/km will decrease to 95 g/km in 2021 (EU, 2016).

In its Integrated Electricity Plan (IEP) for the period 2013–22 (CEB, 2013), the CEB gives an insight into the strategies that will address the short- and medium-term electricity generation challenges of Mauritius. The main objective of the IEP is to match the power system supply side with the forecasted growing demand. It is estimated that the CEB will require USD 300 million over the next 10 years to implement the plan. According to CEB projections, the increase in demand will call for additional electricity generation capacity. The demand forecasts are based solely on an econometric model without taking into consideration potential consequences of various factors, including climate change and grid penetration of RE technologies. The demand predictions also disregard potential savings in electricity production due to energy efficiency programs, which typically constitute a vital resource in energy transition plans. Energy efficiency can help to attenuate demand growth and associated GHG emissions. Planning for additional capacity of power plants by the CEB or IPPs could thus be delayed, giving breathing space to power producers and the economy in terms of capital expenditure and the balance of foreign exchange. The current electricity pricing model of the CEB does not accurately reflect the financial and environmental costs of generating and distributing electricity due to social and economic considerations. In these circumstances, energy efficiency and demand side management (DSM) opportunities do not receive the same interest that their relevance in the energy system mandates. The IEP also underlined the need for additional transmission and distribution infrastructure stemming from the operation of a scheme to encourage small-scale electricity producers to generate their own electricity and sell any excess production or generation to the national grid at preferential rates.

According to the IEP, a new 100-MW coal power plant will be implemented with all the mandatory environmental safeguards from 2015 to prevent power outages. This decision has sparked protests from environmental organizations and residents of the neighborhood of the identified construction site. In the wake of the protests, the government set up a NEC to review national energy requirements and advise authorities on the planning and execution of major projects to address the country's medium- and long-term energy needs. In its report (NEC, 2013), the NEC opined that the IEP of the CEB did not comply with the main objective of the LTES, whereby 35% of the total electricity generation would be achieved through RE technologies by 2025. The NEC also pointed out that it was inconceivable that a plan of such national importance was devised without consultations with all stakeholders. Among NECs recommendations were its advise to the government to commission an additional capacity of 60 MW for an existing power plant running on HFO and liquefied natural gas, increase the use of renewable sources of energy, and integrate energy efficiency and DSM programs.

These developments highlight the limitations in the government's deployment plan for RE by 2025. First, the plan focuses only on the electricity aspect of the energy consumption. Second, the portfolio of technologies considered is restricted and does not taken into account contemporary alternatives. Moreover, it is fraught with inaccuracies. The most obvious one is that the share of bagasse-based electricity will be less than the 17% anticipated in the plan due to the rapidly decreasing area of land under sugarcane cultivation. As such, the focus should be on enhancing the exploitation of readily available RE sources, predominantly solar and wind. The potential of other abundant RE sources, such as tidal and wave, can be tapped in the future but is conditional on resolving some uncertainties about the technologies and their cost. More importantly, further decarbonization of the power grid will be possible through significant emphasis on a more efficient use of the present energy resources and the emergence of cost-effective technologies. The next section describes the key enabling technologies that have the ability to steer the country to not only achieve at least 35% RE by 2025 but also target 100% by 2050.

OPTIONS FOR A 100% RENEWABLE ENERGY SYSTEM BY 2050

Our plan envisions an overarching energy system, encompassing not only electricity but also transport and cooling, which relies entirely on renewable resources. The transition to the 100% RE system must continue to reach the ultimate objective of achieving total energy self-sufficiency and security. Domestic expansion of RE production will underpin this transformation, focusing on technologies appropriate for the resource profile of Mauritius. However, acquiring a formal political commitment by Mauritius to this ambitious endeavor is essential. This will help to garner the support of all stakeholders, including the people, by granting an official mandate for action. Energy efficiency represents another prerequisite for achieving a 100% RE future. Promoting a more efficient energy infrastructure will help to decouple economic growth from increased energy demand, leading to a reduction in additional generation capacity requirements. This section details the renewable technologies that have the potential to turn the 100% renewable future into a reality by 2050. The deployment of these technologies can be backed by financial incentives and concerted efforts to unblock crosscutting nonfinancial obstacles. In addition to financial support from the government, it is also essential for the private sector to be supportive. Although it is also crucial to secure initial support from international organizations in the form of finance, technology development and transfer, and capacity building, the country should strive for economical independence in the long term to sustain the RE transition. The Global Environmental Fund of the United Nations has already earmarked USD 2 million to help remove barriers to effective deployment of solar energy in Mauritius. In 2013, the European Union launched a 15 million Euros program for the development of RE and the improvement of energy efficiency in the islands of the Indian Ocean (IOC, 2016). The support of the Green Climate Fund has been sought for the operationalization of the newly created Mauritius Renewable Energy Agency to oversee the development of RE in Mauritius. The island has also engaged in synergies with regional blocks such as Southern African Development Community, Indian Ocean Commission, and Common Market for Eastern and Southern Africa to promote RE and energy efficiency programs (MoESDDBM, 2015).

Hydrogen Power and Electric Vehicles

As mentioned in the section Primary Energy Requirements, transportation is the largest energy end user in Mauritius, accounting for over half of the primary energy consumption. It is therefore the obvious place to start when attempting to eliminate fossil fuels from the energy mix. There exists a clear evolutionary path for vehicles, from gasoline, compressed natural gas and diesel, through hybrid to electric and hydrogen-propelled. The latter have been touted as future clean alternatives to combustion engines. Although they are presently not commercially viable on a large scale, considerable progress has been made on these technologies during the last decade. Many countries are aiming at a fully electrified vehicle fleet by the end of the next decade. For example, the market share of electric vehicles in Norway increased from 1.5% in 2011 to 23.3% in 2015, or about 36,000 vehicles (IEA, 2016). A corresponding change from 0.0% to 1.0% was observed over the same period in China, representing the introduction of about 215,000 electric vehicles (IEA, 2016). In both cases, electric cars are backed by generous financial incentives with a view of generating a viable market for zero-emission cars in 5 years, when the incentives will gradually phase out. As battery technologies mature over the next decade, a gradual shift toward environment-friendly vehicles will result in a progressive decline in fossil fuel imports. According to a recent report by the United Nations Environment Programme, the cost of batteries for electrical vehicles has dropped dramatically during the past 5 years, from USD 1000 in 2010 to USD 350 per kWh in 2015 (UNEP, 2016). A package of accompanying measures should be deployed at an initial stage to stimulate the uptake of electric and hydrogen fuel cell vehicles. Motorists would benefit from financial incentives to cover the cost difference between clean and conventional vehicles. In July 2016, electric cars have been exempt from the costly excise duty imposed on vehicles in Mauritius. The strategy should also include plans to provide adequate electric charging and hydrogen fueling infrastructure to help develop a network of alternative vehicles throughout Mauritius. Given the small geographic spread of the island, an autonomy of 100–150 km range makes charging at home largely feasible. Over time, as the land transport system is weaned off fossil fuels and becomes electrified, the significance of electricity in energy system of Mauritius will likely increase. Environmental goals make it imperative to decarbonize the electricity infrastructure so that charging and refueling can be powered by clean sources. Electric vehicles, including plug-in hybrid electric vehicles, have the added advantage of providing flexible electrical storage capacity. Flexibility will be critical in future power systems as increased integration of intermittent RE generation sources like wind and solar in the electricity grid calls for additional resources to maintain the delicate balance between energy supply and demand. Ongoing work is aimed at fully capturing the potential synergies between variable generation and electric vehicles (NERC, 2010).

Bagasse Gasification

The use of biomass for electricity and heat generation in sugar mills has considerably mitigated GHG emissions in Mauritius as the fuel is produced in an environmentally sustainable way. The low efficiency of current techniques used for cogeneration indicates that improving the thermal processing of biomass can reinforce its ability to contribute substantially to the growing demands for bioenergy. Conventional cogeneration techniques can produce up to 80 kWh per ton of sugarcane processed, whereas with advanced techniques the generation can reach 200 kWh per ton of sugarcane processed (Pellegrini and de Oliveira, 2011). New technologies are needed to enhance the energy yield from bagasse. Among the options, gasification is one of the most efficient methods for combined heat and power generation due to its potential for higher efficiency cycles (Knoef and Ahrenfeldt, 2005). Gasification refers to the process during which organic or fossil-based carbonaceous materials are converted into carbon monoxide, hydrogen, methane, and carbon dioxide (Speight, 2010). To perform this conversion, the materials are reacted at high temperatures, typically above 1000°C, in the presence of a limited amount of oxygen and/or steam. According to Riehl et al. (2012), gasification technology has been developed over the last decades as an additional option for fuel production, as well as chemical substances, at a more competitive price when compared with crude-oil-based products. Moreover, the payback period of 6 years for incorporating this technology in cogeneration system appears very reasonable (Okure et al., 2006). As shown in Table 15.2, bagasse-based electricity accounted for 510 GWh of the total 2996 GWh electricity production for Mauritius in 2015. According to one study (Autrey et al., 2006) electricity production from higher fiber cane could rise to around 4600 GWh using technologies such as biomass integrated gasification combined cycle. This corresponds to approximately a 10-fold increase in present electricity generation from bagasse and exceeds the current total electricity generation by more than 1.5 times. However, research on making gasification-based power generation commercially viable is ongoing (Asadullah, 2014).

Mauritius has recently begun to tap the potential of landfill gas, primarily composed of methane, to produce electricity. Mauritius is currently disposing about 420,000 tons of solid waste at the unique landfill of the island, situated in the village of Mare Chicose. A landfill gas to energy plant of 3.3 MW capacity has started operation in November 2011 and generates nearly 22 GWh of electricity annually (Sotravic, 2012). The plant is expected to reduce 668,000 tons in GHG emissions throughout its first 5 years of operation, representing about 12% of the island's annual GHG emissions (CAIT, 2015). A study evaluated the present maximum annual energy production capacity of the landfill to be 50.50 GWh (Surroop and Mohee, 2011). Although only 15%–25% of the total gas yield is presently being collected, much higher landfill gas recovery rates in excess of 50% are possible with efficiency gains. The plant will be equipped by mechanisms to enhance the landfill gas collection mechanism by the end of 2016 (Karagiannidis, 2012). These very large volumes of untapped landfill gas have the potential to significantly increase power generation from this source.

Solar Photovoltaic

During the last decade, global solar PV installations have been growing exponentially with the cumulative installed capacity rising at an annual average rate of 49% (IEA, 2013). This growth has been catalyzed by significant reductions in capital costs, ongoing technological advances, and government incentive schemes. Table 15.3 indicates that according to the LTES, solar PV is expected to contribute only 2% of the total electricity required by Mauritius in 2025. This figure seems very low in view of the promising solar potential of the country. According to long-term data from the Mauritius Meteorological Services, the island receives an average daily solar insolation of 5.6 kWh/m^2 and between 2350 and 2850 h of sunshine annually depending on location (MMS, 2015). The underestimation of solar PV potential can be justified by several factors. One relates to the absence of accurate ground radiometric measurement data at the granular spatial resolution needed to build a comprehensive solar radiation database for the country. Such a resource is instrumental in efficient PV planning, design, and implementation. Therefore establishing an accurate solar map should be a priority. To address this deficiency, the United Nations Development Programme is funding a project to deploy a network of solar radiation–measuring equipment in Mauritius and enable the development of a solar resource map in the next few years. The high upfront investment is another factor that limits the wide-scale deployment of solar PV projects. In this context, the government has already worked out a feed-in tariff scheme to support small- and medium-scale solar PV generation. A first large-scale PV farm of 15.2 MW_p capacity, located on the west coast of Mauritius, was connected to the grid in February 2014 to add up to 4 MW_p capacity under the small-scale distributed generation project. PV module prices have dropped substantially over the years and cost less than USD 1 per watt of electricity generated in 2015. Considering these prices, associated components, and typical warranties, PV systems may be able to successfully compete with fossil fuels in sunny regions with high electricity prices without subsidies (Kurtz et al., 2016). Studies have reported that if the decreasing trend in PV prices can be sustained, parity can be achieved in most regions in the United States and China as well as Southern Europe before 2020 (Munoz et al., 2014) and other regions before 2030 (Solar Industry, 2012). Thus Mauritius appears to have the potential to cost-effectively increase its electricity production through the expansion of PV farms in areas where maximum solar radiation is available.

Solar Thermal

Although Mauritius enjoys a subtropical climate, households use hot water extensively, mainly for bathing and cooking purposes. Until the 1990's, water was heated mainly by electric water heaters, which were among the most highly rated electric appliances in homes. These have gradually been replaced by LPG or SWHs. In 2011, 59.4% and 12.1% households were using LPG and solar energy, respectively, to produce hot water, whereas 12.2% used electricity (SM, 2011). SWH represents a cost-effective and clean way to generate hot water for homes. In pursuit of its objective to reduce demand, the government initiated an SWH program in 2008 to provide households with generous subsidies to purchase SWH for their domestic hot water needs. About 60,000 households, representing about 18% of the total households, have already benefitted from this support scheme. It required more than USD 12 millions in terms of grants. Consequently, the SWH market has experienced an average yearly growth of 40% over the past 5 years, such that the installed solar thermal energy installed capacity has increased from 16.3 MW_{th} in 2008 to 109.8 MW_{th} in 2013 (Meister Consultants Group, 2015).

The growing interest in SWH illustrates the enormous potential of solar thermal energy. The current installed capacity value of solar thermal in Mauritius, which stands at 84.7 kW_{th} per 1000 inhabitants (Meister Consultants Group, 2015), is still low when compared with countries like Cyprus and Austria where corresponding values of 542 and 406 kW_{th} per 1000 inhabitants, respectively, were reported in 2011 (Mauthner and Weiss, 2013). Although the subsidies catalyzed the SWH market in Mauritius, several factors indicate that solar thermal growth is sustainable even with limited financial incentives. First, SWHs were mostly imported prior to the introduction of subsidies, but with the increasing demand, many local companies started manufacturing SWHs. Scale economies and enhanced competition among the growing number of local manufacturers caused costs to drop. Consequently, SWH prices have dropped by 30%–50% in the last 15 years. The simple payback period of SWH in Mauritius is presently at 4.3 years (Meister Consultants Group, 2015). Second, the subsidy scheme is aimed mainly at the low-income families. The ubiquity of SWH following the success of the scheme has made them standard in new house construction for medium-income and higher income families. Third, commercial enterprises, particularly large-volume consumers, such as hotels, are adopting SWH on a massive scale. Finally, an aggressive outreach and awareness-raising campaign has helped the public understand the environmental and financial benefits that can be derived from SWH. A survey conducted in 2011 revealed that more than half of the members of the 80.3% of total households that were not equipped with SWH were interested in procuring one (SM, 2011). Equipping these households with SWHs will increase the total installed capacity to about 400 MW_{th}, making the penetration rate comparable to those of countries with mature SWH markets. All of these aspects underline the potential of solar thermal to contribute to the energy transition in Mauritius.

Onshore and Offshore Wind

Although wind power generated in Mauritius is presently negligible, data from the Mauritius Meteorological Services indicates good wind resources in many regions, particularly in the eastern part of the island (MMS, 2015). In fact

Table 15.3 shows that wind is regarded as the most promising RE resource, with its contribution to the total electricity mix forecasted to rise from 0% in 2010 to 8% in 2025. Besides, wind turbines are prevalent in the island of Rodrigues, a dependency located at about 600 km from the east coast of Mauritius. In 2011, wind-generated electricity represented 9% of the total electricity produced in Rodrigues, representing 2.97 GWh (CEB, 2013). Two wind farms of capacities 30 MW and 18 MW are presently being implemented at Plaine Sophie in the southwest and Plaine des Roches in the northeast, respectively (World Bank, 2015). The CEB claims that it can presently accommodate wind-generated electricity up to one-third of the night load, amounting to 60 MW (CEB, 2013). Wind turbines must be specially designed to withstand the wind gusts experienced during tropical cyclones that are common during summer, with pull-down mechanisms and resilient mechanical structures. Onshore wind technology is mature and can be exploited in the short term, whereas offshore wind technology is gradually gathering pace. In 2013, offshore wind power in the European Union represented 14% of the total wind power installations, up by 4% from the previous year (SETIS, 2016). Besides, the installed capacity of offshore wind in Europe is expected to reach 23.5 GW in 2020, more than three times the present capacity (EWEA, 2014). The experience garnered through growth of installed capacity coupled with extensive research and development in this field has enabled better and upscale designs (Voormolen et al., 2016). Although offshore wind turbines involve significantly higher upfront investment than their onshore counterparts, they have numerous benefits, mainly lesser visual and environmental impacts, better levels of reliability, and steadier yields per turbine. For example, the typical annual yield from an offshore 3.6-MW wind turbine is 12 GWh, that is, 50% more than that of an onshore wind turbine of similar capacity (Siemens, 2014). Projections of various energy technologies made by the European Union indicate that capital investment on offshore wind technology is likely to decrease by more than 50%, whereas the technical lifetime of the turbines is expected to increase from 20 to 30 years by 2050 (ETRI, 2014). If as predicted, this technology becomes economically and technically viable, its exploitation will represent an appropriate solution for the highly populated coastal regions of Mauritius that also accommodate most of the energy-hungry hotels.

Ocean Technologies

The marine environment is central to economic development of SIDS because they depend heavily on the fisheries and tourism industries. The ocean also stores tremendous energy in the form of currents, waves, tides, and heat flows that can potentially supplement the power requirements of small islands. In view of the vast exclusive economic zone of Mauritius that extends over a surface area of 1.9 million km^2, the exploitation of the ocean can be envisaged to cater to the substantial energy requirements of coastal regions, where luxury hotels and other tourist facilities are located. Although tidal systems, wave power, and ocean thermal energy conversion (OTEC) are emerging as plausible technologies, many challenges impede their progress. Wave and tidal power technologies can be cost prohibitive and pose environmental threats. OTEC, which exploits the natural thermal gradient at different depths of the ocean as a heat engine, is under serious consideration by local authorities. Results on OTEC from research conducted at the National Energy Laboratory of Hawaii have been promising (Scientific American, 2016). The ancillary benefits of OTEC, including desalinated water, air-conditioning, and marine culture, may make the technology attractive to investors. Provision of cooling to hotels and buildings in the coastal capital city of Port Louis has the potential to serve average and peak load to a significant degree. For this purpose, the Government of Mauritius has already received funding from the African Development Bank for the first phase of the Deep Ocean Water Application (DOWA) project, which consists of implementing a 44-MW Energy Transfer Station to pump deep ocean cold water for building air-conditioning. It is expected that this initial phase of DOWA will curtail the peak power on the grid by about 26 MW (ADB, 2014b).

Biofuels

The global production and use of liquid biofuels increased in 2013 (REN21, 2014). In Mauritius, reengineering of the sugarcane industry following the expiration of sugar quotas has catalyzed the production of sugarcane-based ethanol. Ethanol is obtained from the processing of the juice resulting from the crushing of sugarcane. As such, the production of ethanol does not compete with bagasse. It is estimated that 86 L of ethanol can be produced from 1 ton of sugarcane (Nguyen et al., 2009; Macedo et al., 2008). Based on the local sugar production of 4.5 million tons in 2011, the production capacity of ethanol is estimated at 360 million liters annually. Ethanol is emerging as a low-carbon alternative for fossil fuels in both the transportation and electricity generation sectors. A study conducted by Silalertruksa et al. (2015) revealed that only 390 g of CO_2 are emitted during the production of 1 L of ethanol, whereas producing an equal quantity of gasoline releases 2210 g of CO_2 (IPCC, 2007). The introduction of biofuels in the transportation sector in Mauritius was planned to be started in 2010 (Elahee, 2011). Tests performed in Mauritius have already demonstrated the feasibility of 10% substitution of petrol by ethanol (E10) for use in vehicles. As highlighted in the section Hydrogen Power and Electric Vehicles, the decarbonization of the transportation sector will have to follow an evolutionary path, and the use of E10 can be the starting point of this development. Ethanol can also be used as a substitute for HFO in thermal power plants so that the latter can be retrofitted to exploit the local production of ethanol for electricity generation. In 2010, the first ethanol-fired power plant was successfully launched in the city of Juiz de Fora in Brazil. Having an installed capacity of 87 MW, the plant can supply the power demand of the 150,000 inhabitants (Power-technology, 2015). On the basis of an energy content of 6.53 kWh per liter of ethanol (University of Wisconsin, 2016) and a similar conversion efficiency for the power plant (Ethanol Producer Magazine, 2010), the electricity generation capacity from the total annual ethanol production would amount to approximately 470 GWh, representing 21% of the total electricity generation in 2015 (SM, 2016a).

Flexible Generation and Storage

Some of the renewable sources of energy proposed among the roadmap technologies are intermittent in nature, implying that their availability varies. For instance, the output of PV and wind power plants will depend on the amount of solar irradiance and wind characteristics at the site, respectively. Conventional power systems are used to deal with variability on the load side and balancing supply accordingly. The integration of RE in the grid brings variability on the supply side as well. At high levels of RE integration, the possibility of large fluctuations in generation within small timescales presents new challenges for power systems. To deal with the intermittency of RE resources and maintain reliability, the system must be flexible enough to closely match the electricity demand at all times. Therefore on the supply side, the grid should be upgraded to feature power plants with wide operating capacity ranges and fast ramping rates. Open-cycle gas turbine units are ideal for sudden load because they have the ability to start-up and reach full power output within minutes. On the demand side, efficiency programs and time-of-use pricing (demand response), can also address the balancing issue in an economical way by decreasing or shifting load as needed. Another important measure to reduce the flexibility requirements of the power system is the development of more accurate forecasting and scheduling systems. In recent years, much progress has been made in forecasting intermittent RE output (Kaur et al., 2016).

Adequate storage capacity is also central for enhancing flexibility in power systems with large shares of clean but variable RE. Much research has been dedicated lately to the development of efficient and high-capacity storage technologies, including mechanical storage such as pumped hydro, compressed air energy and flywheels, batteries, supercapacitors, and fuel cells. Due to their fast response time, storage systems are required to provide power during the interval between the falling off of intermittent RE output and the moment that flexible generation capacity is ready to come online. Several of the existing hydropower stations in Mauritius are situated in sites with topographical features that would enable coupling of PV or wind power with pumped storage systems. Convenient storage technologies like batteries, fuel cells, and supercapacitors are still being optimized, and it is anticipated that they will become more affordable and efficient, leading to their widespread deployment. Thus based on the evolution of the performance of electrochemical batteries over the past years, the International Energy Agency expects significant breakthroughs by 2020 leading to decreasing costs (IEA, 2014), while the International Renewable Energy Agency forecasts that lithium ion and flow battery prices will drop by more than 60% and 40%, respectively, by 2020 (IRENA, 2015). The EU's Directorate General for Energy, in its report on the future role and challenges of energy storage, estimated substantial market penetration of supercapacitors, flywheels, compressed air energy, and thermal storage plants over the next decade (EU, 2013). In view of the growing integration of intermittent renewables on the grid, the Mauritian government's budget for the fiscal year 2016–17 provided for USD 10 million (MOFED, 2016) to address this challenge, with a priority on procuring battery storage systems.

Smart Grids

The sine qua non for a green energy revolution in Mauritius is the implementation of a smart grid. The present grid architecture was principally developed around large power plants running on fossil fuels. Communication with electricity customers is still a one-way process involving billing of monthly energy consumption based on manual readings from passive electricity meters. The current grid cannot cope with the fluctuations in electrical supply introduced by intermittent RE sources. As the share of RE grows, the grid must be upgraded with state-of-the-art planning, operational, and security features. Multidimensional communication supporting information flows among numerous devices and stakeholders is necessary to effectively balance variable supply with load. In a smart electricity grid, modern information and communication technologies can be used to coordinate data interchange among sensors, actuators, intelligent metering systems, and controllers to enable real-time management of energy flows. Mauritius has made strides in this context, as it ranks first in Africa and 45th globally in terms of technology and innovation (WEF, 2015). The island is aiming to achieve full broadband fiber connectivity by 2018.

Smart grid pilot and demonstration projects have been deployed worldwide, boosted by government incentives. Many countries are investing substantial effort and financial resources in this area. China, for instance, plans to invest at least USD 96 billion by 2020, while Italy, Japan, and South Korea have earmarked USD 200, 100, and 65 million, respectively, to implement pilot projects (IEA, 2011). In the United States, more than 65 million smart meters have already been installed (USDoE, 2015). Meanwhile, many developing countries are deploying clean and smart micro- and minigrids (Asmus and Gibson, 2011). These are small-scale versions of the conventional electricity grid conceived to serve localized loads and can function in either autonomous or grid-connected modes. The motivations here are mainly to improve reliability and security, diversify energy sources, reduce GHG emissions, integrate distributed RE generation, and lower costs over time. As they encompass all the fundamental building blocks of larger grids, autonomous grids offer a less expensive alternative path for smart grid elaboration. In the same vein, Mauritius can develop a phased planning strategy to facilitate the transition from the existing grid infrastructure to the future smart grid through micro- and minigrids, to enable seamless integration of RE, distributed energy, DSM, and innovative pricing methodologies in the power network.

CONCLUSIONS

This chapter presents the current energy situation of Mauritius and the technologies that have the potential to transform the island and achieve carbon neutrality by the year 2050. In contrast with the existing government energy plan, which is centered on partially decarbonizing the electricity sector through the year 2025, we recommend a comprehensive approach that encompasses energy-intensive transportation and cooling. A revised blueprint would tap the abundant RE resources of the country for energy generation along with energy efficiency and demand-side measures, including pricing. Decarbonization of the transport sector figures high on this proposed plan. Harnessing electricity generation from a diverse range of RE technologies that are most suitable for Mauritius will underpin the energy transformation. We expect that thermal, solar PV, onshore and offshore wind, and bagasse gasification will lead the way in terms of deployment based on resource availability and cost effectiveness. They will be backed by a combination of less mature technologies with high potential for development, such as ocean technologies, biofuels, and flexible generation and storage.

A precondition to the successful implementation of the energy transition is the deployment of a smart grid to facilitate the integration of the various RE technologies. The long-term 2050 horizon entails a range of uncertainties in the planning process. Technological developments, economic growth, government policies, and consumer response will affect the ways in which RE technologies are actually deployed and thus the eventual energy mix. Substantial financial resources, in terms of investments and incentives as well as extensive research, will be required to expand the role of renewables. Moreover, political resolve and public participation are prerequisites for the success of the plan we envision. Extensive changes in the legal and regulatory landscape of the energy sector will also be required to support the growth of RE technologies. Tremendous challenges lie ahead, and undoubtedly mistakes will be made as part of the learning process. The economic, environmental, and social impacts of fossil fuel exploitation mean that we are left with little choice but to embrace this ambitious transition and move boldly toward a sustainable future. Successful implementation in Mauritius will set a landmark for future generations and lead the way for island states.

REFERENCES

ADB, 2014a. Saint Louis Power Plant Redevelopment Project Appraisal Report. African Development Bank. Available at: http://www.afdb.org/fileadmin/uploads/afdb/Documents/Project-and-Operations/Mauritius_Saint_louis_Power_Plant_Redevelopment_ Project_Appraisal_Report.pdf.

ADB, 2014b. SEFA to Support Innovative Energy Efficiency Project in Mauritius. African Development Bank. Available at: http://www.afdb.org/en/news-and-events/article/sefa-to-support-innovative-energy-efficiency-project-in-mauritius-12739/.

Asadullah, M., 2014. Barriers of commercial power generation using biomass gasification gas: a review. Renewable and Sustainable Energy Reviews 29, 201–215.

Asmus, P., Gibson, B., 2011. Microgrid Deployment Tracker 2Q11-commercial, Community, Institutional, Military, and Remote Microgrids: Active Projects by World Region. Pike Research LLC.

Autrey, L.J.C., Kong Win Chang, K.T.K.F., Lau, A.F., 2006. A strategy towards enhanced bio-energy production from cane biomass. Mauritius Sugar Industry Research Institute, Republic of Mauritius.

World Bank, 2015. Private Participation in Energy Database: Mauritius. Available at: http://ppi-re.worldbank.org/snapshots/country/mauritius.

World Bank, 2016. Energy Use per Capita of Countries. Available at: http://data.worldbank.org/indicator/EG.USE.PCAP.KG.OE.

CAIT, 2015. CAIT Climate Data Explorer: Mauritius. World Resources Institute. Available at: http://cait.wri.org/profile/Mauritius.

CEB, 2013. Central Electricity Board. Integrated Electricity Plan 2013–2022. Available at: http://ceb.intnet.mu/.

Clark II, W.W., Bradshaw, T., 2004. Agile energy systems: global solutions to the California energy crisis. Elsevier Press, London, UK and New York NY.

Clark II, W.W., Cooke, G., 2011. Global Energy Innovation. Praeger Press, New York, NY.

Connolly, D., Lund, H., Mathiesen, B., Leahy, M., 2011. The first step towards a 100% renewable energy-system for Ireland. Applied Energy 88, 502–507.

EEA, 2009. Europe's Onshore and Offshore Wind Energy Potential: An Assessment of Environmental and Economic Constraints. Technical report No. 6/2009. European Environment Agency.

Elahee, M.K., 2011. Sustainable energy policy for small-island developing state: Mauritius. Utilities Policy 19, 71–79.

Elliston, B., MacGill, I., Diesendorf, M., 2013. Least cost 100% renewable electricity scenarios in the Australian national electricity market. Energy Policy 59, 270–282.

Enerdata, 2016. Global Energy Statistical Yearbook 2016. Available at: https://yearbook.enerdata.net/world-electricity-production-map-graph-and-data.html.

Ethanol Producer Magazine, 2010. GE Powers Turbines with Ethanol in Brazil. Available at: http://www.ethanolproducer.com/articles/7031/ge-powers-turbines-with-ethanol-in-brazil/.

ETRI, 2014. Energy Technology Reference Indicator Projections for 2010–2050. JRC Science and Policy Reports. European Union.

EU, 2013. The Future Role and Challenges of Energy Storage. The European Union Directorate-General for Energy Working Paper.

EU, 2016. European Commission. Passenger Car Taxation. Available at: http://ec.europa.eu/taxation_customs/taxation/other_taxes/passenger_car/index_en.htm.

EWEA, 2014. The European Wind Energy Association. Wind Energy Scenarios for 2020.

Hansard, July 22 , 2014. Fifth National Assembly, Parliamentary Debates. 3rd Session. Available at: http://mauritiusassembly.govmu.org/English/hansard/Documents/2014/hansardiii0414.pdf.

IEA, 2011. International Energy Agency. Technology Roadmap: Smart Grids.

IEA, 2013. International Energy Agency. World Energy Outlook 2013.

IEA, 2014. International Energy Agency. Technology Roadmap: Energy Storage.

IEA, 2016. International Energy Agency. Global EV Outlook 2016: Beyond One Million Electric Cars.

IEO, 2013. International Energy Outlook 2013, US Energy Information Administration. www.eia.gov/forecasts/ieo/pdf/0484(2013).pdf.

Solar Industry, 2012. Photovoltaic Grid Parity Expected by 2017 in U.S. and China. Available at: http://solarindustrymag.com/e107_plugins/content/content.php?content.10375.

IOC, 2016. Indian Ocean Commission Website. Available at: http://eeas.europa.eu/delegations/mauritius/regional_integration/indian_ocean_commission/index_en.htm.

IPCC, 2007. Climate Change 2007: Synthesis Report. An Assessment of the Intergovernmental Panel on Climate Change. Cambridge University Press, Cambridge.

IRENA, 2015. International Renewable Energy Agency. Battery Storage for Renewables: Market Status and Technology Outlook.

Jacobson, M.Z., Delucchi, M.A., Ingraffea, A.R., Howarth, R.W., Bazouin, G., Bridgeland, B., et al., 2014. A roadmap for repowering California for all purposes with wind, water, and sunlight. Energy 73, 875–889.

Karagiannidis, A., 2012. Waste to Energy: Opportunities and Challenges for Developing and Transition Economies. Springer Science & Business Media.

Kaur, A., Nonnenmacher, L., Pedro, H.T., Coimbra, C.F., 2016. Benefits of solar forecasting for energy imbalance markets. Renewable Energy 86, 819–830.

Knoef, H., Ahrenfeldt, J., 2005. Handbook on Biomass Gasification. BTG Biomass Technology Group BV, Amsterdam, Netherlands.

Krajačić, G., Duić, N., Carvalho, M.G., 2011. How to achieve a 100% RES electricity supply for Portugal? Applied Energy 88, 508–517.

Kurtz, S., Atwater, H., Rockett, A., Buonassisi, T., Honsberg, C., Benner, J., 2016. Solar research not finished. Nature Photonics 10 (3), 141–142.

Levantis, T., 2008. Oil price vulnerability in the Pacific. Pacific Economic Bulletin 23 (2), 214–215.

Lund, H., Mathiesen, B., 2009. Energy system analysis of 100% renewable energy systems - the case of Denmark in years 2030 and 2050. Energy 34, 524–531.

Macedo, I.C., Seabra, J.E.A., Silva, J.E.A.R., 2008. Greenhouse gases emissions in the production and use of ethanol from sugarcane in Brazil: the 2005/2006 averages and a prediction for 2020. Biomass and Bioenergy 32 (7), 582–595.

Mason, I., Page, S., Williamson, A., 2010. A 100% renewable electricity generation system for New Zealand utilizing hydro, wind, geothermal and biomass resources. Energy Policy 38, 3973–3984.

Mauthner, F., Weiss, W., 2013. Solar Heat Worldwide: Markets and Contribution to the Energy Supply 2011. Solar Heating and Cooling Programme. International Energy Agency.

MEF, October 2007. Mauritius Employers Federation. The Business Costs of Traffic Congestion. MEFeedback, Issue Number 3.

Meister Consultants Group, 2015. Solar Water Heating Techscope Market Readiness Assessment: Mauritius and Seychelles. Clean Energy Solutions Center).

MMS, 2015. Mauritius Meteorological Services. Climate of Mauritius. Available at: http://metservice.intnet.mu/climate-services/climate-of-mauritius.php.

MOENDU, 2010. Ministry of Environment and National Development Unit. Mauritius strategy for implementation national assessment report 2010. Available at: https://sustainabledevelopment.un.org/content/documents/1255Mauritius-MSI-NAR2010.pdf.

MoESDDBM, 2015. Minister of environment, sustainable development and disaster and beach management. In: Minister Speech: "toward 100% Renewables in Light of Mauritius Efforts to Lead the Way towards 100% Renewables". Available at: http://environment.govmu.org/English/Documents/speeches/07.12.15%20Speech%20by%20Hon%20Ministe.pdf.

MOFED, 2016. Ministry of Finance and Economic Development. Budget Speech 2016–2017. Available at: http://budget.mof.govmu.org/budget2017/budgetspeech2016-17.pdf.

MPILT, 2004. Ministry of Public Infrastructure and Land Transport. Report on Congestion Pricing in Port Louis. Available at: http://publicinfrastructure.govmu.org/English/Publication/Documents/congest.doc.

MREPU, 2009. Ministry of Renewable Energy and Public Utilities, Republic of Mauritius Long Term Energy Strategy 2009–2025.

Munoz, L.A.H., Huijben, J.C.C.M., Verhees, B., Verbong, G.P.J., 2014. The power of grid parity: a discursive approach. Technological Forecasting and Social Change 87, 179–190.

NEC, October 2013. National Energy Commission, Republic of Mauritius. Making the Right Choice for a Sustainable Energy Future: The Emergence of a "Green Economy". Available at: http://publicutilities.govmu.org/English//DOCUMENTS/NEC%20FINAL%20REPORT.PDF.

NERC, August 2010. North American Electric Reliability Corporation. Flexibility Requirements and Metrics for Variable Generation: Implications for System Planning Studies.

Nguyen, T.L.T., Hermansen, J.E., Sagisaka, M., 2009. Fossil energy savings potential of sugar-cane bio-energy systems. Applied Energy 86 (1), S132–S139.

NREL, 2016. PVWatts Calculator – Solar Resource Data. Available at: http://pvwatts.nrel.gov/pvwatts.php.

Okure, M.A.E., Musinguzi, W.B., Nabacwa, B.M., Babangira, G., Arineitwe, N.J., Okou, R., 2006. A novel combined heat and power (CHP) cycle based on gasification of bagasse. In: Mwakali, J., Wani, T. (Eds.), Proceedings of the First International Conference on Advances in Engineering and Technology (AET2006). Elsevier, Entebbe, Uganda, pp. 465–472.

Pellegrini, L.F., de Oliveira Jr., S., 2011. Combined production of sugar, ethanol and electricity: thermoeconomic and environmental analysis and optimization. Energy 36 (6), 3704–3715.

Power-technology, 2015. Ethanol Power Plant, Minas Gerais, Brazil. Available at: http://www.power-technology.com/projects/ethanol-power-plant/.

REN21, 2014. Renewable Energy Policy Network for the 21st Century. Renewables 2014 Global Status Report. Available at: http://www.ren21.net/Portals/0/documents/Resources/GSR/2014/GSR2014_full%20report_low%20res.pdf.

Riehl, R.R., Shahateet, C.A., de Souza, L.S., Karam Jr., D., 2012. Biomass gasification unit using sugar cane bagasse for power generation. In: Proceedings of the 3rd International Conference on Development, Energy, Environment and Economics (Paris, France).

Scientific American, 2016. Hawaii First to Harness Deep-ocean Temperatures for Power. Available at: http://www.scientificamerican.com/article/hawaii-first-to-harness-deep-ocean-temperatures-for-power/.

SETIS, 2016. Strategic Energy Technologies Information System. Wind Energy. Available at: https://setis.ec.europa.eu/technologies/wind-energy.

Siemens, 2014. Offshore Wind Power as a Pillar of the Energy Transition. Available at: http://www.siemens.com/press/pool/de/feature/2013/energy/2013-08-x-win/presentation-offshore-e.pdf.

Silalertruksa, T., Gheewala, S.H., Pongpat, P., 2015. Sustainability assessment of sugarcane biorefinery and molasses ethanol production in Thailand using eco-efficiency indicator. Applied Energy 160, 603–609.

SM, 2011. Statistics Mauritius. Housing and Population Census 2011. Port-Louis, Republic of Mauritius. Available at: http://statsmauritius.govmu.org/English/Pages/Housing-and-Population-Census-2011.aspx.

SM, 2015a. Statistic Mauritius. Digest of Energy and Water Statistics 2014. Port-Louis, Republic of Mauritius. Available at: http://statsmauritius.govmu.org/English/Publications/Documents/Regular%20Reports/energy%20and%20water/Digest%20of%20Energy%20and%20Water%20Statistics%202014.pdf.

SM, 2015b. Statistic Mauritius. Road Transport and Road Traffic Accident Statistics 2014. Port-Louis, Republic of Mauritius. Available at: statsmauritius.govmu.org/English/Publications/Documents/Regular%20Reports/road%20transport/Digest%20of%20Road%20Transport%20and%20Road%20Accidents%202014.pdf.

SM, 2015c. Statistic Mauritius. Digest of Environment Statistics 2014. Port-Louis, Republic of Mauritius. Available at: statsmauritius.govmu.org/English/StatsbySubj/Documents/Digest/Environment/Digest_Env_2014.pdf.

SM, 2016a. Statistic Mauritius. Energy and Water Statistics - Year 2015. Port-Louis, Republic of Mauritius. Available at: http://statsmauritius.govmu.org/English/Publications/Pages/Energy-and-Water-Statistics-Year-2015.aspx.

SM, 2016b. Statistic Mauritius. Road Transport and Road Traffic Accident Statistics – Year 2015. Port-Louis, Republic of Mauritius. Available at: http://statsmauritius.govmu.org/English/Publications/Pages/RT_RTAS_Year2015.aspx.

Sotravic, 2012. First UN Carbon Credits Granted for Power Station at Mare Chicose. Available at: http://www.sotravic.net/news/12-first-un-carbon-credits-granted-for-power-station-at-mare-chicose.html.

Speight, J.G., 2010. Handbook of Industrial Hydrocarbon Processes. Gulf Professional Publishing.

Surroop, D., Mohee, R., 2011. Power Generation from landfill gas. In: Proceedings of 2nd International Conference on Environmental Engineering and Applications. IACSIT Press, Singapore, pp. 237–241.

UNEP, 2016. Global trends in renewable energy investment 2016. Frankfurt School-UNEP Centre/BNEF.

University of Wisconsin, 2016. Green econometrics: Information and Analysis on the Economics of Solar and Alternative Energies. Available at: http://greenecon.net/ethanol-benefits-and-issues/energy_economics.html.

USDoE, 2015. U.S. Department of Energy. Smart Grid Investment Grant Program – Progress Report II. Available at: https://www.smartgrid.gov/files/SGIG_progress_report_2013.pdf.

Verbruggen, A., Lauber, V., 2009. Basic concepts for designing renewable electricity support aiming at a full-scale transition by 2050. Energy Policy 37 (12), 5732–5743.

Voormolen, J.A., Junginger, H.M., van Sark, W.G.J.H.M., 2016. Unravelling historical cost developments of offshore wind energy in Europe. Energy Policy 88, 435–444.

Walker-Leigh, V., 2012. Small Islands Push for New Energy. Inter Press Service Agency. Available at: http://www.ipsnews.net/2012/09/small-islands-push-for-new-energy/.

WEF, 2015. World Economic Forum. The Global Information Technology Report 2015: ICTs for Inclusive Growth. Available at: www3.weforum.org/docs/WEF_Global_IT_Report_2015.pdf.

Chapter 16

Urban Sustainability and Industrial Migration: The Green Transition of Hefei, China

Benjamin Leffel[a]
University of California Irvine, Irvine, CA, United States

Chapter Outline

INTRODUCTION

Ang (2017) observes that from the early 2000s, local governments along China's coast began increasingly expelling less valuable, high-polluting industries to make room for more valuable, nonpolluting industries, prompting those high-polluting industries to migrate to inland Chinese provinces, as reflected in recorded flows of "beyond-province investment" (*shengwai zijin*), or domestic investment from one Chinese province to another. This study corroborates the connection between beyond-province investment and high-polluting industries by showing that when one ranks the central region provincial capital cities by their relative levels of high-polluting industrial

a. Benjamin Leffel is a Sociology Ph.D. student at the University of California Irvine, holds a Masters of International Affairs in China Studies from the School of International Service at the American University, Washington DC, and is Director of Research for the Tai Initiative, a US–China subnational cooperation organization. His research on Sino-foreign subnational networking and cooperation has informed the work of the US Department of State, the British Government Office for Science, and US local governments.

Sustainable Cities and Communities Design Handbook. https://doi.org/10.1016/B978-0-12-813964-6.00016-1

investment, this ranking matches that of their relative levels of industrial sulfur dioxide (SO_2) emissions and industrial electricity use—with the exception only of Hefei, the capital city of Anhui Province. Explaining this exception is the fact that Anhui has been the leading recipient of high-polluting migrating investment, with its capital city of Hefei being the largest in-province recipient, thus suggesting Hefei was the leading recipient of high-polluting industries expelled from coastal cities. The same is true for Anhui's investment in the local treatment of environmental pollution, for which Hefei is also the largest recipient, making Hefei the region's largest inflows of seemingly opposing investments: high-polluting industries on the one hand, and investment in pollution treatment on the other.

This study frames Hefei leadership decision making as underpinned by urban growth machine politics, in which profit motives are the primary criterion for government decision making, and outlines the series of policy efforts and adjustments undertaken by the Hefei government in the lead up to 2010, through which the city thereafter achieved reduced industrial electricity use and reduced industrial SO_2 emissions. It is argued that Hefei achieved these improvements not only because of high provincial-level public investment in pollution treatment, but also because the city's basic economic growth needs were satisfied through its massive inflows of migrating high-polluting industrial investments, thus overcoming the development demands of urban growth machine politics. The similarly striking impact of per capita income on central capital cities' ability to achieve sustainable transitions is also discussed.

This study is organized as follows: *Green Urban Planning* provides a background on environmental protection-related underpinnings of the 11th and 12th Five Year Plans, trends of green urbanization during these periods in China, and the indices developed to measure Chinese city environmental sustainability. *Dueling Investments* presents inflows of migrating high-polluting industrial investment into the central provinces and relative levels of pollution and energy consumption indicators among central capital cities, and discusses urban growth machine politics in China. *Hefei's Green Transition* describes in detail the environmental protection policy efforts leading up to 2010, the subsequent environmental impacts, and implications for urban growth politics. This study concludes with a discussion of the relationship between interregional migrating high-polluting industries and sustainable reform in conditions of regional development inequality and the future trajectory of sustainable development into China's western region.

GREEN URBAN PLANNING

In 1992 the former Ministry of Construction (today, Ministry of Housing and Urban and Rural Development) began to biennially select and praise "National Garden Cities" nationwide that had perceptibly high standards of sanitation and housing security, the first batch of which was Hefei, Beijing, and Zhuhai

(Zhao, 2011). The National Garden Cities designation was later revised in 2004 by the Ministry of Housing and Urban and Rural Development (MoHURD) with a "National Standard for Eco-Garden Cities" and again in 2005 to incorporate green building and public transport requirements. Since 1997[1] the Ministry of Environmental Protection (MEP) has named several cities and districts as National Environment Protection Model Cities/Districts (Baeumler et al., 2012). In 2001 MoHURD began selecting and praising cities for the "China Habitat Environment Award" in accordance with the United Nations Habitat Award, and in 2007 MEP announced revised Indices for Eco-City and other subnational green designations with a range of socioeconomic and environmental indicators (Baeumler et al., 2012).

This study focuses on China's 11th Five Year Plan (2006–10) and 12th Five Year Plan (2011–15) periods. The 11th Plan included the policy goal of reducing sulfur dioxide (SO_2) and other airborne pollutants (Cao et al., 2009), including a 20% reduction in energy use per unit of gross domestic product (GDP), while doubling per capita GDP. In 2009 before the COP-15 Climate Negotiations in Copenhagen, China expressed its intention to reduce the carbon intensity of its GDP by 40%–45% by 2020 from the levels of 2005 (Baeumler et al., 2012; NDRC, 2006). In 2009 the "China Low-Carbon Eco-City Strategy" was launched by the Chinese Society for Urban Studies under MoHURD, and 12 cities were designated as Eco-Cities, including Tianjin, Shenzhen, and Changsha. In 2010 the National Development and Reform Commission (NDRC) designated eight cities as low-carbon cities, including Tianjin, Shenzhen, and Nanchang (Khanna et al., 2014).

The 12th Plan contained the first explicit targets to reduce carbon intensity per unit of GDP by 17%, reflecting progress on shifting away from "growth at any cost" (Zhang and Barr, 2013) and away from the coal-based industry (Lewis, 2017) and for the first time included climate policy as a key component of sustainable development (Sternfield, 2017). In 2011, the first year of the 12th Plan, the NDRC selected Beijing, Tianjin, Shanghai, Chongqing, Hubei, Shenzhen, and Guangdong Province to carry out a carbon emissions trading pilot scheme (Duan et al., 2014). By 2012 all Chinese prefectural-level cities had announced some form of low carbon or eco-city development strategy (Liu and Wang, 2017). In 2013 MoHURD launched joint international Low-Carbon Eco-City program pilots in collaboration with the US Department of Energy, which were carried out in Hefei, Langfang, Weifang, Rizhao, Hebi, and Jiyuan (MoHURD, 2013).

There is a distinct difference in environmental protection performance between the coastal, central, and western regions, where coastal cities have the highest performance and the western region, being the furthest inland, has the lowest performance. This is broadly reflected in two series of specially

1. Before 2008, MEP was the State Environmental Protection Administration.

developed indices measuring the sustainable development of Chinese cities. First, the Urban China Initiative, affiliated to McKinsey Global Institute, produced a 2011 and a 2013 Urban Sustainability Index for Chinese cities, in which "urban sustainability" was measured by social, economic, environmental, and resource-related categories of sustainability.[2] In both the 2011 and 2013 Indices, each of which reviewed over 100 Chinese cities, coastal cities outperformed central and western cities, in that order. Aside from coastal location, the highest scoring cities had the largest urban population and population density, the highest foreign direct investment and GDP per capita, and the highest percentage of migrant population. Beijing and Zhuhai ranked #1 overall in the 2011 and 2013 Index, respectively (McKinsey, 2012, 2014).

Second, in China's Green Development Index Report, "green development" is measured using a range of economic and environmental indicators[3] in which a higher score reflects a better balance between economic development, natural resources, and the environment than the overall average—for example, lower pollution, more green space, and a higher rate of economic growth. Shenzhen held the top rank, with Beijing and Zhuhai as nearby contenders. As with McKinsey's Urban Sustainability Index, the regional distribution of "green development" in the Green Development Indices was the highest in the coastal region, middle in the central region, and the lowest in the western regions (Li and Pan, 2011, 2012).

DUELING INVESTMENTS

As shown in Fig. 16.1, Anhui maintained a substantial regional lead in domestic investment from the early 2000s until 2011, when Anhui's investment falls to just above the levels of its nearest two competitors, which Ang (2017) explains is because since 2011, Anhui counted only domestic investment projects at or above 100 million RMB in its statistics. Even so, from 2011 forward, Anhui still maintained the regional lead in domestic investment.[4] Anhui Province's regional lead in beyond-province investment is more likely due to geographic than motivational factors, being that by comparison with the other central provinces

2. These included measures of per capita government expenditure on education and health care (social), per capita government investment in R&D and the Gini coefficient (economic), emissions and treatment of air pollution, wastewater and waste (environment), and energy consumption per unit GDP and per capita.
3. Including but not limited to per capita and per unit of land area measures of GDP, emissions and treatment of air and water pollution, water and energy resource consumption, ratio of nonfossil energy consumption to total energy consumption, green space in urban areas and government expenditure on science, education, and health.
4. Annual provincial government work reports were used to extract these data. For example, Guo, G. (Governor), January 8, 2012. 2012 Henan provincial government work report (*2012 nian Henan sheng zhengfu gongzuo baogao*). In: Fifth Session of the Eleventh People's Congress of Henan Province.

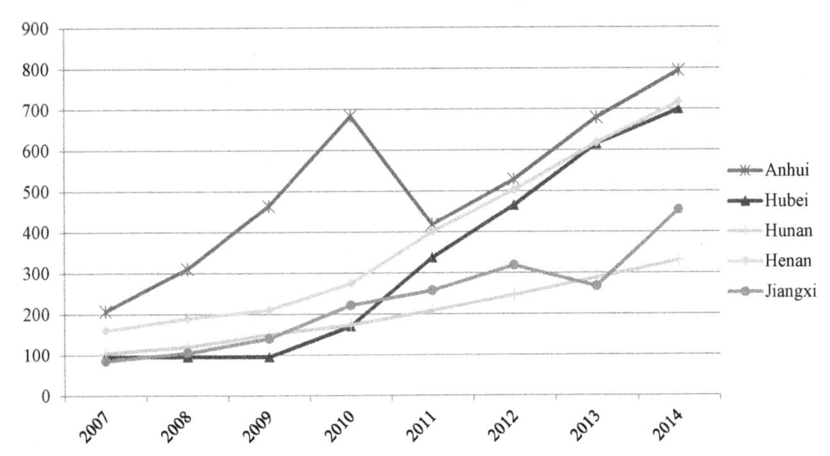

FIGURE 16.1 Beyond-province investment inflows in central provinces, 2007–14 (billion RMB).

(Henan, Hubei, Hunan, and Jiangxi), Anhui has the closest proximity to Shanghai, Jiangsu, and Zhejiang, which are the wealthiest provinces by per capita GDP at purchasing power parity after Tianjin and Beijing (NBSC, 2016). This may suggest that Shanghai/Jiangsu/Zhejiang were the greatest sources of low-end industrial expulsions and subsequent out-migrating high-polluting industries. Among Anhui's cities, Hefei received the largest shares of the beyond-province investment flowing into the province.[5]

As shown in Fig. 16.2, Anhui has also led the region in the provincial investment in the treatment of environmental pollution, the largest individual shares of which went to Hefei. Since financing for public investment has shifted progressively downward to the provincial level and below as fiscal centralization has deepened, including for environmental matters (Wong, 2014), Anhui's spending levels in the treatment of environmental pollution was most likely both the Anhui provincial government's own decision and from its own provincial budget, rather than that of the central government.

5. In 2009, of Anhui's total 464 billion RMB in beyond-province investment, Hefei received the largest share at 77 billion RMB, followed by Wuhu and Anqing, with 57 and 45 billion RMB, respectively. In 2010, of Anhui's total 683 billion RMB of this investment, Hefei received the largest share among the cities of 115 billion RMB, followed by Wuhu and Xuancheng, with 91 and 66 billion RMB, respectively. In 2011, of Anhui's total 418 billion RMB, Hefei received 70 billion, followed by Wuhu with 50 billion, and so forth. In 2012, of Anhui's total 528 billion RMB, Hefei received 87 billion RMB, again the largest share of Anhui cities. In 2013 and 2014, of Anhui's total inflows of 679 and 794 billion RMB, respectively, Hefei received 100 billion and 114 billion RMB. These data were obtained from investment utilization reports. For example, for 2014: Provincial utilization of beyond-province investment and foreign investment, January to December 2014 (*2014 nian 1–12 yuefuen quansheng liyong shengwai zijin he jingwai zijin qingkuang*), *Anhui Province Cooperation Exchange Office*, January 2015.

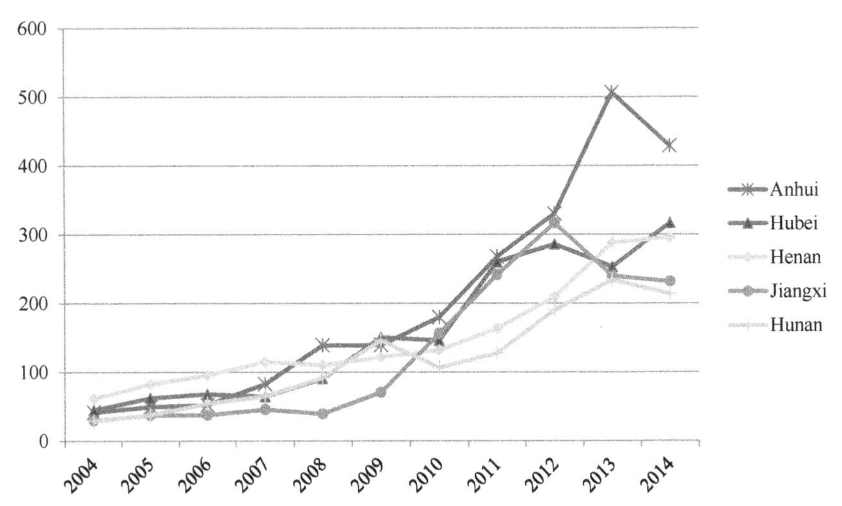

FIGURE 16.2 Total investment of Anhui Province in the treatment of environmental pollution (100 million RMB).

Being simultaneously the central region's largest recipient of beyond-province investment of high-polluting industries and of provincial investment in pollution treatment, Hefei would appear to have both the potential for a severely polluted urban environment *and* the public finances to solve such problems. This begs the question of if and to what extent Hefei has done the latter. The following describes a theoretical basis and an empirical approach that can be used to answer this question.

Regarding a theoretical basis, the concept of a city as a "growth machine" in urban studies describes a coalition between city government officials and parochially interested business elites acting as purveyors for a given urban area's economic growth, whose decisions are motivated by profit maximization. Such a "machine" is commonplace of modern capitalist urban development (Logan and Molotch, 2007; Molotch, 1976), and can also be found in China. In China, however, the local government entities have a significantly stronger position than private sector counterparts because of the state-centric form of governance (Zhang, 2002). This capitalistic propensity toward infinite growth is a principle that underpins institution building in China (Feng, 2016), which is why Chinese city governments normally implement pollution reduction policies by framing them as economically beneficial, emphasizing incentives such as cost savings, employment, and renewable energy opportunities (Koehn, 2016). This study therefore theoretically frames all Chinese cities, Hefei included, as motivated by and subject to the limitations of growth machine politics. These politics can create barriers to progress on environmental protection, but do not obviate the potential for improvements in environmental governance. These conditions do, however, conceptually require that profit-maximization interests be satisfied before environmental reforms can take place and/or that such reforms be accompanied by profit-maximizing opportunities.

Regarding an empirical approach to analyzing Hefei's urban environmental conditions, impacts by industrial processes to Chinese urban environmental conditions are often measured by industrial sulfur dioxide (SO_2) emissions (Pires et al., 2008; Zhang et al., 2016) and industrial coal consumption (Wang and Luo, 2017; You and Xu, 2010), whereas energy efficiency outcomes to sustainable reform efforts are measured using industrial electricity consumption (Shao et al., 2014; Zhang and Cheng, 2009).

Accordingly, the below uses 2004–14 data from Chinese city statistical yearbooks on the annual tonnage of industrial SO_2 emissions, tonnage of industrial coal consumption, and kilowatt-hours of industrial electricity consumption for Hefei as well as each of the provincial capital cities of central region provinces: Changsha (Hunan), Nanchang (Jiangxi), Wuhan (Hubei) and Zhengzhou (Henan).[6] These four cities are included less for direct analytical comparison than for reference purposes, or to provide a regional context for Hefei's indicators. An important caveat is that government-reported air pollution statistics such as these are often manipulated for political purposes in China (Ghanem and Zhang, 2014), so the accuracy of reported SO_2 emissions cannot be certain, although it is important nevertheless to have some baseline of analysis of Hefei's industrial SO_2 emissions. Following Li and Pan (2011, 2012), per capita adjustments are used for city indicators to contextualize variation in the population size for the cities included. This is particularly important for Hefei, given the 2011 decision to place nearby Chaohu City under Hefei's jurisdiction, which suddenly made Hefei responsible for a large new area of over 2 million additional residents along with its industry, energy use, and pollution.[7]

Figs. 16.3–16.5 show industrial electricity consumption, industrial coal consumption, and industrial SO_2 emissions, respectively. An initial observation is that by quantity, the relative levels of beyond-province investment inflows at the provincial level (Fig. 16.1) align with that of their respective capital cities' coal and electricity consumption and of SO_2 emissions, with the exception of Hefei. That is, by measure of relative quantity, all central capital cities but Hefei can be ranked consistently by their respective levels of emissions, coal and electricity consumption (Figs. 16.3–16.5), and beyond-province investment inflows at the provincial level (Fig. 16.1)—in descending order: Zhengzhou, Wuhan, Nanchang, and Changsha.[8] Given that Hefei

6. Owing to data availability limitations, data for annual industrial electricity consumption is only presented for years 2004–13.
7. State Letter no. 84, op. cit.
8. The relative levels by quantity of industrial SO_2 emissions and coal and electricity consumption in central capital cities are more consistent with relative levels of provincial-level beyond-province investment inflows than are city-level population, GDP, and other socioeconomic variables.

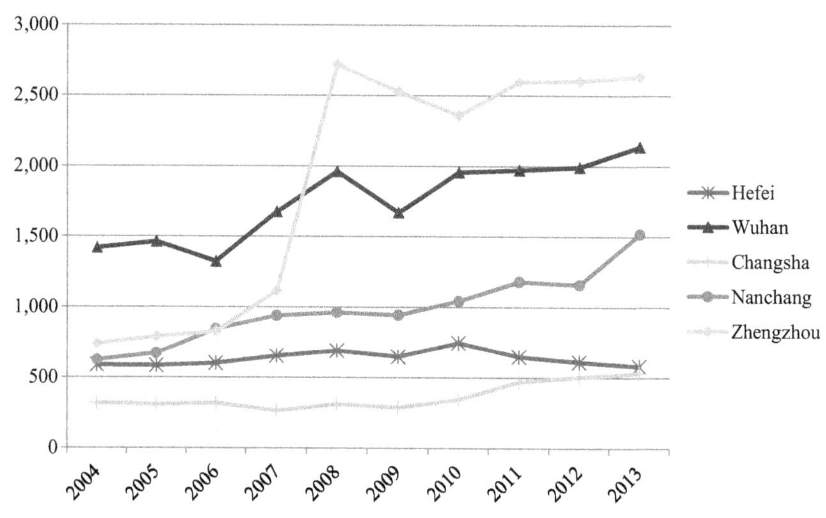

FIGURE 16.3 Industrial electricity use, per capita adjusted (10,000 kWh).

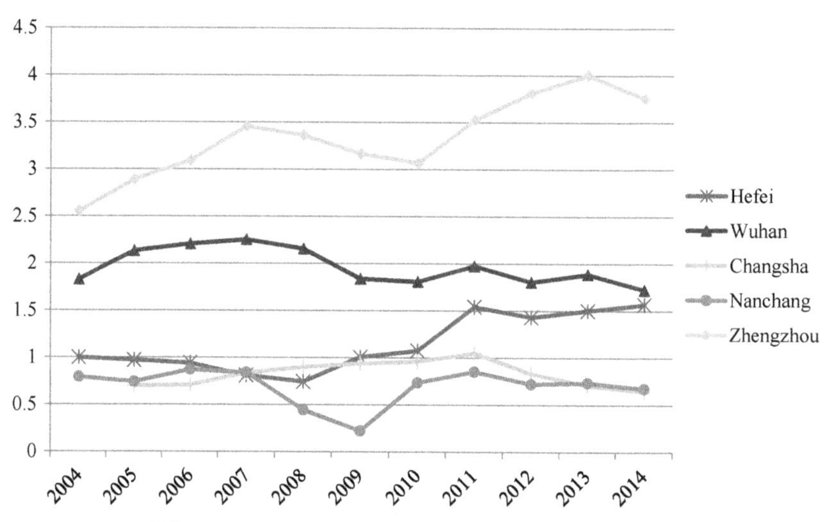

FIGURE 16.4 Industrial raw coal consumption (10,000 tons).

received the largest share of the beyond-province investment inflows at the provincial level in Anhui, it may be reasonably posited that the same is true for the other central provinces and their respective capital cities. Following this assumption, the observed alignment of energy use, emissions, and beyond-province investment would corroborate Ang's (2017) argument that beyond-province investment principally takes the form of high-polluting industries.

That Hefei's relative energy consumption and pollution levels do not match its relative levels of beyond-province investment levels—but rather are

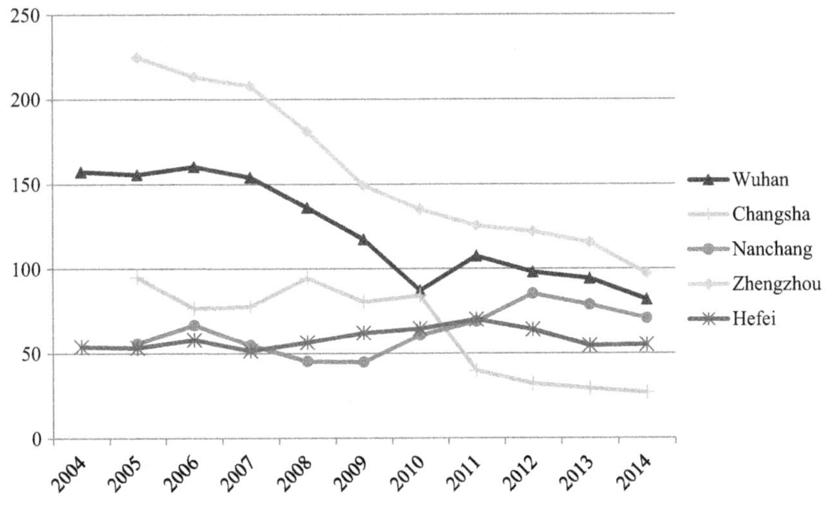

FIGURE 16.5 Industrial sulfur dioxide emissions (10,000 tons).

markedly lower—raises questions. This may be either (1) because the city applies its high-polluting investment inflows in a more environmentally sustainable fashion or (2) because much of its beyond-province investment is not from high-polluting industries. The former would suggest Hefei enforces industrial energy and emission standards more strictly than do the other central capitals, and the latter would beg the question of why the composition of beyond-province investment to Hefei differs from that of the other central capitals in terms of high-polluting industries. This study suggests that the former is the more likely explanation because Hefei engaged in a series of stringent pollution treatment efforts in the lead up to the end of the 11th Five Year Plan (2006–10), which is described in detail in the following sections. These efforts were applied in such a way and in conditions where the urban growth machine politics in which Hefei is theoretically embedded did not prevent successful implementation, thanks in part to Hefei being the largest regional recipient of provincial investment in the treatment of pollution.

Before discussing these policy efforts at length, a few observations about Hefei should be made regarding the above-mentioned figures. First, in 2011 the Anhui Provincial government announced a State Council decision to move nearby Chaohu City from the jurisdiction of Lujiang County to that of Hefei,[9] making Hefei responsible for over 2 million new residents and the industries and pollution that came with them. To ensure that the recorded statistics for

9. State Letter no. 84, "Regarding State Council approval of Anhui Province to revoke the prefecture status of Chaohu City and adjustment of administrative division," Anhui Provincial People's Government, 2011.

Hefei included this new jurisdiction and its population, a review was conducted of a range of Hefei emissions and energy consumption statistics, all of which showed a distinct boost from 2010 to 2011 roughly commensurate with the incorporation of a new jurisdiction of over 2 million people.

Second, Fig. 16.3 shows that only Hefei's industrial electricity use decreased in the post-2010 period among the other central capital cities, and Fig. 16.4 shows that Hefei's industrial raw coal consumption increased in 2011 with its incorporation of Chaohu, and despite a brief drop in 2012, its industrial coal use only increased thereafter. Fig. 16.5 shows that Hefei's 2011 industrial SO_2 emissions increased from the previous year as expected, but the city managed to maintain a decreasing and then a constant emission level in the years following. Hefei's post-2010 improvements in industrial energy efficiency and SO_2 emissions are surprisingly high considering the challenges posed by the sudden expanded jurisdiction and population, but the same cannot be said for industrial raw coal consumption, suggesting that dependence on coal as an energy source in production processes remained constant. As the next section explains, the improvements made are explicable by stringent policy efforts undertaken by the Hefei government during this time period.

Hefei's Green Transition

Hefei's leaders oversaw a substantial period of environmental protection regulatory and capacity improvements during the 11th Five Year Plan period (2006–10), which, although they did not intend this, prepared the city for a 2011 State Council decision that would make the city responsible for over 2 million new residents—and the pollution that came with it. Helping inland China catch up economically with the wealthy coast became a state priority in the early 2000s, beginning with the Western Development Plan, and later the Central Rising Strategy announced in 2004, which sought to redistribute coastal wealth to the central region (Ang, 2017; Lai, 2007). Anhui shares the largest border with Jiangsu, one of the wealthiest coastal provinces in China, and Hefei was presumably deeply involved in these efforts to catch up economically with the coast. In 2005 the Hefei government was trying to lure in as many businesses as possible, having been ranked among the top 50 cities nationwide for investment (KPMG, 2008), following a "growth at any cost" model typical of growth machine politics, and following the coastal development model of previous decades (Ang, 2017). The Hefei municipal government's development model changed during the 11th Five Year Plan period, beginning in 2006. In that year, construction began for Hefei's "Binhu new district" and became a focal point for the city's Eco-City efforts, and its environmental innovations, including green building projects. This project subsequently won national recognition as an Ecological Demonstration Zone (Xia et al., 2010).

Beginning in 2008, the Hefei's Environmental Protection Bureau (EPB) and related government departments markedly increased the number of domestic "study tour" exchanges with other Chinese cities involving environmental management and pollution mitigation, as well as international knowledge-seeking exchanges and partnerships involving the same subject matter.[10] From 2006 to 2009, Hefei government officials increasingly reported their intentions to transition to a substantially more environmentally conscious development model. In 2009 the Hefei EPB implemented a range of new management measures targeting high-polluting industries, such as making several emission sources liable to fines, and subjecting many others to mandatory government inspection.[11] Hefei also held its first environmental protection exhibition that year (Liu, 2009). For example, in 2009 Hefei's EPB was given "a free hand to veto any business projects that are not environmentally viable," resulting in several development projects being turned down by the city. An English translation of a statement by the Hefei mayor was quoted:

> *Hefei will not follow the coastal cities in pursuing an export-oriented and resource-consuming manufacturing business model, which brought them quick growth, but a deteriorating environment. Our business engine should be resource-saving and environmentally sound...overall, the plan is to promote the city as a favorable place to live.*
>
> Si (2009).

A series of stringent environmental governance efforts took place following 2009, in which Hefei officials attempted to achieve environmental improvements while still allowing for economic conditions conducive to high profits. Specifically, Hefei government leaders announced a development model defined by "energy efficient, low-emissions, and high profits," which also would increase market share in alternative-energy vehicles and solar energy products (Zhu, 2011).

In 2010 the final push in the 11th Five Year Plan involved the reduction of SO_2 and other emissions, as well as having 138 enterprises sign environmental

10. For example, starting in 2008, Hefei and its sister city of Columbus, Ohio, USA, engaged in knowledge exchanges pertaining to landfills and automotive companies, which by 2012 added up to 12 exchanges (CSCI, 2013). In 2012, the two cities won an EcoPartnership, a bilateral environmental cooperation partnership program jointly operated by the US Department of State and China's NDRC. Also in 2012, the Hefei government and the Beijing-based China Energy Conservation and Environmental Protection Corporation signed a strategic cooperation agreement (Zong and Wu, 2012).
11. Directive no. 142, "Administrative Measures for the Environmental Protection of Hefei's Service Industry, 2009," Hefei Municipal People's Government; Directive no. 199, "Notice on Deepening the Pilot Project of Enterprise Environmental Supervision," Hefei Environmental Protection Bureau, 2009; and Directive no. 158, "Notice on Special Inspection of Heavy Metal-Polluting Enterprises," Hefei Environmental Protection Bureau, 2009.

commitments on emission reduction (Xia and Liang, 2011; ADHURD, 2010). This led to a range of emissions falling 10% collectively, including SO_2, according to results of an inspection at the end of 2010 by the National Environmental Protection Department of the East China Inspection Center, and by the end of the 11th Five Year Plan period Hefei had carried out 65 emission reduction projects (Zhu, 2011). The Hefei EPB reported putting in place a policy that provided a monetary reward for those who reported enterprises that violated environmental regulations in some discernible form,[12] a range of emission reduction and general pollution prevention projects,[13] banning the burning of straw,[14] and regulations increasing local enterprises' use of automatic air pollution monitoring equipment.[15] These policies were reported to have been put in force alongside central-level emission standards of the same year targeting the magnesium and titanium industries,[16] as well as the industries producing aluminum, lead and zinc, copper and nickel,[17] and standards were established targeting the reduction of nitrogen oxides from thermal power plants.[18]

By the end of 2010, Hefei government leadership proclaimed a successful transition in which they had "steered away from the outdated development path of high energy consumption and high pollution," and transitioned to a low-emission model that sanctioned local enterprises that did not meet new emission-related standards (Zhu, 2011). Overall, a 22% decrease from 2006 to 2010 in Hefei's energy consumption was attributed to the city's new sustainable policy commitments (Tang and Li, 2012), and as shown in Fig. 16.3, energy efficiency improvements reached the industrial sector after 2010. The "energy efficient, low-emissions, and high profits" nature of Hefei's transition (Zhu, 2011) supports the conceptualization of Hefei as driven by urban growth machine forces, given the critical role of profit maximization (Feng, 2016; Logan and Molotch, 2007; Molotch, 1976; Zhang, 2002). That is, environmental reforms were more palatable if accompanied by profit-maximizing opportunities. This apparent green transition for Hefei came at a propitious

12. Directive no. 181, "Hefei city environmental violations reporting reward," Hefei Environmental Protection Bureau, 2010.
13. Directive no. 140, "Second batch of special environmental projects in Hefei 2010," Hefei Environmental Protection Bureau, 2010.
14. Directive no. 65, "Implementation Plan of Straw Banning Law Enforcement Supervision in Hefei in 2010," Hefei Environmental Protection Bureau, 2010.
15. Directive no. 15, "Strengthening operation and management of automatic pollution monitoring equipment," Hefei Environmental Protection Bureau, 2010.
16. Directive no. GB 25468-2010,"Emission standard of pollutants for magnesium and titanium industry," Chinese Ministry of Environmental Protection, 2010.
17. Directive nos. GB 25465-2010, GB 25466-2010, GB 25467-2010, respectively.
18. Directive no. 10, "Technical policy of nitrogen oxides in thermal power plants," Chinese Ministry of Environmental Protection, 2010.

time, as brought the State Council decision for Hefei to expand its jurisdiction to incorporate nearby Lujiang County and its 2 million people.

In the 2011 and 2013 Urban Sustainability Index for Chinese cities, Hefei ranked among the top three most sustainable central region cities, and in the top 10 medium GDP ($5–20 billion) cities showing sustainability improvements from 2008 to 2011 (McKinsey, 2012, 2014). In the 2011 and 2012 Green Development Index reports, among nearly 40 Chinese cities, Hefei scored a middle-level rank overall, but was in the top two for the central region (Li and Pan, 2011, 2012). These indices demonstrate that Hefei's environmental protection performance following its "green transition" has been assessed as high—for the central region—by measures other than those used in this study. Observing the city of Changsha offers additional valuable insights. In all of the above-mentioned Urban Sustainability Index and Green Development Index reports, Changsha was the one central region city outranking Hefei, which is reflective of both Changsha's post-2010 drop in industrial raw coal consumption (Fig. 16.4) and drop in industrial SO_2 emissions, the sharpest among the central capital cities (Fig. 16.5).

Changsha's superior performance may likely be attributable to income. The Urban Sustainability Index reports found that the only city-level indicator that consistently predicted final Sustainability Index scores was per capita income (McKinsey, 2012, 2014), and per capita income was also an important factor determining city sustainability rankings in the Green Development Index reports (Li and Pan, 2011, 2012). Fig. 16.6 shows the urban per capita income levels of the central capital cities, in which Changsha has both the highest relative levels *and* growth rate, and in which Hefei maintains the lowest level

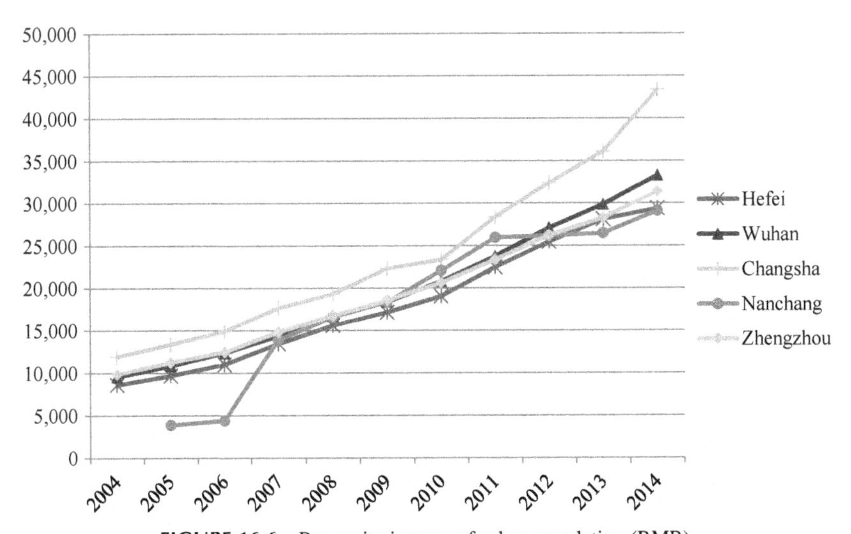

FIGURE 16.6 Per capita income of urban population (RMB).

for most years observed and a modest growth rate. Given Changsha's primacy in the above-mentioned indices, this may suggest that it is only when a city reaches and/or surpasses a certain high threshold of per capita income level and growth rate that income has a significant impact on urban sustainability. Furthermore, Hunan Province has the lowest levels of both in-migrating high-polluting investment (Fig. 16.1) and investment in pollution treatment (Fig. 16.2), suggesting that in the absence of high levels of in-migrating high-polluting industries, high per capita income may provide the sufficient conditions for an effective sustainable urban transition.

DISCUSSION AND CONCLUSION

Through the 2000s, central provinces attempted to catch up economically with the coast by using the same export-oriented, high-polluting model used in the past decades by the coast (Ang, 2017). Just as coastal cities reached a point at which they transitioned away from high-polluting industries, so too did Hefei, as its mayor in 2009 announced explicitly the need to shift away from the old high-polluting coastal model and toward a more sustainable one (Si, 2009). Hefei's transition to a more sustainable, less polluted and more energy efficient development model was enabled only in part by financial resources afforded by the vast quantities of available government financing for the treatment of environmental pollution. The other key element explaining this transition is urban growth machine politics, where profit maximization and economic growth demands guide government decision making (Feng, 2016; Logan and Molotch, 2007; Molotch, 1976; Zhang, 2002). It is due to the persistent presence of such growth-related demand that Chinese city governments normally implement pollution reduction policies by emphasizing their economic advantages (Koehn, 2016), as Hefei did by framing its transition as being toward an "energy efficient, low-emissions, and high profits" development model (Zhu, 2011). However, although this framing helps make sustainable reform in Chinese cities more politically palatable, it does so only in a superficial way.

Rather, for an urban sustainable transition to be theoretically possible in these conditions, it is required that the growth and profit maximization demands of the urban growth machine be substantively satisfied. It is argued here that such conditions were indeed present in Hefei, such that the accumulated growth benefits from years of massive inflows of beyond-province investment satisfied Hefei's economic development and growth needs to the degree that it could afford to incur the economic costs associated with a green transition. Having thus satisfied the urban growth machine demands, and having also the available mass quantities of public investment in pollution treatment, Hefei leaders were both politically and financially capable of exerting stringent policy efforts to reduce emissions and increase energy efficiency. Noteworthy also is that Changsha maintained the highest performance in emissions reduction and energy efficiency among the central capital cities, which appears to be attributable to Changsha having the highest per capita income. This is consistent with findings by the Urban Sustainability Index that per capita income is the strongest predictor of sustainability outcomes in Chinese cities (McKinsey, 2012, 2014), and warrants further research on the impact of income on urban sustainability.

This study's findings may hold lessons for regional inequality regarding sustainability versus economic development. Specifically, how cities in less economically advanced regions of a given nation proceed toward more sustainable models while also being exposed to in-migration of high-polluting industries from cities in more advanced regions. In the case of China's less advanced central region, the city with the most in-migrating high-polluting industries and lowest per capita income (Hefei) transitioned to a more sustainable model by satisfying urban growth demand and by using public investment in pollution treatment, the highest in the region. The starkest reductions achieved in emissions and coal consumption was by the city with the highest per capita income and lowest levels of in-migrating high-polluting industries (Changsha), suggesting such conditions are the most amenable for sustainable transitions. These findings beg the question of if these parameters may similarly explain sustainable transitions in other nations experiencing interregional inequality of economic development and migration of high-polluting industries into the less developed from the more developed regions.

With regard to China, as the cycle of high-polluting industry migration and sustainable urban transitions proceeds further inland over time, China's Western region will be the next (and last) frontier. Several emissions reduction and green urban initiatives of note are paving the pathway to greater sustainable urbanism in China. In 2015 China submitted its intended nationally determined commitment, committing to peak its carbon dioxide emissions by 2030 (NDRC, 2016). The 13th Five Year Plan (2016–20) includes an emphasis on cities peaking their airborne pollutant emissions (NDRC, 2016), and the Chinese government selected 42 pilot cities and provinces to join the national low carbon program at the 2016 US–China Climate Leaders Summit. That same year, 23 Chinese cities and provinces also became members of China's Alliance of Pioneer Peaking Cities to peak carbon dioxide emissions by or before 2030 (Fong, 2016). Such initiatives at present focus predominantly on coastal and central region cities, but just as central region cities' replication of the high-polluting, high-profit development model of coastal cities before them (Ang, 2017) acted as the precursor to urban sustainable transitions, China's western region cities may soon fall into this cycle and thus emerge with a more sustainable development path.

ACKNOWLEDGMENTS

Special thanks to Dr. Ang Yuen Yuen (University of Michigan) and Dr. Benjamin Van Roiij (University of California Irvine) for their invaluable input.

REFERENCES

Anhui Department of Housing, Urban and Rural Development (ADHURD), May 27, 2010. Firm confidence in problem solving and cooperating to reduce emissions (*jianding xinxin po nanti heli zuowei zhua jianpai*). Anhui News. http://www.anhuinews.com/zhuyeguanli/system/2010/05/27/003008589.shtml.

Ang, Y.Y., June 25, 2017. Domestic flying geese: industrial transfer and lagged policy diffusion in China.In: Asian Development Bank Institute Working Paper No. 762.

Baeumler, A., Chen, M., Iuchi, K., Suzuki, H., 2012. Eco-cities and low-carbon cities: the China context and global perspectives. In: Baeumler, A., Ijjasz-Vasquez, E., Mehndiratta, S. (Eds.), Sustainable Low-carbon City Development in China. The World Bank, p. 33.

Cao, J., Garbaccio, R., Ho, M.S., 2009. China's 11th five-year plan and the environment: reducing SO_2 emissions. Review of Environmental Economics and Policy 3 (2), 231–250.

Columbus Sister Cities International (CSCI), 2013. Visits to Date Related to EcoPartnership, Columbus, Ohio USA with Hefei, Anhui Province China (Columbus, Ohio).

Duan, M., Pang, T., Zhang, X., 2014. Review of carbon emissions trading pilots in – China. Energy & Environment 25 (3 and 4), 527–549.

Feng, L., 2016. Ecological values and capitalism. In: Zheng, Y., Fan, L. (Eds.), Chinese Research Perspectives on the Environment, Special Volume, vol. 5. Brill, Leiden, pp. 258–267.

Fong, W.K., June 8, 2016. 23 Chinese cities commit to peak carbon emissions by 2030. World Resources Institute.

Ghanem, D., Zhang, J., 2014. 'Effortless Perfection:' Do Chinese cities manipulate air pollution data? Journal of Environmental Economics and Management 68 (2), 203–225.

Khanna, N., Fridley, D., Hong, L., 2014. China's pilot low-carbon city initiative: a comparative assessment of national goals and local plans. Sustainable Cities and Society 12, 110–121.

Koehn, P.H., 2016. China Confronts Climate Change: A Bottom-up Perspective. Routledge, London.

KPMG, 2008. Hefei Economic & Technological Development Area Investment Environment Study 2008. KPMG Huazhen. https://kpmg.de/media/Hefei_Investment_Environment_Study_2008.pdf.

Lai, H., 2007. Developing Central China: a new regional programme. China: An International Journal 5 (1), 109.

Lewis, J.I., 2017. Green energy innovation in China. In: Sternfield, E. (Ed.), Routledge Handbook of Environmental Policy in China. Routledge, New York.

Li, X., Pan, J. (Eds.), 2011. China Green Development Index Report 2011. Springer, New York.

Li, X., Pan, J. (Eds.), 2012. China Green Development Index Report 2012.. Springer, New York.

Liu, Jun, May 14, 2009. 2009 China Hefei first environmental protection exhibition held in Hefei on June 5 (*2009 zhongguo hefei shoujie huanbao zhan 6 yue 5 ri zai hefei juban*). China Radio Network. http://news.cnr.cn/gnxw/200905/t20090514_505334061.shtml.

Liu, W., Wang, C., 2017. Low-carbon urban development in China: policy and practices. In: Sternfield, E. (Ed.), Routledge Handbook of Environmental Policy in China. Routledge, New York.

Logan, J.R., Molotch, H., 2007. Urban Fortunes: The Political Economy of Place: 20th Anniversary Edition. University of California Press, Los Angeles, CA.

McKinsey & Company, April 2012. The China Urban Sustainability Index 2011. Urban China Initiative, Beijing, China.

McKinsey & Company, April 2014. The China Urban Sustainability Index 2013. Urban China Initiative, Beijing, China.

Ministry of Housing and Urban and Rural Development (MoHURD), July 25, 2013. The Progress of Low-Carbon Eco-City International Cooperation (*woguo ditan shengtai chengshi guoji hezuo qude xin jinzhan*).

Molotch, H., 1976. The city as a growth machine: toward a political economy of place. American Journal of Sociology 82 (2), 309–332.

National Bureau of Statistics of China (NBSC), February 29, 2016. Statistical Communiqué of the People's Republic of China on the 2015 National Economic and Social Development (Beijing, China).

National Development, Reform Commission (NDRC), 2006. China's 11th Five-year (2006–2010) Plan for National Economic and Social Development (Beijing).

National Development, Reform Commission (NDRC), 2016. The 13th Five-year Plan: For Economic and Social Development of the People's Republic of China, 2016–2020. Central Compilation & Translation Press, Beijing, China.

Pires, J.C.M., Sousa, S.I.V., Pereira, M.C., Alvim-Ferraz, M.C.M., Martins, F.G., 2008. Management of air quality monitoring using principal component and cluster analysis-part II: CO, NO_2 and O_3. Atmospheric Environment 42, 1261–1274.

Shao, C., Guan, Y., Wan, Z., Guo, C., Chu, C., Ju, M., 2014. Performance and decomposition analyses of carbon emissions from industrial energy consumption in Tianjin, China. Journal of Cleaner Production 64, 590–601.

Si, T., April 27, 2009. Hefei chooses environment over polluting businesses. China Daily. http://www.chinadaily.com.cn/bw/2009-04/27/content_7717801.htm.

Sternfield, E., 2017. Introduction. In: Sternfield, E. (Ed.), Routledge Handbook of Environmental Policy in China. Routledge, New York.

Tang, S., Li, W., 2012. Hefei's low carbon development. China Today 61 (7), 16–18.

Wang, S., Luo, K., 2017. Atmospheric emission of mercury due to combustion of steam coal and domestic coal in China. Atmospheric Environment 162, 45–54.

Wong, C., 2014. China: PIM under reform and decentralization. World Bank Group.

Xia, Y., Liang, C., February 11, 2011. 2010 Hefei environmental protection 10 major events announced (*2010 nian hefei huanbao 10 jian dashi gongbu*). Hefei Online. http://news.ifeng.com/gundong/detail_2011_02/11/4616878_0.shtml.

Xia, Y., Kong, J., Shi, Q-qian, Song, Ye-hao, Fan, G-hui, Zhou, G-chang, Yi, C., November 11, 2010. Environmental Protection and Eco-city Development (*Huanbao Wei Shengtai Lishi Baojia Huhang*). http://news.ifeng.com/gundong/detail_2010_11/11/3072109_0.shtml.

You, C.F., Xu, X.C., 2010. Coal combustion and its pollution control in China. Energy 35 (11), 4467–4472.

Zhang, T., 2002. Urban Development and A Socialist pro-growth coalition in Shanghai. Urban Affairs Review 37 (3), 475–499.

Zhang, J.Y., Barr, M., 2013. Green politics in China: Environmental Governance and State-society Relations. Pluto Press, London.

Zhang, X., Cheng, X., 2009. Energy consumption, carbon emissions, and economic growth in China. Ecological Economic 68 (10), 2706–2716.

Zhang, J., Zhang, L-Yue, Du, M., Zhang, W., Huang, X., Zhang, Ya-Qi, Yang, Y-Yi, Zhang, J-Min, Deng, S-Huai, Shen, F., Li, Y-Wei, Xiao, H., 2016. Indentifying the major air pollutants base on factor and cluster analysis, a case study in 74 Chinese cities. Atmospheric Environment 144, 37–46.

Zhao, J., 2011. Towards Sustainable Cities in China: Analysis and Assessment of Some Chinese Cities in 2008. Springer, New York.

Zhu, G., January 15, 2011. Eleventh Five-Year Plan Hefei reduction results (*Shiyiwu jianpai hefei chengji zenyang*). Hefei Online. http://news.ifeng.com/gundong/detail_2011_01/15/4295177_0.shtml.

Zong, H., Wu, Jun, March 20, 2012. Hefei and China Energy Conservation and Environmental Protection Group in Beijing signed a strategic cooperation agreement (*hefei yu zhongguo jieneng huanbao jituan gongsi zaijing qian zhanlue hezuo xieyi*). Hefei Daily. http://ah.sina.com.cn/news/hscj/2012-03-20/6583.html.

FURTHER READING

National Development and Reform Commission (NDRC), June 30, 2015. Enhanced Actions on Climate Change (Beijing, China).

Chapter 17

Energy Economics in China's Policy-Making Plan: From Self-Reliance and Market Dependence to Green Energy Independence

Xing Li[1], Woodrow W. Clark, II[1,2]
[1]*Aalborg University, Aalborg, Denmark;* [2]*Clark Strategic Partners, Beverly Hills, CA, United States*

Chapter Outline

INTRODUCTION: FROM SELF-RELIANCE TO DEPENDENCY

The historical transformation from Maoist social-communist China to Dengist capitalist China since the end of the 1970s represents a sharp contrast as to China's national policy objectives, political agenda, economic development, and more importantly development strategy. Dengist China took the capitalist mode of economic development based on privatization of ownership and the means of production and distribution to the marketization and allocation of resources including the total acceptance of economic inequities and political privileges. The basis of this form of capitalism placed emphasis on market-oriented

Sustainable Cities and Communities Design Handbook. https://doi.org/10.1016/B978-0-12-813964-6.00017-3

economics and technology as the essential productive forces, along with the promotion of the interests of the privileged, professional, and entrepreneur classes, to also include the commercialization of welfare and social security. The 1970s was the period in which Western capitalism, promoted by the United Kingdom and the United States, became the norm for over three decades (Economist, 2009). More significantly what China has witnessed in the past three decades is an economic growth path based on increasing demand for energy consumption based on an annual gross domestic product (GDP) growth of 10%. By the end of 2010, China had outpaced the United States in clean technology investments (Scientific American, April 5, 2011). According to M. L Chan, at the end of 2010, China ranked number #1 in clean technology investments, whereas the United States fell to #3 (Chan, 2010).

By the end of 2008, China ranked fourth in the world in wind turbine manufacturing and installation with 12 GW of power (Zhen et al., 2009). Today Chinese companies control over half of the annual US $65 billion wind turbine market (Rosenberg, 2010). Could China become the #1 economy in the world, surpassing the United States? (Time, 2011). With its GDP at $5.88 trillion in 2010, China surpassed the former #1 economy, Japan with a GDP of $5.47 trillion. The GDP per capital of $7518 for the Chinese is much lower than the $33,828 for the Japanese, but some economists predict that by 2030, China will have a GDP of $73.5 trillion, compared with those of the United States of $38.2 trillion and Japan, in third place, of $8.4 trillion. The world economics has changed. Some Americans are skeptical and fearful of the Chinese economic challenge. A recent series of meetings in the US Congress have focused on "China's energy and climate initiatives" with a subtitle "successes, challenges, and implications for US policies" (US Congress, 2011). The presentations were varied and gave no conclusions.

During the socialist period [after World War II (WWII)], due to historical reasons, China took a development course with a commitment and goal by emphasizing human capacity, economic equality, and balance, thereby mobilizing social and economic resources in pursuit of a self-reliance development strategy. No matter how the socialist strategy is interpreted and assessed from today's perspective, it was historically the only possible option if China wanted to sustain its economic development, national security, and independence. The strategy of self-reliance emphasized the primacy of internally generated *independent* development not only at the national level but also at the provincial, regional, and local levels. The national 5-year plans started under Mao continue today. These internally generated *independent* development plans were reflected in the institutional structure of each unit,[1] where the multifunctions of party leadership, production planning, medical service, and

1. A unit, in Chinese "*Dan Wei*," refers to any functional organization, for example, a ministry, a university, a company, a factory, etc. A "Dan Wei" was a self-managed "miniwelfare state" combing supply, demand, and welfare.

resource demand and supplier controls were closely integrated. Self-reliance under such a socioeconomic structure in which politics, economics, service, and supply existed together at the unit level created a sustainable society based on their own resources and needs and under the overall guidance of the state's 5-year plans.

The 5-year plans or *The Five-Year Plan for National Economic and Social Development* are national goal-setting economic and policy position papers derived over a period of 1 or 2 years from high-level and local committees. At the national government level, the 5-year plans outline national key construction projects and infrastructure plans as well as administer the distribution of productive, manufacturing and business sources and growth of academic and educational institutions along with individual sector contributions to the national economy. The plans also provide large sums of national funding for implementation. Aside from giving the nation, business, government, and foreign interest a "road map" of the Chinese policies, the plans map the general direction of future development including specific measureable policies and targets. The last Five-Year Plan was from 2006 to 2010 and officially called the *11th Five-Year Development Guidelines*. The current 12th Five-Year Plan or 12-5 year Plan came out officially in March 2011.

In the context of energy, particularly for homes and buildings, self-reliance also meant that the focus for renewable energy was primarily technologies that could help in the heating and cooling of buildings. Solar thermal systems were developed by state-operated companies in the 1990s and then spun off into private-public-owned firms. Sundra is a case in point with solar thermal systems that appear on homes all over China (Kwan, 2009). Now their market has grown worldwide, as examples from colleges in the United States illustrate (Eisenberg, 2009).

The Chinese choice of a self-reliance and self-sufficiency development path was projected as a potential "ideological threat" by the capitalist Western nations, because the central goal of the socialist politics with specific plans from the central government were seen as an attempt to challenge the capitalist ideology of competition that would reduce costs but unfortunately also lead to an inequitable hierarchy in the world order (Downs, 2000, 2004, 2006; Jiahua et al., 2006; Kaplinsky, 2006; Liu, 2006; Konan et al., 2008; Li, 2010; and today, US Congress, 2011, among others). Seen from the interpretation of world system theory, the socialist self-reliance and self-efficiency policy aimed at transforming the basic logic of capitalism into "social capitalism" (Clark and Li, 2003). However, in reality it was actually designed toward a nation-wide mobilization for industrialization for the purpose of catching up with the core advanced capitalism that rewards people with money, no matter where the funds came from or by what means the funds were acquired. This is because socialist states were still operated within the capitalist world economy, and the dynamics of capitalism was capable of distorting and limiting national economic planning, leading to the constraints of their policy options (Chase-Dunn, 1982, 1989). Nevertheless, such a socialist project based on self-reliance existed outside the US-led capitalist world economic system. In other words, it was merely an ideological challenge without being able to construct a sustainable alternative model (or paradigm based on past 2010–16 historical evidence) to replace the capitalist system.

In addition, the Chinese self-reliance policy was also a response both to the internal constraints of socioeconomic backwardness and the external pressure by the US-led economic blockage and trade embargo after WWII and the Korean War as well as from the failure of China's dependent relations with the Soviet Union since the Republic was founded in 1949. What should be pointed out here is that China's emphasis on self-reliance and independence in the Chinese energy industry was primarily derived from the lessons learned due to the Sino-Soviet split in which China was deprived not only of the Soviet technicians and specialists who were helping China develop its industries but also of around 50% of its oil supplies that were imported from the Soviet Union (Downs, 2000, p. 11–12). The discovery of the Dajing[2] oil field at the end of the 1950s was declared to mark the end of China's external oil dependence both for defense and civilian applications. Such was the case in the modern market-oriented economy after Mao in the 1980s.

The post-Mao transformation of policy orientation from socialism to market capitalism with its objectives has its roots in the change of the regime's perceptions of the external environment of international political economy since the 1980s and especially due to the end of the Cold War in the early 1990s. The perceptions were generated from the conceptualization of international relations that (1) the superpowers, including their respective alliances, were exhausted in their endless competition, leading to a situation in which no major serious conflict was likely in the future even with the almost decade-long US military involvement in the Middle East and the 2011 military warplane presence in Northern Africa.

On the other hand, there would be the emergence of nonconventional security challenges whereby the Chinese government–controlled oil and gas companies bought international oil- and gas-producing and transport companies. (2) Economic development became the key objective for all nations, and economic power thus emerged to become more important and relevant than traditional military strength. Thus, soon, China had to face economic interdependence by increasing its global economic presence through fuel supplies; (3) the post–Cold War US-based and US-controlled world order, which is an American-centric new world order, would likely remain for an unknown period of time. Therefore, China should, in the words of Deng, "observe calmly, secure our position, cope with affairs calmly, hide our capacities and bide our time"; and (4) hence there would be growing global

2. Daqing was the largest oil field in China, and it is located in Heilongjiang province. It was discovered in 1959, and the production started in 1960.

competition for natural resources for energy security. It was this last area that began after the turn of the 21st century, when China's unknown global challenges were still being defined and hence significant tasks needed to be accomplished moving from totally government-controlled industries to ones that were collaboration or joint ventures with foreign companies and often owned by a majority of Chinese workers (Clark and Isherwood, 2010).

Driven by the aforementioned new strategic understanding, China has been pursuing a global foreign economic policy that was directed at creating a stable and peaceful environment for its economic growth through active engagement with the West and with the surrounding Asian nations. This strategy has become China's globalization focus for a new or forthcoming economy (Li and Clark, 2009). China grasped opportunities for increasing international trade and foreign direct investment, and, more importantly, for securing access to natural resources and energy supplies through its own international trade and investment in the resource-rich regions such as Africa, Latin America, and Southeast Asia, and in recent years, Central Asia. China's global policy strategies under an active role of the state have been seen as effectively making it one of "globalization's great winners" (Thøgersen and Østergaard, 2010).

In short, China no longer was an "emerging nation" and founding member of BRIC (Brazil, Russia, India, and China) but had leapfrogged technologically, economically, and politically into the third industrial revolution (3IR), which moves an economy off the dependence on fossil fuels to renewable energy, smart green grids and advanced storage technologies, as well as sustainable communities that reverse carbon emissions and reduce pollution; it started in Asia and the European Union over two decades ago (Clark and Cooke, 2011).

China, by the end of the first decade of the 21st century, moved actively into the 3IR. The Pew Charitable Trusts (SJBJ, 2011, p. 8) was quoted in Apr. 2011 with global data. For example, China reported $54.4 billion in clean tech investments for 2010, which was an increase of 39% from the year before and thus led all nations. Germany was second with $41.2 billion with the United States at the third position at $34 billion or 51% over 2009. Italy was fourth with $13.98 billion or up 124% from 2009, whereas the rest of the European Union ranked fifth with $13,48 billion, which was off 1% from 2009.

China's remarkable achievement in economic growth was made possible by its growing involvement in the capitalist world system. Steven Chan verified this fact as he told the story of SunTech becoming the world's #1 solar manufacturing company in 2010 (Chan, 2011). However, China remained in charge with caution and intense controls from the central government. It did not, for example, experience the deep 2008 global economic recession. In other words, China's economic growth is inseparable from its increasing dependence on global markets, with some estimates suggesting that more than 40% of its GNP is derived from international trade (Chan, 2010). In other words, China's rapid economic growth has been driven by exports with the assistance of foreign investments and joint ventures that have dominated the most dynamic sectors of the economy. Its market-driven growth encourages more concessions to induce capital flows and growth in unlimited possibilities of expansion and more structural changes to meet the demand of the overwhelming pursuit of external markets and resources (Lo, 2011). In addition, its integration with the world market is followed by overdependence of the productive forces on the fluctuations of the world market. The most affected area is China's energy demand and supply. Chart I illustrates the problem in China today with most of its energy demand being met by the supplies of coal, natural gas, and nuclear power. To summarize part of a recent series of presentations to the US Congress (April 2011):

> *China is now the global manufacturing leader of most renewable energy technologies, and the largest user of clean energy. In 2010 alone, clean energy investment in China topped $154 billion, while approximately 77 gigawatts of old fossil fuel power was shut down.*

THE REAL PREDICAMENT: ENERGY-CONSUMPTION-BASED ECONOMIC GROWTH

During the past three decades, China's GDP has enjoyed an average growth rate of 9%–10%. Since 2002, China's energy consumption has been growing at a faster rate than its GDP growth. From 2000 to 2005, China's energy consumption rose by 60%, accounting for almost half of the growth in world energy consumption (Downs, 2006, p. 1). In 2004, China contributed 4.4% of total world GDP, whereas China also consumed 30% of the world's iron ore, 31% of its coal, 27% of its steel, and 25% of its aluminum. Between 2000 and 2003, China's share of the increase in global demand for aluminum, steel, nickel, and copper was, respectively, 76%, 95%, 99%, and 100%. On a global scale, an increase in personal car ownership alone could mean an extra billion cars on the road worldwide within the next 10 years (NYT, 2010). As a report by the Chinese Academy of Social Sciences in 2006 predicted accurately on the mounting worldwide impact of China's resource consumption:

> *China is currently the world's third largest energy producer and the second largest energy consumer. In 2002, China accounted for 10% of total world energy use and is projected to account for 15% of global energy use by 2025. China is the world's largest coal producer accounting for 28% of world coal production and 26% of total consumption. China is the third largest consumer of oil and is estimated to have the world's sixth largest proven reserves of oil. China has roughly 9.4% of the world's installed electricity generation capacity (second only to the United States) and over the next three decades is predicted to be responsible for up to 25% of the increase in global electricity generation. China*

emitted 10.6% of global carbon emission from fossil fuels in 1990 (second only to the United States) and 14.2% in 2003. This share is projected to rise to 22.2% by 2020 (IEA, 2006).

This situation has given rise to problems of net energy imports, environmental pollution and ecological destruction, cross-border pollution, and mounting carbon dioxide (CO_2) emissions:

The domestic environment has deteriorated rapidly, with some 70% of urban population exposed to air pollution, 70% of seven major water systems heavily polluted, over 400 cities short of water, and 3,400 km^2 (equivalent to Japan's Tottori Prefecture) turning into desert every year. Cross-border pollution, notably acid rain and sandstorms, have reached the Korean Peninsula and Japan. Global environmental problems: China is the world's second-largest CO_2 producer after the U.S.

Li (2003, p. 1)

By early 2010, China moved ahead of the United States in being the number emitter of carbon in the world, alto per capital, the US is still the leading nation.

China's escalating energy consumption is placing heavy pressure on the world's energy prices. Chinese energy demand has more than doubled during the past decade. According to the study by Konan and Jian (2008), China will account for about 41% of the global coal consumption and 17% of global energy supply by 2050. Metal prices have increased sharply due to strong demand, particularly from China, which has contributed 50% because of the increase in world consumption of the main metals (aluminum, copper, and steel) in recent years. Due to its rapid growth and rising share in the world economy, China was expected just after the turn of the 21st century to retain its critical role in driving commodity market prices (*World Economic Outlook*, Sep. 2006). It did so, according to almost every study, by the end of the first decade of the 21st century (ACORE, 2011; IEA, 2011; US Congress, 2011). China is willing to offer greater prices for purchasing raw materials than those offered by world markets, which not only attributes great comparative advantages to the developing world but also threatens the West, which seeks to purchase and control those fossil fuel supplies (Tseng, 2008); Historically, the world has been already burdened by the high energy consumption by the West, particularly by the United States. Today China's growing appetite for international trade drives its mounting demand for resources to sustain its economic growth and to fuel its countless development projects. China has already become the world's largest importer of a range of commodities, from copper to steel and crude oil (Silverstein, 2011). The phenomenal rise of commodities prices worldwide in recent years is claimed to be due to China's growing import demand. Some even worry that there would not be enough resources in the world, for example, gas and oil, to satisfy the ever-increasing demand driven by China's economic growth. Furthermore, taking China's neighbor, India, into consideration with its population of 1 billion, it will add double pressures on the demand for the same resources.

After a short period of self-sufficiency in energy supply, especially in coal and petroleum, China became a net importer of petroleum in 1993, and it took only a few years before China became the second largest oil importer and consumer after the United States in 2009 (USDOE, 2011). China's energy profile used to be heavily weighted toward fossil fuel technologies (petroleum and coal) at a time when reductions are urgently needed to stabilize global climate change. Based on the 2008 statistics from the International Energy Agency, the growth rate of China's energy consumption and its share of the global total final consumption are comparably much higher than the rest of the world (Fig. 17.1).

It is foreseeable that China's resource imports of oil and gas will continue to increase if its targeted economic growth is to be maintained and it does not move fast enough into 3IR. The implication is therefore clear that not only global commodity price and international geopolitical power relations will be affected but also China's international politics, such as its foreign policy rationale, international aid objective, arms sale consideration, and compulsory expansion of its long-range naval power projection capabilities, will be closely connected energy economic issues. The rise of China, as a key actor in energy consumption, is forcing the current international energy regime to adjust or modify its established rules of the capital market game. This is because the international energy regime "is influenced not only by economic, political, and social factors of resource-rich countries but also by international political factors, particularly change in the international balance of power, adjustment of relationships among countries and changes to international rules" (Xu, 2007, p. 6). However, events in North Africa and the Middle East in early 2011 may have changed all this. Clearly the price of fossil fuels will rise and remain high.

One of the key challenging questions raised here is whether the global ecological system is on the verge of reaching the limit and whether the

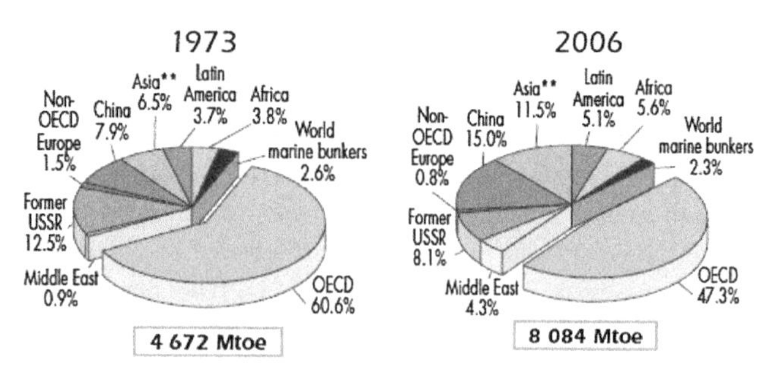

FIGURE 17.1 Shares of total final consumption in 1973 and 2006. *Courtesy International Energy Agency, 2008. Key World Energy Statistics, p. 30.*

expansion of global resource consumption is ecologically possible. The consequences facing China are very severe, and the Chinese growth model could face a fundamental challenge because the peak of resource exhaustion and the imperativeness of ecological sustainability would impose severe limits to its future economic growth, hence fundamental social changes will be inevitable (Li, 2010).

Fig. 17.2A and B calculate and project the peak period of both the world's and China's primary energy supply.

> *World oil production is projected to have peaked in 2008. World natural gas production is projected to peak in 2041. World coal production is projected to peak in 2029. Nuclear energy is projected to grow according to IEA's "Alternative Policy Scenario." Long-term potential of the renewable energies is assumed to be 500 EJ (12,000 million tonnes of oil equivalent). The world's total energy supply is projected to peak in 2029.*
>
> Li Minqi (2010)

Furthermore, China's coal production is projected to peak in 2030; oil production, in 2016; and natural gas, in 2046. China's long-term potential of nuclear and renewable energies is assumed to be 1000 million tons of oil equivalent, China's energy imports are assumed to keep growing from now to 2020 to sustain rapid economic growth. The Chinese economy is assumed to keep growing at an annual rate of 7.5% from 2010 to 2020. By 2020, China's energy imports are projected to grow to near 700 million tons of oil equivalent, comparable to the current US energy imports. After 2020, China's energy imports are assumed to stay at 8% of the rest of the world's total fossil fuel production. China's total energy supply is projected to peak in 2033 (Li, 2010, p. 130–131).

Clark and Isherwood (2010) found this same pattern of short-term fossil fuel energy supplies in China while studying the inner Mongolia autonomous

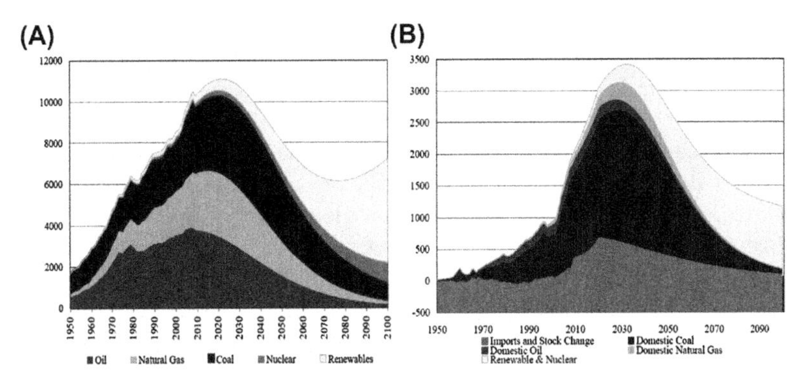

FIGURE 17.2 (A) World primary energy supply (million tons of oil equivalent, 1950–2100). (B) China's primary energy supply (million tons of oil equivalent, 1950–2100). *Courtesy Li, M., 2010. Peak energy and the limits to economic growth: China and the world. In Xing, L. (Ed.), The Rise of China and the Capitalist World Order. Ashgate, Farnham, England, p. 128, 130.*

region (IMAR) for the Asian Development Bank in 2007. IMAR is the second largest coal producing region in China and in the use of renewable energy (solar and wind in particular) to transition from the environmental problems caused by coal. China needs to provide public policies in its next 5-year plan for sustainable development with the financial resources. The 12th Five-Year Plan started that with over half of 1.5 GW of power from global solar installation estimates being produced in China (Chan, 2011). Today, the IMAR is developing such renewable energy resources like wind and solar while controlling its coal production through advanced coal technologies that are "cleaner." Almost every month several megawatts of wind power are being installed and operated in IMAR (Martinot and Li, 2010).

THE RISE OF CHINA IN THE CONTEXT OF ENERGY DEPENDENCY

To keep the economic growth rate, China has to make the access to adequate energy supplies a national priority and to a great extent a national *security* priority (Constantin, 2005; Huliq News, 2008 and Li, 2010). China is perhaps one of the few countries that regard energy security as a vital component of their *national interests*. Currently China is "the world's second largest consumer and third largest producer of primary energy" (Martinot and Li, 2010). There is no sign that China's energy consumption will slow down; on the contrary, it will steadily increase. Thus for energy consumption to keep pace with its targeted economic growth at a moderate rate of 8%–9%, China will have to utilize every fuel source available including investment on renewable energy and the expansion of nuclear power. It is expected that China's import of energy resources will increase at a steady rate particularly from Russian natural gas and liquefied natural gas shipped through Chinese seaports, which are both difficult options (Clark and Isherwood, 2010).

China's growing interest in resource-rich regions such as Africa, Latin America, Middle East, Central Asia, and Southeast Asia is no doubt linked with its energy security consideration (Brautigam, 2008). How will the rapidly increasing demand for energy, raw materials, and other natural resources shape Chinese policies toward its international relations especially with resource-rich countries? Can China afford depending on global energy markets, either via exclusive bilateral deals or via direct investment in resource exploration to sustain its economic growth? What strategies will China use to secure its share of the global resource market? To find the answers to these questions, it is important to take an energy security approach to explore the geopolitical, economic, energy, and environmental implications behind China' rapidly growing energy challenges and to understand the Chinese anxiety and concern with issues of energy security in attempting to search for new sources of energy supply.

China's economic and foreign policy behavior is increasingly influenced by growing energy concerns. Andrew Chung, principal at Lightspeed Venture Partners in Silicon Valley, notes that the Chinese increase in clean (green) tech financing is reflective of the US drop: "First, a lot of the clean technologies are dependent on policy and government support to scale up. In some other parts of the world, you have more consistency in the way these types of funds are appropriated" (SJBJ, 2011, p. 8). As the world's second largest economy and trading nation in 2010, China's search for energy and its global strategies for energy security have led to heavy debates and even in some cases have resulted in political conflict. China is predicted by some economists and members of the US Congress to be the #1 economy and trading nation within the next decade (US Congress, 2011).

The Western nations have been expecting that ideally China's energy vulnerability might drive it toward cooperation with rival oil-consuming nations through participation in multilateral organizations and other forums. Since energy security is no doubt playing a more decisive role in Chinese foreign policy, Beijing's relations with both the existing major energy-consuming powers and energy-exporting countries will shape its motivation and justification on energy issues as well as on nonenergy issues.

In recent years, China's "going-out" economic and foreign policy encouraged its national oil companies (NOCs) to try to acquire some Western oil companies, but still secure the control over access to some overseas energy supplies including purchasing equity stakes in foreign oil companies (US Congress, 2011). This strategy has been regarded as "mercantilist" in the West and particularly in the United States where the attempt of a Chinese NOC to buy out the American oil corporation UNOCAL in 2005 triggered political backlash in the US Congress causing the final withdrawal of the Chinese company. The incident indicates the lack of trust of the United States in China's energy diplomacy, because the US politicians felt that the Chinese move was to undermine American energy security. Hence the US oil and gas company Chevron bought UNOCAL.

In the studies of China's energy security with its economic and foreign policy, a number of geopolitically vital areas cannot be disassociated with China's efforts to maintain both energy security and good international relations within these regions and with the major Western powers. China's energy diplomacy with the Middle East, Russia and Central Asia, the Asia-Pacific, Africa, and Latin America has become a global topic, where Beijing's efforts toward greater energy security through multilateral organizations is discussed. It is still too early to predict whether the world is to witness the evidence supporting the liberal hypothesis that economic interdependence promotes international cooperation, or confirming the realist conviction that competition and power accumulation will eventually lead states to conflict and war, as history has shown. Energy demand is seemingly accelerating China's "peaceful" rise to global prominence, and moderating the conflicting aspects of Chinese foreign policy, while China establishes hundreds of new solar, wind, and other renewable energy companies. Chan (2011) stated that of the global installation of 2.5 GW of solar energy in 2011, over half will be in China.

China has been struggling to develop and promote good relationships with underdeveloped regions that contain potential energy reserves, such as Africa and Latin America, through its unique international aid system linking development aid and trade with energy suppliers. Recently, China has aimed to prepare for technological advances and changes of climatic circumstances that will bring maritime transport in the Arctic waters to make possible the linking of North Atlantic and the North Pacific into closer commercial relations. Some policy makers expect that China will increasingly strengthen its political economy of international relations in the Arctic region and speed up its research through its polar research bases in the Antarctica. In addition, China is adopting different policy strategies and objectives to different regions.

Africa and Latin America

Compared with other regions in the world, these two regions are seen as relatively stable markets as energy suppliers. China's central policy objectives in Africa and Latin America are stated clearly in its policy papers *China's African Policy Paper* (2006) and *Latin American policy paper* (2008). The policy objectives are aimed at strengthening diplomatic and political ties with these two resource-rich regions while at the same time securing and diversifying energy supplies and other raw material resources including the opening of these regions' commodity markets. Currently, China is one of the key investors in Africa, and its trade and investment relations in Latin America are going to accelerate in the coming years (Hanergy, 2011). China's increasingly dynamic economic relations with these regions through long-term financing of infrastructures, renewable energy technologies, and smart grid systems are seen by some Western critiques as challenging the traditional ties between these regions and their historical colonial ties with the Western powers. Intensification in China–African and China–Latin American trade relations also accelerated the "neo-colonialist" argument claiming that China is imposing the regions with a renewed "colonial" relationship. However, despite the criticism of China's energy-oriented policy in its economic and political relations with the two regions, the Chinese style of approach and engagement, especially its aid policy and practice, has indeed a far-reaching long-term and permanent realignment of power relations in the conventional international aid system and has already changed the system in many ways (Opoku-Mensah, 2010).

Middle East and Central Asia

These two regions are world's most unstable energy markets. China has gradually emerged as one of the regions' main partners. China's emerging presence in the regions is seen as one of the major geopolitical changes in the aftermath of demise of the Soviet Union's power and the consolidation of China's new power (Peyrouse, 2007). The two regions are now set to play a major part in China's energy policies and in the war against the separatists in China's northwestern region. The economic and political rise of the Chinese has a great implication in the two regions in terms of the reinforcement of China's policy objectives and the reinforcement of the geopolitical alliance embedded in the strategic calculation of energy security. Currently the Iran nuclear issue is testing China's foreign policy orientation in the context of its energy security consideration. China–Iran partnership has grown out of mutual need for products, ranging from arms and technology to consumer goods and China's soaring need for energy supplies (Dorraj and Currier, 2008, p. 70). Thus it has been a painful foreign policy decision for China to lend support to the US-led United Nations' sanctions against Iran's nuclear program, fearing the grave consequence that this might lead to loss of one of its major energy suppliers.[3] China is being torn between the imperative need for energy on the one hand and the US pressure on its role as "responsible stakeholder" and "strategic reassurance."[4]

The enlarging discrepancy between the economics of energy demand and domestic supply is driving China to reply on a number internal and external policy choices to keep the planned growth rate, a tendency that makes it politically vulnerable to economic setbacks. First, China has to provide huge investments into discovering oil and natural gas resources in the Western part of the country despite the burdens of massive capital investment, high production costs, infrastructures, as well as environment and geological risks. Second, China has to depend on the unstable Persian Gulf areas and other crude oil suppliers in Africa and Central Asia where civil wars, geopolitical risks, and sociopolitical conflicts are unavoidable. Even if the energy supply sources are secure, the transportation issue is becoming another headache for China. In the case of its neighbors, China can construct an oil and natural gas overland pipeline from Central Asian and Russia. Cross-land pipeline options were already put on the highest negotiation table between China, Russia, and other Central Asian countries.

In connection with its rising energy import, the transport of energy products has been the lifeline of China's economic development. China's coastal line areas are the heart of its economic growth and the frontier of its international trade. China's growing maritime ambitions, which already

---

3. According to data released by the General Administration Agency, Iran supplied 11.3% of China's energy consumption in 2009 (adapted from *People's Daily Online*, February 10, 2010).
4. "Strategic reassurance," coined by James Steinberg, Deputy Secretary of State, in a conference sponsored by the Center for a New American Security, states that "China must reassure the rest of the world that its development and growing global role will not come at the expense of security and well-being of others."

boasts of being the world's fourth largest merchant fleet, contributes 6.8% to global tonnage (UNCTAD, 2005). However, in the aftermath of 9/11, the security landscape of international trade and maritime transport changed significantly. The challenges facing global maritime security are increasingly of a nontraditional nature, such as terrorist acts against shipping, trafficking in weapons of mass destruction, armed pirate robbery, as well as smuggling of people and arms. Pirates and Islamist terrorist groups have long operated in those water areas, including the Arabian Sea, the South China Sea, and in waters off the coast of western Africa. Since 2008, the Chinese government has dispatched warships to the waters off Somalia to protect Chinese vessels and crews from pirate attacks. The Chinese fleet would join warships from the United States, Denmark, Italy, Russia, and other countries in patrolling the Gulf of Aden, which leads to the Suez Canal. Currently this is the quickest route from Asia to Europe and the Americas. This is a remarkable foreign policy change from a home-based passive defense to an offshore-based "preventive defense," which is directly linked and coordinated with the Western developed nations for their collective energy security. The aftermath of the 2011 conflicts in the Middle East and North Africa will be far more significant.

From an internal perspective, energy security has become the essential premise for China to achieve its national goal of quadrupling its GDP in 2020. There is a genuine consensus among Chinese leaders and scholars that energy has become a key strategic issue for China's economic development, social stability, and national security and that the realization of China's key national interests[5] depends highly on the access to sufficient energy resources (Liu, 2006; Zhang, 2006). China's "market economy" had locked itself in a "tiger-riding dilemma," i.e., any slowdown in economic growth would put the country in a risky situation, leading to social unrest and political illegitimacy (Li and Clark, 2009). China's government fears that domestic energy shortage and rising energy cost could undermine the country's economic growth and thus seriously jeopardize job creation (Lo, 2011). Beijing increasingly stakes its political legitimacy on economic performance and rising standards of living for its people. Consequently, the threat of economic stagnation due to energy shortage represents real risks of social instability, which could in turn threaten the continued political authority of the state and the Communist Party. Energy security, hence economic stability, and sustainable development are basic strategic political concerns for the leadership.

In fact, some scholars of energy politics point out that state-led pursuit of energy supplies is often seen as the source of international conflicts

5. China's national interests are defined by the government as including sustained economic growth, the prevention of Taiwanese independence, China's return to a global power status, and the continuous leadership of the Chinese Communist Party. Today, energy security is defined as a core part of China's national interests.

(US Congress, 2011). However, behind it, other sources of conflict, nationalism, geopolitical competition, competing territorial claims, are most likely to have been at the root cause of those conflicts (Constantin, 2005). One Chinese scholar of strategic studies clearly explains the reason why energy security has become a core component of China's national interest:

> *With external trade accounting for almost 50 percent of China's economy, China is now highly interdependent with a globalized market. This shift also includes hard social, political and geopolitical choices that deeply impact matters of national security. The more developed China becomes the greater its dependence grows not only on foreign trade but also on the resources to fuel the economy. With these complex and expanding interests, risks to China's well-being has not lessened but has actually increased, making China's national security at once both stronger and more vulnerable.*
>
> Zhang (2006)

China's sensitivity on the confluence of geopolitics and resource politics is also derived from the fact that historically China has been a weak sea power. One of China's key weaknesses through centuries of its development and into the modern age is its lack of a strong navy to safeguard its global interest, and this is perhaps one of the major factors leading to China's massive investment on raising and modernizing its naval capabilities. China therefore has good reasons for acquiring an aircraft carrier to enable it to protect its national interests (Cole, 2006). China has territorial disputes in the South China Sea over the Spratly Islands with neighboring countries and is also worried about the security of the major maritime transportation routes through which it transports the majority of its foreign trade, as well as its oil imports, upon which it is totally dependent. Based on the historical lessons, China has a clear understanding on the linkage between its energy security and international geopolitics, which is spelt out clearly by one scholar:

> *The history of capitalism and its spread globally have shown that it is often accompanied by cruel competition between nation states. Those countries that lose out are not necessarily economically or technologically underdeveloped or those with a low level of culture. Rather, they are most often those nations who forgo the need to apply their national strength to national defense and therefore do not possess sufficient strategic capability.*
>
> Zhang (2006, p. 17)

Today the rise of China is due in large part to its rapid emergence as a major force in world energy markets and energy geopolitics (Chan, 2011; Lo, 2011). Beijing's booming energy consumption and heavy investment for energy security have raised a new range of contentious issues between China and other world powers that are adding a new layer of issues to the already complex and dynamic relationships. China's economic growth is supported by three primary pillars: (1) export-led growth, (2) real property growth,

(3) government spending; among them export has been the key engine driving its economic growth. The current global financial crisis (2008–09) has already indicated a concern for the first pillar because European and especially American consumers can no longer consume at the debt-supported levels as they had in the past (Economist, 2009). One of the perplexing questions is whether the sustainability of China's export-oriented development strategy can be counted on to be sustainable and reliable into the future. Current data suggest that China has weathered this storm and become the new financial center for economic markets. Hence businesses are coming to China not only to invest but also to seek investments. The new Chinese export has become investment "capital" and financing. (SJBJ, 2011; Lo, 2011).

NEW POLICY THINKING: CHANGE OF ECONOMIC GROWTH STRATEGY AND PROMOTING SUSTAINABLE ENERGY

In Nov. 2005 Chinese Premier Wen Jiabao declaimed at the Plenary of the Chinese Communist Party that "energy use per unit of China's GDP must be reduced by 20% from 2006 to 2010," and this declaration was turned into a policy goal set up by China's current 11th Five-Year Plan (2006–10).[6] In this national policy planning China's energy policies are defined to be "from a growth at any cost model" to "a sustainable, energy-secure growth path." To deal with the rising energy intensity, Chinese government has introduced a number of energy- and emission-saving policies as well as administrative plans and legal frameworks to strengthen energy conservation work. According to HSBC in late 2010, the next 12th Five-Year Plan (enacted in Mar. 2011) would focus on three key issues:

1. "Achieving more balanced and sustainable growth is the key."
2. Requiring "real reforms of income distribution, industrial regulations and fiscal system" and
3. Taking "steps towards financial reforms ... (that will) unleash the power of consumers and inland regions."

As Lo (April, 2011) reports from the Central Government and his corporation in Shanghai, China has aggressively begun doing just these three target areas plus more. Lo adds three other key elements to the 12th Five-Year

---

6. Five-Year Plan, shortened for *The Five-Year Plan for National Economic and Social Development*, and even shorter for the plans are 11th Five-Year Plan, is a national goal-setting policy paper. At the macrolevel, it disposes national key construction projects, administers the distribution of productive forces and individual sector's contributions to the national economy and maps the general direction of future development including specific policies and targets. The current 5-year plan for 2006–10 is also called the *11th Five-Year Development Guidelines*. The 12th Five-Year Plan or 12-5 year Plan came out in March 2011.

Plan, however: (4) "strengthen environmental protection," (5) "enhance innovation," and (6) "improve living standards" by changing the focus of Chinese "export-oriented" economic model to a "domestic demand–oriented" economic model (Lo, 2011, pp. 7–9) Some concrete policy measures were implemented in line with these macro policy goals. In Dec. 2007 China's Information Office of the State Council issued the country's first ever white paper on its energy conditions and policies—*China's Energy Conditions and Policies*. China's National Energy Administration (NEA) was set up in 2008 to coordinate energy issues concerning various ministries, commissions, and state-owned energy companies. To promote the development of emerging energy industries and meet the carbon emissions reduction targets of 2020, the NEA has compiled a development plan for emerging energy industries from 2011 to 2020 that will require direct investments totaling 5 trillion yuan according to *China Daily*.[7]

Consider the growth of renewable power systems in China. The Chinese government is to launch a series of policies to support new energy development through China's 12th Five-Year Plan, or 12th Five-Year Plan for National Economic and Social Development from 2011 to 2015 (Lo, 2011), which is focusing on new energy, including wind, solar, and nuclear power, and the plan is under final review. The new energy policy will increase China's proportion of nonfossil energy in overall energy consumption from 12% to 13% by 2015, according to China's Energy Research Institute. This development trend is noticed by an international consulting organization with the following data:

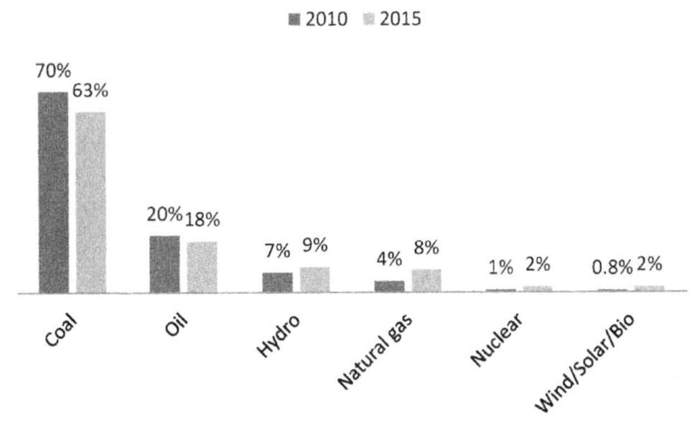

Energy security and environmental problems in China should be resolved not only through self-reliance efforts but also through international economic cooperation. To diversify access to energy supplies and reduce dependency on

7. *People's Daily* online, Jul. 22, 2010, available from http://english.peopledaily.com.cn/90001/90778/90862/7076933.html.

certain exporters, China is taking many political and economic measures and providing economic aid to strengthen its cooperative relations with resource-rich countries (Tseng, 2008; Ziegler, 2006). However, the dynamic debates on energy security are still going on in China (Downs, 2004). The debates indicate that many of China's analysts and policy makers are not fully convinced of the benefits of reliance on world energy markets. The political consensus today is to move toward "green" (renewable energy power generation) and integrated energy infrastructure systems that are sustainable.

In addition, China has been alert regarding the soaring demand in global energy in recent years and at the possibilities of long-term global energy shortage, called "peak oil," which now includes "gas" (US Congress, 2011). Hence, China's energy security will be one of the most important parts of its broader foreign policy in the years to come. The world will soon focus on China's new economic and energy policies, its energy market reform, and its new strategies in meeting the political challenges of rising energy costs and environmental pollutions (Clark and Li, 2003). Much global attention has seen China move toward technology development and innovation in generating clean coal and natural gas power along with new institutional developments. Chinese energy policies and each of China's steps and practices bear significant implication on greenhouse gas emissions and climate change.

Already being burdened with serious environmental problems and energy shortage, the ongoing global economic downturn presents China with a historic opportunity to rethink its growth strategy to move toward a more stable and sustainable path. Today, a promising optimism is that China seems to be firmly committed to the creation of a largely self-sustaining innovation system as part of a knowledge-based economy of the future. China is sparing no effort to meet its 11th Five-Year Plan's energy conservation goals, in which China will cut its per unit of GDP energy consumption by 20% from 2005 levels by the end of 2010 (Fig. 17.3).

China's policy determination in clean and renewable energy can be clearly seen from its ambitious plan published in 2007—"Middle and Long-term Development Plan of Renewable Energies," which was approved by the People's Congress in 2008 as *The Renewable Energy Law*. The new policy is determined at moving the country toward renewable energy to reduce energy consumption and cut the surging carbon dioxide emissions. The policy expects to derive 10% of China's energy supply from renewable sources by 2010 and 15% by 2020. To meet the 2020 goal the total expected investment will be two trillion Chinese yuan (US $133.3 billion). If successful, by the end of 2010, China would emit 600 million tons less carbon dioxide a year, and by 2020 the annual reduction in carbon dioxide emissions would reach 1.2 billion tons. The target of this policy plan reflects another of China's policy concerns in coping with the environmental and economic challenges of climate change. The linkage of energy policy and climate change policy can be read from the policy document—*China's National Climate Change Programme* 2007, prepared by

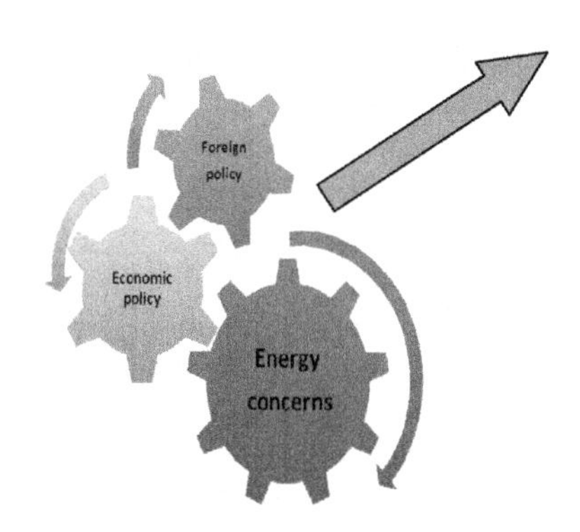

China	$34.6 billion
United States	$18.6 billion
United Kingdom	$11.2 billion
Rest of EU-27	$10.8 billion
Spain	$10.4 billion
Brazil	$7.4 billion
Germany	$4.3 billion
Canada	$3.3 billion
Italy	$2.6 billion
India	$2.3 billion

(Pew Charitable Trusts, 2010:7)

FIGURE 17.3 China's ongoing and future policies connected with energy concerns. *Courtesy Pew Charitable Trusts, 2010. Who's winning the clean energy race? In G-20 Clean Energy Fact Book. The Pew Charitable Trusts, Washington, p. 7*

one of China's key government institutions, the Development and Reform Commission. Some examples of China's successes can be seen in communities that are becoming sustainable (Wang and Li, 2009; Kwan, 2009).

Through the legal framework stipulated in the new laws, the Chinese government has set efficiency goals and imposed taxes and regulations designed to curb demand and reduce emissions of greenhouse gases. In addition, the government energy and environment institutions are imposed with defined guidelines and responsibilities. The new policy toward alternative energy is supported by financial incentives including direct subsidies and innovative policy measures, tax-related incentives, custom duties, and pricing incentives. Some concrete policy incentives are: (1) connecting "intermittent"[8] sources of electricity like wind or solar to the national grid; (2) connecting utilities mandated to open transmission lines to renewable generators, with ratepayers bearing part of the extra costs; (3) feed-in tariffs guaranteeing renewable energy producers a steady, high price for electricity so as to enable them to compete with coal producers; as well as (4) tax breaks, preferential loans, and other financial incentives encouraging investors to support renewable ventures (China FAQs, 2010).

China, however, has focused its efforts in the 11th Five-Year Plan to grow the renewable energy sector primarily for export. As is pointed out by Lo (2011) for the 12th Five-Year Plan and SGCC's report on the 11th Five-Year Plan: "The strategy of building a world-leading strong and smart grid with ultra-high voltage grid as its backbone and subordinated at various voltage levels featured as being IT-based, automated, interactive, based on independent

8. It refers to energy generation installations, which are not state-owned.

innovation...Since 2009, SGCC has started 228 demonstration projects of 21 categories in 26 provinces and municipalities." (SGCC, 2010, p. 1)

It is likely that China will meet and even exceed its renewable energy development targets for 2020 by applying other alternative energies including hydro, wind, biomass, and solar PV power. It is expected that more than one-third of China's households could be using solar hot water by 2020 if current targets and policies are continued (Martinot and Li, 2010). China expects the policy objectives to be reached through the integration of a number of relationships: the responsibility of the state and the obligation of the public, institutional promotion and market mechanism, current demand and long-term development, and domestic practice and international experience.

In recent years, China has won the global recognition for its achievement in the development and application of alternative energy. China overtook the United States for the first time in 2009 in the race to invest in wind, solar, and other sources of clean energy. American clean energy investments were $18.6 billion last year, which were a little more than half the Chinese total of $34.6 billion. Just a few years ago, China's investments in clean energy totaled just $2.5 billion (*Los Angeles Times*, Mar. 25, 2010). In recent years, it is increasingly recognized that China's "green leap forward" policy has made it become the world's largest makers of wind turbines and solar panels surpassing Western competitors in the race for alternative energy. As one of the key US newspapers points out:

> *China vaulted past competitors in Denmark, Germany, Spain and the United States last year to become the world's largest maker of wind turbines, ... China has also leapfrogged the West in the last two years to emerge as the world's largest manufacturer of solar panels. And the country is pushing equally hard to build nuclear reactors and the most efficient types of coal power plants. These efforts to dominate renewable energy technologies raise the prospect that the West may someday trade its dependence on oil from the Mideast for a reliance on solar panels, wind turbines and other gear manufactured in China.*
>
> New York Times (January 30, 2010)

Clean renewable energy strategy emphasizes a sustainable growth path based on equity is leading the transition to knowledge and information economy. When referring to China's alternative renewable energy policy, some studies have shown that China is facing both opportunities and challenges. The potential opportunities are plenty, such as solar energy, wind energy, biomass energy, small hydropower, geothermal energy, and ocean energy, whereas the challenges are apparent as well, such as the lack of coordination and policy consistency, weakness and incompleteness in incentive system, lack of innovation in regional policy, immature financial system for renewable energy projects, and the limited investment in research and development of renewable energy (Zhang et al., 2007). There is still a long way to go before China's renewable energy market becomes mature and socially and culturally embedded.

CONCLUSION REMARKS: CHALLENGES AND OPTIMISM AHEAD

The objective of the chapter is to provide a framework of critically understanding China's transformation from a self-reliance development path to a "market-driven" dependent growth strategy to now a "social capitalist" economic system. The chapter's emphasis is on the economic and ecological consequence of China's insatiable demand for energy driven by the growth-based industrialization policy in the past decades. This chapter argues that since the beginning of the 21st century, energy has become a key concern on the agenda of China's economic and foreign policy-making calculations as it moves rapidly into the 3IR. Among China's core national interests—securing energy resources, generating national renewable companies and systems, gaining market access, and political recognition—energy and economic security is the top priority for developing a sustainable nation. It is expected with the rapid economic development and the improvement of people's living standard that the energy demand in China will unavoidably continue to increase, which will be inseparable with its environmental problems, such as the emission of sulfur dioxide, carbon dioxide, and particulates among other issues of pollution and waste.

Above all, the Western definition of "market-driven" economies in energy is questionable in China such that different definitions and meanings are need for "market" and therefore "capitalism." And that is what China has done. It has redefined capitalism so that it has a societal focus, direction, and set of policy along with financial strategies. For example, the rapidly emerging renewable energy industry in China has created a new market finance mechanism for long-term debt, which involves the Chinese business financing the entire sale, installation, and operations along with maintenance of the renewable energy technologies and products. In short, China may have discovered a new form of The World Bank.

Chinese policy makers understand the fact that due to a growing need for and even competition over energy resources and maritime transportation security, global resource-based competition and geoterritorial claims on sea areas and shelves will become harsh and could lead to armed confrontations. To understand the implication of the underlying dynamics of international political economy in resource-rich regions, it is of great importance to understand the source of international economic competition and the political conflict for access to energy and natural resources to understand the interactions between individual national interests. The international geopolitics and geoeconomics in the acquisition and distribution of states' wealth and power is manifested in the respective country's economic and foreign policies.

China's soaring demand for energy in connection with its export-oriented economy poses a variety of new challenges for its economic and foreign policy. Hence the country will be more and more dependent on the purchase of natural resources abroad for sustaining its economic development. Any crisis to China's access to overseas resource and maritime shipping routes will have a negative impact on China's growth and trade-dependent economy. China will endeavor to protect the strategic areas concerning its national interest. It has no choice. In recent years China's energy diplomacy in the context of the political economy of global energy developments has drawn the attention of the West, especially in connection with the sensitive regions, such as the Middle East, Central Asia, Latin America, and Africa. As one Chinese scholar bluntly states, "The determining factor shaping the rise and fall of a country ultimately is not just the size of its total economic volume but also the strategic ability of the country; that is, the ability to use national forces to achieve political goals" (Zhang, 2006, p. 22).

However, despite the global reality described earlier by this realist perception, China's deep sense of its energy insecurity and vulnerability is changing its development policy toward clean and renewable energy. China is accelerating research and development on renewable energy supply and advanced energy conservation–based techniques and products; it is making necessary structural changes in industrial and agricultural sectors moving to non–energy-intensive industries. Furthermore, China is trying to rely primarily on domestic resources while strengthening mutually beneficial international energy cooperation. The optimism that China is presenting to the world is not groundless. China not only is one of the world's leading producers of renewable energy but also is overtaking more developed countries in exploiting valuable economic opportunities, creating green-collar jobs, and leading development of critical low-carbon technologies.

Such optimism in China's own "green revolution" is also confirmed by the front page of a recent report by Climate Group (2009), "As one of the world's major economic powers, China will have to be at the forefront of this journey. This report shows that it can be." Nevertheless, China still has a long way to meet its policy objectives on energy and environmental sustainability. Due to its size and population the consequences of failure in China's case are much more serious than in many other counties. China should not be left struggling alone on the road to optimism, and the whole world must pay more attention to China. World peace and a sustainable planet depend on global harmony and collaboration beyond conventional competition over supply and demand.

CHART 1

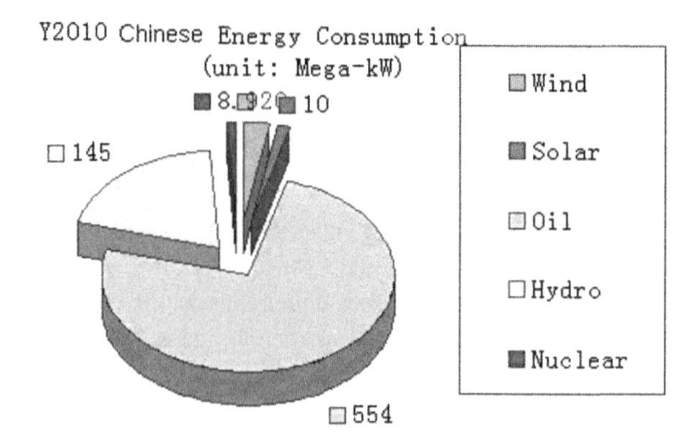

ACKNOWLEDGEMENTS

The authors express special thanks to Jerry Jin, PhD, who can be reached at jerryjin88@yahoo.com; David Nieh, environmental economist in Shanghai, who can be reached at: david.nieh@shuion.com.cn; and ML Chan, PhD, who can be reached at: mlchan@juccce.com

REFERENCES

ACORE, March 2011. American Council on Renewable Energy. US-China Quarterly Review.

Brautigam, D., April 2008. China's Africa Aid. The German Marshall Fund of the United States.

Chan, M.L., 2010. Green White Paper Series. Technical Advisor. JUCCCE, Shanghai, China.

Chan, S., President of SunTech, March 2011. Global solar industry prospects in 2011. In: SolarTech Conference, Santa Clara, California. www.SunTechSolar.com.

Chase-Dunn, C., 1982. Socialist States in the World-System. Sage, Beverly Hills.

Chase-Dunn, C., 1989. Globalization: a world-systems perspective. Journal of World-Systems Research 5, 165–185.

APCO, 2010. China's 12th Five-Year Plan: How it Actually Works and What's in Store for the Next Five Years. APCO Worldwide.

Clark II, W.W., Cooke, G., 2011. The Third Industrial Revolution. Praeger Press.

Clark II, W.W., Isherwood, W., Winter 2010. Report on energy strategies for inner Mongolia autonomous region. Asian Development Bank, December 2007. In: Clark II, W.W., Isherwood, W. (Eds.), Utility Policy Journal. Special Issue on Line as China: Environmental and Energy Sustainable Development. https://doi.org/10.1016/j.jup.2007.07.003.

Clark II, W.W., Li, X., 2003. Social Capitalism: transfer of technology for developing nations. International Journal of Technology Transfer 3, 1–11.

Cole, B.D., 2006. Chinese Naval Modernization and Energy Security, a Paper Prepared for the Institute for National Strategic Studies. Washington, D.C.

Constantin, C., 2005. China's Conception of Energy Security: Sources and International Impacts. Working Paper No. 43. University of British Columbia.

Hanergy Corporation. www.hanergy.prc.

IEA, 2011. International Energy Agency. Technology Roadmap: Smart Grids.

The Development, Reform Commission, 2007. China's National Climate Change Programme 2007. PRC, Beijing, China.

Dorraj, M., Currier, C.L., 2008. Lubricated with oil: Iran-China relations in a changing world. Middle East Policy XV, 66–80.

Downs, E.S., 2000. China's Quest for Energy Security. RAND, Santa Monica (CA).

Downs, E.S., 2004. The Chinese energy security debate. The China Quarterly 177, 21–41.

Downs, E.S., 2006. China – executive summary. The Brookings Foreign Policy Studies Energy Security Series The Brookings Institution.

Economist, July 16 , 2009. Special issue and articles on "collapse of modern economic theory".

Eisenberg, L., 2009. "Los Angeles Community College District (LACCD)" and "Appendix a". In: Sustainable Communities. Springer Press, pp. 29–44.

China FAQs, 2010. Renewable Energy in China: An Overview. World Resources Institute.

Jiahua, P., et al., 2006. Understanding China's Energy Policy. A Background Paper Prepared for Stern Review on the Economics of Climate Change, the Chinese Academy of Social Sciences (CASS). Beijing.

Kaplinsky, R., 2006. Revisiting the revisited terms of trade: will China make a difference? World Development 34, 981–995.

Konan, D.E., Zhang, J., 2008. China's quest for energy resources on global markets. Pacific Focus 23, 382–399.

Kwan, C.L., 2009. Rizhao: China's green Beacon for sustainable Chinese cities. In: Sustainable Communities. Springer Press, New York, pp. 215–222.

Li, M., 2010. Peak energy and the limits to economic growth: China and the world. In: Xing, L. (Ed.), The Rise of China and the Capitalist World Order. Ashgate, Farnham, England, pp. 117–134.

Li, X., Clark, W.W., 2009. Globalization and the next economy: a theoretical and critical review". In: Li, X. (Ed.), Globalization and Transnational Capitalism: Crises, Challenges and Alternatives. Aalborg University Press, Aalborg-Denmark, pp. 83–107.

Li, Z., March 2003. Energy and environmental problems behind China's high economic growth – a comprehensive study of medium- and long-term problems, measures and international cooperation. IEEJ. http://eneken.ieej.or.jp/en/data/pdf/188.pdf.

Liu, X., September 2006. China's Energy Security and its Grand Strategy. Policy Analysis Brief. The Stanley Foundation.

Lo, V., April 25, 2011. China's 12th five-year plan. In: Speech by Chairman. Shui on Land and President, Yangtze Council. Asian Society Meeting, Los Angeles, CA.

Martinot, E., Li, J., 2010. Powering China's development: the role of renewable energy. Worldwatch Institute, Worldwatch Annual Reports.

NYT. New York Times, January 30, 2010. China Leading Global Race to Make Clean Energy.

Huliq News, 2008. China's Energy Demand Increases Global Pressure to Seek Out New Sources. Available at: http://www.huliq.com/597/73705/chinas-energy-demand-increases-global-pressure-seek-out-new-sources.

Opoku-Mensah, P., 2010. China and the international aid system: transformation or cooptation? In: Li, X. (Ed.), The Rise of China and the Capitalist World Order. Ashgate, Farnham, England, pp. 71–85.

Pew Charitable Trusts, 2010. Who's winning the clean energy race?. In: G-20 Clean Energy Fact Book. The Pew Charitable Trusts, Washington.

Peyrouse, S., 2007. The Economic Aspects of the Chinese-Central Asia Rapprochement, Silk Road Paper, Central Asia-Caucasus Institute & Silk Road Studies Program – A Joint Transatlantic Research and Policy Center. Johns Hopkins University.

Rosenberg, M., December 17, 2010. Global Renewables War is on. U.S. Missing in Action. EnergyBiz.

SJBJ, April 1, 2011. Clean energy financing jumps to record $243B. San Jose Business Journal 8. The News, Data from Pew Chartable Trusts.

Scientific American, April 5, 2011. Clean tech rising. www.scientificamerican.com/technology.

SGCC, August 18, 2010. State Grid Road Map of China. State Grid Corporation of China. Framework and Roadmap for Strong and Smart Grid Standards.

Silverstein, K., 2011. Energy avenues open with China. EnergyBiz. http://www.energybiz.com/article/11/01/energy-avenues-china-open.

The Climate Group, 2009. China's Clean Revolution II. The Climate Group, London.

Thøgersen, S., Østergaard, C.S., 2010. Chinese globalization: state strategies and their social anchoring. In: Li, X. (Ed.), The Rise of China and the Capitalist World Order. Ashgate, Farnham, England, pp. 161–186.

Time, February 28, 2011. World: Now No. 2, Could China Become No. 1?, p. 17.

Tseng, Yu-H., 2008. Chinese foreign policy and oil security. Internationals Asian Forum International Quarterly for Asian Studies 39, 343–362.

UNCTAD, 2005. Review of Maritime Transport 2005. UNCTAD.

US Congress, April 5, 2011. China's energy and climate initiatives: successes, challenges, and implications for US policies. Energy Strategy Institute (EESI) and World Resources Institute (WRI).

USDoE, 2015. U.S. Department of Energy. Smart Grid Investment Grant Program - Progress Report II.

Wang, W., Li, X., 2009. Ecological construction and sustainable development in China: the case of Jiaxing municipality. In: Sustainable Communities. Springer Press, New York, pp. 223–241. www.WorldWatchReports.org.

Xu, Q., 2007. China's Energy Diplomacy and its Implications for Global Energy Security. FES Briefing Paper 13. Friedrich-Ebert-Stiftung, Berlin, Germany.

Zhang, W., Summer, 2006. Sea power and China's strategic choices. China Security 17–31.

Zhang, P., et al., 2007. Opportunities and challenges for renewable energy policy in China. Renewable and Sustainable Energy Reviews 13, 439–449.

Zhen, Yu Z., Hu, J., Zuo, J., 2009. Performance of wind power industry development in China: a Diamond Model Study. Renewable Energy Journal 34, 2883–2891. Elsevier Press.

Ziegler, C.E., 2006. The energy factor in China's foreign policy. Journal of Chinese Political Science 11, 1–23.

FURTHER READING

Chan, M.L., May 2009. Building strong grid policy, state grid corporation. In: International Conference on UHV.

IEA, 2010. International Energy Agency. Key World Energy Statistics. IEA, France.

Qu, H., Sun, Jun W., October 2010. Economics: From Quantity to Quality of Growth, China's Next Five Year Plan: HSBC Global Research (revised March 2011). www.research.hsbc.com.

Shoumatoff, A., May 2008. The Arctic oil rush. Vanity Fair. Available at: http://www.vanityfair.com/politics/features/2008/05/arctic_oil200805.

Spears, J., 2009. China and the Arctic: the awakening snow dragon. ChinaBrief IX, 10–13.

Sundra Solar Corporation, 2011. Beijing. www.sundrasolar.com.

The Chinese government Reports: (2008) The Renewable Energy Law (2008) China's Policy Paper on Latin America and the Caribbean (2006) China's African Policy Paper.

US DOE. US Department of Energy. Annual Global Reports. www.usdoe.usa.org.

US DOE. US Department of Energy. US-China Clean Energy Research Center. http://www.whitehouse.gov/blog/2010/09/03/us-and-china-advancing-clean-energy-research-throughcooperation.

Valsson, T., 2007. How the World will Change with Global Warming? University of Iceland Press, Reykjavik.

Chapter 18

Energy Strategy for Inner Mongolia Autonomous Region

Woodrow W. Clark, II[1], William Isherwood[2]
[1]*Clark Strategic Partners, Beverly Hills, CA, United States;* [2]*Asian Development Bank, Manila, Philippines*

Chapter Outline

THEME 1: ENERGY BASE, SUSTAINABLE DEVELOPMENT, AND FINANCE

We need to define clearly the *energy sector* in terms of an energy base and how it is organized, financed, and regulated. The Inner Mongolia autonomous

Sustainable Cities and Communities Design Handbook. https://doi.org/10.1016/B978-0-12-813964-6.00018-5

region (IMAR) can be a "Strategic Energy Base" for demonstration sites in various locations. For example, Belgium is considering an "energy base" for the EU hydrogen highway. And California is implementing its Hydrogen Highway with 220 fueling stations.

An important concept is "sustainable development" (SD). For most policy makers, SD includes environment, energy, and economics. Energy includes conventional fossil fuels, nuclear, clean coal, renewables, and others.

Among the important economic terms is "stranded costs," referring to capital costs for system equipment, structures, and facilities that constitute long-term repayment commitments, but become unrecoverable because of being superseded by new technologies or competition. For example, a conventional coal power plant built today will strand costs that need to be recovered over a 20- to 30 year-period, as the plant becomes replaced with cleaner technologies before the obligations are paid off.

Other critical areas include "infrastructures," which include water, waste, energy, environment, transportation, education, and medicine. Related to this are "hybrid" systems that can utilize technologies in combination with one another. This approach works for vehicles and is now available for homes and buildings.

THEME 2: DIFFERENCES THAT EXIST BETWEEN REGIONS, CITIES, AND NATIONS

Inherent differences may pose serious conflicts between a *nation and its regions, cities, and states*. Nations also have vast differences between regions and states/provinces. Scotland, for example, has aggressive renewable energy programs in wind, wave, and solar power. This region of the United Kingdom was prominent for its coal generation for almost a century, but the region needed to shift dramatically from labor-intensive and polluting forms of energy generation into something entirely new and climate friendly.

THEME 3: LONG-TERM COMMITMENTS

Along with government policy, financial programs must be created and applied over a long period of time. Finance, public policy, and regulations are the keys to any energy base. Government and business alike need to have policies in place along with funding so that they can plan. The People's Republic of China's (PRC's) Five-Year Plans are very useful in this regard. Such plans do not exist in most Western nations, since commitments depend on political elections. IMAR is uniquely positioned to be able to meet both the national and regional concerns for leadership and long-term commitments to turn its policies into education, environmental protection, and economic development of the energy base.

RECOMMENDATIONS AND CONCLUSIONS

Applying Key Case Study Features

Vision

Economic growth with balance and harmony for China (CPRC) and Inner Mongolia (IMAR).

Objectives

1. *For the PRC as a whole, as provided in the Five-Year Plan for National Economy and Social Development:*
   - Approved: October 2005
   - Authorized by: the 16th Central Committee of the Communist Party of China
   - Period covered: 2006–10

   Quantitative near-term national targets:
   - Double the per capita gross domestic product (GDP) from the 2000 level ($800) to $1600
   - Reduce energy consumption per GDP to 20% less than that in 2006 by 2010
   - Keep population growth below 1%

   Midterm (by 2020) national objectives:
   - Quadruple the per capita GDP from the 2000 level—from $800 to $3000
   - Have GDP average growth of 7.2% per annum
   - Create a "moderately well-off society"

   Long-term (by 2030) national objectives:
   - Achieve full-scale industrialization
   - Solid infrastructure
   - Wide-scale urbanization
   - Decrease percentage of high-energy-consuming industries (Fig. 18.1)
2. *For IMAR:*
   - To become a major player in making China energy independent
   - Take advantage of IMAR's unique strengths to make it a strategic energy base in the near term
   - Expand development and commercialization of renewable resources
   - Revise and reinvigorate coal production through the conversion of coal to liquid and/or gaseous fuels with most modern clean technologies with near-zero emissions of greenhouse gases (GHG)
   - Provide infrastructure to both amply supply IMAR with its own energy needs and export to both China and the world
   - Develop and expand on-site or distributed energy generation for communities with available and future energy generation sources

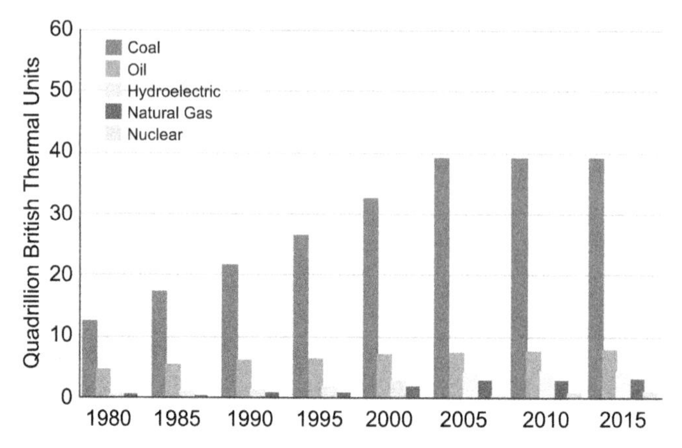

FIGURE 18.1 Overall China energy production over a 30-year period with predictions from the next Five-Year Plan. Framework of PRC polices and plans with data from Rob Koepp (Asia Capital Group).

- Create and develop hybrid technologies, such as coal or biomass gasification into hydrogen; on-site solar with energy storage, geothermal and hydrogen, natural gas and solar thermal, among others
- Develop and leverage a workforce capable of becoming one of the world leaders in energy development and sustainable technologies
  - Create collaborations between universities and industry
  - Develop an energy base and science parks
  - Provide and promote international exchanges and partnerships in public education, government, and private sector businesses

Unique IMAR and PRC Opportunities

Because the IMAR and PRC are rapidly expanding their energy infrastructures, they can invest in new technologies without abandoning existing facilities before the end of their useful lives (stranding recent costs). Clark and Li (2003) describe the PRC system as "social capitalism." Concern for society can be matched with new business enterprises for improving the environment and climate.

Old, inefficient, or environmentally dirty facilities can be replaced by newer technologies that are cleaner and more efficient, and this can be done in harmony with the design of new facilities that look forward to the future energy needs of the country and region (see California Commission for the 21st Century, Energy Chapter, 2002).

In this way, IMAR can immediately take steps to meet the needs of the future. National priorities support the targeting of environmental protection investments, now proposed in the 11th Five-Year Plan, as shown in Table 18.1. Environment protection as an investment is targeted in the 11th Five-Year

TABLE 18.1 Targeting of Investment in Environmental Protection in the 11th Five-Year Plan

	Billion RBM	Billion US$
Urban environmental infrastructure	660	82
Remediate industrial pollution	210	26
Environmental protection for new projects	350	43
Ecological protection	115	14
Nuclear safety	10	1
Environmental monitoring	30	4
Total	1375	170

China Economic Information Network.

Plan. The economic consequences for both growth and environmental "energy harmony" are expressed in a similar vision and goal in California (Clark and Sowell, 2001).

The issue, as seen in all countries and expressed later in the discussion, is that public policy by decision makers *must* proceed with both financial and implementation programs that have defined mechanisms. The policies set forth by the PRC and IMAR governments provide stability for the public sector in terms of resource allocation, planning, and future development. Additionally, stability in established public policies encourages private sector partnerships through commitment of resources (capital, physical, and personnel) and both debt and equity investments.

Predictions indicate that China will continue to use coal for about two-thirds of its energy consumption by 2010. If this happens then major upgrades will be necessary for the coal industry to become the basis for SD of the national economy. However, as rail service now creates a bottleneck in transporting coal, new pipelines for liquid and/or gaseous fuels produced near the mine mouth can avert the need for large investments to expand rail capacity. Fig. 18.2 shows he portion of world energy consumption used by the PRC.

One obvious problem arises from further and continued competition for these now globally scarce energy resources. In the United States, for example, this problem in the petroleum industry has been called "peak oil," meaning that there are limits to oil reserves and that production will be "peaking" soon. By some industry and government estimates, world-side peak petroleum production will come within the next 5–10 years. This peak can only be postponed by higher oil prices, as new fields will be more difficult and costly to develop.

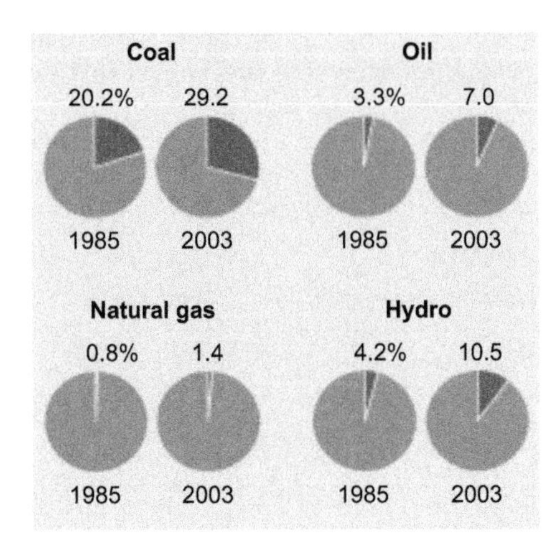

FIGURE 18.2 Percentage of various world energy sources used by the People's Republic of China.

Specific suggestions follow, first by commercial technology, next by some of the potential barriers to successful implementation, and then by the policies and financial mechanisms that might best encourage and implement those technologies.

Commercial Technological Approaches Available Today (2006) and Areas for Collaboration With Research and Development and Universities for the Future

1.Expand renewable energy generation such as wind and solar use.

a. Provide the infrastructure to integrate these seamlessly with the grid

b. Use these resources especially where they can be colocated with local loads, and ideally as "hybrid systems"

c. Investigate and utilize appropriate energy storage technologies where appropriate [e.g., compressed air storage, superconducting magnetic energy storage, high-tech electromechanical batteries, or conversion into another energy carriers such as hydrogen (by electrolysis)]. If sufficient energy is converted to hydrogen, develop new local energy centers for distribution

d. Take the following steps toward establishing "agile energy systems" (see the Glossary of Terms (Clark, 2017) and the following points for definition and further discussion)

i. Establish and implement Energy Infrastructure Plans based on the Agile Systems Model for 5–10 years.

ii. Learn from the environmental and energy mistakes in other nation states such that future technological advances can be incorporated and implemented to assure energy security and environmental protection.

iii. Facilitate central grid connections with on-site (e.g., distributed energy) power generation.

iv. Use the "civic market" approach (public and private partnerships where the "civic" or public has the majority controlling interest) to create and implement plans.

v. Create short-term (3–5 years) incentives (see Glossary) that encourage renewable energy generation now rather than fossil and conventional power generation systems.

vi. Provide long-term incentives including contracts, grants, and cofinancing for renewable energy generation.

vii. Seek "hybrid" or combined technologies that merge renewable energy such as wind or solar with storage and delivery devices, such as hydrogen and fuel cells.

viii. Encourage and collaborate with other infrastructures, especially water, waste, transportation, and education.

ix. Focus on long-term goals, which should include a "hydrogen economy" in the next 10–20 years or sooner, with demonstrations.

2. Develop advanced technologies, such as clean coal and coal mine methane (CMM), as "transition" energy supply approaches.
   a. Use underground coal gasification where feasible, in combination with CO_2 sequestration in unmineable coal seams.
   b. Employ CMM and ventilation air methane (VAM) technologies to working mines, as a source of natural gas and as a mine safety measure.
   c. Search for opportunities to extract abandoned mine methane and coal bed methane from virgin coal seams, and apply these where feasible.
   d. Continue with construction of additional high-tech coal washing facilities, to remove excess ash, increase thermal content, and for desulfurization processes.
   e. For power generation alone, continue to seek the most efficient technologies, e.g., supercritical fluidized bed combustion or integrated coal gasification combined cycle.
   f. Follow and develop potential new technologies such as direct carbon fuel cells and integrated coal gasification fuel cell—these could be important in the not too distant future.
3. Investigate the potential to build mine mouth polygeneration plants for coal, producing town gas for cooking and heating. DME for heating and transportation, and methanol and F-T liquids for transportation, with electricity as a coproduct.
   a. Design these plants with the ability to switch portions to hydrogen production as fuel cells become cheaper and more accepted.

   b. Find locations for and install underground sequestration programs for CO_2 and SO_2.
   c. Blend methanol with gasoline and F-T liquids or DME with diesel via China's current infrastructure.
4. Follow closely and build on the experiences of South Africa and Chinese pilot plants to scale-up liquefaction of coal for transportation fuels.
5. Build infrastructures to support the agile energy systems model with the concern for short-term (3–5 years) use of the aforementioned alternative fuels (see Glossary).
   a. Upgrade transmission lines, using best available high-voltage DC technologies, to transmit electricity to eastern load centers.
   b. Complete electrification of rural areas with a combination of connections to the central grid and local on-site energy generation from solar, wind, biomass polygeneration, and/or biodigester facilities.
   c. Replace coal with town gas for industrial and urban residential heating.
   d. Introduce DME as a motor vehicle fuel, starting with buses and trucks, and shift (especially in the public sector) to hybrid cars (e.g., electric and gas) or compression-ignition engines.
   e. In the longer term, introduce renewably or cleanly generated hydrogen fuel availability for fuel cell vehicles.
6. Explore other fuels from polygeneration plants in IMAR, especially liquid fuels for the transportation sector and DME for cooking and heating.
7. Build or expand current pipelines for transporting coal-derived fuels to sites closer to load centers in the east to utilize cost-competitive alternative fuels.
8. Upgrade power transmission capability with latest technologies, e.g., high-voltage DC lines, inverters, ultracapacitors, and efficiency and conservation devices.
9. Encourage other environmentally improved energy approaches, such as combined heat and power, ground source heat pumps, demand-side management, and "hybrid" technologies such as wind and pump storage or fuel cells.

Potential Barriers to Implementation in Inner Mongolia Autonomous Region

Note that some nations and various states within the United States have solved many of these problems but still have high energy demands, conservation, and environmental issues to deal with.

1. High capital costs, uncertain markets, and intermittency for renewables.
   - Wind is a good model on how to overcome these barriers, whereby today it is cost competitive with all fuels (especially natural gas) and financial accounting for intermittency has been resolved (see evidence

in California, 2003; and the USA Federal Energy Regulatory Commission, 2003).
- Combine renewable energy generation with advanced technologies such as fuel cells, fly wheels, and biodigesters to reduce costs.
- Cost accounting for comparison of power from different technologies should include the costs of "externalities," such as health costs borne by society caused by traditional coal emissions.
- Devise a scheme under which cleaner technologies that reduce "externality" costs can be subsidized by the savings from these reduced costs.
- Create educational and training programs at all levels of the work force, including career advancement, advanced education in research and development, and education of public officials.
- Policies are needed to overcome barriers, including methods to amortize high up-front costs and assure access to markets, such as civic markets (public–private partnerships), with incentives, grants, rate rebates, and tax advantages.

2. Changes in coal infrastructure can cause social disruption, safety problems, and initial up-front stranded costs.
   - Extensive retraining programs can improve the status of the workforce, while new production facilities will create new jobs. Other policies can be tailored to overcome other social disruptions.
   - New safety and health issues may arise (but can be addressed), but existing safety and health issues may be reduced.
   - Need to weigh carefully these continued costs with alternative advanced technologies and potential for environmentally sound energy generation.
3. New alternative fuels (see Glossary) may be slow to be accepted, but must be started for an orderly and definable "transition" to meet the future vision for energy.
   - New vehicle fueling stations must be in place before new fuels will be accepted. The government must ensure financial incentives, which could require initial subsidies.
   - New pipelines and other infrastructure must be built in parallel with local and on-site fuel generation (e.g., biogas, ethanol, hydrogen) for meeting demand and the availability of new fuels. Some incentives for industrial, commercial, and domestic users to convert to new fuels may also be necessary.
   - The public must be made aware of the need for "cleaner" energy sources and be willing to pay for it today, as has been in the case in Germany, the United Kingdom, and the United States. (Many of these costs can be covered by reduced environmental and health costs borne by the society.)

- Planning for the future must set goals and implement projects to achieve them, so the public sees and benefits from the changes.
- Mass transit (especially as a partial remedy to major fuel consumption issues) must be improved and remain part of the public lifestyle.
- Harmony with the social, cultural, and political aspects surrounding a region and the IMAR in particular must be a continuous concern.

Solutions to Barriers for Inner Mongolia Autonomous Region

Nations, states or provinces, and regions (collections of nations and states) must develop proactive strategies for the long term and use their remaining fossil fuels as efficiently as possible without accelerating global warming further. This requires development of many options during this transition. Of the available options, renewables have the most environmental appeal, and in many cases are already, or soon will be, cost effective and available. Clean coal technologies can also play a significant role in the near and intermediate time frames.

Renewable generation technologies tend to be underutilized due to higher costs today, but they are rapidly decreasing. Wind power, for example, only 8 years ago was expensive, but today (since 2003), it can be cheaper than natural gas. Issues such as intermittency have been resolved through government policy, tariffs and financing as well as hybrid technologies, metering, and storage devices. Given the approaching worldwide "peak oil and gas" crises, along with energy security issues, development and use of renewable energy generation is far more strategic and cost competitive.

Long-term contracts based on stated policy visions, goals, and regulations along with finance mechanisms can reduce both the risk and costs for alternative sources of energy. Other strategies include more effective use of energy, such as cogeneration, other uses of "waste" heat, and more efficient processes and appliances.

Fig. 18.3 illustrates the history and projection of where the world got and will get its energy over 2½ centuries, by the International Institute of Applied Systems Analysis. This projection shows coal use phasing out by 2100, which we believe is unnecessarily soon. If new clean coal technologies are employed, coal can play a more major role in the transition from fossil fuels, which needs to take place.

Policies to Move the Inner Mongolia Autonomous Region Energy Structures to Meet Its Future Needs

- Enact specific public (civic) policies and plans to establish regions or areas within IMAR as "energy bases" for IMAR and the PRC.
  - Eliminate import duties for equipment for wind, solar, or clean coal-related equipment.

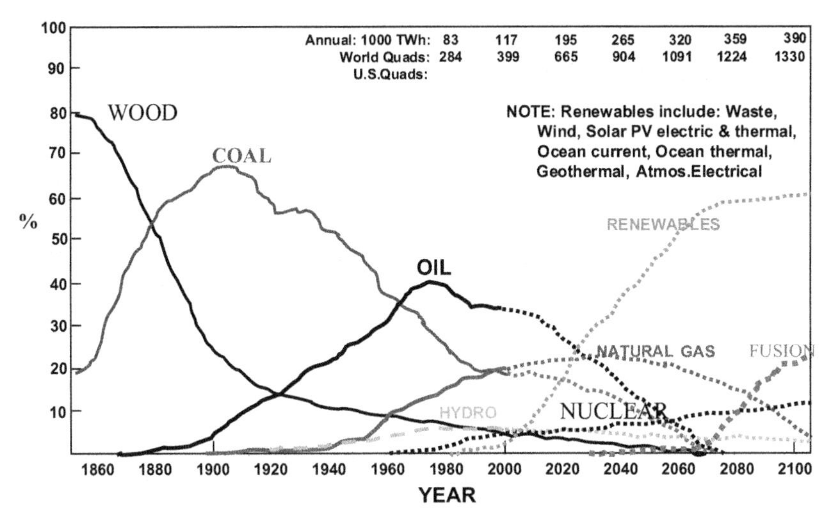

FIGURE 18.3 The projection of trend in energy sources for the future.

- Expand joint ventures with companies in these industries and coordinate them with education, training, and academic programs.
- Provide funds for specialized training programs in IMAR (e.g., at IMAR University of Technology) on advanced coal technologies and renewable energy.
- Establish research and development centers, especially in areas such as hybrid technologies, that are not found in the other chapters in the book on other nations.
- Become a leader in GHG capture and storage technologies (sequestration), including oxyfuel combustion, by initiating demonstration programs.
- Support large-scale demonstration projects in the renewable and clean energy generation areas.
- Establish global business partnerships with commercial joint venture programs.
- Create long-term financing, funding, and contract mechanisms such as preferential feed-in laws.

- Establish IMAR as a "Strategic Energy Base" for demonstration sites in various locations. For example, in Europe today a number of countries have established "science parks" (see Clark, 2003a,b) with a focus on certain technologies or interrelated groups of technologies. More recently, Belgium is considering establishing an "energy base" for hydrogen in an industrial area that needs both a vision (similar to IMAR) for the future and technologies that create businesses with education, training, and jobs.

Several nations and states in the Case Studies (particularly the United Kingdom, Germany, the United States, and especially California) historically

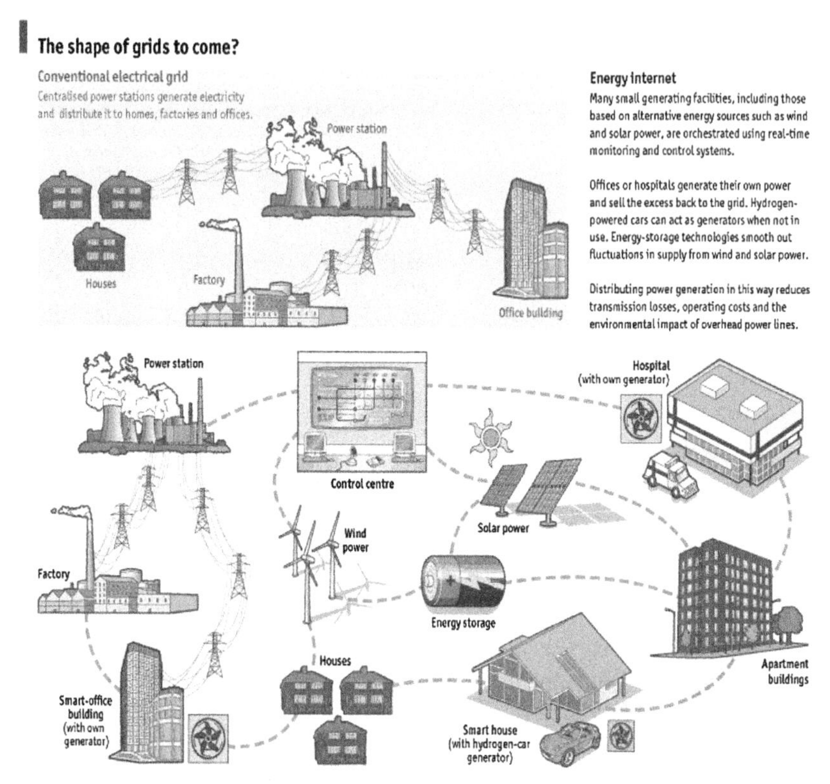

FIGURE 18.4 Agile Energy Infrastructures: central grid and on-site generation.

have had energy base locations. As shown in Fig. 18.4 (Agile Systems expressed as a model like the "Internet") and in the United States case study (among others), there are regions within nations and communities that are well endowed with special energy (and other) resources.

The southwest United States has abundant sunshine, so concentrated solar and photovoltaic energy generation provides the opportunity for an energy base. Likewise the mountains of the western United States have both hydroelectric and geothermal energy bases. Along with the energy base, related economic and educational opportunities exist that supplement and support these resources.

- Establish and set public policy(s) with standards, protocols, rules, and regulations, which need to include decision making, ownership, and accountability. Central government policy and rules are important but are often meant as a guide and goal. Local and regional governments can exceed those rules and go further. These policies can facilitate transitions as new emerging technologies come into the marketplace. Leadership and meeting the vision for IMAR is doing just that.

- Offer long-term contracts to provide commitment to the policy(s) and rules for the IMAR. Both the private and public sectors need such long-term commitments for planning, financing, and sustainable energy systems.
- Make the government the initial "market driver" through the procurement of materials, buildings, transportation, and vehicles among other public sector supplies. Set standards that require continuous change and building on a solid base. These standards could include renewable portfolio standards, fuel efficiency standards, and emission standards (see later). See the recent California Public Utility Commission rulings on solar systems for buildings (2006).
- Create, implement, and monitor an Energy Plan, to include both the central grid and distributed generation systems (that is, agile energy systems).
- Provide policy goals (not limits but goals to meet and exceed) such as Kyoto carbon emission and Renewable Portfolio Standards.
- Establish related energy program goals, such as for recycling products and goods with financial support to companies.
- Develop "take back" laws where companies are required to reuse materials used in their products.
- Institute additional programs modeled after existing programs such as green tags, efficiency ratings or requirements, and standards, which have market value. These values can be traded globally and nationally, to offset emissions from one area for finance credits (see California Climate Registry, 2006 and Chicago Climate Exchange, 2006).
- Follow up on current and recent demonstration projects in China, e.g., coal liquefaction and gasification. Support the inclusion of lessons learned in new high-efficiency mining operations using clean coal and CMM technologies.
- Training:
  - Set up a special energy program at Inner Mongolia University of Technology.
  - Develop joint ventures with foreign partners that are expert in technologies for which China does not yet meet the international standard; e.g.,VAM capture.
  - Provide additional training for blue-collar workers who might have their jobs transformed by new technologies.
- Provide incentives to reduce and conserve energy use per GDP:
  - Rebates for efficient energy use
  - Programs for manufactures to produce efficient appliances and machines
  - Notification for consumers of new conservation life style products for use in their daily lives
  - Establish "Energy Star" (low energy use and high efficiency values) ratings for appliances and product programs with rebate incentives

The US case study mentioned and cited (Fig. 18.4) "agile energy systems" as being similar to the "Internet." That is, like information on the Internet, energy does not rely on one central computer terminal or generating station for the distribution of power. Instead, the system for energy, like that for the Internet, needs to be flexible and dispersed. This approach also makes economic sense and reduces vulnerability.

The Energy Internet analogy as diagramed in Fig. 18.4 comes from the Economist (May, 2004), a publication from the United Kingdom, which reflects the changes in energy infrastructures and systems occurring there now: the outcome of the privatization and liberalization of the energy system in the United Kingdom (30 years later) has been a diverse and flexible energy system combining the central grid and distributed power generation.

As defined in the California Case study and the Glossary, the experiences of energy crises worldwide show that local or on-site energy generation from solar, wind, biomass, and other renewable resources are critical for providing secure, cost-controlled, and reliable energy generation. Grid connected or central power transmission must also include renewable energy generation sources so as to mitigate and lessen waste and impacts on the environment, and the atmosphere. In short, such public policies can significantly reduce global warming.

Fig. 18.4 illustrates the idea of "agile energy systems" as they are being created globally when compared with the old "conventional electric grid," which had been developed in all the case studies for the last 100 years. The old model, analogous to the Internet and computers, has a central computer with all the lines/transmission going to and from that central "grid." The new model (agile energy systems) instead uses the central grid that must exist in a minor role, whereas energy needs to be generated at the local level. Renewable energy is cost effective and environmentally preferable today, and local energy sources must be developed and used to meet regional needs in ways that are environmentally benign and sound.

From the US perspective, the lessons from the California energy crisis (Clark and Bradshaw, 2004) represent a similar pattern of change for the "modern energy system development" evolving into a flexible energy system conforming with regional needs and strengths of the power supply. Like the United Kingdom, the United States and individual states "deregulated" (basically the same approach as "privatization"), which resulted in an energy crisis, due in large part to market manipulation by central grid power generators.

Isherwood et al. (2000a,b) made a similar argument in the analysis of remote Alaskan communities regarding the need for central grid energy along with distributed or on-site power generation. In fact, these articles also argued or suggested that conventional energy sources like clean goal with gasification (e.g., hybrid energy systems) could be converted into hydrogen for storage and energy for use on demand. Similar lessons learned would be useful for IMAR.

Note also that intermittent renewable energy can be used to electrolyze water into hydrogen, in combination with a number of environmentally benign technologies. The example, the US Case Study, shows a hydrogen fueling station from Honda in Torrance, California, USA, that uses renewable energy (solar) to convert water into hydrogen for the Honda hydrogen fuel cell cars, expected in the mass market place within 3–5 years, according to industry reports in February 2006.

IMPLEMENTING MEASURES

Form an Energy Advisory Council for IMAR

- This should be a small permanent group with the ability to bring in experts in various areas, as necessary
- One task of the Council would be to recommend roles and actions by the various players, including: the IMAR government, state-owned enterprises, private industry, nongovernmental organizations, universities, consumers, investors, and experts (both domestic and foreign)
- A specific analysis should be made of the important externalities for the energy industry, especially health impacts of pollution, mine safety, and climate change
- These externalities need to be factored into the economics of comparing clean energy technologies with conventional ones; the Council should recommend how to do this (a strong central government has the advantage of being able to overcome the myopic perspective of individual companies whose actions are controlled by the own internal accounting.)
- The Council should examine current feed-in laws, pricing schemes, taxation/credits, etc., and take necessary actions to manage other potential impediments to constructing the infrastructure required for the new systems required for the energy transition described in our reports
- Seek direct foreign investment with joint ventures and partnerships
- Create an investment fund or "bank" that provides favorable financing based on vision and goals
- Establish an international and reciprocal program for visiting scholars, scientists, engineers, entrepreneurs, etc.
- Set up regional "science parks" or hubs that are focused on an energy and environmental base

CONCLUSIONS

New and alternative sources of energy can replace fossil fuels, conventional coal combustion, and other "dirty" sources as new power generation approaches come into the marketplace and become more cost effective. The policy(s) and plans from the central government and IMAR will promote and implement these alternative strategies, perhaps based on lessons learned in other industrialized nations or states like California in the United States. These programs can begin immediately, with near- and long-term impacts.

A "Green Electricity Strategy" could become policy, such as in Germany, Japan, and California, with solar/photovoltaic policy and laws with financial incentives for related industries. Some attainable objectives are as follows:

- Place the burden of increased costs on the business and wealthier consumer sectors.
- Exempt certain sectors of consumers, such as the agriculture sector and the poor, from increased electricity tariffs.
- Define a long-term price that adds investor certainty in the market.
- Spur growth, for example, in wind power generation (already increased in PRC; 40% in 2004 and 2005), and also for other renewables like solar and photovoltaic, geothermal, biomass, hydroelectric, and related areas.
- Provide special incentives for hybrid energy industries and companies.
- Create university training and educational programs for workers and entrepreneurial companies.

The IMAR should try to set energy goals (in conjunction with the PRC central government) that could attract global attention. Such goals could be based on demonstrations of future technologies that are becoming available today (2006), and their applications to green power and transportation, green buildings, and include models and demonstrations of the IMAR's new energy infrastructures.

The national PRC energy and environment goal, like the "blue skies" (pollution free) on 80% of the days in Beijing by the year 2008, could be adopted for the IMAR, as the coal and energy industry is one of the largest sources of atmospheric pollution. Many cities in the nations covered in the Case Studies set specific goals for air pollution from their industries. Because energy production has such high potential for causing pollution in the downwind provinces, specific emission goals have special importance to the IMAR. An international organization called "Green Cities" holds annual meetings to share goals, objectives, experiences, and ideas. Annual awards are given for both clean cities and those that have mitigated atmospheric pollution among other environmental concerns.

All governments, like those in the Case Studies, must have enforcement power and successes behind the public policies that they enact. These enforcement and regulatory powers include (1) mandatory catalytic converters for cars; (2) factories moved out of cities; (3) "dirty coal" replaced by natural gas or "clean" coal, and (4) major efforts for renewables.

Finally, regions, cities, and communities, as part of the "agile energy" systems strategy, must form seamless combinations of the central gird and on-site power generation that meet their needs and are based on available and plentiful energy resources. Today, the costs for producing energy from these resources have been greatly reduced.

Hence, governments at all levels might start with the development of a vision and then enact policies like: (1) restrict private car access to the city centers, (2) create enforcement incentives, (3) adopt green building (efficiency) standards, (4) invest in public transport, (5) increase peripheral green belts to counteract geological forces, (6) attempt to deemphasize the auto industry (individual vehicle ownership) in favor of mass transportation systems, and (7) use transitional cleaner fuels, like ethanol and biofuels, methanol, natural gas, propane, DME, and hydrogen, for transportation fuel and building power. The key, however, is to limit the capital costs for these systems to short and definable time frames.

SUPPLEMENTAL RECOMMENDATIONS: RURAL INNER MONGOLIA

Introduction

Over half the population of IMAR lives outside the large cities. They face unique energy problems including lack of reliable grid electric power and poor access to heating and cooking fuels. Because of IMAR's ample energy resources, these problems should be solvable in a way to improve the lives of the rural inhabitants and those in small villages, as well as those in urban areas, for the benefit of all.

The Problem: Electricity

In some cases, no grid power is available. In these situations, local renewable resources can be used to provide power for small clusters of homes or even nomadic encampments. However, because wind and solar are inherently intermittent, energy storage is critical. Such systems have been conceptualized and analyzed for other rural settings and villages in similar circumstances (see Isherwood et al., 2000; Andersen and Lund, 2006; among other publications).

One general conclusion of earlier studies is that stand-alone systems can compete economically in remote locations with the costs of installing power lines and infrastructure from a distance. The chart (Energy Internet, 2004) demonstrates a comparison of the old central grid—connected energy "paradigm" with the new one, which is more "on-site" and distributed like the "Internet" (see Clark and Bradshaw, 2004).

The basic costs of wind generation can compare favorably with other energy sources. Solar photovoltaic systems can be practical, as well, but cost more per unit of energy delivered. The initial costs of advanced energy storage technologies such as electrolysis of water and fuel cells remain fairly high today but are rapidly becoming commercially viable. Hence in this transition phase (see Transition is Uncertain Chart, Shell, 2001), external financing options may be necessary to initiate such projects with a focus on public and government as the "market" driver. Conventional batteries can suffice with low initial costs, but storing enough energy to provide power for extended periods will cost much more on a life cycle costing basis.

New technologies such as anaerobic digestion (gasification) of waste can potentially provide power, along with heat and other fuels. Because of greater efficiency and a major reduction of indoor air pollution over the traditional burning of biomass, this process (describe more in further discussion) should be vigorously pursued. In general, combined heat and power has great potential in small village settings, especially when combined with some small industry.

As grid connections become upgraded, attention should be paid to systems that accommodate distributed generation, integrating locally generated power with the larger system. Here the principles of "agile energy systems' proposed by Clark and Bradshaw (2004) can be applied.

Nonelectric Energy for Heating and Cooking

Solar hot water heaters already play an important role in China, and can be further expanded for rural IMAR. Similarly, solar cookers have been employed around western China with some success.

Much of rural IMAR depends on biomass (e.g., animal dung and wood) and coal for their cooking. This results in some of the worst indoor air quality in China. Shifting some of the cooling load to solar cookers can help a little, but much more is necessary. The two main avenues are developing alternative fuels from processing animal waste where some new technologies such as biodigesters are available from new companies who would welcome start-up and new business opportunities in China. Also, by distributing clean cooking fuels such as DME or natural gas along with IMAR's coal gasification polygeneration can be developed.

Heating continues to be a major need throughout much of the year. Building design can be improved (especially if indoor burning of biomass for cooking is reduced). Home efficiency principles applicable to local situations could be prepared by IMAR researchers at the Inner Mongolia Technical University and distributed to rural districts. The distribution of DME as a heating fuel could also follow the path developed for liquified petroleum gas, and could be made common in small villages and clusters of permanent homes.

Transportation

Besides animal power, rural residents also use conventional gasoline and diesel power vehicles. Although their contribution to greenhouse gases is small, the availability of polygenerated DME and/or alcohols can be part of a general push to develop these infrastructures in lieu of expanding the already inadequate conventional petroleum-based fueling system.

Toward an Integrated Energy, Water, Waste, and Transportation Infrastructure Strategy

Distributed "on-site" power systems now being examined for urban areas contain many features with more in common with small rural systems than with the traditional urban, large single power plant, model of the past.

We outline a strategy for the Los Angeles Community College District (nine campuses) that is being implemented along the lines of thinking discussed earlier concerning on-site and central grid energy generation. This kind of combination of energy needs to be integrated with other basic infrastructures such as water, waste, and transportation with a strong concern for education and training.

Energy Internet

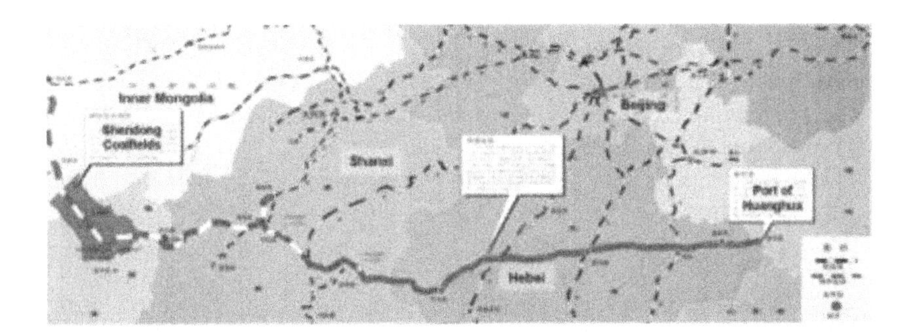

Energy Generation is in Transition

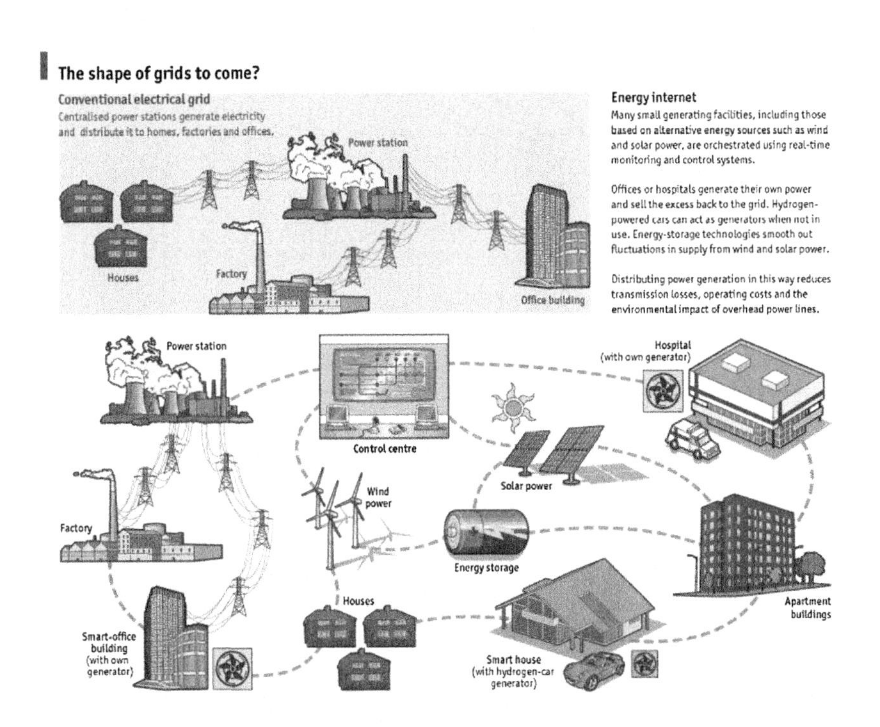

References

Clark, W.W., Bradshaw, T., 2004. Agile Energy Systems: global lessons from the California energy crisis. Elsevier Press, Oxford, UK.

Clark, W.W. II, 2006. Partnerships in Creating Agile Sustainable Development Communities, Journal of Clean Production. (Elsevier Press).

Clark, W.W. II, Lund, H., 2006. Sustainable Development in Practice. Journal of Clean Production. (Elesvier Press).

Isherwood, W., Smith, R., Aceves, S., Berry, G., Clark, W., Johnson, R., Das, D., Goering, D., Seifert, R., 2000. Remote power systems with advanced storage technologies for Alaskan villages, Energy, The International Journal 25, 1005–1020.

Andersen, A., Lund, H., 2006. New CHP Partnerships offering balance of fluctuating renewable electricity productions. Journal of Cleaner Production. (Elsevier Press).

Coal Report and Data (March 06) www.harbour.sfu.ca/dlam/Taskforce/energy-rpt.htm.

The transition is uncertain...

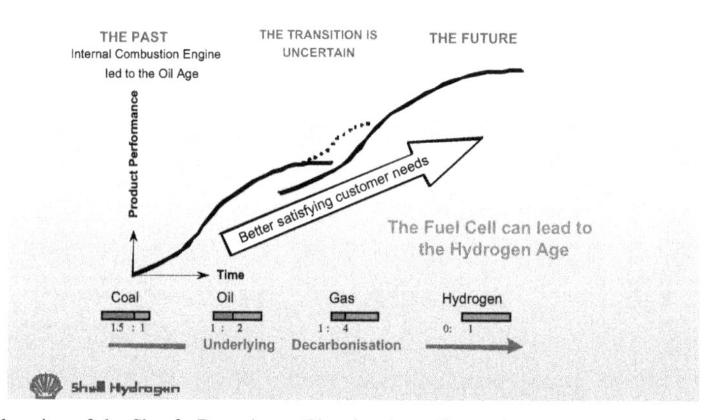

The location of the Shenfu Dongsheng (Shendong) coalfields. Click to enlarge. The map is of the much larger Shenhua Project, which is designed to increase coal production, rail transshipment, power generation, and export. The ultimate target is the production and transport of 100 million metric tons of coal per year.

REFERENCES

California Climate Registry (2006), Los Angeles, CA. http://www.epa.gov/climateleaders/.

California Commission for the 21st Century, "Infrastructure Report, Building Better Buildings: A Blueprint for Sustainable State Facilities (Blueprint)", Chairs: Maria Contreras-Sweet (Secretary of Business, Housing and Trade Agency) and Guy Bustemonte (Lt. Governor of California), Governor in Executive Order D-16–00, February 02.

California Public Utility Commission, January 12, 2006. Order Instituting Rulemaking Regarding Policies, Procedures and Incentives for Distributed Generation and Distributed Energy Resources. Rule #47: RO$-03-017, San Francisco, CA. http://www.cpuc.ca.gov/Cyberdocs/AgendaDoc.asp?DOC_ID=214778.

Chicago Climate Exchange, 2006. Overview and Programs. http://www.chicagoclimateexchange.com.

Clark II, W.W., 2003a. Science parks (1): the theory. International Journal of Technology Transfer and Commercialization 2 (2), 179–206. Inderscience, London, UK.

Clark II, W.W., 2003b. Science parks (2): the practice. International Journal of Technology Transfer and Commercialization 2 (2), 179–206. Inderscience, London, UK.

Clark, W., 2017. Agile Energy Systems. Elsevier.

Clark, W.W., Bradshaw, T., 2004. Bradshaw. Agile Energy Systems: Global Lessons from the California Energy Crisis. Elsevier Press, Oxford, UK.

Clark II, W.W., Li, X., December 2003. Social Capitalism: transfer of technology for developing nations. International Journal of Technology Transfer 3. Inderscience, London, UK.

Clark II, W.W., Sowell, A., 2002. Standard Practices Manual: life cycle analysis for project/program finance. International Journal of Revenue Management (London, UK: Inderscience Press).

Economist, March 13, 2004. Building the Energy Internet.

Isherwood, W., Ray Smith, J., Aceves, S., Berry, G., Clark II, W.W., Johnson, R., Das, D., Goering, D., Seifert, R., 2000a. Economic Impact on Remote Village Energy Systems of Advanced Technologies. University of Calif., Lawrence Livermore National Laboratory. UCRL-ID-129289: January 1998. Published in Energy Policy.

Isherwood, W., Smith, J.R., Aceves, S., Berry, G., Clark, W.W., Johnson, R., Das, D., Goering, D., Seifert, R., 2000b. Remote power systems with advanced storage technologies for Alaskan villages. Energy, The International Journal 25, 1005–1020.

FURTHER READING

Clark, W.W., 2002a. The California Challenge: energy and environmental consequences for public utilities. Utilities Policy 10. Elsevier, UK.

Clark II, W.W., 2002b. Entrepreneurship in the commercialization of environmentally sound technologies: the American experience in developing nations. In: Kuada, J. (Ed.), Culture and Technological Transformation in the South: Transfer or Local Development. Samfundslitteratur Press, Copenhagen, DK.

Clark II, W.W., 2002c. Innovation and capitalisation. International Journal of Technology Management 24 (4), 391–418. Inderscience, UK.

Clark, W.W., Morris, G., 2002. Public-private partnerships: the case of intermittent resources. Energy Policy (Elsevier).

Clark, W.W., Lund, H., 2001. Civic markets the case of the California energy crisis. International Journal of Global Energy Issues 16 (4), 328–344 (Inderscience, London, UK).

Technology transfer of renewable energy from developed to developing countries. In: Clark, W.W., Chong, R.K. (Eds.), 2000. Public Funds for Technology Project. Framework Convention for Climate Change, United Nations. Landmark Study of the Economics of Environmentally Sound Technologies from Six Countries.

Energy Flow Chart, Lawrence. Lawrence Livermore National Laboratory, US Department of Energy, Livermore, CA

Lamont, A., Clark, W.W., Barry, G., Watz, J., 2000. Renewable Energy Systems: Hydrogen for Energy Storage and Transportation Fuels. Working paper. Lawrence Livermore National Laboratory, US Department of Energy, Livermore, CA.

Lund, H., 2005. Renewable energy strategies for sustainable development. In: Dubrovnik Conference. Aalborg University, Denmark.

Lund, H., Clark, W.W., 2002. Management of fluctuations in wind power and CHP: comparing two possible Danish strategies. Energy Policy 27. Elsevier Press.

Livermore National Laboratory, 2004. US Department of Energy. University of California, Oakland, CA.

Santa Barbara County, 2005. Community Environment Council, "Fossil Free by 2033" Plan. Santa Barbara, CA.

Standard Practices Manual (SPM), 2002. A Five Year Energy Efficiency and Renewable Investment Plant for Public. Buildings, prepared by the Interagency Green Accounting Working Group.

Williams, R.H., 2001. Toward zero emissions from coal in China. Energy for Sustainable Development V (4).

Additional Sources

Renewable Energy Sources: IMAR

www.efchina.org.

www.nrel.gov/china.

American Bar Association: Renewable Energy Program

www.abanet.org/environ/committees/renwableenergy/home.html.

Coal Report

www.harbour.sfu.ca/dlam/Taskforce/energy-rpt.htm.

Chapter 19

Business Ventures and Financial Sector in the United Arab Emirates

Robert Rumiński
University of Szczecin, Szczecin, Poland

Chapter Outline

INTRODUCTION

Sustainable development is widely defined as "Development that meets the needs of the present without compromising the ability of future generations to meet their own needs."[1] The United Nations has supplemented this definition with the following statement: "At the heart of operationalising sustainable

1. Brundtland Commission, 1987, Our Common Future, Chapter 2: Towards Sustainable Development From A/42/427. Report of the World Commission on Environment and Development available at: http://www.un-documents.net/ocf-02.htm#I.

Sustainable Cities and Communities Design Handbook. https://doi.org/10.1016/B978-0-12-813964-6.00019-7

development is the challenge of evaluating and managing the complex interrelationships between economic, social and environmental objectives."[2] Meeting the challenges of sustainability is vital both globally and locally.

According to the UN (2002), "... critical objectives for environment and development policies that follow from the concept of sustainable development include:

- reviving growth;
- changing the quality of growth;
- meeting essential needs for jobs, food, energy, water, and sanitation;
- ensuring a sustainable level of population;
- conserving and enhancing the resource base;
- reorienting technology and managing risk; and
- merging environment and economics in decision making."

This chapter is devoted to the economic dimension of sustainability, some of the macro- and microeconomic processes and interrelationships between different financial and nonfinancial entities executing well-balanced public policies leading to sustainable growth of the UAE. It refers to the effects of the UAE policies rather than the policy itself.

The sustainable growth is supported by the following initiatives:

- issuing effective laws and legislations to encourage the business environment and develop national industries and exports;
- strengthening investments and promoting the small and medium businesses sector;
- protecting consumer rights and intellectual property rights;
- diversifying trade activities under the leadership of qualified national resources while adhering to international standards of excellence and the tenets of knowledge economy, thus ushering in balanced and sustainable growth for the UAE.[3]

The UAE has a superb location for international trade, which makes it a natural gateway into the other Gulf Cooperation Council (GCC) Countries of Saudi Arabia, Qatar, Bahrain, Kuwait, and Oman. The investment environment built a global reputation in an unprecedented length of time and demonstrates how global business thrives with visionary leadership and commercial cooperation thanks to sustainable growth policy. The UAE remains the region's most attractive destination for foreign investment. The country ranked 27th among 142 countries in the Global Competitiveness Index (2011−12), where it was the only Arab economy categorized as innovation-driven, and 33rd among 183 countries in the ease of doing

2. The United Nations Department of Economic and Social Affairs (2002).
3. Ministry of Economy, United Arab Emirates: http://www.economy.ae/English/Pages/VissionAndMission.aspx.

business (Doing Business 2011)[4]; 35th among 183 countries in the Index of Economic Freedom (2012)[5]; and 28th among 178 countries in transparency and accountability (Corruption Perceptions Index 2011).[6] In addition, the UAE is a contracting party to the General Agreement on Tariffs and Trade (GATT) since 1994 and a member of the World Trade Organization (WTO) since 1996. It is also a member of the Greater Arab Free-Trade Area in which all GCC states participate.[7]

The country is committed to maintaining the policy of economic openness, actively seeking to develop economic/business projects that are in harmony with the changes taking place in the world. The improved credit and loan provision as well as major public investment project expansion measures contribute to building the country's positive economic growth, despite the current unfavorable economic environment.

To attract strategic foreign investments and to add value to the existing business community as well as to provide a stable and attractive business environment, the UAE government continues to launch initiatives and incentives to boost its investment environment. Nevertheless, the volume of investment might contract as a result of the restructuring of global businesses.

The UAE economy has been slowly recovering from the 2009 crisis and its balance sheets have improved. The authorities strengthened the banking sector through liquidity support, recapitalization, and deposit guarantees, and the emirate of Abu Dhabi provided financial support to the emirate of Dubai. The Dubai Financial Support Fund was called to support troubled entities in the emirate and has now almost exhausted its funding of 20 billion USD. There are several factors that might undermine economic recovery. They are the following:

- *Massive property projects and uncertainty regarding their size*: The excess supply of property in Dubai, which will further increase (as unfinished projects come to completion), will continue to weigh on property prices and growth prospects. Abu Dhabi's strategy to increase its housing supply may also pose risks by placing additional pressure on the property market.
- *Government Related Entities (GREs)*[8] *debt rescheduling*: According to the International Monetary Fund, GREs have an estimated 32 billion USD of debt due in 2011–12 (of which at least 5 billion USD is in the real

4. http://www.ukiet:doingbusiness.org/rankings.
5. http://www.heritage.org/index/ranking.
6. http://cpi.transparency.org/cpi2011/results/.
7. http://www.polishbusinessgroupuae.com/page.aspx?l=1&pg=8&md=pagedetail.
8. Government Related Entities.

estate sector).[9] For that reason, Dubai may continue to face significant rollover risks in the short term, which may raise the cost of borrowing. The Dubai World (DW) debt restructuring was completed (March of 2011—final DW agreement to restructure 14.7 billion USD of its debt with all of its creditors; DW will divide its liabilities in two tranches—with 4.4 billion USD to be repaid in 5 years and the remaining 10.3 billion USD in 8 years), but several other troubled GREs are still in the process of restructuring.

- *International sanctions on Iran*: Iran is one of UAE's largest trading partners, and as a result, sanctions on Iran could weaken the UAE's recovery.
- *The political unrest in the region poses downside risks to the outlook*: It may result in more difficult market conditions as evidenced by the sharp drop in equity markets during Q1-2011. On the other hand, there are also indications that the UAE may benefit from increased tourism and investments. Moreover, the higher oil prices are also benefiting the UAE as it exports oil and hydrocarbons.[10]

In 2011 the economic recovery continued. Although the construction and real estate sectors still remained subdued in the aftermath of the crisis, real gross domestic product (GDP) growth reached an estimated 4.9%, supported by high oil prices and production in response to disruptions in Libya. Nonhydrocarbon growth strengthened to around 2.7%, backed by strong trade and a buoyant sector of tourism. The current account surplus increased significantly, to around 9% of GDP. Average inflation remained at 0.9%, largely due to declining rents. The UAE economic recovery was supported by an expansionary fiscal policy and the nonhydrocarbon primary deficit rose to nearly 42% of nonhydrocarbon GDP in 2011 (36% in 2010), because of Abu Dhabi's increased current and development expenditures and its support for the weakened real estate sector. Nevertheless, the overall fiscal balance, backed by high oil prices, improved significantly. The UAE economy is proving to be extremely resilient in a difficult global economic climate.

SMALL AND MEDIUM ENTERPRISES AND THE GOVERNMENT FINANCIAL SUPPORT

Small and medium enterprises are recognized as an engine of economic growth and a source of sustainable development. Within this sector micro- and

9. A renewed worsening of global financing conditions could make it more difficult to roll over some of the GREs' maturing external debt and would raise the overall cost of their borrowing from international markets.

10. KAM CO Research, United Arab Emirates (UAE), Economic Brief and Outlook 2011, April 2011 pp. 5–6.

small enterprises are of special importance because they are considered as the cradle of entrepreneurship, particularly in environments facing high unemployment. Entities of that size play a vital role in both developing and well-developed economies and are perceived as those creating new jobs and enhancing competition. Small and medium-sized enterprises (SMEs) control 99% of the business sector worldwide, account for 50% of the global GDP, and employ 85% of the world's labor force.

The UAE private sector has been the major vehicle driving the UAE economy, providing the major part of investments and the biggest contribution to growth rate. This would have not been possible without the efforts and support of the government resulting from the policy. The government has been focusing on identifying a number of ways to support SMEs to boost economic growth and employment in line with the **Country Vision 2021**.[11] The government defines a small business as a company that employs less than 50 employees and a medium business as a firm employing between 50 and 100 employees.

SMEs in the UAE are considered as the backbone of the economy, representing 85%−95% of the total business sector. The SMEs' contribution is a critical element of the UAE economy, driving growth and prosperity and providing diversification, as well as employing large numbers of people in the country. The Ministry of Economy is currently drafting a new SME Law to regulate investment in this sector.

To facilitate the development of the SME sector, the government launched a loan scheme. Abu Dhabi Emirate established **Sheikh Khalifa Fund**[12] for SMEs, which offered 246 million AED in funds (2008) for 154 new enterprises (industry and services). The fund also conducted training programs for young men and women on ways to design and choose projects. It aims to create a new generation of Emirati entrepreneurs by enriching the culture of investment among young people as well as supporting and developing small to medium-sized investments in the Emirate.[13]

In 2009 His Highness Sheikh Mohammed bin Rashid issued Law No. 23 of 2009 concerning the **Mohammed Bin Rashid Establishment** for SME's Development, one of the Department of Economic Development establishments, recently renamed the Mohammed Bin Rashid Establishment for **Young Business Leaders**. One of the Establishment's key objectives is to help address the challenges of the funding gap for SMEs and make capital available to them. This comprehensive financing program is to ease the UAE entrepreneurs' financial challenges and facilitate the start-up.

11. http://www.uaeinteract.com/docs/Cabinet_releases_UAE_Vision_2021_(full_text)/39555.htm.
12. http://www.khalifafund.gov.ae/En/AboutUs/Pages/MessagefromtheChairman.aspxhttp://www.khalifafund.gov.ae/En/AboutUs/Pages/MessagefromtheChairman.aspx.
13. Khalifa Fund was launched in 2007 with a total capital investment of 2 billion AED.

The Establishment has set up a dedicated 700 million UAE Fund based on Islamic banking principles in conjunction with the Dubai Islamic Bank to provide access to capital at preferential terms for new entrepreneurs. The application procedure is simple and repayment terms are highly favorable. For existing small and medium business owners, the Establishment provides loans at preferential terms via a network of affiliated banks. Repayment terms are highly favorable. Moreover, the Establishment performs as an advisor to the entrepreneurs seeking financing. The key benefits of the program are the following:

- dedication to small and medium enterprises,
- simple and streamlined application process,
- competitive interest rates,
- highly favorable financing terms (repayment),
- financing both existing and new entrepreneurs.[14]

The new law (Law No. 23 of 2009) confirms the commitment of the Dubai Government to support the development of the SMEs. This sector represents 98.5% of registered businesses in Dubai and 61% of the work force employed. The law aims to strengthen Dubai's position as a center for entrepreneurship and enterprise development based on innovation and intellectual property.

The emirate of Ras Al Khaimah is focusing on SMEs by encouraging them to invest in untapped areas. It increases their management skills and ensures high sales of their products. The **Saud bin Saqr Programme** for SMEs supports business ventures in the field of Information Technology where young investors can succeed. There is high demand for these services from the market.

HSBC introduced the **SME International Fund**[15] in line with the government's efforts to enhance the role of SMEs in the country. A fund of 100 million USD is dedicated to support UAE-based international SME business. The Fund is a part of HSBC's global strategy to support internationally focused SMEs, helping them to grow and conduct business internationally.[16]

Apart from the above-mentioned programs supporting SMEs, there are significant government policies in place[17]:

- SME Development Support Policies, e.g., Access to Finance, IP, Exports and Internationalization, Bankruptcy, and Company Closures;

14. http://www.sme.ae/english/index.html.
15. http://www.ameinfo.com/hsbc-launches-dhs1bn-international-trade-sme-300518.
16. Inverstor's Guide to the UAE 2010 – 2011, Your one - stop information resource, Ministry of Economy, 2011.
17. PKF – Doing business in the UAE – Financial Reporting and Auditing, PKF International Limited, UAE 04.2012.

- Industry-Specific Policies, e.g., Information and Communications Technology, Creative Business Sectors, Emerging Businesses;
- SME Professionalization and Capability Development Program covering skills and leadership development, corporate governance, accounting, financial and investment management, business continuity, operational and productivity excellence, etc.

EASE OF DOING BUSINESS IN THE UAE

Doing Business[18] report is one of the key instruments providing quantitative measures of regulations for:

- starting a business,
- dealing with construction permits,
- getting electricity,
- registering property,
- getting credit,
- protecting investors,
- paying taxes,
- trading across borders,
- enforcing contracts
- resolving insolvency.

The above-mentioned issues are concerned with the domestic SMEs. This internationally recognized report assesses countries on how easy it is for SMEs to conduct business.

In 2010 the UAE climbed 14 places in the Doing Business report compiled by the World Bank and its International Finance Corporation. The UAE rose to the 33rd position in the global rankings for regulatory reform, partly as a result of the government's decision to abolish a 150,000 AED minimum capital requirement for some start-ups.

Two other key reasons for the country's rise was a streamlining of the process involved in obtaining construction permits and the improving of capacity at Dubai ports.[19]

The following important improvements were carried out in 2010:

- business start-up was eased by simplifying the documents needed for registration,
- abolishing the minimum capital requirement,

18. http://www.doingbusiness.org.
19. Vine P. (Ed): *UAE 2010 Yearbook*, Trident Press Ltd and The National Media Council. Mayfair, London 2010.

- removing the requirement that proof of deposit of capital be shown for registration,
- the time for delivering building permits was shortened by improving its online application processing system,
- greater capacity at the container terminal,
- elimination of the terminal handling receipt as a required document, an increase in trade finance products.

Doing Business 2010 Report

	Doing Business 2010 Ranking	Doing Business 2009 Ranking	Doing Business Change in Rank
Doing business	33	47	+14
Starting a business	44	118	+74
Dealing with construction permits	27	54	+27
Employing workers	50	45	−5
Registering property	7	7	0
Getting credit	71	68	−3
Protecting investors	119	114	−5
Paying taxes	4	4	0
Trading across borders	5	13	+8
Enforcing contracts	134	135	+1
Closing a business	143	143	0

Based on http://www.doingbusiness.org/~/media/giawb/doingbusiness/documents/profiles/country/ARE.pdf.

In 2012 the following reforms were introduced (according to the DB report):

- the UAE made starting a business easier by merging the requirements to file company documents with the Department for Economic Development, to obtain a trade license, and to register with the Dubai Chamber of Commerce and Industry,
- the country improved its credit information system through a new law allowing the establishment of a federal credit bureau under the supervision of the central bank,
- the access to credit was enhanced by setting up a legal framework for the operation of the private credit bureau and requiring that financial institutions share credit information,
- the UAE streamlined document preparation and reduced the time to trade with the launch of Dubai Customs' comprehensive new customs system (Mirsal 2).[20]

20. http://www.doingbusiness.org/reforms/overview/economy/united-arab-emirates#.

INVESTMENT BODIES—THE KEY PLAYERS AND CONTRIBUTORS

According to the UN Conference on Trade and Development, between 2003 and 2008 the UAE was the third largest recipient of foreign direct investment in West Asia, behind Saudi Arabia and Turkey.

Investments in overseas markets have been integral to the country's strategic plan to create a security net for future generations who will face the prospect of a depletion of the hydrocarbon reserves. The major international investment bodies in the Emirates significantly contribute to the sustainable growth of the country. They are the following:

- **Abu Dhabi Investment Authority (ADIA)**[21]: Its mission is to secure and maintain the prosperity of the emirate through the management of its investment assets. ADIA is a leading international investor and for the past 35 years has established itself as a trustworthy and strong investor (supplier of capital). The Authority supervises a substantial global diversified portfolio of assets across varying sectors, regions, and asset classes (private equity, property, public listed equities). ADIA does not seek active management of the entities it invests in, only long-term sustainable financial returns.
- **Abu Dhabi Investment Council (ADIC)**[22]: It is responsible for investing part of Abu Dhabi's surplus financial resources. It executes a globally diversified investment strategy focused on gaining positive capital returns across a range of asset classes.
- **Invest AD**[23]: It is a subsidiary of ADIC; a government investment vehicle similar to ADIA, established in 1977. Invest AD allows outside investors to put their money in it (alongside). Invest AD invests on behalf of the government and attracts capital from external investors.
- **The Investment Corporation of Dubai**[24]: It invests to create stability and foster diversification. It owns 60% of **Borse Dubai**,[25] a holding company that performs as a holding company for Dubai Financial Market (DFM) and NASDAQ Dubai.
- **Dubai Holding**[26]: It is one of Dubai's major holding companies, divided between the Dubai Holding Commercial Operations Group (DHCOG) and the Dubai Holding Investment Group (DHIG). It was formed in 2009 when Dubai Group and Dubai International Capital (DIC) were combined. Dubai

21. http://www.adia.ae/En/home.aspx.
22. http://www.adcouncil.ae/.
23. http://www.investad.com/.
24. http://www.icd.gov.ae/.
25. http://www.borsedubai.ae/.
26. http://dubaiholding.com/.

Holding has reorganized its companies into property, business park, hospitality, and investment units.

- **DHCOG**[27]: Property developers such as Dubai Properties Group, Sama Dubai, and Tatweer fall under DHCOG. Moreover, DHCOG holds the hotel operator Jumeirah Group and the business park operator TECOM Investments.
- **DHIG**[28]: This was formed after combining previously separate entities of Dubai Group and DIC. It controls six financial companies that are under the responsibility of Dubai Group, including Dubai Capital Group, Dubai Financial Group, Dubai Investment Group, Dubai Banking Group, Dubai Insurance Group, and Noor Investment Group. Focused on the private equity asset class, the DIC operates through global buy-outs specializing in secondary leveraged buyouts in Europe, North America, and Asia. DIC owns stakes in the Travelodge hotel chain, the Middle-Eastern operations of the property consultancy CB Richard Ellis, and the UK engineering company Doncasters.
- **DW**[29]: This is a holding company that has been at the forefront of Dubai's rapid growth. It operates in different industrial segments and invests in four main sectors:
  - transport and logistics,
  - dry docks and maritime,
  - urban development,
  - investment and financial services.

  The DW portfolio includes some of the world's best known companies:
  - DP World (maritime terminal operator);
  - Drydocks World and Dubai Maritime City;
  - Economic Zones World (operates several free zones around the world);
  - Nakheel, the property developer behind The Palm Islands and The World;
  - Limitless (the international real estate master planner);
  - Leisurecorp, a sports and investment group;
  - Dubai World Africa;
  - Istithmar World.

LEGAL ASPECTS OF CONDUCTING BUSINESS ACTIVITY

The UAE legal system is essentially a civil law jurisdiction heavily influenced by French, Roman, and Islamic laws. The increasing presence of

27. http://dubaiholding.com/media-centre/news/2012/Dubai-Holding-Commercial-Operations-MTN-Limited-Bond-Coupon-Payments/.
28. http://www.dubaigroup.com/aboutus/dubaiholding_en_gb.aspx.
29. http://www.dubaiworld.ae/.

international law firms from Common Law jurisdictions has demonstrated the application of Common Law principles in commercial contracts. This indirectly has further influenced the UAE legal system.[30] Establishing business in the UAE is subject to licensing requirements as well as foreign investment restrictions. Businesses can be set up in the following two investment locations:

1. mainland UAE, and
2. Free Trade Zones ("FTZs").

To regulate matters such as commercial transactions, commercial agencies, civil transactions, labor relations, maritime affairs, intellectual property, and commercial companies, a number of codified federal laws have been passed. In addition, a number of local laws have also been passed in various areas by individual Emirates. There are two main types of laws in the UAE:

- federal—applicable to the UAE as a whole and issued either by the legislative body or by the Ministers of each Ministry (Ministerial Order);
- local—decrees and orders apply only to a particular Emirate (passed by the Ruler or Crown Prince of a particular Emirate and issued by a member of the Royal Family of that Emirate).

All emirates have brought their judicial systems into the UAE Federal Judicial Authority except Dubai and Ras Al Khaimah. Dubai has retained its own independent courts and judges, which are not a part of the UAE Federal Judicial Authority. Dubai's courts apply federal laws (e.g., Companies Law or the Civil Code) as well as the laws and decrees enacted by the Ruler of Dubai, in case the federal law is absent.

In 2004 the **Dubai International Financial Center Courts** (DIFC Courts) were founded and they are an independent common law judiciary based in the Dubai International Financial Centre (DIFC), with jurisdiction governing civil and commercial disputes.

The DIFC Courts comprise international judges from common law jurisdictions such as England, Malaysia, and New Zealand and their procedural rules are largely modeled on English civil procedure rules. The official language of the DIFC Courts is English. All proceedings are also conducted in English.

The Dubai government has expanded the jurisdiction of the DIFC Courts, which allow any parties (even not incorporated within the DIFC free zone) to use the DIFC Courts to resolve commercial disputes.[31] Currently, regional

30. Doing business in the UAE, A business and tax profile, PKF Accountants & business advisers, April 2012.
31. Previously, only companies based in the DIFC or those that had an issue related to the DIFC could use the DIFC Courts.

and international parties can agree to use the DIFC Courts in the event of a dispute (the parties should then agree to incorporate the jurisdiction of the DIFC Courts into their contracts before taking the dispute to the DIFC Courts).

The expansion of DIFC Courts jurisdiction represents an important policy shift and will give the business community an unprecedented access to the DIFC Courts. The move is likely to be welcomed by both the legal and business communities, since international parties may be more likely to wish to resolve their disputes in a more familiar forum that uses the common law English model.[32]

One of the crucial decisions that entrepreneurs need to make to set up their business entity is to choose the appropriate legal form of conducting business activity. The way the business is structured and operated will most likely influence the access to future financing opportunities.[33] It is essential for the potential business owner to understand how each legal structure works and then pick up the one that best meets the entrepreneur's needs. Each business form has its own unique legal and financial ramifications; therefore it is advisable that the choice is made in conjunction with a lawyer and an accountant. The choice of the form of business organization can have a great impact on the success of the business. It may influence:

- how easy it is to obtain financing;
- how taxes are paid;
- how accounting records are kept;
- whether personal assets are at risk in the venture;
- the amount of control the owner has over the business, etc.

Foreign investors can choose between several types of cooperation and partnerships for conducting business in the UAE. Entrepreneurs can also conduct business activity through the UAE branch office. Limited Liability Companies (LLCs) are more commonly used by the foreign investors.

The UAE company law determines a total local equity of not less than 51% in any business entity. Moreover, it defines seven categories of business organization that are allowed to be established in the UAE:

1. **General Partnership**—formed by two or more partners who will be jointly liable to the extent of all their assets for the company liabilities.
2. **Simple Limited Partnership**—formed by one or more general partners liable for the company liabilities to the extent of all their assets, and one or more limited partners liable for the company liabilities to the extent of their respective shares in the capital only.

32. See footnote 30.
33. Sitarz D.: The complete book of Small Business Legal Forms, third Edition, Nova Publishing Company 2001, Carbondale, Illinois, USA, pp. 21.

3. **Joint Venture**—a company concluded between two or more partners to share the profits or losses of one or more businesses being performed by one of the partners. Local equity participation must be at least 51%.
4. **Public Joint Stock**—any company whose capital is divided into equal value negotiable shares and a partner therein is only liable to the extent of his share in the capital.
5. **Private Joint Stock**—not less than three founder members may incorporate. Shares are not offered for public subscription. The founder members will fully subscribe to the capital (not less than 2 million AED).
6. **Limited Liability Company**—An LLC can be formed by a minimum of 2 and a maximum of 50 persons whose liability is limited to their shares. Most companies with expatriate partners have opted for this LLC, because this is the only option that will give maximum legal ownership, i.e., 49% to the expatriates for a trading license; 51% UAE nationals participation is the general requirement (normal share holding pattern is: local sponsor, 51%; foreign shareholder, 49%). There is no minimum capital requirement for establishing a company. The foreign equity capital cannot exceed 49%, but the profit and loss distribution can be mutually agreed. The management of an LLC may be performed by foreign and national partners or third parties. The formation of a company takes approximately 1–2 weeks from the date of receipt of all the documents.
7. **Share Partnerships**—a company formed by general partners who are jointly liable to the extent of all their assets for the company liabilities and participating partners who are liable only to the extent of their shares in the capital.

Foreign direct investments (FDIs) are encouraged through branches and representative offices of foreign companies and 100% foreign ownership is permitted in the FTZs. Partnerships etc. are generally open only to UAE nationals.

As mentioned before, LLCs are more commonly used by the foreign investors and the following documents are required for LLC:

- certificate of capital contribution from a bank;
- auditor's certificate for shares of all kinds;
- all other items requested in the application form.

There are five steps to set up an LLC:

1. Approval of the company name and activity from the relevant office of Economic Development, Municipality and Chamber of Commerce;
2. Articles of Association must be notarized according to the requirements of each emirate(s);

3. Application package must be delivered to the Department of Economic Development or the Municipality as appropriate;
4. Following approval, the new company will be included in the Commercial Register and the Articles of Association published in the Bulletin of the Ministry of Economy;
5. A license will then be issued by the Department of Economic Development (Dubai and Sharjah) or the Municipality or the Chamber of Commerce of the other emirates.

The Commercial Companies Law allows for setting up branches and representative offices of foreign companies. They may be 100% foreign owned, provided a local service agent[34] is appointed.

A **branch office** (regarded as part of its parent company) is a full-fledged business, permitted to realize contracts or conduct other activities similar to those of its parent company as specified in its license. Activities are approved on a case-by-case basis.[35]

A **representative office** is limited to promoting its parent company's activities (e.g., gathering information and soliciting orders and projects to be performed by the company's head office). Representative offices are also limited in the number of employees that they may sponsor.

Local service agents (also referred to as sponsors) are not involved in the operations of the company but assist in obtaining visas, labor cards, etc. The time required for setting up a branch of a foreign company is approximately 3–4 weeks from the date of receipt of all the documents. To prove the credibility of the company, a business plan and current profile, as well as financial statements of the last 2 years, should be submitted.

In the case of **professional firms**, 100% foreign ownership, sole proprietorships, or civil companies are permitted. They may engage in professional or artisan activities. A UAE national must be appointed as a local service agent.

The basic requirement for all types of business activities in the UAE is one of the following three categories of **licenses**:

1. *commercial license*—covering all kinds of trading activity,
2. *professional license*—covering professions, services, craftsmen, and artisans;
3. *industrial license*—establishing industrial or manufacturing activity.

34. Only UAE nationals or companies 100% owned by UAE nationals may be appointed as local service agents.
35. Expanding your horizons? A guide to setting up business across the Middle East and North Africa region, KPMG International Cooperative, 2010.

To set up a company to engage in certain activities (including financial institutions), the official approval is required from the appropriate government ministry or department. After receiving the license, companies are also required to register with the local Chamber of Commerce. These institutions are very important and investors should regard them as effective resources of information (databases, business literature) to establish their projects. Moreover, the Chambers provide an extensive range of basic and more sophisticated services for entrepreneurs.

Another opportunity for the foreign investor is to set up a company in one of the **FTZs**. They are special economic areas established to promote foreign investment and economic activities within the UAE. FTZs are governed by an independent Free Zone Authority (FZA) responsible for issuing FTZ operating licenses and assisting entrepreneurs with establishing their businesses. The procedures for establishing a business are relatively simple. Entrepreneurs (investors) can either register a new company in the form of a Free Zone Establishment (FZE) or simply establish a branch or representative office of their existing or parent company based within the UAE or abroad. An FZE is a limited liability company governed by the rules and regulations of FTZ in which it is established. There are currently 36 FTZs with many more in the pipeline[36]:

1. Masdar City
2. Abu Dhabi Ports Company
3. Abu Dhabi Airport Free Zone
4. Khalifa Industrial Zone
5. ZonesCorp
6. twofour54
7. Dubai Airport Freezone
8. Dubai Silicon Oasis
9. Jebel Ali Free Zone
10. Dubai Multi Commodities Center
11. Dubai Internet City
12. Dubai Media City
13. Dubai Studio City
14. Dubai Academic City
15. Dubai Knowledge Village
16. Dubai Outsource Zone
17. Enpark
18. Intl Media Production Zone

36. http://www.uaefreezones.com.

19. Dubai Biotech Research Park
20. Dubai Auto Zone
21. Gold and Diamond Park
22. Dubai Healthcare City
23. Dubai Intl Financial Centre
24. Dubai Logistics City
25. Dubai Maritime City
26. Dubai Flower Centre
27. Intl Humanitarian City
28. Sharjah Airport Free Zone
29. Hamriyah Free Zone
30. Ahmed Bin Rashid FZ
31. Ajman Free Zone Authority
32. RAK Investment Authority
33. RAK Free Zone
34. RAK Maritime City
35. Fujairah Free Zone
36. Fujairah Creative City

The companies operating in the FTZs are treated as being offshore or outside the UAE for legal purposes. The major advantages in setting up in a free zone are:

- 100% foreign ownership of the enterprise,
- 100% import and export tax exemptions,
- 100% repatriation of capital and profits,
- exemption from all import and export duties,
- corporate tax exemptions for up to 50 years,
- no personal income taxes,
- assistance with labor recruitment,
- inexpensive energy and workforce,
- additional support services (e.g., administration, sponsorship and housing),
- companies at FTZ can operate 24 h a day.

The FTZ also has some limitations:

- a company set up in the FTZ is not allowed to trade directly with the UAE market,
- a company can undertake the local business only through the locally appointed distributors,
- a custom duty of 5% is applicable for the local business.[37]

37. http://www.polishbusinessgroupuae.com/page.aspx?l=1&pg=8&md=pagedetail.

BANKING SECTOR—THE STRUCTURE AND RECENT DEVELOPMENTS

Islamic finance is a 400 billion USD industry, growing at a rate of over 151% per annum with an expected high growth rate for the next 15–20 years. Each Islamic market has developed relatively independently, setting its own regulations and standards, developing a wide variety of products with different benchmarks and pricing techniques.[38]

The UAE financial services sector has served as an important element of growth toward the diversification of the UAE's strategy. Abu Dhabi and Dubai financial sectors constitute the majority of the UAE's financial system. There is a sound, modern, and competitive banking industry. In 2011, 23 locally incorporated banks and 28 branches of foreign banks were operating in the country. The international financial crisis has had a relatively mild influence on the banking sector, thanks to the government interventions (the improvement of banks' liquidity), which gave a boost to economic activity (the UAE Central Bank AED 50 billion facility to support local lenders and the UAE Ministry of Finance AED 70 billion liquidity support scheme). The crisis witnessed in Dubai in 2008 led its neighbor Abu Dhabi to intervene and provide financial aid.

In consequence, the capitalization of the UAE's banking system remains sound as the capital adequacy ratio increased from 13.3% at the end of 2008 to 21% in 2010. Moreover, the banking sector has a strong deposit base (it increased from 923 billion AED in 2008 to 967 billion AED in 2010). A strong capital base has allowed banks to increase lending in the UAE, even during the crisis.[39] The sector remains resilient to shocks, backed by solid capital base, including money injected by the government, and strong earnings, despite the doubling of nonperforming loans since the global financial crisis struck.

A renewed worsening of global financing conditions could make it more difficult to roll over some of the GREs' maturing external debt, and would raise the overall cost of their borrowing from international markets. About 32 billion USD of sovereign and GRE debt is estimated to mature in 2012, of which $15 billion is in Dubai.[40]

The UAE banking sector's capitalization remains sound and is sufficient to absorb debt problems faced by Dubai GREs. Banks' credit portfolios in the UAE have extensive exposures to trade, real estate, and construction sectors. Bank lending is reviving but credit growth remains sluggish. Lending was up

38. http://www.difc.ae/.
39. INVESTOR'S Guide to the UAE 2010–11.
40. International Monetary Fund, Middle East and Central Asia Department, United Arab Emirates, 2012 article IV consultation concluding statement, March 19, 2012.

1.4% in 2010, reaching 972.1 billion AED[41] (264.6 billion USD), which indicates the bank's general cautiousness in lending to the economy. Lending to the government maintained its upward trend, which is in line with the government's continued expansionary policies [100 billion AED (27.2 billion USD) in 2010]. Lending for construction decreased 2.6% in 2010, whereas the credit extended for real estate mortgage loans increased 15%. Moreover, stress tests on aggregate banking data indicate resilience to shocks in spite of nonperforming loans doubling since the crisis.

Effective bank governance is fundamental to support financial sector soundness. In light of the government's control of banks and the banks' high exposure to GREs, a clear governance framework is needed to safeguard the banks' financial integrity.

Selected Monetary and Banking Indicators (in AED)

	December 2009	March 2010
Total Bank Assets (net of provisions)[a]	**1519.1**	**1533.1**
Certificate of Deposits held by Banks	71.9	63.3
Bank Deposits	982.6	967.0
Loans and Advances: (net provisions)[a]	1017.7	1022.0
Personal Loans	209.8	212.2
Letters of Credit	102.8	104.2
Total Private Funds (Capital + Reserves)[b]	231.4	252.8
Specific Provisions for Nonperforming Loans	32.6	34.4
General Provisions	10.7	13.7
Total Investments by Banks		
Capital adequacy ratio—Banking System	**19.2%**	**20.3%**
Banking Institutions (Total Numbers)		
UAE Incorporated Banks		
Head Offices	24	23
Branches	674	687
Electronic Banking Service Units	25	25
Pay Offices	71	73
Foreign Banks		
Head Offices	28	28
Branches	82	82
Electronic Banking Service Units	43	45
ATMs	**3599**	

[a]Net of interest in suspense, specific provisions, and general provisions.
[b]Excluding current year profit.
Based on United Arab Emirates: Selected Issues and Statistical Appendix, IMF Country Report No. 12/136, June 2012.

The Central Bank of the UAE has also made important progress in strengthening its financial stability approach, reorganizing the regulatory framework and developing macroprudential policies.[42]

41. 2010 data from the Central Bank of the UAE.
42. See footnote 30.

As mentioned earlier, the number of locally incorporated commercial banks stood at 23 during 2011, whereas the number of their branches increased from 732 at the end of Dec. 2010 to 768 at the end of December 2011, and the number of their electronic/customer service units remained at 26. The ratio of nonperforming loans of national banks stood at 6.2% in 2011, a sharp increase from 2008 precrisis levels, whereas that of Dubai banks was higher at 10.6%. The stress tests showed that the domestic banking system could absorb a significant increase in nonperforming loans.

The following national banks are currently registered in the UAE[43]:

1. Abu Dhabi Commercial Bank[44]
2. Abu Dhabi Islamic Bank[45]
3. Al Hilal Bank[46]
4. Arab Bank for Investment and Foreign Trade (Al Masraf)[47]
5. Commercial Bank of Dubai[48]
6. Commercial Bank International[49]
7. Dubai Bank[50]
8. Dubai Islamic Bank PJSC[51]
9. Emirates NBD Bank[52]
10. Emirates Islamic Bank[53]
11. First Gulf Bank[54]
12. Mashreq Bank[55]
13. Noor Islamic Bank[56]
14. RAK Bank[57]
15. Sharjah Islamic Bank[58]
16. United Arab Bank PJSC[59]
17. Union National Bank[60]

---

43. http://www.centralbank.ae/en/index.php.
44. http://www.adcb.com/general/chargesandfees/chargesandfees.asp.
45. http://www.adib.ae/savings-account.
46. http://www.alhilalbank.ae/.
47. http://www.al-masraf.ae/.
48. http://www.cbd.ae/cbd/index.aspx.
49. http://www.cbiuae.com/.
50. https://www.dubaibank.ae/?item=%2fcontent%2fdefault&user=extranet%5cAnonymous&site=website.
51. http://www.dib.ae/support/schedule-of-charges.
52. http://www.emiratesnbd.com/en/.
53. http://www.emiratesislamicbank.ae/default.aspx.
54. http://www.fgb.ae/en/.
55. http://www.mashreqbank.com/personal/service-charges/overview.asp.
56. http://www.noorbank.com/english/help/contact-us.
57. http://rakbank.ae/rakbank/personalbanking/personalbanking.jsp.
58. http://www.sib.ae/en/retail-banking/products-services/fees-and-charges.html.
59. http://www.uab.ae/cms/index.php?option=com_content&view=article&id=60&Itemid=92.
60. http://www.unb.ae/english/inner.aspx?p=1&mid=403.

The UAE National Banks and Distribution of Their Branches (2011)

No	Name of the Bank	Head Office	Abu Dhabi	Dubai	Sharjah	Ras Al Khaimah	Ajman	Umm Al-Quwain	Fujairah	Al Ain	Total Number of Branches	Pay offices	Electronic Banking Service Units
1.	National Bank of Abu Dhabi	Abu Dhabi	39	18	10	2	1	1	3	12	86	42	0
2.	Abu Dhabi Commercial Bank	Abu Dhabi	20	11	3	1	1	0	2	7	45	5	1
3.	ARBIFT	Abu Dhabi	3	4	1	0	0	0	0	1	9	0	0
4.	Union National Bank	Abu Dhabi	19	14	8	2	2	1	1	7	54	10	0
5.	Commercial Bank of Dubai	Dubai	3	17	1	1	1	0	1	1	25	5	0
6.	Dubai Islamic Bank PJSC	Dubai	9	32	12	4	2	1	2	6	68	0	5
7.	Emirates NBD Bank	Dubai	15	83	7	3	1	1	2	3	115	18	0
8.	Emirates Islamic Bank	Dubai	4	17	5	1	1	1	1	3	33	1	0
9.	Mashreq Bank PSC	Dubai	13	33	9	2	3	1	2	3	66	0	7
10.	Sharjah Islamic Bank	Sharjah	1	3	20	0	0	0	1	1	26	1	0
11.	Bank of Sharjah PSC	Sharjah	1	1	1	0	0	0	0	1	4	0	0
12.	United Arab Bank PJSC	Sharjah	2	3	3	2	1	0	1	1	13	0	0
13.	InvestBank PLC	Sharjah	2	2	4	1	1	0	1	1	12	0	0
14.	The National Bank of R.A.K	RAK	5	12	4	7	1	0	0	1	30	1	4
15.	Commercial Bank International	Dubai	3	5	2	3	1	1	1	1	17	1	0
16.	National Bank of Fujairah PSC	Fujairah	2	4	2	0	1	0	5	1	15	0	0
17.	National Bank of U.A.Q PSC	U.A.Q	2	6	2	1	2	2	1	1	17	1	7
18.	First Gulf Bank	Abu Dhabi	7	3	2	1	2	0	1		18	0	0
19.	Abu Dhabi Islamic Bank	Abu Dhabi	28	11	8	3	2	1	2	11	66	0	0
20.	Dubai Bank	Dubai	4	13	3	1	1	0	1	1	24	0	0
21.	Noor Islamic Bank	Dubai	3	9	2	0	0	0	0	1	15	0	2
22.	Al Hilal Bank	Abu Dhabi	10	8	1	1	0	0	0	2	22	0	0
23.	Ajman Bank	Ajman	3	2	1	0	4	0	0	1	11	2	0
	Total		198	311	111	36	28	10	28	69	791	87	26

Based on www.centralbank.ae.

The complete list of commercial banks and representative offices in the UAE is the following:

1. ABN Amro Bank NV
2. Abu Dhabi Commercial Bank
3. Abu Dhabi Islamic Bank
4. Al Ahli Bank of Kuwait
5. Al Rafidain Bank
6. Al Hilal Bank
7. Arab African International Bank
8. Arab Bank for Investment and Foreign Trade
9. Arab Bank PLC
10. Bank Melli Iran
11. Bank of Baroda
12. Bank of Sharjah
13. Bank Saderat Iran
14. Blom Bank France SA
15. Banque Du Caire
16. Calyon Bank
17. Al Khaliji France
18. BNP Paribas Bank
19. HSBC Bank Middle East
20. CitiBank NA
21. Commercial Bank International
22. Commercial Bank of Dubai
23. Dubai Islamic Bank
24. Dubai Bank PJSC
25. El Nilein Bank
26. Emirates Bank International
27. Emirates Islamic Bank
28. First Gulf Bank
29. Habib Bank Limited
30. Habib Bank AG Zurich
31. Invest Bank
32. Janata Bank
33. Lloyds Bank TSB
34. Mashreq Bank
35. National Bank of Abu Dhabi
36. National Bank of Bahrain
37. National Bank of Dubai
38. National Bank of Fujairah
39. National Bank of Oman
40. National Bank of RAK
41. National Bank of UAQ

42. Noor Islamic Bank PJSC
43. Sharjah Islamic Bank
44. Standard Chartered Bank
45. Union National Bank
46. United Arab Bank
47. United Bank Limited

Representative offices are as follows:

1. Bank of Bahrain & Kuwait
2. Barclays Bank
3. Doha Bank
4. ED & F Investment Products Ltd.
5. ICICI Bank
6. Korea Exchange Bank
7. Standard Bank London Ltd.
8. Union Bancaire prive'e (CBI-TDB)
9. Westdeutche Landesbank

In 2011 two licenses were granted to wholesale banks, namely, Deutsche Bank AG and Industrial & Commercial Bank of China. Moreover, two investment banks commenced operation in the country, Arab Emirates Invest Bank and HSBC Financial Services (Middle East) Limited.

Commercial Banks Operating in the UAE (2010 and 2011)

	2010			2011	
	December	March	June	September	December
National Banks					
Head Offices	23	23	23	23	23
Branches	732	736	745	757	768
Electronic Customer Service Units	26	29	27	27	26
Cash Offices	86	86	86	87	87
GCC Banks					
Main Branches	6	6	6	6	6
Additional Branches	1	1	1	1	1
Other Foreign Banks					
Main Branches	22	22	22	22	22
Additional Branches	82	82	82	82	82
Electronic Customer Service Units	50	48	47	47	50
Cash offices	1	1	1	1	1
Number of ATMs	3.758	3.846	3.963	4.053	4.172

Based on www.centralbank.ae.

The number of GCC banks in 2011 remained unchanged at 7, whereas the number of other foreign banks remained unchanged at 22, the number of their branches remained at 82, and the number of their electronic/customer service units remained at 50.

The number of Automated Teller Machines (ATMs) in the UAE increased from 3758 ATMs at the end of 2010 to 4172 at the end of 2011.

There are 11 foreign banks operating in thc UAE. They are as follows:

1. Bank of Baroda[61]
2. Barclays Bank[62]
3. Bank Saderat Iran[63]
4. Citi Bank[64]
5. Doha Bank[65]
6. Habib Bank A.G. Zurich[66]
7. HSBC Middle East Limited[67]
8. Lloyds TSB Bank PLC[68]
9. National Bank of Bahrain[69]
10. National Bank of Oman[70]
11. Standard Chartered Bank[71]

The number of licensed foreign banks' representative offices operating in the UAE reached 110 at the end of 2011. The new representative offices licensed during 2011 are the following:

1. AXIS Bank Ltd.
2. Falcon Private Bank Ltd.
3. Doha Bank
4. Bank of Montreal
5. SBI Funds Management Private Ltd.
6. Bank of the Philippine Islands
7. Liechtensteinische Landesbank (Liechtenstein) Ltd.
8. ABN Amro Bank N.V.
9. Banque Privee Edmond De Rothschild SA
10. Fairbairn Private Bank

---

61. http://www.bankofbaroda.com/.
62. http://www.barclays.ae/accounts/savings-account/.
63. http://www.banksaderat.ae/OpenAcc.html.
64. http://www.citibank.com/uae/homepage/index.htm.
65. http://www.dohabank.ae/en/Personal/Accounts/SavingsAccount.aspx.
66. http://www.habibbank.com/.
67. https://www.hsbc.ae/1/2/.
68. http://www.lloydstsb.ae/personal-banking/savings-investment/instant-access.html.
69. http://www.nbbonline.com/default.asp?action=category&ID=48.
70. http://www.nbo.co.om/.
71. http://www.standardchartered.ae/personal/Important-Information/en/service-charges.html.

Foreign Banks in the UAE (2011)

Sl. No	Name of the Bank	Head Office	Abu Dhabi	Dubai	Sharjah	Ras Al Khaimah	Ajman	Umm Al-Quwain	Fujairah	Al Ain	Total Number of Branches	Electronic Banking Service Units Pay Offices
1.	National Bank of Bahrain	Abu Dhabi	1	0	0	0	0	0	0	0	1	0
2.	Rafidam Bank	Abu Dhabi	1	0	0	0	0	0	0	0	1	0
3.	Arab Bank PLC	Abu Dhabi	1	2	1	1	1	0	1	1	8	0
4.	Banque Misr	Abu Dhabi	1	1	1	1	0	0	0	1	5	0
5.	El Nilein Bank	Abu Dhabi	1	0	0	0	0	0	0	0	1	0
6.	National Bank of Oman	Abu Dhabi	1	0	0	0	0	0	0	0	1	0
7.	Credit Agricole Corporate and Investment Bank	Dubai	1	1	0	0	0	0	0	0	2	0
8.	Bank of Baroda	Dubai	1	2	1	1	0	0	0	1	6	7
9.	BNP Paribas	Abu Dhabi	1	1	0	0	0	0	0	0	2	2
10.	Janata Bank	Abu Dhabi	1	1	1	0	0	0	0	1	4	0
11.	HSBC Bank Middle East United	Dubai	1	3	1	1	0	0	1	1	8	16
12.	Arab African International Bank	Dubai	1	1	0	0	0	0	0	0	2	0
13.	Al Khaliji (France) S. A.	Dubai	1	1	1	1	0	0	0	0	4	0

14.	Al Ahli Bank of Kuwait	Dubai	1	1	0	0	0	0	0	0	2	0
15.	Barclays Bank PLC	Dubai	1	1	0	0	0	0	0	0	2	3
16.	Habib Bank Ltd	Dubai	1	4	1	0	0	0	0	1	7	0
17.	Habib Bank A G Zurich	Dubai		5	1	0	0	0	0	0	8	1
18.	Standard Chartered Bank	Dubai		7	1	0	0	0	0	1	11	3
19.	Citi Bank NA	Dubai	1	2	1	0	0	0	0	1	5	7
20.	Bank Saderat Iran	Dubai	1	3	1	0	1	0	0	1	7	0
21.	Bank Meli Iran	Dubai	1	2	1	1	0	0	1	1	7	1
22.	Blom Bank France	Dubai	0	1	1	0	0	0	0	0	2	1
23.	Lloyds TSB Bank PLC	Dubai	0	1	0	0	0	0	0	0	1	5
24.	The Royal Bank of Scotland N.V.	Dubai	1	1	1	0	0	0	0	0	3	3
25.	United Bank Ltd	Dubai	3	3	1	0	0	0	0	1	8	2
26.	Doha Bank	Dubai	0	1	0	0	0	0	0	0	1	0
27.	Samba Financial Group	Dubai	0	1	0	0	0	0	0	0	1	0
28.	National Bank of Kuwait	Dubai	0	1	0	0	0	0	0	0	1	0
Total			27	47	15	6	2	0	3	11	111	51

Based on United Arab Emirates: Selected Issues and Statistical Appendix, IMF Country Report No. 12/136, June 2012.

The list of investment banks, wholesale banking, and companies that conduct finance and investment activities is relatively short:

- Arab Emirates Invest Bank P.J.S.C., Dubai
- HSBC Financial Services (Middle East) Limited, Dubai
- Deutsche Bank AG, Abu Dhabi
- Industrial & Commercial Bank of China, Abu Dhabi
- Mubadala GE Capital P.J.S.C., Abu Dhabi

The UAE banking sector has served as an important element of growth toward the diversification of the UAE's growth strategy. It appears to be sound, modern, and very competitive. Nevertheless, "Dubai SME," part of the Department of Economic Development, revealed that 86% of the SMEs have not sought bank finance. It reflects how difficult it is for this sector to get a bank loan. This shows the big gap between the government directives and the banks' policies. Bank financing is usually the only option, and is the predominant source of external financing for most SMEs. However, banks consider SMEs to be relatively high risk, as most of their businesses are service activities, and this leads to charging higher interest rates.

Most of the commercial banks are keen on funding the working capital needs of businesses, but less on funding start-ups. Thus there is a need for dedicated banks, working on commercial principles but devoted to financing of the SME start-ups. Currently, not only the UAE but also the entire GCC lack institutions that specialize in funding SMEs. Challenges of the banks in serving the SME sector include:

- quality of financial reporting (usually poor, especially in the case of microenterprises),
- lack of credit history,
- inadequacy of collateral (especially in case of start-ups),
- informal management (especially in micro- and small enterprises),
- short-term planning horizon,
- weak cash flow management.

Despite the structural challenges, innovation is widespread in the banking community.

OTHER KEY PLAYERS OF THE FINANCIAL MARKET

Dubai International Financial Centre[72] (DIFC) is a financial center that offers a convenient platform for financial institutions and service providers (financial intermediaries). The center was established as part of the vision to position Dubai as an international hub for financial services and as the regional

72. http://www.difc.ae/.

gateway for capital and investment. DIFC's ambition is to become the global hub for Islamic finance.

The city of Dubai has an established track record of realizing projects in safe environment and is perceived as one of the fastest growing cities in the world. It has a well-diversified economy based on international trade, banking and finance, information and communication technology, tourism, and real estate. In 2005 oil contributed less than 6% of Dubai's GDP and in 2010, it was less than 1%. This economic diversification is continuing with the establishment of new industries, private sector growth, and increased regional economic integration.

DIFC seems to fill the gap between the financial centers of Europe and South East Asia, the region comprising over 42 countries with a combined population of approximately 2.2 billion people. This region, stretching from the western tip of North Africa to the eastern part of South Asia had (until 2004) been without a world class financial center.

DIFC aims to meet the growing financial needs and requirements of the region while strengthening links between the financial markets of Europe, the Far East, and the Americas. DIFC's mission is to be a catalyst to facilitate the mobilization of capital for regional economic growth, development, and diversification by performing as a globally recognized financial center. The center attracts regional liquidity back into investment opportunities. It has attracted international firms such as Merrill Lynch, Morgan Stanley, Goldman Sachs, Mellon Global Investments, Barclays Capital, Credit Suisse, Deutsche Bank, and many other leading international financial institutions. DIFC intends to be the regional gateway for investment banks as well as other financial institutions who wish to establish underwriting, M&A advisory, venture capital, private equity, foreign exchange, trade finance, and capital markets operations to service this large and relatively untapped market.

The objectives of DIFC are the following:

- to attract regional liquidity back into investment opportunities within the region and contribute to its overall economic growth;
- to facilitate planned privatizations in the region and enable initial public offerings by privately owned companies;
- to give impetus to the program of deregulation and market liberalization throughout the region;
- to create added insurance and reinsurance capacity, 65% of annual premiums are reinsured outside the region;
- to develop a global center for Islamic finance (serving large Islamic communities stretching from Malaysia and Indonesia to the United States).

DIFC focuses on the following main financial services sectors:

- banking and brokerage,
- capital markets,

- wealth management,
- reinsurance and captives,
- Islamic finance,
- ancillary services.

Benefits of establishing an institution in DIFC include:

- 100% foreign ownership,
- 0% tax rate on income and profits,
- modern office accommodation and sophisticated infrastructure,
- double taxation treaties available to the UAE incorporated entities,
- no restriction on foreign exchange,
- world class English language court system based on the common law,
- freedom to repatriate capital and profits (no restrictions),
- high-standard laws, rules, and regulations,
- high-standard operational support.

DIFC offers a wide range of investment opportunities, such as:

- mutual funds,
- exchange traded funds,
- open- and closed-ended investment companies,
- index funds,
- hedge funds,
- consultant wrap accounts,
- Islamic compliant funds.

Moreover, DIFC supports the operational needs of financial institutions and provides an ideal environment and a highly skilled work force to asset management firms and private banks. These services include accounting and legal practices, actuaries, management consultants, recruitment firms, and market information providers, among others.

There is an independent regulator, the Dubai Financial Services Authority (DFSA) supervising DIFC and the independent status of the center is further enhanced by the DIFC Courts providing the highest international standards of legal procedure.

The DFSA's primary functions include[73]:

- policy development,
- enforcement of legislation and authorization,
- supervision of DIFC licensees.

It manages companies offering asset management, banking, securities trading, Islamic finance, and reinsurance and it regulates the Nasdaq Dubai exchange.

73. www.dfsa.ae.

Dubai Financial Market[74] was established as a public institution with its own independent corporate body[75] and started the operations in March 2000. DFM is operating as a secondary market for the trading of:

- securities issued by public joint-stock companies,
- bonds issued by the Federal Government or any of the Local Governments and public institutions in the country,
- units of investment funds;
- other financial instruments (local or foreign) that are accepted by the Market.

As decided by the Executive Council Decree in Dec. 2005, DFM is set up as a public joint stock company in the UAE.[76] It operates on an automated trading system (screen based) that offers a major advantage over traditional floor trading in terms of transparency, efficiency, liquidity, and trading of prices. The Market has collaborated with renowned international experts in trading systems. Safeguarding the market efficiency and integrity of trading requires DFM to conduct regular monitoring and controlling. The Market Control Section monitors the compliance with the trading rules and regulations of the ESCA.[77] Brokers Licensing and Inspection Section monitors brokers' conduct to ensure integrity of the brokers' activities and that the best service is given to investors.

In 2010 DFM launched a global service with the iVESTOR card, a revolutionary and innovative solution that enables retail investors to instantly receive DFM dividends directly into their investor card. Investors no longer have to wait for dividend checks or have the hassle of depositing checks into their bank accounts. The card will enable DFM to credit any future dividends directly into the cardholders' balance. In addition, the cardholder can easily withdraw cash from ATMs.

Finance Companies[78] undertake one or more of the following major financing activities:

- extending advances and/or personal loans for personal consumption purposes,
- financing trade and business, opening credit and issuing guarantees in favor of customers,
- subscribing to the capital of projects and/or issues of stocks, bonds, and/or certificates of deposit.

Contribution of the financing company to the capital of projects, issues of stocks and/or bonds, or certificates of deposit should not exceed 7% of its own

74. http://www.dfm.ae/Default.aspx.
75. Resolution from the Ministry of Economy in 2000.
76. http://www.dfm.ae/pages/default.aspx?c=801.
77. ESCA - the UAE Securities and Commodities Authority.
78. Central Bank of the United Arab Emirates, Report 2011 at www.centralbank.ae.

capital. The paid-up capital of a finance company should not be less than 35 million AED and national shareholding should not be less than 60% of total paid-up capital, without prejudice to provisions of Federal Law No. 8 of 1984 and any subsequent amendments.

The number of finance companies licensed to operate in the UAE increased from 23 in 2010 to 24 in 2011, because of a license given to AMEX (Middle East) BSC.

Finance Companies in the UAE (2011)

1.	Osool "A Finance Company" LLC, Dubai
2.	Gulf Finance Corporation, Dubai
3.	HSBC Middle East Finance Co. Ltd., Dubai
4.	Maf Orix Finance P.P.C., Dubai
5.	Finance House P.J.S.C, Abu Dhabi
6.	Dubai First P.P.C., Dubai
7.	Reem Finance P.J.S.C., Abu Dhabi
8.	Majid Al Futtaim JCB Finance L.L.C., Dubai
9.	A1 Futtaim GE Finance P.P.C., Dubai
10.	Dunia Finance L.L.C., Abu Dhabi
11.	Abu Dhabi Finance P.P.C., Abu Dhabi
12.	Amlak Finance P.J.S.C., Dubai
13.	Tamweel P.J.S.C., Dubai
14.	Al Wifaq Finance Company P.P.C., Abu Dhabi
15.	Mashreq Al Islamic Finance Co. P.P.C., Dubai
16.	Islamic Finance Co. P.P.C., Dubai
17.	Aseel Finance "Aseel" P.P.C., Abu Dhabi
18.	Mawarid Finance P.P.C., Dubai
19.	Abu Dhabi National Islamic Finance P.J.S.C., Abu Dhabi
20.	Islamic Finance House, Abu Dhabi
21.	Emirates Money Consumer Finance, Dubai
22.	Abu Dhabi Commercial Islamic Finance Company P.P.C., Abu Dhabi
23.	Siraj Finance Company P.P.C., Abu Dhabi
24.	Amex (Middle East) B.S.C., Dubai

Based on www.centralbank.ae.

One new investment company was issued a license in 2011, namely, Masdar Investment,[79] thereby increasing the number of licensed investment companies from 21 in 2010 to 22 in 2011.

Financial Investment Companies (2011)

1.	Oman & Emirates Investment Holding Co., Abu Dhabi
2.	Merrill Lynch International & Co. C.V., Dubai
3.	Emirates Financial Services, Dubai
4.	Shuaa Capital P.S.C., Dubai
5.	The National Investor, Abu Dhabi

79. http://www.masdar.ae/en/home/index.aspx.

6.	Islamic Investment Co. P J.S.C., Dubai
7.	Abu Dhabi Investment House P.J.S.C, Abu Dhabi
8.	Al Mal Capital P.S.C.
9.	Injaz Mena Investment Company P.S.C., Abu Dhabi
10.	National Bonds Corporation P.S.C., Dubai
11.	Noor Capital P.S.C., Abu Dhabi
12.	Unifund Capital Financial Investment P.S.C., Abu Dhabi
13.	Daman Investment P.S.C., Dubai
14.	Allied Investment Partners P.J.S.C., Abu Dhabi
15.	Gulf Capital P.S.C., Abu Dhabi
16	CAP M Investment P.S.C., Abu Dhabi
17.	Royal Capital P.P.C., Abu Dhabi
18.	Al Bashayer Investment Company L.L.C., Abu Dhabi
19.	Dubai Commodity Asset Management L.L.C., Dubai
20.	ADIC Investment Management P.P.C., Abu Dhabi
21.	ADS Securities L.L.C., Abu Dhabi
22.	Masdar Investment L.L.C., Abu Dhabi

Based on www.centralbank.ae.

BUSINESS COOPERATION BETWEEN THE UAE AND POLAND

Legal and treaty foundations of the cooperation between the UAE and Poland arise from the following bilateral agreements[80]:

1. Double Taxation Avoidance Agreement between Poland and the UAE of 1993.
2. Agreement Between the Government of the United Arab Emirates and the Government of the Republic of Poland for the Promotion and Protection of Investments, 1993r.[81]
3. Civil Aviation Agreement of 1994.

The above-mentioned agreements, along with the liberalization of the regulations concerning foreign investments in Poland, create favorable conditions for capital cooperation and capital influx from the UAE to Poland. In 1994, the Polish party submitted a draft trade agreement to the UAE; however, its negotiation was unnecessary because both Poland and the UAE are

80. The market guide for entrepreneurs - The United Arab Emirates, Polish Agency for Enterprise Development, Warsaw 2010.

81. (1) Desiring to create favorable conditions for greater economic cooperation between and particularly for investments by investors of one Contracting State in the territory of the other Contracting State. (2) Recognizing that the encouragement and reciprocal protection under international agreements of such investments will be conductive to the stimulation of business initiative and will increase prosperity in both Contracting States.

signatories of a multilateral GATT 94/WTO Agreement, which regulates the rules of international trade in a comprehensive and obligatory manner.[82]

Prospects for Polish Enterprises in the Persian Gulf States

The Persian Gulf states feature a high level of money reserves. The last financial crisis forced some Polish enterprises to undertake a more intensive expansion abroad, which for some of them may contribute to establishing cooperation with the countries in that part of the world. Several months have already passed since the Arab Spring occurred. In some of the states, where a government change had taken place, certain shifts in politics started, yet pundits' opinions are divided on whether the revolutions in a few countries of North Africa can actually affect the economic relations with those countries.

Polish Exports in 2010 (mln PLN)

Saudi Arabia	**580.3**
Bahrain	43.8
Iraq	151.7
Iran	370.1
Oman	58.1
UAE	742.8
Qatar	48.8
Kuwait	84.1

Based on Central Statistical Office, http://www.stat.gov.pl/gus/index_ENG_HTML.htm.

Currently, it is chiefly construction companies that seek to win foreign contracts. A significant drop in the number of public tenders in Poland gave the impetus for undertaking actions in that direction. In the course of 2009 and 2010, the value of tender offers realized for GDDKiA[83] exceeded 40 billion PLN; however, in 2011 orders stood at 4 billion PLN. Yet, a large part of Polish construction companies market has foreign shareholders, who vie for contracts in the Near East on their own behalf. They do not welcome any competitors from Poland; hence, they may oppose the expansion of affiliated enterprises into the Arab region.

The Persian Gulf Markets

Saudi Arabia, Qatar, and Kuwait register significant budget surpluses (amounting to 20.5% GDP, second highest in the world), so do Iran, Oman, and the UAE. Only Bahrain and Iraq follow the example of the Western world and spend more than their budget income.

---

82. http://www.obserwatorfinansowy.pl/2012/02/20/przedsiebiorcom-z-polski-marza-sie-kontrakty-z-1001-nocy-jest-potencjal/.
83. http://www.gddkia.gov.pl/en/1618/News.

Budget Income and Spending in the GCC (bln USD)

Country	Budget Income	Budget Spending
Saudi Arabia	293.1	210.6
Bahrain	7.93	8.297
Qatar	68.79	37.88
Kuwait	92.23	61.16
Iraq	69.2	82.6
Iran	130.6	92.22
Oman	23.75	23.21
UAE	120.8	102.9

Based on CIA The World Factbook at https://www.cia.gov/library/publications/the-world-factbook.

Buying something from the Arab states is much easier (chiefly oil and oil products) than selling anything in those markets. All the Persian Gulf states show a positive, and distinctly so, balance of trade. For instance, Saudi Arabia ranks as the 15th global exporter, but only the 33rd importer. The largest global economies, such as the United States, Japan, China, South Korea, Germany, Italy, Great Britain, and India, are the trade partners of the Persian Gulf states. Thailand is another major importer of goods from Oman and Singapore from Saudi Arabia, whereas Turkey plays an important role in trade with Iran and Iraq. Iraq is the fifth most important export market for Turkey (5.3%), whereas Iran is the seventh in respect of the value of import volume.

Imports and Exports of the GCC Members (bln. USD)

Country	Import	Export
Saudi Arabia	106.5	350.7
Bahrain	16.8	20.23
Qatar	25.33	104.3
Kuwait	22.41	94.47
Iraq	53.93	78.38
Iran	76.1	131.8
Oman	21.47	45.53
UAE	185.6	265.3

Based on CIA The World Factbook at https://www.cia.gov/library/publications/the-world-factbook.

Poland's trade with the Persian Gulf states has been small so far. It amounts to hundreds of millions of PLN in the case of Saudi Arabia, whereas with any of Poland's top 10 trade partners it is counted in tens of billions of PLN (with Germany even in hundreds of billions of PLN). Polish export to, e.g., Saudi Arabia is chiefly furniture, and also medical equipment and lamps. Enterprises of certain industries managed to establish their position in those markets. Annual trade with the United Arab Emirates amounts to nearly 320 million USD.[84] The UAE is one of Poland's biggest trade partners in the area of North Africa and the Near East. Higher total turnover in trade between Poland and the

84. Central Statistical Office, http://www.stat.gov.pl/gus/index_ENG_HTML.htm.

UAE is only recorded in the exchange of trade with Israel and Saudi Arabia, but the UAE ranks first in respect of the volume of Polish exports in this area.

The number of foreign investments in the UAE is rising month by month, mostly in the largest metropolises, Dubai and Abu Dhabi. The amount of Polish capital is also on the increase. It is allocated on highly favorable conditions, inter alia into real estate, purchasing of which guarantees a relatively quick and certain profit.

Polish Imports From the GCC Members, 2010 (bln PLN)	
Saudi Arabia	636.9
Bahrain	150.3
Qatar	8.1
Kuwait	5.3
Iraq	0.0175
Iran	205.9
Oman	24.7
UAE	214.8

Based on Central Statistical Office, http://www.stat.gov.pl/gus/index_ENG_HTML.htm.

Polish entrepreneurs do not limit their expansion exclusively to the largest countries of the Persian Gulf. "Inglot,"[85] a producer of cosmetics for women, has already opened 21 shops in the Near East.[86] It sells its cosmetics in the UAE (six outlets in Dubai alone, in addition to shops in Abu Dhabi and Sharjah), Saudi Arabia, Qatar, and Bahrain, with two outlets in each of these countries, and in Kuwait and Oman with one outlet in each. The company is also present outside of the Gulf states in Lebanon.

Mr. Wojciech Inglot, the founder and president of the cosmetics company, is very satisfied with the company's operations in the Near East. "The region is inhabited not only by Arabs, but also by people coming from all over the world, mostly from the south-east Asia. In some countries, e.g. in the UAE, there are more foreigners than there are native residents. We are planning further expansion in the region: in Saudi Arabia and Kuwait," says Wojciech Inglot.[87]

Entry Barriers for Polish Entrepreneurs on the Example of Polish Companies of Inglot and Can-Pack/Arab Can

Poles find it hard to establish their footing in the Arab markets, and practitioners offer a number of theories explaining why it is so. Stanisław Waśko, the Vice President of Can-Pack,[88] a manufacturer of packaging, claims that the largest problem encountered in the Near East countries is that it is impossible

85. http://inglotcosmetics.com/.
86. http://biznes.newsweek.pl/wojciech-inglot–wladca-kolorow,79347,1,1.html.
87. http://www.ekonomia24.pl/artykul/864257.html#.
88. http://www.canpack.eu/index.php?lang=en.

to hold a majority package in the share capital of local companies and the need to have a local partner. It is a significant discomfort, particularly in case of production companies, which by their very nature need to invest a lot. Inglot solved that problem with the help of a franchise partner.

In 1999 Can-Pack founded Arab Can in the UAE, in 2004 it started up a second plant in Dubai, in 2009 it established a joint venture company of Can-Pack Linco[89] in Cairo, and in 2010 it opened a beverage can production facility in Casablanca in Morocco.

In turn, it is more difficult for construction companies to succeed in the markets where cheap local companies along with well-recognized international brands already operate. Polish construction companies are unable to compete against the local ones, and they have not yet achieved the position in the market that would enable them to compete against international concerns. Finding qualified workers in Arab states is also a problem. In practice, workers are recruited from among immigrants coming from India, Pakistan, the Philippines, etc., and assembling an efficient production facility team requires a lot of effort.

Can-Pack is one of the few Polish companies that have found out what it is like to operate—and not only sell—in the Near East. The fear of the situation in Iraq is too great for most companies. Companies in Poland demonstrate great interest at the stage of talks. However, when it comes to concrete actions, only a small percentage of companies decide to go there. Fear and a misconception of the situation in Iraq constitute the largest problem. There are many entities in Poland that are interested in cooperation, but news of explosions and other types of dangerous events refrain them from action.

If an entrepreneur is determined to take subsequent steps aimed at starting cooperation with partners in the Persian Gulf region, then the next—highly significant stage—is going there. Arab states do not conduct business at a distance. Some companies tried to start operations remotely, but without a physical presence locally, not much could be achieved. Some experts claim that the time of Polish enterprises in the Near East is about to come. Companies without references that want to implement contracts in the Near East must start as subsuppliers or subcontractors. That is the strategy adopted by Spanish companies, which in the last 30 years have achieved an unprecedented global expansion. Nevertheless, to enter new foreign markets, Polish companies need to take a greater advantage of the financing of national institutions that support and promote such types of undertakings (banks or insurance corporations). For instance, a Swedish equivalent of KUKE S.A.[90] (Export Credit Insurance Corporation Joint Stock Company, which establishes the conditions for the safe and stable functioning of Polish enterprises by securing export and domestic transactions and by facilitating access to external

89. http://www.canpack.eu/index.php?lang=en&action=1.
90. http://www.kuke.com.pl/home.php.

financing) signed contracts to the tune of 25 billion EUR over a year period, whereas KUKE, only to the amount of several hundred million PLN.[91]

Factors Determining the Expansion of Polish Enterprises in the UAE

The most important factors contributing to the cooperation and expansion of Polish enterprises in the territory of the UAE include:

1. political stability of the UAE—lack of significant influence of fundamentalist movements on the internal policy (a consistently implemented vision of a modern state, in which old Arab traditions and the fundamental principles of Islam combine with modernity and the progress of civilization),
2. good economic situation (high GDP growth, significant budget surpluses, surpluses in foreign trade balance),
3. no restrictions on profit transfer or capital repatriation,
4. low import duties (less than 5% for virtually all goods), nonexistent in the case of items imported for use in the free zones,
5. competitive labor costs; corporate and personal taxes are nil, numerous double taxation agreements and bilateral investment treaties are in place,
6. excellent infrastructure as well as a stable and safe working environment,
7. development of new branches of industry in the UAE (chemical, metallurgical, machine industries, trade, and construction),
8. rich deposits of natural resources (approximately 10% of global oil deposits and the fifth largest gas deposits),
9. growth recorded in certain areas outside of the oil sector (apart from industry, in construction, trade, services, and banking),
10. substantial financial reserves (estimated at over 250 billion USD), invested in foreign markets (Western Europe, the United States, the Near East),
11. dynamic development of financial services sector (e.g., the establishment of DFIC), constituting a bridge that links the financial markets of the East and the West),
12. increase in trade between Poland and the UAE,
13. most advanced telecommunication technologies,
14. access to numerous trade fairs and international conferences,
15. high-quality office and residential space,
16. security of energy supplies,

91. http://www.obserwatorfinansowy.pl/2012/02/20/przedsiebiorco-z-polski-marza-sie-kontrakty-z-1001-nocy-jest-potencjal.

17. presence of Polish citizens in the UAE (chiefly engineering personnel, architects, medical personnel),
18. successful development of cultural and scientific contacts.

The factors limiting the cooperation with the UAE include high competition of local and international companies as well as moderate interest demonstrated by the local enterprises in conducting business with Poland.

Market Chances and Threats

As already mentioned, the UAE is without a doubt a highly attractive trade partner, especially for exporting of goods. There are no financial or administrative barriers restricting access to that market. Moreover, Poland and the UAE share a rich tradition of trade and the UAE has an extensive, modern infrastructure at its disposal. A varied group of consumers in the Emirates alone creates favorable conditions for export expansion, but above all, it provides wide-ranging possibilities of reexport of goods from other states of the Persian Gulf region.

Apart from the oil and gas sector, crucial for the UAE, the following branches of industry can be seen as promising from the point of Polish investment export: chemical industry, machine-making sector, power engineering, industrial and residential construction, as well as food processing. Numerous duty free zones operating in the UAE are a chance for Polish (production and trade) enterprises (the UAE ranks third, after Hong Kong and Singapore, with respect to the size of a reexport center in the world). The fact that in general the country uses a zero rate corporate tax and personal income tax seems to be a significant investment incentive.

Arabic Investments in Poland

Apart from conducting business with the Near East, it is worthwhile drawing capital from there. The Persian Gulf states are classified as the strongest players in respect of their capital power. Saudi Arabia invested 18 billion USD abroad in 2010, Kuwait 39 billion USD, and Qatar 23.5 billion USD. By way of comparison, on average Poland receives 10 billion EUR of FDIs in a year, and its accumulated value stands at 150.4 billion EUR.

A bond issued by the ministry of finance or a large city addressed to Islamic investors could be a good step. Several years ago a German land of *Sachsen-Anhalt* issued bonds worth 100 billion USD. Thanks to presentations of the region during a road show, contracts for sale of agrofood products were concluded that were worth three times more than the bond issue itself. To enable the operation of Islamic banks in their markets, some countries of Western Europe introduced a provision in their regulations permitting the application of the Quran law (the Sharia). Above all, it concerns a ban on calculating interest on loans and credits. The economic principle of Muslim banking instruments is largely based on market patterns of traditional banking. Interest is replaced with other, equivalent financial constructions.

THE SWOT ANALYSIS OF THE UAE MARKET—BUSINESS PERSPECTIVE

Taking into consideration all of the above recent and future UAE market developments, the country characteristics including the market data, as well as the policies realized within the last decade, appropriate conclusions can be drawn by making the SWOT analysis from the potential foreign investor's perspective. The market strengths, weaknesses, opportunities, and threats are as follows[92]:

Strengths

- Political neutrality and stability of the UAE—lack of significant influence of fundamentalist movements on the internal policy (a consistently implemented vision of a modern state, in which old Arab traditions and the fundamental principles of Islam combine with modernity and the progress of civilization),
- Booming economy—good economic situation (high GDP growth, significant budget surpluses, surpluses in foreign trade balance),
- No restrictions on profit transfer or capital repatriation,
- Low import duties (less than 5% for virtually all goods), nonexistent in the case of items imported for use in the free zones,
- Competitive labor costs; corporate and personal taxes are nil, numerous double taxation agreements and bilateral investment treaties are in place,
- Excellent infrastructure as well as stable and safe working environment,
- Development of new branches of industry in the UAE (chemical, metallurgical, machine industries, trade, and construction),
- Rich deposits of natural resources (approximately 10% of global oil deposits and the fifth largest gas deposits),
- Growth recorded in certain areas outside of oil sector (apart from industry, in construction, trade, services, and banking),
- Substantial financial reserves (estimated at over 250 billion USD), invested in foreign markets (Western Europe, the United States, the Near East),
- Dynamic development of the financial services sector (e.g., the establishment of DFIC), constituting a bridge that links the financial markets of the East and the West),
- Increase in trade between Poland and the UAE,
- Most advanced telecommunication technologies,

92. Hong Ju Lee and Dipak Jain. Dubai's brand assessment success and failure in brand management – Part 1, Place Branding and Public Diplomacy, 234–246 (August 2009).

- Access to numerous trade fairs and international conferences,
- High-quality office and residential space,
- Security of energy supplies,
- Numerous and well-balanced policies leading to sustainable growth of the country.

Weaknesses

- Negative image of the Middle East,
- Barren desert, lack of natural resources (other than oil and gas),
- Only 20% of UAE nationals,
- The lack of fundamental infrastructure (e.g., transportation, water)
- Luxuries might appeal too small a segment.

Opportunities

- Increasing oil price,
- Increasing job opportunities for immigrants and natives,
- Growing luxury market,
- Increase in foreign investment,
- Proactive attitude,
- Well-developed Meetings, Incentives, Conventions & Exhibitions (MICE) environment.

Threats

- Strong competitors: within the region: Abu Dhabi, Qatar; outside the region: Singapore, Hong Kong,
- Oil running out in 30 years,
- Terrorism and war could further the negative image of the Middle East,
- Limited media coverage.

CONCLUSIONS AND SUMMARY

The UAE financial services sector and SMEs have served as important elements of growth toward the diversification of the UAE's strategy. There is a sound, modern, and competitive banking industry and innovation is widespread in the banking community. Abu Dhabi and Dubai financial sectors constitute the majority of the system.

The country's sustainable growth is supported by issuing effective laws and legislations (encouraging the business environment and developing national industries), strengthening investments, and promoting the SME sector (numerous incentives for foreign investors, e.g., FTZs), as well as protecting consumer and intellectual property rights and diversifying trade activities.

All these government initiatives and incentives adhere to international standards of excellence and the tenets of knowledge economy. They aim to attract strategic foreign investments and add value to the existing business community, providing a stable and attractive business environment. The UAE market strengths (according to the presented SWOT analysis) are numerous. The country is committed to maintaining the policy of economic openness, actively seeking to develop projects that are in harmony with the changes taking place in the world.

Despite the numerous policies that are in place, there are some deficiencies regarding SMEs' debt financing. "Dubai SME," part of the Department of Economic Development, revealed that 86% of the SMEs have not sought bank finance. It reflects how difficult it is for this sector to get a bank loan. This shows the gap between the government directives and the banks' policies. Bank financing is the predominant source of external financing for most SMEs. However, banks consider SMEs to be relatively high risk as most of their businesses are service activities. Most of the commercial banks are keen on funding the working capital needs of businesses, but less on funding start-ups. There is a need for dedicated banks, working on commercial principles but devoted to financing the SMEs, especially start-ups. Currently, not only the UAE but the entire GCC lacks institutions that specialize in funding SMEs.

Chapter 20

ECO-GEN Energy Solutions: Tahiti Story

Cheryl Stephens
ECO-GEN Energy Inc., Van Nuys, CA, United States

Chapter Outline

BACKGROUND

Tahiti, French Polynesia, is an island in the South Pacific that many associate with lovely vacations, volcanoes, black sandy beaches, French artist Paul Gauguin, and tourism. With a population of 268,645 inhabitants (2012 census), it is divided into two major parts, consisting of 12 communes, with the most populous one being the capital of Papeete. Tahiti imports its petroleum and has no local refinery or production (French Polynesia Population, 2014). The Pacific Islands are home to some of the countries that are most at risk from the effects of climate change. French Polynesia, whose per capita gross domestic product is the second highest among Pacific Islands, is also above average in its per capita electricity consumption. Electricity per inhabitant is 2468 (kWh). Tahiti has a goal of reducing its fossil fuel consumption and shifting to renewable sources of electricity generation by 50% by 2020, yet it only had 26% in renewable energy in 2014 (Hourçourigaray et al., 2014). Tourism is a $10 billion industry, and close to 200,000 visitors annually visit Tahiti, which also creates a need for a solution for energy consumption and efficiency at peak usage without surges (ADEME, 2014).

Distributed generation is of vital importance in these areas because they do not have the infrastructure for large-scale transmission and distribution. Often

Sustainable Cities and Communities Design Handbook. https://doi.org/10.1016/B978-0-12-813964-6.00020-3

dirty and costly diesel generators remain the best solution for primary and backup electrical generation. We know that the affordability of fossil fuels comes at a heavy cost to the environment. The extraction of natural gas leads to fracking. Transportation releases methane that significantly traps more heat than CO_2. Current renewables are not good enough. Typical solar produces power 4–6 h a day at best, and typical wind, 8–10 h per day. Photovoltaic and concentrated solar or farms require large areas, are usually away from urban areas where power is needed, threaten wildlife, and lack inertial stability to support the grid. In Tahiti, mountainous islands, often forested and wet, develop daytime cloud cover linked to forest evaporation, and the southern flanks are badly oriented to the sun's path (Hourçourigaray et al., 2014).

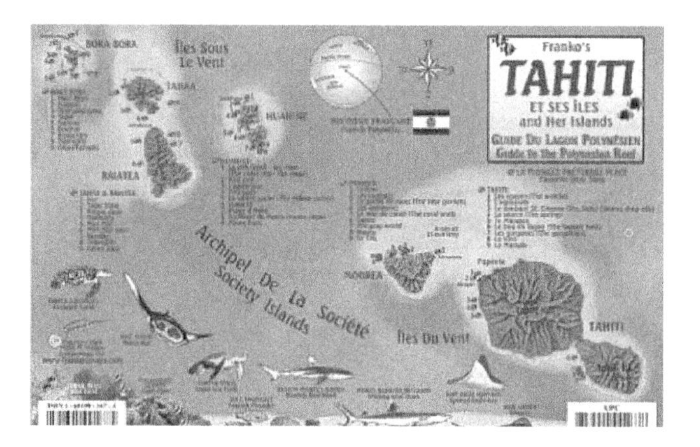

THE SOLUTION

ECO-GEN Energy developed the exclusive, advanced technology to produce clean, green, energy with the patented and Underwriters Laboratories-tested JouleBox Hybrid Generator that is more efficient than anything else on the market today. The JouleBox has a UCC Manufacturer's Component Warranty and Performance Guarantee from Lloyds of London. Using different and proven technologies, in tandem, the developers were able to secure a new intellectual property (IP). Our development partners, Green Power Partners, with a power purchase agreement (PPA), are creating local jobs as well.

The JouleBox makes electricity from solar, turbines, hydraulics, lithium ion battery bricks, high-efficiency switched reluctance motors and generators, and N50 neodymium permanent magnets. Each JouleBox has its own IP satellite address and can be remotely and continuously monitored for 45 different components. If a unit goes out, it could be replaced within an hour with a simple plug-n-play swapping of the unit.

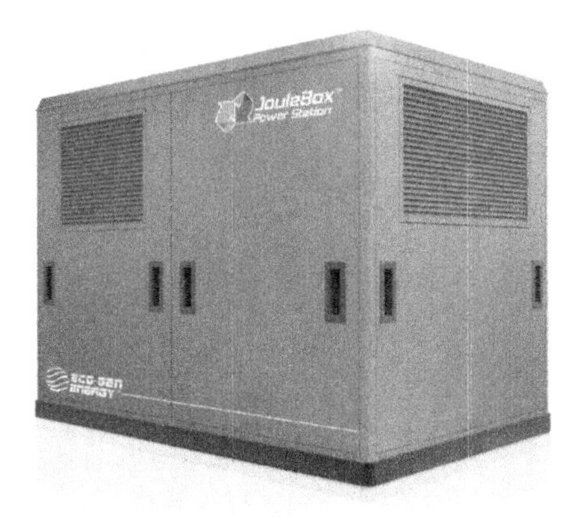

The Technology

Since lithium ion batteries cannot be shipped fully charged, ECO-GEN installs four to eight solar panels per JouleBox, enclosed in a metal box that is approximately 7 X 5 X 5 feet. The solar panels and batteries continue to produce additional electricity that can be used to keep the building cool or to sell back to the grid. French Polynesia has renewable incentives, similar to those with the State of California Go Solar program other solar electric generating technologies, for which ECO-GEN is approved (gosolarcalifornia).

Serial and parallel wiring of lithium ion battery bricks to charge and discharge simultaneously means the batteries will not use their 5000 charge/discharge life cycle and should last for 20.9 years. Solar panels and the DC generators have the ability to keep the battery brick fully charged at all times to constantly run the turbines.

Wind/hydraulic turbines powered by proprietary, permanent magnet motors are used to turn the generators. Turbines are an accepted and proven source of power to generate electricity. A Vortex Reduction Chamber increases the torque. The key is to keep the turbines at 1800 RPM (1500 for 50 Hz). There is minimal energy required to maintain the speed. High-efficiency, gated flux, permanent magnet switched reluctance motors typically produce energy for the next 30–40 years, with baseload power 24/7/365. The unit does not need to be turned off except for annual maintenance once a year to change the oil in the gear box!

N50 neodymium permanent magnets, capable of picking up 1000 lb each, are used between coils to create dual flux and redirect the back electromotive force (EMF), producing more horsepower using less watts. The Star Motor uses patented software processes to redirect the back EMF so that the power is reused and not lost. The unit complies with the laws of thermodynamics wherein energy cannot be created or destroyed.

Inside the JouleBox: How the JouleBox Power Station Works

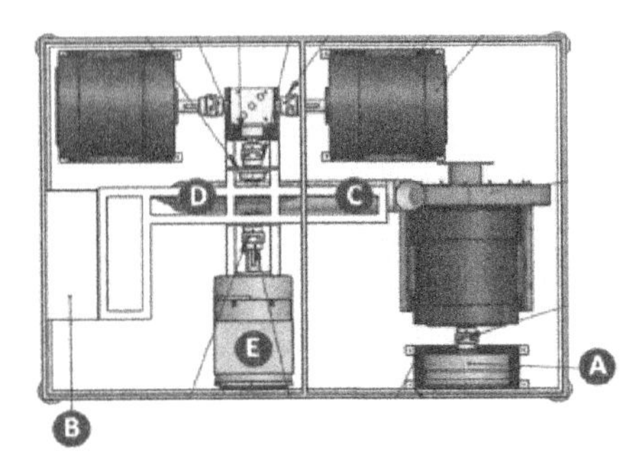

(3) Solar Panels (not shown) and DC Generators A constantly charge batteries

Lithium Ion Batteries B with patented serial and parallel wiring charge and discharge. (When not in use, the batteries will last 20 years.)

High pressure manufactured wind is sent through a Vortex Reduction Chamber C to increase force with reduced energy to turn the Power Turbine. D

The Power Turbine D turns the AC Generator E

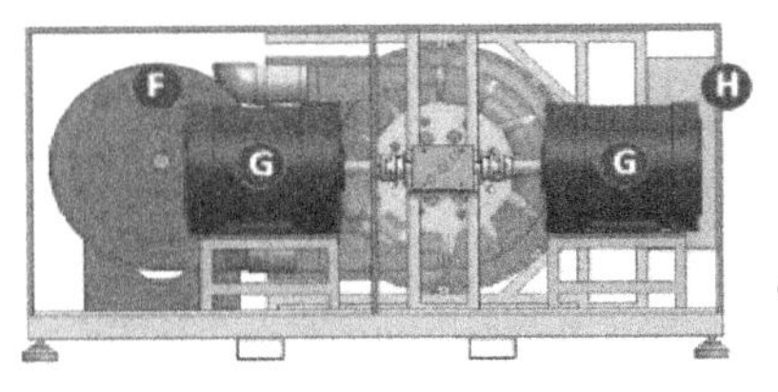

- High efficiency Speed Control DC Motors F and DC Blower Motors G and controllers result in greater durability and reliability. The motors use less than 1/3rd the energy of conventional motors. Recycle back EMF
- Power Q controls H the power quality output to produce clean electricity, stable Hz, synced sine waves, and harmonics, and no power surges.

60-kW JouleBox Hybrid Solar Generator Produces 24/7/365
Annual kWh produced: 525,600
Annual kWh′ is estimated as unit size (60 kW) × 24 h per day × 365 days in a year

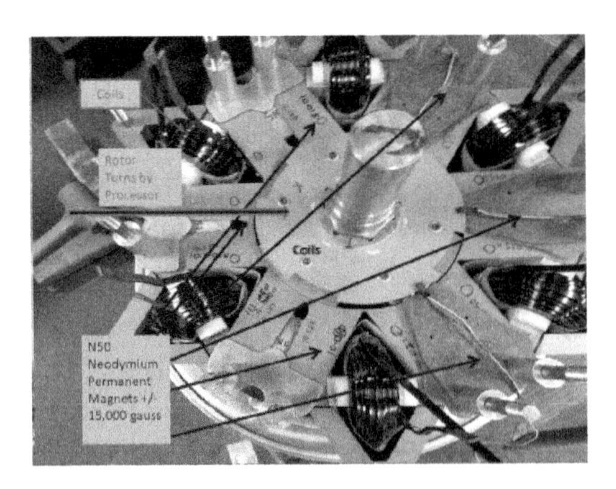

NOT THE ONLY WAY!

HP= Torque X RPM / 5252

- Typical motors use 30kw to create 40 HP.
- We use less than 8kw for each 40HP.
- Redirect and reuse the Back EMF for 98% efficiency!

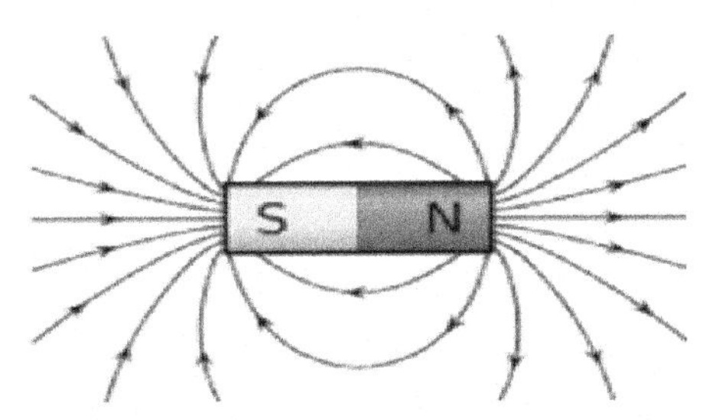

- We increase HP by positioning N50 Neodymium permanent magnets between coils.
- Each one can pick up over 1,000 lbs.
- Combined with the magnet flux to create dual flux, when charge to the coil is turned on.
- Dual Flux creates twice the Torque without using more Watts.

As per Lenz's Law, many people have tried to destroy the back EMF (Electrical). It is not possible because energy cannot be created or destroyed; however, with strong enough magnets the back EMF can be redirected and reused (Khanacademy).

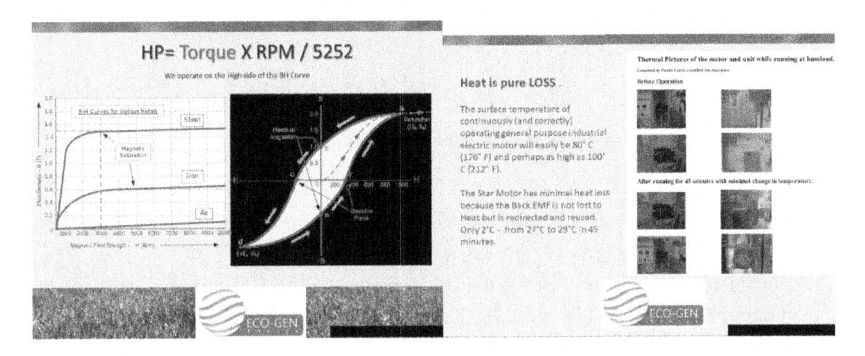

See reference section for independent heat loss testing study (Laurie, 2015).

The JouleBox Power Station Plants Solution

The JouleBox Power Station Plants Solve All These Problems

The 60-kW units are configured in parallel clusters, with double-stacked containers in a climate-controlled, sound-proofed facility. The buildings have a small environmental footprint compared with other renewable energy options: 3 MW (1007.81 m^2) to 50 MW (13,601 m^2).

Since these plants are modular, they can be scaled larger; installing them next to a substation, under transmission lines, or any electrical right of way, makes distribution efficient and cost effective and produces power where it is needed. There is no theoretical limit to the number of units that could be clustered together. A 3-MW system would produce 26,280,000 kWh per year or enough electricity to run 2200 US homes. Six JouleBox Power stations can power a large factory. Seven JouleBoxes fit into a shipping container and can produce 420 kW of power. The JouleBox equates to approximately 4 MW of solar, 3 MW of wind, and 2 MW of hydro. By producing power from a cluster of units that are not dependent on each other to operate, the reliability of production output is greatly enhanced and the grid is safer.

Intertie and grid protection is vitally important. Our interconnection system was designed by a senior member of the IEEE who sits on the (1) IEEE Power Systems Relaying Committee (PSRC), Main Committee; (2) IEEE PSRC Rotating Machinery Protection Subcommittee as Chair Emeritus; and the (3) IEEE PSRC Rotating Machinery Protection Subcommittee as Committee Member. With each installation a JouleBox facility makes the grid safer and more secure. These facilities can operate as a microgrid or in islanding mode. All units are synchronized and electronically locked down and combined to be delivered at medium voltage (11 KV to 69 KV). Even if one JouleBox is shut down, the other JouleBoxes continue to run.

Since the JouleBox provides baseload power 24/7/365 the question of peak demand has been raised. Waste to energy gasification can provide power for peak demand while eliminating landfill on islands with limited space.

Substation

Transmission Right of Way

THE TAHITI SOLUTION

ECO-GEN Energy is creating a solution for Tahiti and the world as demand grows. Modern life would flounder without lighting, air-conditioning, computers, and the untold number of devices and services that rely on electricity. Pollution, climate change, toxic waste, and disasters require increased use of renewable energy as a solution. Where energy is available, the economy naturally will grow.

By forming 20-year Power Production Special Purpose Entities, which receive income from the sale of electricity (kWh) via PPA with the local utility internationally, ECO-GEN helps create jobs domestically and in foreign countries, stabilizing power for the masses. We are also committed to returning 10% of the profits to local charities. This solution can be easily repeated for many island and developing nations. No one technology can solve all the problems, but by working together, all the existing diesel-powered electricity generation can be replaced with new, clean, green, renewable energy.

REFERENCES

ADEME, 2014. Agency for the Development and Management of Energy.
https://www.electrical4u.com/lenz-law-of-electromagnetic-induction/.
Major Cities in French Polynesia Population, 2014. worldpopulationreview.com.

http://www.gosolarcalifornia.ca.gov/equipment/other.php.
Hourçourigaray, J., Wary, D., Bitot, S., Agence Française de Développement, 2014. Renewable Energy in the Pacific islands.
https://www.khanacademy.org/science/physics/...law/.../lenzs-law.
Laurie, P., 2015. Bremara Power Inc. Study.

Chapter 21

Japanese Smart Communities as Industrial Policy

Andrew DeWit
Rikkyo University, Tokyo, Japan

Chapter Outline

JAPANESE SMART COMMUNITIES: PLENTY, BUT POORLY PROMOTED

The scale of Japan's smart community policies and projects remains little known or understood, even among experts in smart cities. Only recently has this lack of awareness begun to change: on April 27, 2017, the smart city market analysts at ABI Research declared that Japan, China, and South Korea were leaders in Asian smart city initiatives. ABI Research Industry Analyst Raquel Artes highlighted the countries' robust wireless communication networks as the principal factor in their ability to quickly and efficiently diffuse smart cities. She also argued that the region's projects are likely to accelerate over the coming 5 years and pointed to the role of "catastrophe management" in Japan's projects (ABI Research, 2017).

The main reason for the dearth of information on Japan is that the Japanese themselves are notoriously poor at communicating. This problem was evident,

Sustainable Cities and Communities Design Handbook. https://doi.org/10.1016/B978-0-12-813964-6.00021-5

to use just one illustrative example, at the March 28, 2017, International Energy Agency's (IEA) annual workshop of the IEA's Renewable Energy Working Party.[1] The workshop was titled "Scaling-up renewables through decentralised energy solutions," and included a panel on "Drivers for change – the role of cities, industry and smart solutions." Stockholm and Seoul sent experts to describe their combination of good governance and smart technology. By contrast, the Japanese dispatched an engineer from Hitachi, also employed as a professor at the Tokyo Institute of Technology. He unwisely opted to showcase his firm's kit in his presentation on "Our experiences and activities of smart communities and renewables." Not surprisingly, the IEA praised Stockholm and Seoul in its workshop summary article (IEA, 2017), with no mention of what Japan is doing.

Moreover, the flagship "Japan Smart Community Alliance" (JSCA), a consortium of 255 firms established on April 6, 2010, is remarkably poor at communicating, even in Japanese. The JSCA declares (in Japanese and English) that its role is to promote smart communities in Japan and globally. It defines the smart community as "a community where various next-generation technologies and advanced social systems are effectively integrated and utilized, including the efficient use of energy, utilization of heat and unused energy sources, improvement of local transportation systems and transformation of the everyday lives of citizens" (JSCA, n.d.). Yet this definition has not changed to reflect the powerful role of disaster resilience following March 11, 2011 (3-11). Neither has the JSCA Japanese language listing of domestic microgrid projects been updated to include, for example, the Higashi Matsushima City Smart Disaster Prevention Eco Town that we examine later in this chapter.

As of this writing (May 2017), the JSCA's most recent English language material is a June 24, 2015, document "Smart Community: Japan's Experience."[2] This material is not only woefully out of date, but also fails to include Higashi Matsushima in a curiously incomplete list of post-3-11 projects that stress disaster resilience. The entire document is also replete with inexcusably poor English translations, such as the following paragraph on page 2: "Smart Community being addressed in Japan has the concept involving smart grid. Whereas smart grid refers to the state being smarter by information and communication technologies (ICT) for electric power system, Smart Community is the effort of changing social system of a defined area into smarter state with technologies not only for electric power system but also for a variety

1. The IEA workshop outline and contents are available at the following URL: https://www.iea.org/workshops/scaling-up-renewables-through-decentralised–energy-solutions-.html.
2. The JSCA English language pamphlet "Smart Community: Japan's Experience" can be accessed at the following URL: https://www.smart-japan.org/english/vcms_lf/Resources/JSCApamphlet_eng_web.pdf.

of public infrastructure including heat supply, water and sewerage, transportation and communications."

The above examples reflect shoddy governance in coordinating Japan's domestic and international communication of what its smart communities are and how 3-11 has changed them. The JSCA and its member firms, along with the New Energy and Industrial Technology Development Organization and Ministry of Economy, Trade and Industry (METI), which serve as secretariat and support organizations for the JSCA, have clearly allowed the crucial role of PR to fall through the cracks. One great irony here is that Japan excels at good governance of the smart community itself, both at the project level and when it comes to integrating policy supports.

OPENING THE WINDOW ON JAPAN'S SMART COMMUNITIES

In fact, smart communities are written in Japan's 2016 "Energy Innovation Strategy" (hereafter, EIS) and its "Society 5.0" industrial policy (Kashiwagi, 2016; METI, 2016). They are also deeply embedded in its disaster resilience and spatial planning (EcoNet Tokyo 62, 2016, p. 13). Japan's smart community industrial policy aims to reshape the energy economy rather than merely swap solar and other green energy for coal, nuclear, and other brown energies. The policy includes a shift away from reliance on the conventional power grid and other sprawling and vulnerable critical infrastructure.

Some of this ambition is evident in Fig. 21.1, adapted from a METI publication. The figure offers a stylized portrayal of a typical Japanese city, fronted by the sea and with moderately high mountains at the back. The residential, commercial, and industrial districts are marked by rings, outlining their respective deployments of distributed energy and related network infrastructures. The figure not only shows that the locus of Japan's distributed energy transition is the smart community but also clearly states that this transition is the primary purpose of the smart community.

Hence, in Japanese usage, the smart community can be a residential district, public facilities, a factory cluster, a commercial sector, or any other zone where there is sufficient density of demand for smart energy and disaster resilience. The Japanese smart community is where disaster resilience, energy efficiency, and the uptake of renewable energy endowments are maximized by smart energy systems, including virtual power plants; power microgrids; district heating and cooling (DHC) networks; home energy management system (HEMS) as well as building/factory/community/mansion energy management systems; advanced energy storage; light-emitting diode (LED) lighting; waste heat recovery; and other infrastructure. All these components represent the ongoing convergence of energy, new materials, and ICT/robotics.

In addition, Japanese policy makers are quite explicit that their smart energy approach aims to maximize the diffusion of renewable energy and

The Purpose of the Smart Community:

- Large-scale deployment of distributed energy systems centred on renewable energy
- Use of IT, storage and cogeneration to balance demand and supply in the community. Networking residences, offices and other buildings, use renewable energy, storage and other assets to realize a disaster-resilient energy system with a high level of autonomy.

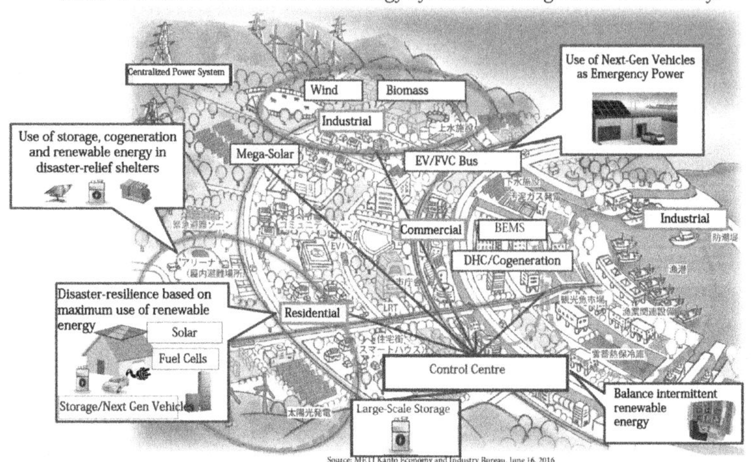

FIGURE 21.1 Japanese smart communities.

efficiency, while encouraging densification. Their aim is to increase disaster resilience, economic revitalization, energy security, socioeconomic equity, and related public goods (Kashiwagi, 2016). Also, they stress that the network integration of these discrete smart community districts is the basis for building the overall smart city (Murakami, 2017).

Yet there is no authoritative count of Japanese smart communities. Tallying cases requires extensive searches through national and subnational government materials as well as among the major home builders (e.g., Sekisui House, Panahome, Misawa Home) and other sources. Recent quantitative and comparative research, including the ABI Research announcement cited earlier, suggests that Japan's initiatives are plentiful and robust. Additional evidence is available from the Smart Energy Group of the Japan Economic Center (JEC), a market research firm established in 1966, which regularly surveys the Japanese and international markets. The JEC's 2016 surveys indicated that Japan is a global leader in deploying the smart energy management systems that are central to the smart community. Table 21.1 displays the JEC survey results and projections for HEMS sales. The table reveals that 75,000 units were installed globally in 2011, and that a total of 15,000 (20% of the global total) were in Japan. By 2015, the number of HEMS had increased to 870,000 globally, with 150,000 (17.2% of the global total) in Japan. For 2020, the JEC projected a worldwide diffusion of 1.632 million HEMS, with 240,000 (14.7% of the global total) in Japan. The results for Japan are roughly consistent with an

TABLE 21.1 HEMS, Global and Japan, 2011–20 (Units: 1000)

Year	Global	Japan
2011	75	15
2012	156	30
2013	243	45
2014	448	80
2015	870	150
2020	1632	240

Adapted from Japan Economic Center, September 23, 2016. 2016 Survey on Current and Project Smart House Markets. Japan Economic Center (in Japanese).

August 2016 Fuji Keizai survey of the Japanese domestic market for energy management systems (Fuji Keizai, 2016).

Also impressive was the JEC survey concerning the "smart house" market, another segment that is a core element of the smart community. Table 21.2 shows that the JEC's results indicated that Japan represented over half of 2011's total global value of JPY 150 billion. The survey also revealed that Japan had held this share in 2015, when the total global sales had doubled to JPY 301 billion. Projections for 2020 indicated that Japan's share would decline to just below 40% as the total market enlarged to JPY 990 billion.

TABLE 21.2 Smart House Markets, Global and Japan, 2011–20 (Units: JPY Billion)

Year	Global	Japan
2011	150	80
2012	180	95
2013	220	115
2014	260	140
2015	310	160
2020	990	390

Adapted from Japan Economic Center, September 23, 2016. 2016 Survey on Current and Project Smart House Markets. Japan Economic Center (in Japanese).

These data are not especially surprising. University of California (Berkeley) Professor Dana Bruntrock, an expert on Japanese and other green building technologies, points out that Japanese home builders have long been involved in energy conservation and disaster resiliency (Bruntrock, 2017). The shock of 3-11 provided them with enormous incentives to build smart homes and smart communities (Honda, 2014).

Interestingly, Japan is generally not seen as a significant player in the microgrid market, which is dominated by North America. Indeed, the US power microgrid and DHC lobbies joined forces on May 18, 2016, when the International District Energy Association (IDEA) and the Microgrid Resources Coalition merged to push for a more rapid diffusion of their resilient, efficient, and distributed infrastructure. As IDEA President and chief executive officer Rob Thornton argued on the day of the merger, "we are witnessing a paradigm shift from remote central station power plants toward more localized, distributed generation for enhanced reliability, resiliency and energy efficiency, especially in cities, communities and campuses" (IDEA Industry News, 2016).

Yet Hitachi, Toshiba, and other Japanese firms are among the industry's leaders, powerfully incentivized by the impact of 3-11. Japanese firms dominate their domestic market and have an increasing presence in the US and other markets. In fact, Hitachi's smart community project at Kashiwanoha (in Chiba Prefecture) came to include the firm's first microgrid because 3-11 impelled a "rethink on the design of the country's energy infrastructure" (Wood, 2015).

Certainly Japanese smart communities include the digital signage and other ICT-based amenities that feature prominently in smart cities elsewhere (Maras, 2017). However, Japan's focus is different, and perhaps more apt as an industrial policy. Fig. 21.1 tells us that smart energy networks and decarbonizing energy inputs are the core of Japanese smart communities. Such networks are critical to modern industrial policy. Smart networks are, qua networks, comparable to the roads that were core networks in the Fordist economy and the railroads that marked the steam-based economy. As Nicholas Stern, the leading economist on climate change and energy, argues in his 2015 book *Why Are We Waiting? The Logic, Urgency, and Promise of Tackling Climate Change*:

> *[e]conomic history tells us that networks, be they power grids or railways, played a central role in past economic transformations: grids enabled great surges of creativity and innovation and led to opportunity and growth across the economy...More effective temporal and spatial management of the energy system, for instance with smart technologies or increased flexibility of the energy markets, could aid in the management of low-carbon generation, reduce the need for extra infrastructure, and unlock the potential for renewable energy to meet both base and peak demand for energy*
>
> Stern (2015, pp. 48–49).

HIGASHI MATSUSHIMA CITY SMART DISASTER PREVENTION ECO TOWN

As noted earlier, one example of Japan's approach is the Higashi Matsushima City Smart Disaster Prevention Eco Town. Fig. 21.2 locates Higashi Matsushima City on the northeast coast of Japan's main island, Honshu, the focal point of 3-11. The figure also notes that the population prior to 3-11 was 43,142. Over 1130 residents died in the disaster, which saw 65% of the city inundated by the sea. The disaster led to the official opening of a microgrid-based smart community on June 12, 2016. The microgrid allows the community's power system to "island" from the regional power grid in the event of a disaster.

Fig. 21.3 illustrates how the Higashi Matsushima Eco Town's renewable generation, complete with battery storage, is linked through the microgrid. Smart meters on all local facilities provide real-time tracking of energy generation, storage, and consumption. The project has 460 kW of solar capacity, of which 400 kW is an array that overtops a regulating pond. This regulating pond is itself part of disaster resilience, as it was built to accumulate rainwater and thus alleviate flood risks. In addition to the solar capacity, there is

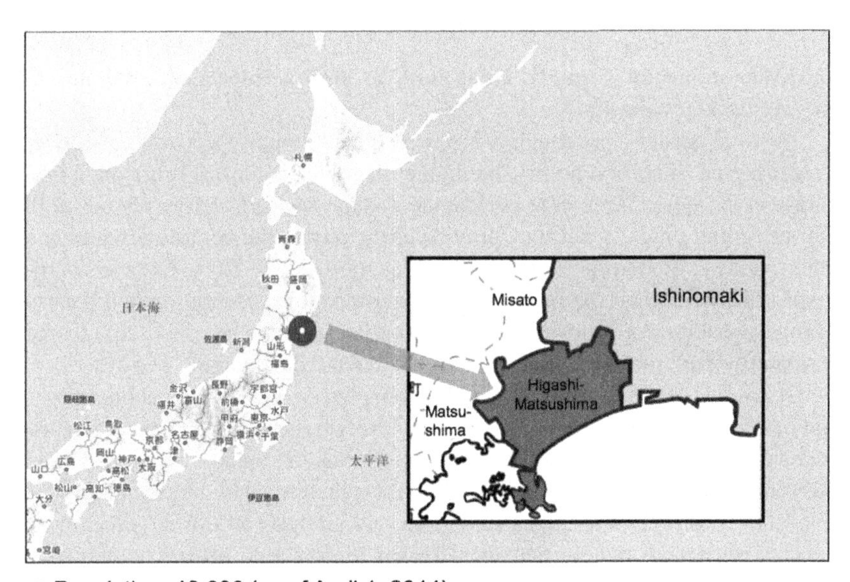

■ Population: 40,090 (as of April 1, 2014)
(Population before earthquake: 43,142)

FIGURE 21.2 Overview of Higashi Matsushima city. *Adapted from Japan Future City Initiative, 2015. Higashi Matsushima City. http://doc.future-city.jp/pdf/forum/2015_malaysia/Case_Studies_of_Japan_01_04_01_Ctiy_of_Higashi-matsushima.pdf.*

Overview of Smart Disaster Prevention Eco Town

Public Housing, 4 Hospitals, and 1 public facility linked by a microgrid

Composed of 460 kW solar (PV), 480 kWh large-scale battery storage, 500 kW and biodiesel emergency generation capacity

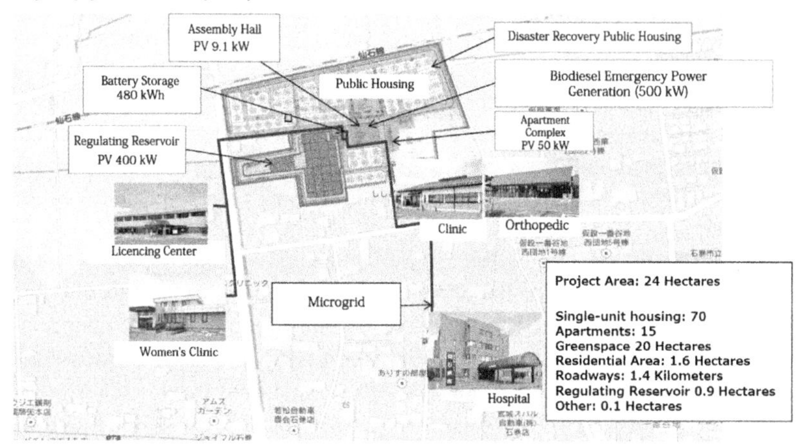

FIGURE 21.3 Higashi Matsushima disaster-ready smart eco town. *Adapted from Higashi Matsushima City, 2016: 15.*

480 kWh of battery capacity along with a 500-kW biodiesel generator to provide backup capacity.

The total capacity is enough to cover all the community's power needs over a short period of several hours, should the main grid go down. If the main grid outage is for more than a short period, the system suspends delivering power to the residential units. It can thus provide electricity to the community hall and other facilities (including four medical) for up to a few days. Even in a protracted disaster, when the biodiesel is exhausted, solar generation and battery storage capacity are able to provide essential supplies of electricity to the community hall and critical care facilities (MOE, 2016a, pp. 17–21).

Of course, Japan is often dismissed as replete with wasteful public works and top-down policy making. And this kind of criticism has been directed at its smart communities, particularly by Japanese fans of "small is beautiful" and market fundamentalists alike. Both dislike planning and industrial policy, respectively, privileging citizen initiatives and market-conforming incentives. So it is important to note that the Higashi Matsushima project is neither a white elephant nor the product of administrative fiat from the central agencies. Quite the contrary. Higashi Matsushima exemplifies Japan's smart investment and collaborative governance, involving an impressive diversity of stakeholders in planning projects that maximize benefits to the national and local communities.

For one thing, the city itself was a first mover rather than a passive participant. Higashi Matsushima Mayor Abe Hideo[3] was a practitioner of citizen-centered city management even before 3-11. After the disaster, Abe led the rebuilding of his city under the slogan "Never Forget That Day" and with the ambition of energy autonomy. Higashi Matsushima residents were fully in support, having direct experience of the death and prolonged suffering that results when critical infrastructure fails. The depth of community spirit was evident in how quickly and efficiently they got rid of the wreckage. Higashi Matsushima's residents, local businesses, and other stakeholders recycled an incredible 99% of 1.098 million tons of ruins from the disaster, achieving an extremely low JPY 18,000/ton cost (MOE, 2016a, p. 6). The community also organized in the "Higashi Matsushima Organization for Progress and Economy, Education and Energy (HOPE)," formally inaugurated as a company on October 1, 2012. Under this leadership, the city undertook a citizen-centered project on relocating the 1400 households whose residences were destroyed by 3-11.

In her January 2017 Japanese book on "community energy," the expert environmental journalist Kono Hiroko (2017) reveals both the extensive community engagement and the crucial role of the central government and business interests. Kono's careful research, the best yet on Higashi Matsushima's project, relies on in-depth interviews as well as concrete fiscal and administrative details. Kono relates that Japan's Ministry of the Environment (MOE) recognized Higashi Matsushima's local capacity and, in the spring of 2014, suggested city officers apply for a distributed energy subsidy. The city got the funds and was able to use them (along with other monies) to cover three-fourths of the JPY 500 million cost of the eco town's power system (Kono, 2017, p. 63).

Higashi Matsushima City Smart Disaster Prevention Eco Town also includes large corporate interests. Sekisui House, one of the world's largest home builders and a developer of 16 smart communities in Japan that were energy self-sufficient and disaster resilient (Sekisui House, 2016, p. 25), was brought in to construct the housing units. The business consortium, "Smart City Project," inaugurated on September 9, 2009, provided design advice. The Smart City Project includes 27 leading Japanese firms, including home builders such as Mitsui, Sekisui, and Shimizu; ICT firms such as NTT Communications and NEC corporation; smart energy systems firms like Azbil Corporation, Toshiba, Hitachi, Mitsubishi, and Tokyo Gas; and urban planning specialists such as Nikken Sekkei and Kokusai Kogyo Group.[4]

3. In this chapter, Japanese names are rendered surname first, in accordance with Japanese usage.
4. An English language introduction to the Smart City Project and its members is available at the following URL: http://www.smartcity-planning.co.jp/en/index.html.

The Higashi Matsushima project is indeed a flagship initiative emblematic of Japan's inclusive governance and commitment to advanced technology centered on the city post 3-11.

THE POWER BUSINESS

Moreover, the Higashi Matsushima eco town's 460 kW of solar power capacity is just a fraction of the city's renewable portfolio. As shown in Fig. 21.4, the deployment of renewable energy in Higashi Matsushima city (not just the eco town) multiplied by nearly 20 times between 2011 and 2015, rising to 35% of the city's power consumption. The city aims to achieve a target of 120% renewable energy by 2026 (Kono, 2017, pp. 85–88).

Japan's policies on smart communities promote distributed energy as local economic revitalization in addition to disaster resilience. The ongoing deregulation of Japan's retail power, gas, and district-heating markets is deliberately being used to foster local initiatives. Reflecting this thinking, from April 2016, Higashi Matsushima began operating a local power company, "HOPE Electricity." As of October, 2016, HOPE Electricity had a client base of 34 city and 98 private sector facilities and was delivering just over 8 MW of power at 1.5% below the cost of Tohoku Power, the regional utility. HOPE Electricity's power sales are expected to bring in JPY 10 million of profit, to be reinvested in the community. HOPE also aims to increase its power sales to

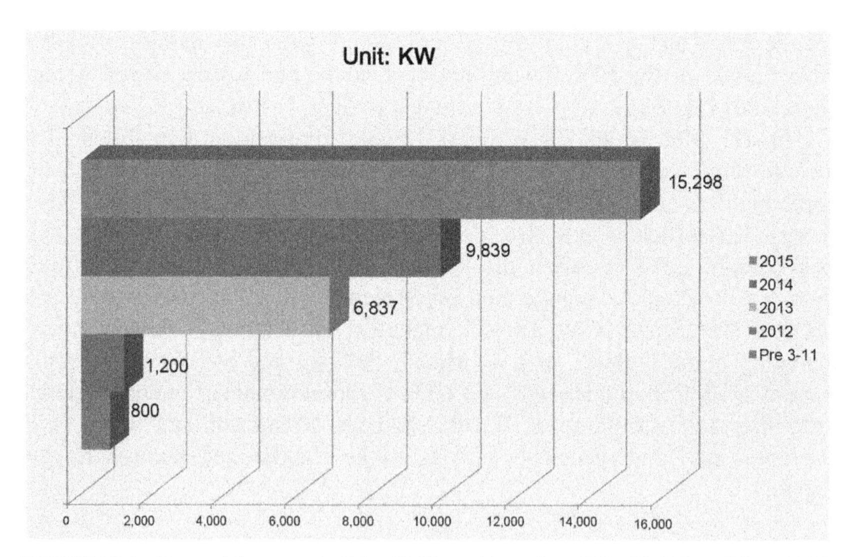

FIGURE 21.4 Renewable energy in Higashi Matsushima city, 2011–15. *Adapted from MOE, October 6, 2016a. Town Planning for Local-production/local-consumption and Disaster Recovery. Ministry of the Environment, (in Japanese). http://www.env.go.jp/press/y0618-05/mat05_1.pdf, p. 11*

over 20 MW. This goal seems achievable, given HOPE Electricity's advantage of being local and its lack of competitors in the region (Kono, 2017, p. 77).

The Higashi Matsushima project is a greenfield development, rather than the renovation of an existing, or "brownfield," community. All the same, its essential elements form a template being replicated nationwide. This template links local energy endowments (such as rooftop and utility solar) with smart energy management systems, to decarbonize while maximizing disaster resilience. The template also seeks to encourage as much local involvement as possible, to bolster local organization (key to resilience) and business opportunities. In short, Japan's approach emphasizes smart governance as much as it does the deployment of smart infrastructure.

JAPAN'S INCENTIVES

Geography, geology, and demography are among the principal reasons Japan places a heavy stress on maximizing energy autonomy and disaster resilience in compact, smart communities. Japan is able to act because it is still the world's third largest economy, at JPY 505 trillion (USD 4.7 trillion) in 2016. The country boasts a formidable endowment of scientific, financial, and other resources.

Japan is also compelled to act: although Japan's 127 million citizens made it the world's 11th most populous country in 2015, the country is depopulating and aging more rapidly than any other Organisation for Economic Co-operation and Development (OECD) country (Below, 2016). The grim facts on demography give Japan powerful incentives to plan communities that are sustainable and resilient.

Extreme dependence on conventional energy provides yet another incentive to stress smart and distributed solutions. In 2015, Japan was the world's fifth largest energy economy, using 435 million tons of oil equivalent, totaling roughly JPY 40 trillion in direct fuel costs and ancillary costs. Japan was also the fourth largest electric power market, consuming 921 TWh in 2015, a market worth just under JPY 20 trillion (Enerdata, 2016). Japan imported virtually 100% of the oil, coal, and natural gas it consumed in 2015. Reflecting its economic size, extreme dependence on imports, and de facto withdrawal from nuclear power after 3-11, Japan was the world's largest importer of liquefied nitrogen gas and the third largest importer of coal and oil in 2015 (EIA, 2017). Table 21.3 shows that Japan imported over 90% of its primary energy supply as fossil fuels. The increasing geopolitical and other risks (e.g., water, disinvestment) associated with nearly complete reliance on imported fossil fuels is an unsustainable position for a global power.

As for geology, Japan is also a 3000-km long, narrow and highly seismic archipelago, with comparatively high mountains (2000–3500 m) running its length. No point in Japan is more than 150 km from the sea. Moreover, the mountainous terrain is subject to significant rain, which occurs as increasingly

TABLE 21.3 Japan's Total Primary Energy Supply, by Source, 2015 (Units: %)

Oil	42.9
Coal	27.5
Gas	23.3
Hydro	1.7
Other renewable	4.0
Nuclear	0.6

Adapted from IEA, 2016. Energy Policies of IEA Countries, Japan. https://www.iea.org/publications/freepublications/publication/EnergyPoliciesofIEACountriesJapan2016.pdf.

intense and damaging downpours. Thus, Japan's typical community is on the coast with mountains behind and is run through by plenty of short rivers that have very steep gradients. Most Japanese communities are subject to periodic and potentially catastrophic flooding and landslides. Japan also has unparalleled typhoon, earthquake, and tsunami risks (Fudeyasu, 2016).

This chapter emphasizes Japan's hazards because Japanese state managers do, in thousands of plans and reports (DeWit, 2014b). In addition, the hazards are undeniable and worsening. Japan's pragmatic, science-based public debate also understands seismic and climate hazards. This makes the intellectual context for climate change and smart communities very different in Japan than it is in Anglo-America.

Disasters, aging, dependence on conventional energy, and the like all threaten the viability of Japan's local communities. The challenges therefore all have a common focus. And the vulnerability of local communities undermines the fiscal, administrative, and other capacities of the national community, because Japan is a unitary state with massive intergovernmental redistribution (DeWit, 2017a).

In addition to the incentives listed so far, 3-11 hit Tokyo especially hard, giving the capital a powerful shaking as well as protracted power outages. This direct blow to Tokyo was in sharp contrast to previous disasters such as the 1995 earthquake that leveled much of Kobe but left Tokyo untouched (Edgington, 2011). The threat to the capital galvanized policy entrepreneurs. The scale of 3-11, and its energy—climate context, has fostered impressive and increasing policy integration in energy environmental policy, urban planning, science and technology policy, health and welfare policy, intergovernmental finance, and related policy domains.

The 3-11 natural and nuclear disasters led to collaborative planning, focused on smart communities and creating a distinctively Japanese path to greater energy security and disaster resilience. Japan is not Germany, with its *energiewiende* (energy transition) contingent on continental energy trading infrastructures as well as a very activist civil society. Nor is Japan one of the Anglo-American regimes, again with international energy linkages, plus generally plenteous resource endowments, competitive party politics, growing populations, and a tendency to rely on market-led solutions rather than planning and explicit industrial policy.

We have seen that, compared with its peer countries, Japan has very poor conventional resource endowments and strikingly adverse geography and geology. Japan also has the developed countries' lowest levels of foreign investment, immigration, and other indicators of internationalization. Crucially, Japan also has no direct international energy connections through power grids, gas pipelines, and other energy networks. The country is also very distant from its principal sources of energy supply. It thus makes sense to collaborate on fostering public goods. It also seems advisable to deploy distributed energy and its networks, maximizing efficiency, energy security, and local benefits, rather than immediately spend enormous sums on building an international ("Asia Super Grid") or even an integrated national network.

The smart community distributed energy solution is also attractive because Japan has a wealth of underground infrastructure corridors for communication cables and other critical infrastructure. The replacement of thick conventional communication cables with thin fiber-optic wires left considerable unused space in these utility corridors. Smart community experts point out that this space could be used for the deployment of heat pipelines. They argue that the business opportunities stemming from installing distributed heat and power networks would be very productive public works. Pertinent in this respect is that highly efficient combined heat and power systems are crucial to cutting energy use and carbon emissions (DeWit, 2016b; UNEP, 2015). And these "district energy" systems are necessarily local, as the efficiency of thermal distribution drops off dramatically after a few kilometers. These factors dovetail nicely with Japan's other reasons to build smart and compact communities (Kashiwagi, 2015).

JAPAN'S NEW INSTITUTIONS

Japan shows us that broad societal incentives (e.g., energy security, interregional equity, economic revitalization) can lead to action when they have institutions and policy entrepreneurs empowered by a crisis. Since 3-11 Japanese governance has become increasingly collaborative, integrated, and focused on patent seismic, climate, energy, and other risks. Indeed, Japan may be an outlier in evolving inclusive governance and long-term planning, in the form of an integrated and adaptive industrial policy. Japanese policy makers and others have built an impressive cluster of powerful new institutions that group central agencies, subnational governments, businesses, and academics and other actors. These institutions are well funded and administratively empowered, providing venues for addressing the multiple crises that threaten Japan's built and natural environments.

One example of the scale of Japan's post 3-11 collaboration is the Japan Academic Network for Disaster Reduction (JANET-DR).[5] The JANET-DR groups 54 academic associations, crossing multiple disciplinary boundaries (including energy and spatial planning). It also cooperates with the prestigious Science Council of Japan, whose coordinating role in Japan is roughly equivalent to that of the American Association for the Advancement of Science. The JANET-DR was formalized on January 9, 2016, building on an ad hoc 30-association liaison that emerged in May 2011, shortly after 3-11. The liaison played a large role in shaping Japan's resilience debate, including smart and compact cities, through 11 major events and several publications. On November 1, 2016, the JANET-DR held their second disaster resilience symposium, analyzing the worsening threat of typhoons and intense rain (JANET-DR, 2016). They also published a detailed specialist volume exploring water and landslide hazards in the context of climate change (Ikeda et al., 2016).

Another example is seen in the fact that there are 12 separate collaborative subsidy programs for fostering the deployment and development of renewable energy and smart energy systems (e.g., heat and power microgrids). No fewer than 6 of the 12 collaborations included METI and the MOE. And 9 of the 12 projects include 3 or more central agencies.[6]

This degree of cooperation has already had a powerful impact on smart community policy making. Japan's EIS, mentioned earlier, was developed through extensive consultation with the business community's peak associations (METI, 2016). The EIS aims to increase the diffusion of distributed energy alternatives and efficiency in the context of smart communities. The EIS explicitly relies on a coordinated, strategic approach, rather than market mechanisms. Its governance includes all levels of the state, business, academics, and civil society, and is backed up by ample fiscal and regulatory action. The policy also seeks to exploit potential synergies between sectors. This approach includes bringing the "Internet of things" directly into the energy economy, fostering even greater efficiencies and the uptake of an array of renewables and hitherto wasted heat. Moreover, the EIS expressly commits policy to diffuse smart communities. This objective reflects an expanding policy of bolstering local government resilience through smart energy systems and their capacity to exploit local energy resources (METI, 2016).

Furthermore, Japan's overall policy integration on smart communities is achieved within the National Resilience Plan (NRP). Fig. 21.5 illustrates how

5. The website of the JANET-DR is available at the following URL: http://janet-dr.com.
6. See (in Japanese) "On the 12 Collaborative Projects to Deploy Renewable Energy Over the Next 5 Years," New Energy Net, April 14, 2017: http://pps-net.org/column/34210.

FIGURE 21.5 National resilience as an umbrella plan. *Adapted from Japan Cabinet Secretariat, n.d. Building National Resilience. http://www.cas.go.jp/jp/seisaku/kokudo_kyoujinka/en/e01_panf.pdf, p. 9.*

Japan's NRP serves as an "umbrella plan." The figure shows that the NRP has de jure administrative authority over other national plans, including energy, environmental, and spatial plans. The NRP appears to be far more authoritative and inclusive than comparable planning initiatives among the OECD countries.

The NRP is based on the National Resilience Law passed by the Japanese Diet (parliament) on December 4, 2013. National resilience itself is under the authority of a State Minister, a new position announced during December 26, 2012, inauguration of the first cabinet of Liberal Democratic Party (LDP) Prime Minister Abe Shinzō. The NRP was then worked up into a plan by the Cabinet Secretariat, advised by the National Resilience (Disaster Prevention and Reduction) Deliberation Committee (NRDC). The NRDC first met on March 5, 2013, and continues its deliberations as of this writing. Its 14-person membership is drawn primarily from the top ranks of Japan's academic, government, and business communities. Japan's top energy policy advisor, and its strongest advocate of smart communities, Kashiwagi Takao, advises on energy. Other members, including board-level executives from Toyota and NTT, make recommendations on aging, primary industries, local communities, local administration, risk communication, industrial structure, the environment, disaster prevention, finance, national lands, and information services.

SMART COMMUNITY POLICY ENTREPRENEURS

Japan's National Resilience initiative started in early 2013 with only three items on the table: the linear shinkansen, cross-laminated timber, and methane hydrates (Furuya, 2017). However, the involvement of a broad range of energy and disaster expertise has progressively expanded the initiative's ambit. Notably, planning and financing smart energy systems became central to resilience in large part due to the role of Kashiwagi Takao. As noted, Kashiwagi is not only Japan's foremost energy policy expert but also the central figure in a powerful intellectual community, one whose policy making influence has greatly increased since 3-11. After 3-11, Kashiwagi's main contribution, as a policy entrepreneur, has been to forge a broad public–private coalition that links decarbonizing smart energy systems (heat and power grids) with disaster resilience, spatial planning, and local economic development.

Kashiwagi has long been a dogged champion of smart communities. He designed Japan's first smart community, a 100% renewable microgrid project, for the 2005 Aichi World's Fair (Kashiwagi, 2011). He has consistently argued that the focus of Japan's public works should shift from roads and bridges to resilient, smart energy systems that maximize efficiency and the uptake of renewable energy (Kashiwagi et al., 2001; Kashiwagi, 2009, 2010, pp. 10–15, 2015a,b, 2016, pp. 178–179).

In addition, Kashiwagi was ideally positioned, organizationally and intellectually, when the 3-11 crisis erupted. He had credibility with the business community as well as a thorough understanding of evolving energy paradigms and the crucial role of integrated policy. As an academic, Kashiwagi was professor at the Tokyo Institute of Technology from 2007, and in 2012 , he became specially appointed professor and head of the International Research Center of Advanced Energy Systems for Sustainability (AES Center).[7] The AES Center was inaugurated in September 2009, with funding from some of Japan's largest firms. Its 200 specialist researchers in environment and energy are focused on solar and fuel cells. It also collaborates with the smart energy divisions of the leading power, gas, and other firms in Japan's energy business, along with blue-chip firms in construction, home building, engineering, auto making, and other areas. Moreover, the AES Center includes 15 local governments (prefectures and cities), many of which are exemplar smart communities. The AES Center also clearly understood the crisis to be an opportunity to accelerate the deployment of smart communities that maximize renewable energy, local leadership, resilience, and other priorities (Gotou, 2017).

Kashiwagi has long been a top-rank policy advisor. In the years preceding 3-11, he was member, and often chair, of METI's main energy policy-making

7. The English language version of the AES Center's website is available at the following URL: https://aes.ssr.titech.ac.jp/english.

councils. These councils include the Advisory Committee for Natural Resources and Energy (ACNRE), the Strategic Policy Committee of the ACNRE, and the Energy Efficiency and Renewable Energy Committee of the ACNRE, in addition to numerous other national and subnational bodies. From 2011, he became Chairman of the Green Investment Promotion Organization (GIO), a quango set up by METI in September 2010 to manage low-carbon subsidies and promote such investment. The GIO members and cooperating businesses include several of Japan's major insurance firms together with a large number of business associations in energy (conventional and renewable), ICT, engineering, and electrical equipment.[8]

Right after 3-11, Kashiwagi served as the chair of Research Commission on Urban Planning and Integrating the Effective Use of Heat Energy, one of the emergency environmental and energy research efforts that METI undertook. The Commission included several other specialists, and its proceedings were observed by several national and Tokyo Metropolitan Government infrastructure-, environment-, and energy-related bureaucracies. It began its deliberations in May 17 2011, and then quickly met six additional times, delivering a report on August 1, 2011.[9] The report underscored the merits of integrating energy and urban planning, so as to realize the diffusion of smart communities and distributed energy. It also advised that smart heat and power networks be deployed as the core of a larger, strategic initiative to maximize disaster resilience, efficiency, decarbonization, the uptake of local energy resources, and local energy security (METI, 2011).

When the LDP returned to power in the December 2012 election, it was quick to call on Kashiwagi's expertise. For example, the Ministry of Internal Affairs and Communications (MIC), which—in a decidedly fortuitous combination of responsibilities—both oversees Japan's 1719 local governments and the country's ICT infrastructure, wanted to use the crisis as an opportunity to expand local governments' energy initiatives. Hence, the then-MIC minister, Yoshitaka Shindō (2012–14), brought Kashiwagi into the MIC as chair of a special research commission on diffusing smart energy systems as one means of revitalizing local areas (DeWit, 2014a).

Kashiwagi was also named to several major committees. For example, he chaired METI's Hydrogen/Fuel Cell Strategy Council, from its first meeting on December 19, 2013.[10] He was also named as a member of the Cabinet

8. The GIO's website (in Japanese) is available at the following URL: http://www.teitanso.or.jp/index.
9. The membership, minutes, and materials studied by the Research Commission on Urban Planning and Integrating the Effective Use of Heat Energy are available (in Japanese) at the following URL: http://www.meti.go.jp/committee/kenkyukai/energy/nestu_energy/001_giji.html.
10. The membership, minutes, and materials studied by the Hydrogen/Fuel Cell Strategy Council are available (in Japanese) at the following URL: http://www.meti.go.jp/committee/kenkyukai/energy_environment.html.

Office's Specialist Deliberation Committee on Important Issues, whose first meeting took place on October 11, 2013.[11] This committee is one of the main advisory organs for the Council for Science, Technology and Innovation, chaired by the prime minister. And from November 18, 2013, Kashiwagi also became chair of the Energy Strategy Conference, one of the Specialist Deliberation Committee's working groups.[12] The Energy Strategy Conference groups key scholars and business interests involved in smart energy and climate change. It evaluates the role of energy within Japan's overall "Society 5.0" innovation strategy, paying increasingly close attention to smart networks and resilience. Akin to Germany's Industry 4.0 initiative, Japan's Society 5.0 project seeks to harness the technologies of the fourth industrial revolution. However, incentivized by disasters, demographics, and other challenges, Japan's effort transcends Germany's smart factories and aims to deploy smart systems throughout the entire society (Sayer, 2017; Kashiwagi, 2017).

In addition to his prominence in academe and policy making, Kashiwagi is also directly involved in business circles that embody the ongoing revolution in smart and distributed energy. Particularly noteworthy is his chairmanship of the Advanced Cogeneration and Energy Utilization Center (ACEJ). The ACEJ is dedicated to promoting cogeneration systems (DHC as well as fuel cells) and the use of renewable energy. In September 2011, it revised its name from "Japan Cogeneration Center" to reflect this larger purpose. After April 2014, the ACEJ's membership also expanded, to encompass not just energy firms but also electronics, construction, design, and other firms, reflecting the cogeneration's increasing sophistication and diffusion in Japan (ACEJ, 2016).

JAPAN'S NATIONAL SPATIAL STRATEGY

Kashiwagi and his circle also helped shift Japan's priorities on energy policy and infrastructure via Japan's new National Spatial Strategy (NSS), which was adopted in August 2015. Like the NRP, the NSS was produced by an inclusive process. Indeed, the OECD described the NSS as "an intensive exercise in inter-ministerial co-ordination and consultations extending beyond the government itself under the aegis of the National Land Council, which brings together parliamentarians, academic experts, representatives of the private sector, elected officials from the cities and regions, and others." Like the NRP, the NSS was both distinctive from Japan's previous top-down planning strategies and had regional versions. This broad range of consultation gave the

11. The membership, minutes, and materials studied by the Specialist Deliberation Committee on Important Issues are available (in Japanese) at the following URL: http://www8.cao.go.jp/cstp/tyousakai/juyoukadai/index.html.

12. The membership, minutes, and materials studied by the Energy Strategy Conference are available (in Japanese) at the following URL: http://www8.cao.go.jp/cstp/tyousakai/juyoukadai/wg.html.

NSS a legitimacy that transcended its predecessor documents. On top of that legitimacy was legal authority: at least 20 other national laws are obligated to refer to the NSS (OECD, 2016, pp. 79–80). In turn, the NSS is obliged to reference the NRP.

Compared with its 2008 iteration (not to mention earlier national plans), the 2015 NSS devoted much attention to how Japan's energy, demographic, climate, and other challenges can be alleviated through smart communities, renewable energy, resilience, and related initiatives. In spatial terms, the 2015 NSS stressed the need to build (in ascending order of size) "small stations" (*chiisana kyoten*) in rural areas, "sustainable residential areas" (*teijuu jiritsu ken*) within cities, and "collaborative core urban areas" (*renkei chusuu toshi ken*) across cities. All of these spatial forms underscore the multiple benefits of concentrating public services (such as medical, welfare, and elderly care facilities) in compact nodes and clusters. Such densification encourages greater energy efficiency and disaster resilience.

The 2015 NSS paid careful attention to smart and distributed energy because its design team included Kashiwagi as its first-ever energy expert and he made a point of emphasizing smart communities and smart energy systems (Kashiwagi, 2015). Considering Japan's vulnerability when it comes to energy, it is rather surprising that its previous spatial planning initiatives did not include an energy expert. However, in fact, few countries integrate these policy domains, which is why the European Commission set up the Spatial Planning and Energy for Communities In All Landscapes project from March 2013 to March 2016 (SPECIAL, 2013).

POLICY INTEGRATION IN THE COMPACT CITY

Policy integration on building compact cities is an additional development that is amplifying collaboration and the effective use of scarce fiscal, human, and other resources. Fig. 21.6 outlines the compact city and role of the "Locational Optimization Plan." As in virtually all Japanese smart community projects, policy integration between energy and spatial planning has encouraged robust policies to foster densification, to help cope with depopulation and aging in the context of accelerating climate change and other hazards.

The administrative agency to achieve this integration is also in place, via the Compact-City Design Assistance Team (CCDAT). The CCDAT was developed in March 2015 under the auspices of the Comprehensive Strategy for Regional Development, which itself was given Cabinet assent on December 27, 2014. The CCDAT is centered on the Ministry of Land, Infrastructure, Transport and Tourism (MLIT), but it also includes representation from 10 other agencies, including METI and MOE. This broad representation is deliberate because designing the compact city requires addressing energy, disaster resilience, interregional cooperation, urban farming, education, health

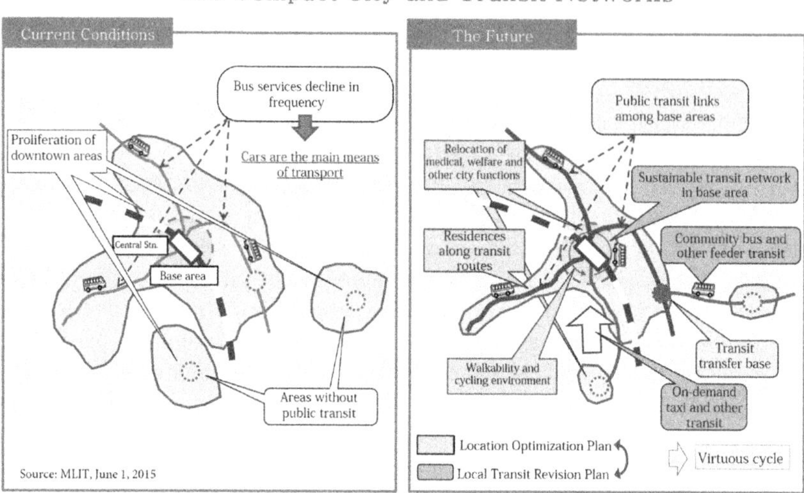

FIGURE 21.6 Japan's compact city approach.

and welfare, urban revitalization, local transport, revising local facilities, residential areas, and schools. The CCDAT's role is to deliberate with the specific local government concerning such matters as issues relating to the relocation of local facilities. The CCDAT then integrates how to relate policies (e.g., energy, transport, disaster resilience) into an overall package of institutional reforms and fiscal measures to achieve greater densities in tandem with better livability.

These plans are already guiding the relocation of hospitals, schools, elderly care, and other public services, a relocation that in turn increases the cost–benefit performance of smart energy networks and inputs at the same time as it reduces energy use. That reduction in energy use is achieved via the reduced need for motorized (especially single-car) transport in favor of public transit, cycling, and walking. The reduced spatial footprint of the community also leads to less energy used to move water around, plus lower per capita costs to maintain roads and other critical infrastructure, in addition to other energy savings. This integration of spatial planning with energy has been underway over the past 3 years and has linked most of the central agencies together, enhancing the effectiveness of planning and spending through reducing overlap and other sources of administrative inefficiency.[13] And the policy support has led to a rapid increase in plans: between December 2014 and

13. An explanation of the Compact-City Design Assistance Team and related institutions is available (in Japanese) at the following URL on the MLIT's website: http://www.mlit.go.jp/toshi/city_plan/toshi_city_plan_tk_000016.html.

March 2017, the number of subnational governments undertaking these spatial plans increased from 62 to 348.[14]

FISCAL SUPPORT FOR SMART COMMUNITIES

Smartly targeted subsidies for critical infrastructure also help drive the smart community projects. One example is the METI "Subsidy for the Promotion of Local Production-Local Consumption Style Renewable Energy Areal Use Projects." This subsidy began in fiscal year 2016 with a total value of JPY 4.5 billion and funded 28 separate local projects in the first round for 2016. METI's outline of the program's purposes described it as aimed at fostering the diffusion of distributed energy. This is because METI portrayed the 3-11 nuclear and natural catastrophes as having led to an increased understanding of the risks of reliance on centralized generation systems. It stated that in consequence Japan needed to promote the diffusion of decentralized energy, particularly systems centered on renewable energy. METI added that the use of energy management and other technologies, in tandem with the spatial deployment of energy systems, could help maximize the effective use of local energy resources. Moreover, the local production/local consumption model could lead to significant cuts in energy use and costs in normal, nondisaster circumstances. The system's disaster resilience role is described as providing the community with a source of energy in emergencies.

The METI cautions that these systems remain relatively costly. Hence the subsidy program aims at facilitating the diffusion of these advanced energy systems, commensurate with local conditions. The goals include reducing the unit costs of microgrids and other energy management systems through greater economies of scale, the creation of new business services linked to demand response and other energy-related services, and the development of energy systems that can be deployed nationwide.

The METI subsidy program period is 5 years, from 2016 to 2020, and the primary criterion for assessing the performance of supported projects is whether overall system efficiencies of 20% or over are achieved. The METI also points out that the renewable generating capacity eligible for inclusion is not to be covered by Japan's feed-in tariff (FIT). As in other countries, such as Germany, the end of Japan's FIT is in sight, and the subsidy aims to foster the nonsubsidized diffusion of distributed renewable energy.

METI's subsidy is only one of many. From fiscal year (FY) 2015, the MIC began implementing a similar fund, the "Distributed Energy Infrastructure Project," for encouraging the deployment of heat and power grids. This program was developed by a special MIC study group, one chaired by Kashiwagi

14. The data and other relevant information concerning Locational Optimization Plans are available (in Japanese) at the following URL on MLIT's website: http://www.mlit.go.jp/toshi/city_plan/toshi_city_plan_fr_000051.html.

Takao from November 2014 to the spring of 2015. Similar to the METI subsidy, MIC's program seeks to foster renewables, particularly biomass, geothermal, and other 24/7 "baseload" energy, with the local community as the lead agent in the project.

Additional finance for related smart energy systems includes cogeneration-related subsidies managed by the MLIT and MOE. In the FY 2015 supplementary budget and the FY 2016 initial budget, these subsidies total JPY 167.94. Moreover, the governing LDP (through its Study Commission on Resources and Energy Strategies) maintains a comprehensive list of 46 national-level distributed energy subsidies. The LDP categorizes the subsidies by their respective central agency funder and publishes them on its website, indicating a high level of interest in diffusing smart communities and smart energy.[15]

It is not possible to calculate the total monetary value of these individual subsidies, as many of them are part of much larger programs that are not solely focused on energy. To take one example of this, the MLIT offers fiscal support for local governments that want to harness waste heat energy in their sewage networks. Japan's potential for waste heat capture in the best areas of its 460,000 km of sewerage has been assessed at 15 million households' worth of heat energy use, so this program is potentially quite significant. However, the MLIT's support was part of the JPY 898.3 billion comprehensive disbursement for social infrastructure (*shakai shihon seibi sougou koufukin*). There is no indication of how much of the total disbursement is to be spent on waste heat recovery. Moreover, the MLIT supplements this particular initiative on waste heat recovery from sewers with the offer of sending expert staff to advise local governments and other actors (such as private firms and public–private collaborations), to assist the latter in working up project proposals and other pertinent items. This deployment of expert assistance has a monetary value that also cannot be quantified.

In addition to the aforementioned finance, the NRP is also in part devoted to smart communities, and is a very well-funded initiative. Its initial budget for FY 2017 is JPY 3.7 trillion, a sum that represents an increase over previous years (NRPO, 2017). And as of March 2017, all of Japan's 47 prefectural governments, along with 65 cities and towns, had adopted local versions of the NRP. The local versions differed in their respective lists of hazards and energy inputs, reflecting local circumstances. However, they were all consistent in their emphasis on exploiting local energy resources to bolster resilience. The large number of subnational plans, less than 3 years after formal passage of the NRP, suggests its legitimacy among subnational policy makers. The increasing number of plans also adds to the fiscal support for smart and compact communities.

15. The LDP list (in Japanese) is available at the following URL: https://www.jimin.jp/policy/policy_topics/energy/131871.html.

Supplementing direct finance, Japanese policy makers also use numerous tax incentives to incentivize smart and compact communities. Policy makers in MIC and MLIT are focusing tax reform on increasing urban density (to raise land values) and foster the diffusion of smart energy systems (e.g., LEDs, district heating, light rail, fuel cell vehicles buses, etc.). The property tax is one mechanism in this initiative. Several municipal governments explicitly aim at densification to raise property values and thus increase property and other tax revenues while cutting expenditures for infrastructure maintenance and other costs. The use of "special tax measures" (*sozei tokubetsu sochi*) is also notable. After 3-11, Japan's special depreciation tax measures for energy and the environment mushroomed from YEN 800 million in FY 2011 to JPY 552.5 billion in 2013. FY 2013 special depreciation tax measures for energy and the environment were thus well over half of FY 2013's YEN 949.3 billion in total special depreciation allowances (Sakamoto, 2015). From FY 2011, these special measures include LEDs, biomass, small hydro, waste heat recovery (from sewerages), batteries, cogeneration, and other energy-producing and storage systems.[16] Most of this investment is clustered in compact and smart cities.

The smart communities fostered by the resilience paradigm are also important to Japan's plans to increase infrastructure exports from JPY 10 trillion in 2010 to JPY 30 trillion in 2020. As seen in Fig. 21.7, Japan's infrastructure exports in 2010 totaled just under JPY 10.3 trillion. JPY 3.8 trillion of this total was energy related, most of it being JPY 2.2 trillion in coal plant (listed in the figure as "conventional power"). The distributed energy and other systems of the smart community composed JPY 800 billion, more than the JPY 300 billion in nuclear sales and the JPY 500 billion in gas and oil plant exports. The Japanese government aims to triple infrastructure exports to JPY 30 trillion, by 2020, with energy infrastructure more than doubling, to JPY 9 trillion. It achieved JPY 19 trillion in 2014, suggesting that the 2020 target is realistic. What remains to be seen is how much the relative shares of smart community exports increase versus the proportions for coal and other fossil fuel plant as well as nuclear plant. One powerful determinant of the shift is likely to be the domestic deployment of smart communities, including microgrids, energy management systems, and the array of distributed energy inputs being developed.

Aiding in this effort to export smart communities was harnessing them to the Japan International Cooperation Agency, Japan Overseas Infrastructure Investment Corporation, and other export promotion networking and finance facilities.[17]

---

16. For a listing of the items included in energy and environmental special tax measures, see (in Japanese) "Special Tax Measures Law, Tax Deductions" at the following URL: http://www.zeiken.co.jp/25kaisei/soz1-3.html.
17. On this objective, see (in Japanese) the Japanese Government's revised 2016 "Infrastructure Export Strategy" at the following URL: http://www.kantei.go.jp/jp/singi/keikyou/dai24/kettei.pdf.

Sector			Current Approximations 2010	2014	Projections Estimates 2020
Energy	Power (Conventional)		22,000	56,000	90,000
	Nuclear Power		3,000		
	Oil/Gas Plant		5,000		
	Smart Community		8,000		
	Subtotal		38,000		
Transport	Rail		1,000	10,000	70,000
	Next-Gen Auto		10		
	Adv. Safety Vehicles		–		
	Roads		2,000		
	Harbours	Construc.	500		
		Operation	500		
	Aviation	Airports	500		
		Control	1		
	Subtotal		4,511		
ICT	Subtotal		40,000	91,000	60,000
Infrastructure	Industrail Parks		100	18,000	20,000
	Construction		10,000		
	Subtotal		10,100		
Living Environment	Water		2,000	4,000	10,000
	Recycling		1,000		
	Subtotal		3,000		
New Sectors	Medical		5,000	11,000	50,000
	Agriculture and Food		1,000		
	Space		200		
	Marine Infra/Ships		1,000		
	Postal		150		
	Subtotal		7,350		
Overall Total			102,961	190,000	300,000

FIGURE 21.7 Japan's infrastructure exports, by sector (units: JPY 100 million). *Adapted from MLIT, July 7, 2017. The government's approach. In: Materials Presented to 8th Meeting of Deliberation Commission on Overseas Port Facilities Distribution Project. Ministry of Lands, Infrastructure, Transport and Tourism, Japan (in Japanese), p. 6*

The Japanese also began working through International Council for Local Environmental Initiatives on this goal. Further to this end, Japan's MOE set up the "Asia Low-Carbon Cities Platform" (MOE, 2016b). It lists 25 of Japan's smart communities, with overviews of their programs for smart energy systems and other critical infrastructures. The site does not seem to provide a comprehensive list of Japan's smart communities. Rather, it summarizes the core infrastructures of the Japanese smart community and shows potential customers how to arrange consulting, financing, and other assistance.

Another important influence on export possibilities is that the main proponent of national resilience, Nikai Toshihiro, was appointed LDP Secretary General on August 3, 2016. Nikai is influential and an internationalist. He has long emphasized cooperating with regional countries, particularly China and Korea. Nikai has made it clear that he is committed to leveraging Japan's expertise on disaster resilience and renewable energy. He has called for using it to expand external engagement and exports, combining domestic security and economic goals (in Kashiwagi, 2016, pp. 177–178). At Hawaii University on May 4, 2017, Nikai argued for the deployment of renewable energy in Pacific Island states as one measure to bolster their resilience against climate change (*Kyoto Shimbun*, May 4, 2017). The evidence thus suggests that smart community and associated exports are increasingly prioritized in Japan's infrastructure export strategy.

POST 3-11 STAKEHOLDER SUPPORT FOR SMART COMMUNITIES

A further consequence of the 3-11 disaster is greater subnational government and popular support for energy alternatives and smart communities. Before 3-11 Japan did have distributed energy initiatives aimed at increasing local energy autonomy through biomass, geothermal, and other projects. Yet these "local production and consumption" programs gained minimal traction due to the ambivalence of local communities, the disinterest (or outright opposition) of the regional power monopolies, the lack of incentives for local leaders, byzantine regulatory regimes, and other hurdles. However, after 3-11 virtually all public and private sector stakeholders, together with most of civil society, were able to agree on the need to bolster resilience against hazards.

For example, a March 2014 Japanese METI survey of smart communities showed that 82.2% of surveyed local governments listed resilience against disasters as their top priority for undertaking a smart community project, with energy autonomy second, at 73.3% and the creation of new local services and businesses third at 71.1% (Oguro, 2014). Moreover, Japan's annual and authoritative "Environmental Consciousness Survey," released in September of 2016 by the National Institute for Environmental Studies, showed that the country's strongest level of consensus for any initiative related to energy and the environment was the 77.8% support for using public funds to build resilience in the face of climate change. And 68.1% supported using ODA to build resilience in developing countries (NIES, 2016, p. 20). In short, Japanese local governments and the public were quite amenable to changing the built environment as an adaptation response. They were also willing to foster resilience in developed countries.

HIROSAKI SMART CITY

The aforementioned evidence on policy integration, finance, and other factors indicates that the diffusion of Japan's smart communities is primed to accelerate further in FY 2017. We can see this play out in Hirosaki city, an urban center of 176,590 in Aomori Prefecture. The city's location is marked in Fig. 21.8.

Fig. 21.9 outlines the Hirosaki city community energy system, which links up heat, power, and information networks as well as an array of decarbonizing inputs. As of 2017, the city's initial smart city plan is 4 years old. Its authors emphasize that it requires updating because of massive and rapid technological changes coupled with the fact that the plan's phase 2 is set to begin in 2017.

Fig. 21.10 shows that phase 1 of Hirosaki city's project ran between FY 2013 and FY 2016. As described in the figure, the first phase centered on the deployment of extant technologies and disaster resilience. To these ends, the city installed LED lighting, some solar, energy management, advanced waste treatment, and other technologies. Fig. 21.10 shows that the next phase of the smart city plan is to deploy the "community energy system" whose elements have been undergoing test-bedding throughout the country as well as overseas. This system is to link the city's projects.

The Hirosaki Smart City plan portrays the community energy system as key to major gains in reducing energy (thermal and electricity) consumption as well as replacing reliance on electricity and fuels brought into the city. The incentives to do this include the fact that in 2010 JPY 36.3 billion flowed out of the city to regional utilities and other suppliers of energy and fuels. For comparison, the city's total revenues in FY 2013 were JPY 85.16 billion, of which local taxes totaled JPY 20 billion and national subsidies amounted to JPY 36.9 billion.

FIGURE 21.8 Hirosaki city. *Adapted from Hirosaki City. http://www.en-hirosaki.com/access.html.*

Hirosaki City's Smart City Vision, 2017

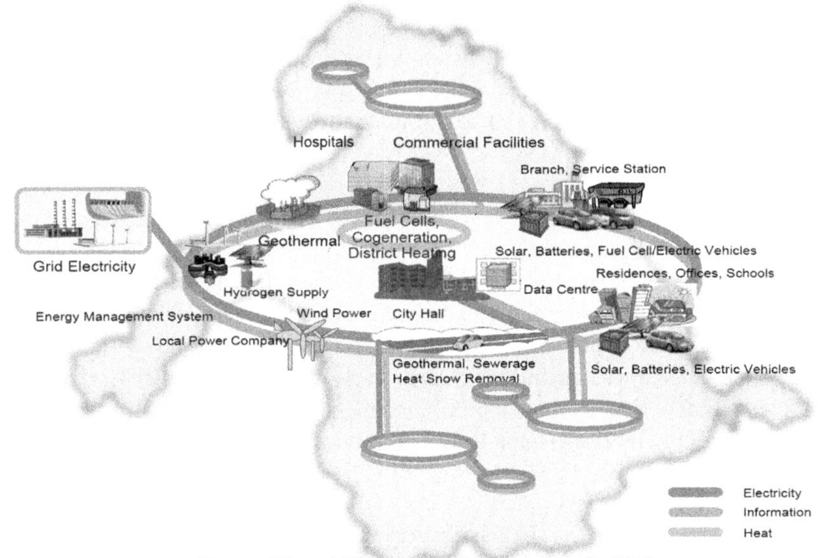

Source: Hirosaki Smart City Concept, March, 2017:
http://www.city.hirosaki.aomori.jp/jouhou/kocho/p_coment/sc-kousou-kaitei-an.pdf

FIGURE 21.9 Hirosaki smart city 2017.

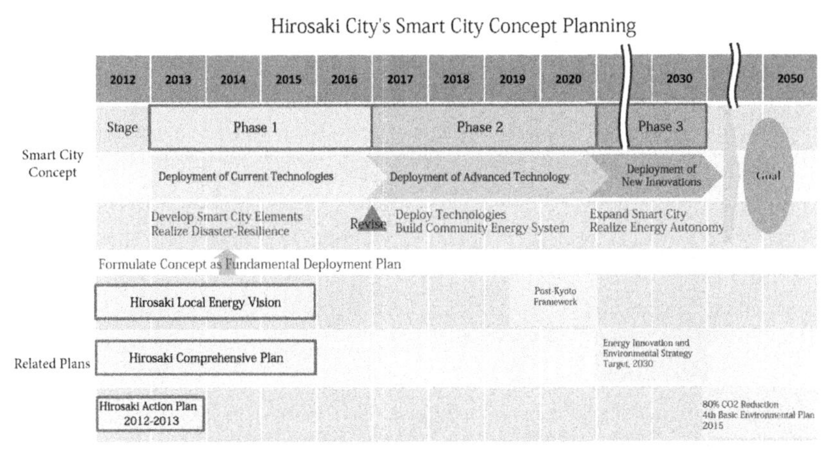

Source: Hirosaki Smart City Concept, March, 2013:
http://www.city.hirosaki.aomori.jp/jouhou/keikaku/sc_koso.pdf

FIGURE 21.10 Hirosaki smart city concept planning.

Similar to Higashi Matsushima, Hirosaki city's smart city plan thus includes the institution of a local power firm, as one means to cut this outward flow of money and keep it in the community. The central government agencies are in support of this ambition, just as they are in all the other smart community initiatives. The fiscal and other support for local-community-led power and thermal initiatives is crucial to the long-term collaborative planning that has evolved in post-3-11 Japan.

Fig. 21.11 shows Hirosaki city's spatial planning, or the "Locational Optimization Plan." This plan is explicitly part of the city's smart city plan. As in virtually all Japanese smart community projects, policy integration between energy and spatial planning has led to robust policies to foster densification, to deal with depopulation and aging in the context of accelerating climate change and other hazards. These plans are already guiding the relocation of hospitals, schools, elderly care, and other public services, a relocation that in turn increases the cost−benefit performance of smart energy networks and inputs at the same time as it reduces energy use. That reduction

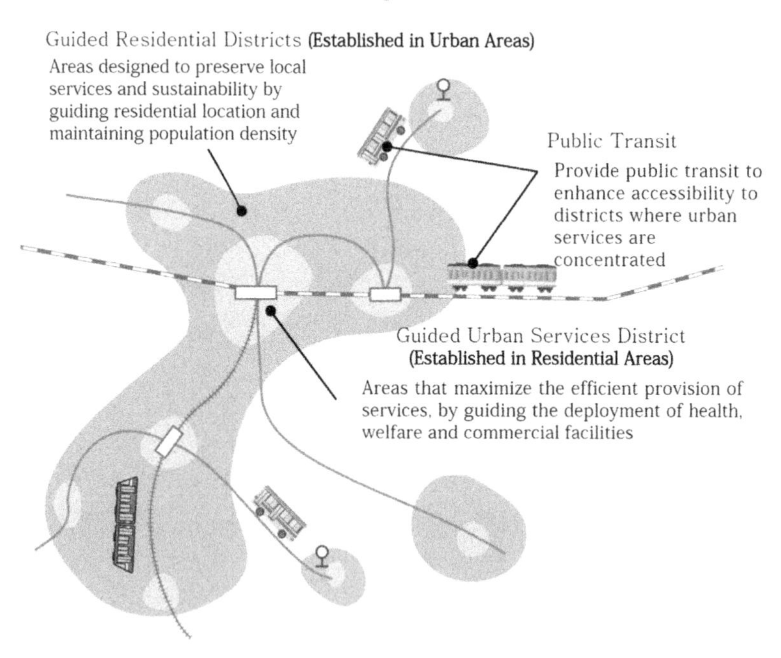

Source: Hirosaki City Locational Optimization Plan, Hirosaki City, October 4, 2016: http://www.city.hirosaki.aomori.jp/oshirase/jouhou/soannzennpenn.pdf

FIGURE 21.11 Hirosaki city locational optimization plan.

in energy use is achieved via the reduced need for motorized (especially single-car) transport in favor of public transit, cycling, and walking. The reduced spatial footprint of the community also leads to less energy used to move water around, plus lower per capita costs to maintain roads and other critical infrastructure, in addition to other energy savings. This integration of spatial planning with energy has been underway over the past three years and has linked most of the central agencies together, enhancing the effectiveness of planning and spending through reducing overlap and other sources of administrative inefficiency.

CONCLUSION

We have seen that Japanese policy makers foster the diffusion of such communities through "whole of government" planning, focused on resilience. The smart communities discussed in this chapter are just a few examples of a paradigm unfolding throughout Japan. Japan's smart communities encompass residential districts, industrial clusters, roadside stations (*michi no eki*), and other areas designated for disaster resilience and local revitalization. These districts center on distributed energy systems, smart and inherently local network infrastructures—for power as well as DHC—that maximize energy efficiency and the uptake of local renewable resource endowments, including solar, wind, biomass, geothermal, and waste heat. The smart city is the linkage of these districts. Japan's projects also stress strong stakeholder engagement, smart communities being a key focal point for Japan's robust and integrated policies for intensifying local revitalization, disaster resilience, compact cities, and the deployment of alternative energy.

Japan's combination of challenges, including disaster threats, dependence on conventional energy, unprecedented demographic change, and declining economic competitiveness, greatly outweighs what other developed countries face. Japanese technocrats are fully aware of these crises and how they interact. Because Japan is rich, technologically sophisticated, and a unitary state, it has the capacity to respond and maximize positive externalities. The competitiveness and resilience that follow from focusing on smart communities could be key to a Japanese economic renaissance.

REFERENCES

ABI Research, April 27, 2017. China, Japan, and South Korea Lead Asian Smart City Initiative Deployments. ABI Research Press Release. https://www.abiresearch.com/press/china-japan-and-south-korea-lead-asian-smart-city-/.

ACEJ, June 2016. Cogeneration White Paper, 2016. Advanced Cogeneration and Energy Utilization Center, Japan (in Japanese).

Amari, J., July 2016. Will smart cities save Japan? The Journal of the American Chamber of Commerce in Japan (ACCJ). https://journal.accj.or.jp/metro-wise/.

Below, B., April 11, 2016. The Case of the Shrinking Country: Japan's Demographic and Policy Challenges in 5 Charts. OECD Insights.

Bruntrock, D., 2017. Prefabricated housing in Japan. In: Smith, R.E., Quale, J.D. (Eds.), Offsite Architecture: Constructing the Future. Routledge, New York.

DeWit, A., December 1, 2014a. Japan's radical energy technocrats: structural reform through smart communities, the feed-in tariff and Japanese-Style 'Stadtwerke'. The Asia-Pacific Journal 12 (48), 2. http://apjjf.org/2014/12/48/Andrew-DeWit/4229.html.

DeWit, A., December 22, 2014b. Japan's "national resilience plan": its promise and perils in the wake of the election. The Asia-Pacific Journal 12 (51), 1. http://apjjf.org/2014/12/51/Andrew-DeWit/4240.html.

DeWit, A., August 1, 2016a. Hioki's smart community and Japan's structural reform. The Asia-Pacific Journal 14 (15), 10.

DeWit, A., July 1, 2016b. "Are Asia's energy choices limited to coal gas or nuclear? The Asia-Pacific Journal 14 (13), 5.

DeWit, A., March 2017a. Japan's fixed-assets tax in context (2). Rikkyo Economic Review 70 (4).

DeWit, A., 2017b. Energy transitions in Japan. In: Lehmann, T.C. (Ed.), The Geopolitics of Global Energy: The New Cost of Plenty. Lynne Rienner, Boulder.

EcoNet Tokyo 62, March 2016. Guidelines for the construction of smart communities, revised edition. EcoNet Tokyo 62 (in Japanese). http://all62.jp/saisei/sc_h27/09_guideline%20refined.pdf.

Edgington, D., 2011. Reconstructing Kobe: The Geography of Crisis and Opportunity. UBC Press, Vancouver.

EIA, February 2, 2017. Country Analysis Brief: Japan. US Energy Information Administration (EIA).

Enerdata, 2016. Global Energy Statistical Yearbook, 2016. https://yearbook.enerdata.net.

Fudeyasu, H., December 1, 2016. Worsening damages from typhoons and intense rain: the approach of the Japan atmospheric association. In: Paper Presented to Second Symposium of the Japan Academic Network for Disaster Reduction. Tokyo. (in Japanese). http://janet-dr.com/07_event/161201/1612100_s1all.pdf.

Fuji Keizai, August 8, 2016. Survey of Energy Management System Markets. Fuji Keizai Press Release (in Japanese). http://www.group.fuji-keizai.co.jp/press/pdf/160808_16064.pdf.

Furuya, K., March 15, 2017. Remarks to the Annual Meeting of the Association of Resilience Japan (in Japanese).

Gotou, M., 2017. Local Economic Development and the Diversification of Energy Supply. AES News No. 8 Winter, (in Japanese).

Honda, May 21, 2014. Sekisui House, Toshiba and Honda Embody 2020 Lifestyle of the Future with Real-world Smart House. Honda News Release. http://world.honda.com/news/2014/c140521eng.html.

IDEA Industry News, May 18, 2016. International District Energy Association and Microgrid Resources Coalition Announce Merger to Promote Wider Adoption of Microgrids. http://www.districtenergy.org/blog/2016/05/18/international-district-energy-association-and-microgrid-resources-coalition-announce-merger-to-promote-wider-adoption-of-microgrids/.

IEA, April 6, 2017. Cities Lead the Way on Clean and Decentralized Energy Solutions. International Energy Agency Press Release. https://www.iea.org/newsroom/news/2017/april/cities-lead-the-way-on-clean-and-decentralized-energy-solutions.html.

JANET-DR Japan Academic Network for Disaster Reduction, December 1, 2016. Worsening typhoons and intense rain: damages and counter-measures. In: Second Symposium of the Japan Academic Network for Disaster Reduction. Tokyo. (in Japanese). http://janet-dr.com/07_event/event13.html.

Japan Cabinet Secretariat, n.d. Building National Resilience. http://www.cas.go.jp/jp/seisaku/kokudo_kyoujinka/en/e01_panf.pdf.

Japan Economic Center, September 23, 2016. 2016 Survey on Current and Project Smart House Markets. Japan Economic Center (in Japanese).

Japan Future City Initiative, 2015. Higashi Matsushima City. http://doc.future-city.jp/pdf/forum/2015_malaysia/Case_Studies_of_Japan_01_04_01_Ctiy_of_Higashi-matsushima.pdf.

JSCA, n.d. Smart Community Development. Japan Smart Community Alliance. https://www.smart-japan.org/english/index.html.

Kashiwagi, T., July 2009. Japan's low-carbon strengths. Journal of Energy Conservation 61 (7) (in Japanese).

Kashiwagi, T., February 2011. The smart city: achieving both economic development and environmental measures. MLIT Shinjidai 71 (in Japanese).

Kashiwagi, T., 2015. Smart Communities: Compact, Networks and Local Revitalization. Jihyo Books, Tokyo (in Japanese).

Kashiwagi, T., 2016. The Super Smart Infrastructure Revolution. Jihyo Books, Tokyo (in Japanese).

Kashiwagi, T., January 27, 2017. Reform of the energy system and the role of district heating. In: Special Presentation to Annual Symposium of Japan Heat Supply Business Association. Tokyo, (in Japanese).

Kashiwagi, T., Hashimoto, N., Kanaya, T., 2001. The Micro-power Revolution. TBS Britannica, Tokyo (in Japanese).

Kashiwagi, T., September 2010. Towards Achieving Both Economic Development and Environmental Countermeasures: The Smart City Perspective. In: New Era, vol. 71. Ministry of Land, Infrastructure, Transport and Tourism, Japan (in Japanese).

Kashiwagi Takao, 2015. The Cogeneration Revolution. Nikkei BP Consulting (in Japanese).

Kono, H., 2017. Community Energy. Chuokoronsha, Tokyo (in Japanese).

Maras, E., May 2, 2017. Digital signage powers smart cities. Digital Signage Today.

METI, August 1, 2011. Report of the Research Commission on Urban Planning and Integrating the Effective Use of Heat Energy. Ministry of Economy, Trade and Industry, Japan (in Japanese).

METI, April 2016. The Energy Innovation Strategy (in Japanese).

MLIT, July 7, 2017. The government's approach. In: Materials Presented to 8th Meeting of Deliberation Commission on Overseas Port Facilities Distribution Project. Ministry of Lands, Infrastructure, Transport and Tourism, Japan (in Japanese).

MOE, October 6, 2016a. Town Planning for Local-production/local-consumption and Disaster Recovery. Ministry of the Environment (in Japanese). http://www.env.go.jp/press/y0618-05/mat05_1.pdf.

MOE, 2016b. Asia Low-carbon Cities Platform. Ministry of the Environment, Japan. http://lowcarbon-asia.org/english/portal.html.

Murakami, K., January 27, 2017. The advent of deregulation: rethinking the attractiveness of district heating. In: Keynote Speech to Annual Symposium of Japan Heat Supply Business Association. Tokyo. (in Japanese).

NIES, September 2016. Environmental Consciousness Survey. National Institute for Environmental Studies, Japan (in Japanese). https://www.nies.go.jp/whatsnew/2016/jqjm10000008nl7t-att/jqjm10000008noea.pdf.

NRPO, August 2017. An Outline of the Fiscal 2018 Budget Request for National Resilience. National Resilience Promotion Office, Japanese Cabinet Secretariat (in Japanese).

OECD, 2016. Territorial Reviews: Japan 2016. OECD, Paris.

Oguro, Y., August 15, 2014. Smart Community: The Sustainable City. Daiwa Institute of Research (in Japanese). http://www.dir.co.jp/research/report/esg/esg-place/esg-municipality/20140815_008861.pdf.

Sakamoto, S., October 19, 2015. Special depreciation. In: Presentation to Japanese Tax Research Centre (in Japanese).

Sayer, P., March 19, 2017. Japan looks beyond Industry 4.0 towards Society 5.0. PC World.

Sekisui House, 2016. CSV Strategies, 2016. https://www.sekisuihouse.co.jp/english/sr/datail/__icsFiles/afieldfile/2016/07/29/p22-36.pdf.

Ikeda, S., Toshimitsu, K., Kenshi, B., Tsuneyoshi, M. (Eds.), 2016. Water and Landslide Hazard and Countermeasures under Climate Change. Kindai Kagakusha, Tokyo (in Japanese).

SPECIAL, April 2013. Energising Europe's Planners: New European Energy Project Springs into Action. Press Release, Spatial Planning and Energy for Communities In All Landscapes.

Stern, N., 2015. Why Are We Waiting? The Logic, Urgency, and Promise of Tackling Climate Change. MIT Press.

UNEP, February 25, 2015. Modernizing District Energy Systems Could Reduce Heating and Cooling Primary Energy Consumption by up to 50% Finds New Report. http://www.rona.unep.org/news/2015/modernizing-district-energy-systems-could-reduce-primary-energy-consumption-heating-and.

Wood, E., December 17, 2015. Hitachi moves into the north America microgrid market with 100-year plan. Microgrid Knowledge. https://microgridknowledge.com/hitachi-north-america-microgrid/.

FURTHER READING

Hirosaki City, 2017. Hirosaki-style Smart City Concept (in Japanese). http://www.city.hirosaki.aomori.jp/jouhou/keikaku/2014-1212-0934-50.html.

Kashiwagi Takao, 2014. Smart Communities: A Smart Network Design for Local Government Infrastructure. Jihyo Books (in Japanese).

MLIT, June 1, 2015. Concerning the Revised Law on Special Measures for Urban Renewal. Ministry of Lands, Infrastructure, Transport and Tourism, Japan (in Japanese).

Morse, R.A., 1981. Introduction: Japan's energy policies and options. In: Morse, R.A. (Ed.), The Politics of Japan's Energy Strategy: Resources, Diplomacy, Security. Institute of East Asian Studies, University of California. http://digitalassets.lib.berkeley.edu/ieas/IEAS_03_0002.pdf.

Nakazato, K., December 13, 2013. National Resilience Begins, with the Passage of the Basic Law. Daiwa Institute of Research (in Japanese). http://www.dir.co.jp/research/report/capital-mkt/20131213_008009.pdf.

Tomigahara, N., July 29, 2016. Infrastructure exports in the purview of the JPY 30 trillion target by 2020. Market Weekly (886) (in Japanese).

Chapter 22

Sustainable Communities in Costa Rica

Gerardo Zamora
Anthem Software Solutions & Services (S^3), San Jose, Costa Rica

Chapter Outline

INTRODUCTION

Searching for Hope

Millions of acres are destroyed (deforested) every year around the world. About 77 million acres a year. The main cause, in my opinion, is "greed," corporate or simply human greed. Businesses have to thrive, some people have to make fortunes, and others are bound to make a mere living wage if not less. Our waste is enormous, and carelessness knows no limits; the raping of the Earth is a concerted effort by all of us! Forgive me, for saying "us," but it is true. To some degree we all participate. Some more than others who are indifferent and choose to be blind. Others, because we cannot possibly isolate ourselves from the products of the modern world and somehow indirectly, at least, we help by our mere existence to destroy parts of the Earth. I am one of them! There are billions of people like me. That sure is not a good thing, but is a fact. The concerted destruction by irresponsible corporations and

Sustainable Cities and Communities Design Handbook. https://doi.org/10.1016/B978-0-12-813964-6.00022-7

governments, I think, produces the most damage as it is fuelled by vast amounts of money, power, and consumerism.

Luckily, there is some oasis of hope in the world. There are some entire countries and some very large areas within countries where people have chosen to protect and preserve for the benefit of future generations of humans, animals, plants, and the earth itself. I wish each country of the earth was in this list (perhaps they all are?), but I am going to concentrate on Costa Rica as it has been one of the leaders in sustainable living for many years. Costa Rica is one of the best democracies in the world, primary education is obligatory, there is free education at preschool and high-school levels, there is no army and no recreational hunting, and the sanitary standards are high.

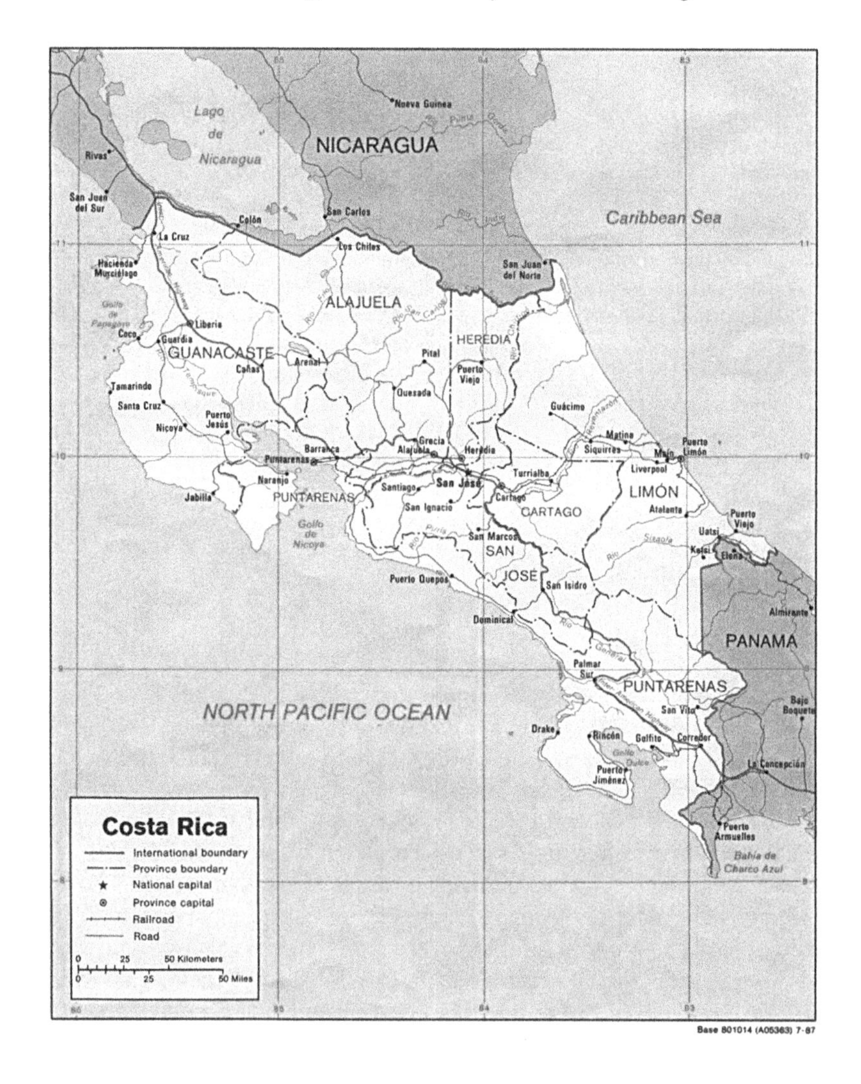

Costa Rica is located in the Central America isthmus, with Nicaragua to the north, Panama to the south, the Pacific Ocean to the west, and the Atlantic Ocean to the east.

The above figure shows the Coat of Arms. The colors of the Costar Rican flag are shown below.

A Jewel in Latin America

Costa Rica stands out as one of the developing tropical countries for its commitment toward environmental and natural resource issues. In my opinion one of the best things the government of Costa Rica has done is to decentralize the decision-making process for sustainability to improve the management of biodiversity conservation by local groups. This decentralization has helped expedite work of projects needed by the local regions, minimized corruption as the local regions are very much interested in the benefits for that area, and has made the formation of alliances with foreign organizations more easy as less "red tape" is needed and the foreign organizations may be very much interested in just that one region. Despite the fact that hotel constructions are rampant in some areas and harming the environment, Costa Rica is a good example for developing countries that economic well-being can be compatible with forest preservation.

The Importance of Sustainability

Historically deforestation has been a cancer in Costa Rica. It started after 1945 and then peaked during the 1960s when the US government offered cattle ranchers millions of dollars in loans to produce beef for US consumption. Then in the 1990's Costa Rica had one of the worst deforestation rates in Central America. Costa Rica has successfully managed to diminish deforestation from one of the worst rates to zero achieved in 2005 (note that this report differs from the table report for 2005). I do not believe that zero is an attainable goal, but small numbers would be ideal. Unfortunately, the table even shows an increase from the year 2000!

Year	Forest Cleared (acres)
1977	128,495
1983	107,614
1985	103,784
1987	79,074
1991	44,348
1996	44,479
2000	7,495
2005	11,705

The response of Costa Rica, whether private local groups or government, or even foreign organizations interested in helping the environment obtain sustainable living conditions, has been incredibly good and effective. The programs are particularly ambitious, and with the country having a high level of biodiversity and ecozones that are relatively small much progress has been made to reforest it. One program that has helped minimize deforestation is the National Bamboo Project of Costa Rica founded in 1986. What they do is to plant indigenous giant bamboo, which is then used instead of timber as the primary building material at low cost for Costa Rica's rural poor. Another area that has helped and is helping Costa Rica in sustainable living is "ecotourism," that is, the ability to raise revenue through tourism while at the same time protecting the forests, lakes, rivers, and seas. There are various programs such as recycling, trash pick up, walking only in trails, limiting the number of people that can visit a given area, some of which involve participation of the tourist in preserving the environment.

A Model for Other Areas

The Arbofilia Asociacion Protectora de Arboles, one of the many groups that keeps an eye and takes action to protect the environment, has an ambitious project to plant trees along the "corridor." This "corridor" stretches from the coastline close to the Carara National Park up to the Potenciana Mountain Range and peaks with Turrubares Mountain. They plant trees in a relatively small region, about 12 miles of amazing biodiversity.

What follows is a report on many areas of interest in Costa Rica, some for their projects in sustainability and the others for the attractions that they pose to the tourists. By all means this is not a complete list. There are many more areas in this relatively small country.

SUSTAINABLE COMMUNITIES AND AREAS OF INTEREST

Northern Areas

Cano Negro refuge is a 10,000-acre area that is a remote tropical everglade with a large network of rivers and wildlife. Hundreds of bird species, river turtles, caimans, jaguars, ocelots, giant anteaters, monkeys, and many more animals are found here.

Northwestern Areas

One of the major ecotourism destinations is Monteverde. It is located in Cordillera de Tilaran at about 4- to 5-h drive northwest from San Jose.

Monteverde Cloud Reserve otherwise known as El Bosque Eterno de los Niños (The Eternal Rainforest of the Children), Bosque Nuboso, Bajo Del Tigre, and Reserva Santa Elena. It is interesting that a major portion of this area was founded by American Quakers, and that El Bosque Eterno de los Niños was founded by a project from schools and children from all over the world. Who else but the kids can we dedicate this gardens to? This area is visited by thousands of tourists each year, who not only take care not to destroy the environment, but also enjoy seeing hundreds of different species of thousands of animals and plants. The University of Georgia owns acres in this region and runs a satellite campus. University of Georgia Costa Rica has study-abroad programs, ecological and forestry research, ecotourism via its on-campus lodging, various conservation and sustainability initiatives, and reforestation efforts. Fortunately for humankind Costa Rica lends itself very friendly to sustainable living. There are many kinds of models for all, and people can create their ideal living achieving more with less to realize their dreams. There are many sustainable homes, businesses, and communities with infrastructures to support them. A trend in Costa Rica real state is the creation of communities dedicated to green living. Most of them are in rural areas helping to provide healthy lifestyles for residents to live in harmony with the natural surroundings and with the local culture. Some are low-budget commune style with simple installations, whereas others are expensive with elaborate homes and infrastructures. What makes these communities different is the commitment to save energy and use renewable energy sources. These communities work on growing organic foods; have free range animals; support local merchants by buying locally; participate in the social and economic life of nearby communities, supporting schools, and local projects; etc. One can say that these communities are organized by like-minded spiritual, religious, and philosophical individuals, but the most important shared value is the commitment to protecting the environment and sustainable living in their daily lives.

Pacha Mama is found in Guanacaste, just North of Nosara. It is considered a spiritual ecocommunity for inner transformation. There are programs for meditation, emotional healing, physical rejuvenation, ancestral wisdom, biodiversity and sustainability, as well as musical journeys. It is a nonprofit center for international gatherings. This spiritual eco-community is spread over an area of 500 acres; whole families live there and there is a school for children.

Northeastern Areas

Siquirres is a town east of San Jose and close to Limon. This is of importance because it leads to other parks and reserves such as Tortuguero National Park to the north by boat. This whole region is dotted with ecological farms.

The city of Turrialba is on the way to Siquirres, and Limon is a beautiful countryside town with many ecofarms. There is an abundance of cattle ranches, milk processing plants, and acres of land for growing arabica coffee, sugarcane, and deliciously large *Macadamia* nuts. A very unique feature of the town is that it manufactures Rawlings baseballs used in major league games.

Parque Nacional Tortuguero has high rainforests to marshy lowlands, long stretches of beach, and canals, and it is the most important breeding grounds for the green sea turtle.

Central Areas

Rancho Mastatal is located $2^1/_2$ hours west of San Jose in the mountains of Puriscal. It spans an area of 300 acres and is considered "an education center, working permaculture farm, lodge and community rooted in environmental sustainability, meaningful, place-based livelihoods, and caring relationships." This is a place to learn about sustainable living by just being there and participating. Living in cooperation with the local community they write, "The people of Mastatal are involved in every phase of our projects. They are greatly responsible for its success. We are honored to work and live amongst and share with such special people." I am sure they mean it, for only people who want to live in concert with nature, and taking care of it, do love to be in cooperation with people of their communities.

La Ecovilla is about 1 h west of San Jose. Several houses have been built with local wood. It is a community with people from many different countries. The houses are designed for living with low impact on the environment. They are solar powered, with solar water heater, black water biodigestor, gray water irrigation, habitat for wild life and share community resources. This is a good example of perhaps expensive homes but with ecology conservation in mind.

Tacotal collective is located west of San Jose in the Avocado Mountains of Costa Rica. It is a permaculture-based community ecovillage consisting of diverse and passionate people who are artists, musicians, builders, tradesmen, scientists, and educators. Their land is host to permaculture courses, natural building workshops, biological research groups, bird watchers, and community of musical gatherings. From their website, "...The linear growth model that currently dominates our societies has falsely oriented us in time, and created an illusion of separation and independence from the natural world. Understanding that our survival depends on a healthy Earth ecosystem, a sustainable community works to re-integrate with nature, and acts as a part of the whole, a cell in the Earth organism. We all have indigenous roots, ancestors who lived sustainably for countless generations, but many of us have completely lost contact with our ancestors' traditional knowledge. The Indigenous peoples around the world that have maintained their cultures and traditions present inspiring models of sustainable living, in which all physical and spiritual nutrition is sourced from the natural environment...our search is for those projects and communities that are working towards sustainability in a holistic way, knowing that these models and examples can help catalyze a greater movement towards sustainable living on a mass scale."

FINCA
TACOTAL

Southwestern Areas

Portasol is a sustainable community located in the mountains of the Central Pacific of Costa Rica. It is close to Manuel Antonio and Dominical beaches. This is a luxurious gated community built causing least damage to the environment and helping the local economy.

Tierra Verde is in the southern Pacific rim south of Quepos and 55 km from San Isidro del General (Perez Zeledon). It a "model of sustainable ecological development and design." The designers are careful to slow water runoff and allow percolation to put water back into the aquifer. As for the land, Tierra Verde is using ecological landscaping and habitat restoration programs. They are undergoing a reforestation project to restore the land to its natural state over time. The designers of Tierra Verde are well aware of preserving the biodiversity that exists in this region for the enjoyment of the residents.

Southern Areas

Las Alturas Del Bosque Verde is a 33,000-acre preserve near Panama border and the slopes of the Talamanca Range. This area possesses a rich biodiversity where monkeys, large cats, tapirs, hundreds of bird species, insects, and plants abound. There are a few indigenous people living here, and some work at the preserve. Las Alturas is about 60% virgin rain forest continuing with the conservation efforts set forth by the owner and the property manager. Wild cats thrive at Las Alturas. The Las Alturas Research Facility is run in cooperation with the Stanford University and others throughout the world. It supports graduate and postgraduate research on tropical ecosystems.

Punta Mona Center for Sustainable Living and Education is an 85-acre beachfront, family-owned, environmental education center, botanical collection, established organic permaculture farm, and an ecolodge, dedicated to regenerative ways of living. It is located in the Caribbean side, close to Manzanillo and the border with Panama.

THE FUTURE OF SUSTAINABLE COMMUNITIES IN COSTA RICA

Cultural Movements and Governmental Programs

Costa Rica as a nation is well aware that waste and rampant deforestation is no good for the economy, the country's morale, or the its growth. They have experienced this first hand. In 1945 75% of the country was covered with luscious forests. In 1987 this number dropped to 21%, and soon thereafter the leaders began to see that the replacement of "unproductive" land (the forest) for "productive" land (agriculture) was making the land infertile, leaving the farmers and investors scrambling. The clean air and water was becoming polluted. The leaders began to see the benefits of retaining healthy ecosystems to promote tourism, recreation, and health care. They implemented policies that promoted conservation and ecotourism, established more national parks, and encouraged organic farming. At any time and in different communities throughout Costa Rica you will see many programs taking place to better the environment, social conscience, and economic health.

A brief description of some of the major programs in Costa Rica is as follows:

> Certification for Sustainable Tourism (CST) aims at helping Costa Rican businesses take a long-term perspective on maintaining the country's environment, culture, and communities. The CST program rates businesses based upon how well they comply with certain sustainable practices.
> The Bandera Azul Ecological Program helps Costa Rica's communities and the natural areas that surround them to stay healthy, safe, and clean. This program encourages environment-friendly practices and active civic engagement.
> The Payments for Environmental Services program provides payments to landowners who maintain healthy, robust land. The National Emergency Commission is a branch of the central government that coordinates the prevention of risks and the response to emergency situations.

CONCLUSION

A Strong Light in the Horizon

That a country, as a whole cultural movement, cares about itself and the future of its citizens, and in addition takes steps to do better as far as sustainable living is concerned is to me a very strong hope. The right- and left-wing sentiments are battling for the future of a nation—the ones who want to exploit nature for the money they can make and the ones who want to protect the natural resources to enjoy them and live a good life. We can all join the ones who care, want to cooperate, and take into account all resources be they human, animal, plant, or mineral. There is always hope that the right-winged mind-sets see the light more clear.

Healing in All Directions

"At the deepest level of ecological awareness you are talking about spiritual awareness. Spiritual awareness is an understanding of being imbedded in a larger whole, a cosmic whole, of belonging to the universe." Fritjof Capra. There is no bigger whole to which we can relate than our beloved Earth. Since day one when creatures and plants appeared it has given us nutrients and sustained our lives (sometimes destroying us!). Some affectionately call it Mother Earth. We are one with her. There is no other home. As we seek healing when feeling ill or in disarray, the Earth heals in many ways. As ignorant children, sometimes we chose not to care for her. It is not much we have to do to care for her. All it takes is that we be aware that this planet is precious beyond anything we know and take steps to live simpler caring lives for all inhabitants be they humans, animals, or plants. Say no to the insidious habit of consumerism.

Ticos y Ticas Mobilizing

The ticos and ticas as they call themselves in Costa Rica are very proud of their country, and collectively they have taken steps to become more environmentally aware. There is much work to be done, but culturally there is this motivation to do better and care at least for the country's environment, if not the whole Earth. The people are clean and care for themselves; they want to eliminate waste and pollution. When you come to Costa Rica, become a Tico or Tica, feel it in your bones! Remember the land, air, and waters are from the Earth...you are at home! Pure Life!!

FURTHER READING

Wikipedia (Costa Rica).
Preserveplanet.org.
Rainforest Alliance.
Earth Action Network.
Arbofilia.net.
Cedarena.org.
Mongabay.com.
Pachamama.com.
Ranchomastatal.com.
Laecovilla.com.
Tacotal.org.
Portasol.cr.
Tierraverdeuvita.com.
LasAlturas.com.
Anywhere.com/costa-rica/.
Capra, F., 1988. Uncommon Wisdom—Conversations With Remarkable People.

Chapter 23

Sustainable Development Cases in Africa

Samantha Bobo
Rice University, Houston, TX, United States

Chapter Outline

AFRICA

Egypt

Located in Northern Africa, Egypt occupies an area of 1,001,450 km^2 bordering the Mediterranean Sea between Libya and the Gaza Strip. With a population of just under 100 million people as of July 2016, Egypt is the most populous country in the Arab world, and the third most populous in Africa. However, given the country's terrain consisting of mostly desert, 95% of the total population lives along River Nile, occupying only 5% of the total land area. Having such limited space for a population growing at about 2.5% annually requires significant and innovative infrastructure and sustainable development, especially in the urban centers such as Cairo and Alexandria (CIA World Fact Book, 2017a).

Sustainable development in its broadest sense requires the interplay of the environment, people and the economy. Looking back at Egypt's development we see efforts to promote social and economic welfare, although their attempts to promote environmental welfare have fallen flat. Historically it was not the lack of environmental legislation or planning that hindered environmental policy in Egypt, but rather the weak regulatory compliance, weak enforcement, and lags in execution. In the postrevolutionary period of the 1960s, there was antigrowth sentiment within the country that resulted in the socialization of the Egyptian economy. The government's laws created an inwardly oriented

Sustainable Cities and Communities Design Handbook. https://doi.org/10.1016/B978-0-12-813964-6.00023-9

development strategy that resulted in nationalization of private enterprises throughout the economy. The government worked to increase domestic industry and production, as well as to increase social programs, distribution of wealth, and education. The economic and social development programs of the time included price controls and a social focus on issues such as health, housing, water and sanitation, employment, and social security. However, the push for extensive industrial development resulted in significant environmental degradation in spite of extensive legislation.

> *State owned industrial enterprises with massive investments in Soviet-style industrial plants were given water and electricity virtually free of charge. These distortions in the prices of production inputs (water, energy, wastewater, etc.) provided little economic incentives for State dominated enterprises to rationalize the use of their resources, and resulted in wasteful polluting industrialization.*
>
> Wahaab (2003).

To both the government and the people, the most pressing problems were poverty and meeting basic human needs. The modern concept of environmental awareness was nonexistent if it did not have a direct impact on health (Wahaab, 2003).

In the 1970s, Egypt began economic liberalization and encouraging private entities and foreign investment. At the same time, it maintained socialist welfare policies and their policy of import substitution to benefit domestic production. During this time, there was rapid urbanization within Egypt, and as the population exploded, so did the demand for transportation, housing, food, services, and industrial production. As the country worked to keep up with both urbanization and the services required by the growing population, environmental management proceeded similar to that of the 1960s, in spite of the new regulations enacted by the government. The general thought was that the waste generated would be assimilated into the environment easily and without consequence (Wahaab, 2003).

Egypt began taking steps toward sustainable development in 1980 with the establishment of the Ministerial Committee for Environmental Affairs as well as the Egyptian Environmental Affairs Agency (EEAA) to be the coordinating body for environmental policy making. By 1985, the country was in the middle of an economic crisis spurred by rising interest rates and oil price collapse. Soaring inflation and unemployment from decreased capital inflows, as well as pressure from the World Bank and the International Monetary Fund, led the government to begin sweeping economic reforms in 1992.

> *The economic reform program included: deregulation of interest rates and the foreign exchanges regime, reduction of government spending through gradual removal of subsidies, implementation of a privatization program, introduction of a new capital market, the abolishment of investment licensing, and the revision of the trade regime.*
>
> Wahaab (2003).

In spite of the hard economic times, Egypt formally announced its intentions to achieve sustainable development and reformulation of its environmental policy. Realizing that rapid growth in both population and urbanization had severely damaged the country's limited natural resources the government worked to ameliorate the previous lapses in regulatory power by passing legislation, Law 4/1994, granting the EEAA a broad base of regulatory controls over areas from air and water pollution to discharge control and hazardous substance/waste management. In addition to regulatory power the EEAA was also granted the power to gather environmental information and implement environmental education programs. In addition to specific standards for air quality, waste management, and coastal zone management the same legislation also laid out a specified adjustment period for existing industrial firms to become compliant with the said standards and the last-resort penalties for those who failed. Egypt finally had a well-outlined policy for environmental protection and a strongly supported regulatory entity to execute it. Although it was not without its flaws, Law 4/1994 marked the first step in policy toward the goal of sustainable development (Wahaab, 2003).

When the United Nations (UN) updated their Millennium Development Goals in 2015 to include a broader sustainability agenda and address the root causes of poverty, Egypt launched Vision 2030 with the statement:

> *The new Egypt will possess a competitive, balanced and diversified economy, dependent on innovation and knowledge, based on justice, social integrity and participation, characterized by a balanced and diversified ecological collaboration system, investing the ingenuity of place and humans to achieve sustainable development and to improve Egyptians' life quality.*
>
> Egypt Vision 2030 (2016).

Egypt's Vision 2030 is structured in the same way as the UN's Sustainable Development Goals (SDGs) (formerly Millenium Development Goals), addressing the social, economic, and environmental dimensions that characterize sustainable development (UNDP, 2016). The following charts outline the three pillars (People, Planet, Profit) that characterize Egypt's Vision 2030, as well as the subcategories within each pillar and the stated goals pertaining to the subcategories. First, within the People pillar of sustainable development, Egypt turns its focus to social justice, health, education and training, and culture (Egypt Vision 2030, 2016).

People Pillar	**Goals by 2030**
Social Justice	Society that: • Is fair and interdependent • Characterized by equal economic, social, and political rights and opportunities realizing social inclusion • Supports citizens' right in participation • Provides protection and support for marginalized and vulnerable groups

Health	Health care that is: • Integrated, accessible, high quality, and universal • Offers early intervention and preventative coverage • Allows Egypt to be a leader of health care services and research in the Arab world and Africa
Education	An education and training system: • Available to all without discrimination • With an efficient, just, sustainable, and flexible institutional framework • That empowers students and trainees to think creatively • That contributes to the development of a proud, creative, responsible, and competitive citizen who accepts diversity and differences
Culture	A Culture that: • Respects diversity and differences • Allows citizens access to knowledge that will build their capacity to interact with modern developments while recognizing their history and heritage • Gives citizens freedom of choice and cultural creativity • Adds value to the national economy by representing Egypt's soft power at regional and international levels

Reproduced from Egypt Vision 2030, 2016. Sustainable Development Strategy: Egypt Vision 2030. Retrieved from: http://sdsegypt2030.com/?lang=en.

The second pillar in sustainable development, and in Egypt's Vision 2030 pertains to the planet. Under this pillar, the Egyptian government outlines its goals for the environment and urban development.

Planet Pillar	**Goals by 2030**
Environment	• Preservation of natural resources and efficient use and investment to preserve resources for posterity • Clean, safe, healthy environment leading to diversified economic activities, supporting competitiveness, providing new jobs, eliminating poverty, and achieving social justice
Urban Development	• Balanced spatial development • Management of land and resources to accommodate the population and improve the quality of life

Reproduced from Egypt Vision 2030, 2016. Sustainable Development Strategy: Egypt Vision 2030. Retrieved from: http://sdsegypt2030.com/?lang=en.

Last is the Profit/Economy pillar that focuses on economic development, energy, scientific research and innovation, and transparency and efficiency of government.

Profit/Economy Pillar	**Goals by 2030**
Economic Development	An economy that is: • Balanced, knowledge-based, competitive, diversified, and market driven • Characterized by a stable macroeconomic environment

	• Capable of achieving sustainable, inclusive growth • An active global player, responsive to international developments • Generating jobs, maximizing value added, and increasing GDP per capita
Energy	An energy sector that: • Meets national sustainable development requirements • Maximizes the efficient use of traditional and renewable resources • Contributes to economic growth, competitiveness, social justice and environmental preservation • Is a leader in renewable energy and efficient resource management • Is innovative, adaptable, and compliant with the UN SDGs
Scientific Research, Knowledge, Innovation	• To be a creative and innovative society producing science, technology, and knowledge • To ensure the developmental value of knowledge and innovation using the outputs to face challenges and meet national objectives
Transparency, Efficiency of Government Institutions	A public administration sector that is: • Efficient and effective at managing state resources with transparency, flexibility, and fairness • Subject to accountability, maximizing citizens' satisfaction and is responsive to their needs

GDP, gross domestic product; *UN SDG*, United Nations Sustainable Development Goals.
Reproduced from Egypt Vision 2030, 2016. Sustainable Development Strategy: Egypt Vision 2030. Retrieved from: http://sdsegypt2030.com/?lang=en.

The two repeating themes within all parts of the vision are the focus on human capital and economic innovation. There is constant emphasis on increasing access to education and training, as well as job creation, with a concerted effort made for youth and women empowerment. Success in this area will create qualified planners, civil servants, and professionals that are integral to realizing Vision 2030. Focus on sustainable youth development is vital for a country where 25% of the population is under the poverty line and 60% of the population is under the age of 30 (CIA). Egypt's problems with youth unemployment have historically been linked to lack of training, which has caused complaints from the private sector regarding quality of training and skills of Egyptian workers. Access to education and upgrading that education are necessary to reduce the disenfranchised youth, while nongovernmental organizations (NGOs) can help bridge the representation gap between the youth and the government (Mansour and Mansour, 2016). The year 2016 was declared the Year of the Youth, and it kicked off with several projects targeting at-risk youth and families. The first was the Social Housing Programme to improve affordability of housing for low-income families. The government has also instituted Takaful and Karama (Solidarity and Dignity) social protection programs that provide cash transfers to impoverished families with children who are both enrolled in school and have regular medical checkups

(UNDESA, 2016). Another proposed project that will provide both social and economic benefits is the "King Salman bin Abdulaziz Bridge" that would serve to connect Egypt and Saudi Arabia across the Strait of Tiran. Construction of the causeway could result in job creation for the Egyptian population as well as potential for Saudi Arabian investment in the Egyptian economy. Construction of this magnitude is not without environmental concerns, especially for the reef ecosystems and the threatened Red Sea dugong population; however, environmental NGOs such as the Hurghada Environmental Protection and Conservation Association have stated that they would be placated by a thorough environmental impact assessment, which has been required by both the Egyptian government and the EEAA since passage of Law 4/1994 (Wahaab, 2003; Walker, 2013; Al Jazeera, 2016).

Vision 2030 also emphasizes economic growth and innovation. The government is emphasizing a customizable, knowledge-sharing, digital economy and sees science, technology, and innovation as key to economic growth. To accomplish this, it is starting from ground-up by creating a legal and regulatory framework and financial instruments to support this futuristic, sustainable economic model. And to fuel the new economy, Egypt is maintaining the strategy set forth in 2008. This mandates that 20% of generation be from renewable sources by 2020 with wind power representing 12% of the total electricity generation (~7200 MW). The 7200 MW capacity will be reached through two paths:

1. "State-owned projects implemented by the NREA with total capacity of 2375 MW (represents 33% of total installed capacity), financed through governmental agreements.
2. Private sector projects with total capacity of 4825 MW (represents 76% of total installed capacity). Policy of increasing the participation of private sector will include two phases:-
   a. Phase I: Adopting Competitive Bids approach as the Egyptian Electricity Transmission Company will issue tenders internationally requesting private sector to supply power to build, own, operate wind farms and selling electricity for the company with price agreed upon between the company and the investor.
   b. Phase II: Application of Feed-in-tariff system, taking into consideration the prices and experience achieved in phase I (NREA, 2016)."

The government has also established incentives to stimulate private sector participation, particularly in wind generation, which includes the aforementioned competitive tender and bilateral agreements, and long-term Power Purchase Agreements (PPAs) (20–25 years). Additionally, the government of Egypt will guarantee all financial obligations under the PPA and investors will benefit from selling certificates of emission reduction resulting from project implementation. Other incentives include a carbon tax credit and exemption of project equipment and spare parts from customs duties and sales tax (NREA, 2016; Fulbright, 2013).

Egypt itself has significant renewable energy potential. Most of the country's hydroelectric potential is already built out with the Aswan Dam providing about 15,000 GWh per year in generation. In contrast, the use of solar energy has been disproportionately low for the available resource. Egypt receives between 2000 and 3200 kWh of solar radiation per square meter annually. In spite of the abundance, solar energy has been slow to develop it was more expensive than traditional thermal generation. In 2010, there was only one large-scale solar project in operation, an installation as part of a combined cycle power plant, where solar produced 20 MW of the 140-MW plant in Kuraymat. The Egyptian government clearly views the renewable sector within the country being driven by wind energy, with notably good resources in the Gulf of Suez and on both banks of River Nile. The Zafarana district is Egypt's most developed wind region with a total installed capacity of 550 MW. A second, large-scale onshore wind installation is in the commissioning phase near the Gulf of El Zayt. With financing from the European Union, the 200-MW project consisting of 100 turbines will be the largest wind farm on the African continent and reduce carbon emissions by 400,000 tons annually. Currently, the project is scheduled to be completed by 2018; however, investor uncertainty remains in the wake of Jan. 25 Revolution, so development has slowed. Even so, the government incentives for private investment and a push for transparency in legal and regulatory frameworks under Vision 2030 are good steps to spur the development of a successful renewable energy sector (Fulbright, 2013; EU, 2017; ESI, 2015).

Zooming into the urban centers, the capital city of Cairo is ranked 99th overall in the ARCADIS Sustainable Cities Index, primarily due to its social welfare institutions. Cairo has now begun to focus on the other areas within the sphere of sustainable development, particularly solid waste management (SWM). Historically, Cairo and the other urban centers in Egypt have struggled with effective and sustainable waste management. Since the 1950s, waste management was handled by the poor working class people referred to as the Zabbaleen. They would take the waste to their homes and turn it into anything from quilts, to rugs, paper, pots, livestock food, compost, and recycled plastic products. The Zabbaleen were experts in efficiency and recycling, reusing around 85% of what they collected. To put that in some perspective, Western recycling systems, under optimal conditions, are able to recycle 70% of the material.

In spite of the Zabbaleen, as well as a government waste management system, Egypt still had a waste problem. In the early 2000s, only about 60% of Egypt's total generated waste was being collected by the two systems (40% by the Zabbaleen and 20% by government waste management). The waste problem was so bad that it was affecting the country's much needed tourist revenue, so in 2003, the government announced a plan to "modernize" Cairo's waste management. To execute the modernization, the government contracted three European waste management companies to displace both the previous government system as well as Zabbaleen's waste collection, which they viewed as "unsanitary and backwards" due to their practices of hand sorting even biohazardous waste. The "modernized" system collapsed after a year primarily due to the Zabbaleen's competitive and comparative advantages in waste collection. The European compacting trucks meant that only 20% of the waste collected could be recycled, which was significantly less efficient than the Zabbaleen. After having sunk US $50 million into the failed modernization of Cairo's waste management and as waste began to once again accumulate on the city streets, the government realized that the proper system for the area would not be adopting the techniques of a foreign entity, but rather a unique solution that would necessarily include the Zabbaleen (DAC, 2014).

In 2013, the Egyptian government decided to establish an Integrated Solid Waste Management Sector under the Ministry of State for Environmental Affairs, which would be responsible for overseeing and implementing the National Solid Waste Management Program. The goal of the new program is "the protection of public health, environment and quality of the living environment for Egyptian citizens through sustainable development of waste management practices," through reform of the SWM sector by way of both policy and related infrastructure. The national waste management policy itself will be characterized by the following:

- "Self-sufficiency: A network of services and facilities is required to ensure that all wastes generated are properly managed;
- Waste management hierarchy: Certain waste management practices should be prioritized over others;
- Proximity principle: Waste should be managed as close as possible to the source of its generation;
- Principle of recognition: Waste management and recycling is an important professional sector, and major future employer of skilled, semi-skilled and unskilled workers; and
- Polluter pays principle: Those who manufacture products which lead to waste, and those who generate waste should be responsible for paying the costs for its appropriate management."

To execute these wide-sweeping reforms the Government of Egypt has acknowledged the need for involving all stakeholders in the SWM sectors, especially the formal and informal private sectors, specifically the Zabbaleen. The following table outlines all the key players within the Cairo SWM landscape.

The Central Government								
National				Local				
Ministry of State for Environmental Affairs	Egyptian Environmental Affairs Agency	Governorates	Municipalities	Cleaning and Beautification Authorities	International Private Companies	National Private Companies	The Zabbaleen	NGOs
	Ministries of Agriculture & Land Reclamations, Housing, Utilities & Urban Development, Trade & Industry, Local Development, Interior and Finance				Private formal SWM Sector		Private informal SWM Sector	

SWM, solid waste management.
Reproduced from El Gamal, M., 2012. Municipal Solid Waste Management in Egypt-Focus on Cairo. Retrieved from: https://www.academia.edu/4805143/MUNICIPAL_SOLID_WASTE_MANAGEMENT_IN_EGYPT_-_Focus_on_Cairo.

These key players would all be managed under the institutional and organizational framework outlined below:

National Government		
Ministry of State and Environmental Affairs		Governorates
Egyptian Environmental Affairs Agency	Egyptian Solid Waste Management Authority	SWM units
Environmental Regulation	SWM Strategy, Policy & Legislation	Planning services & Infrastructure
Environmental Strategy, Policy & Legislation	Investment Programming	Implementation
	Support to Governorates & new housing communities	Service Provision

Reproduced from SWEEPNET, 2014. Country Report on the Solid Waste Management in Egypt. Retrieved from: www.sweep-net.org/sites/default/files/EGYPT%20RA%20ANG%2014_1.pdf.

The new reform addresses some of the previously insurmountable problems faced by Cairo's SWM sector. The new system takes into account the Zabbaleen and NGOs, as well as other previously ignored stakeholders in the policy-making process. The objective of the reform is clearly stated, as well as the primary principles under which the policy will be carried out. However, there are still some constraints that need to be addressed. As seen with implementing environmental legislation in general, Egypt has a history of ineffective enforcement, and the SWM sector is no different. There has historically been a lack of personnel to enforce the laws, a lack of funds to assist with enforcement, a lack of necessary infrastructure, and a legal base fraught with corruption. Another problem has been one of financing, especially attracting private sector investment domestically and abroad. The instability and ineffectiveness of the institutional frameworks within Egypt has previously disincentivized private investment, especially in the SWM sector. Only time will tell if the new policies will effectively address these historical roadblocks (El Gamal, 2012).

That being said, active incorporation of the Zabbaleen in the municipal waste management strategy is a huge step forward for Cairo and its sustainable development. As seen previously, one of the main foci of the Egyptian government has been social welfare and alleviation of poverty. Although they are not technically contracted by the government, allowing the Zabbaleen to perpetuate their livelihood of waste collection is mutually beneficial to the quality of life of the Zabbaleen and to the waste management and socioeconomic goals of the Egyptian government. Through local government programs, as well as international donors such as the World Bank, the Zabbaleen have been able to invest in technologies that help them even more efficiently recycle waste and turn a profit. In addition, the establishment of the Recycling School for Boys within Mokattam, also known as the Garbage City, has helped hundreds of children to learn relevant job skills and also to become literate.

The curriculum includes "literacy, numeracy, business math, personal and environmental hygiene, income generation and recycling, computer literacy, principles of project management, bookkeeping and simple accounting, along with recreational theatre arts (SWEEPNET, 2014)." Providing the Zabbaleen the skills needed to improve their business model as well as incorporating them into the overall SWM strategy will not only help alleviate the pressures of poverty in the Garbage City and the surrounding areas but also improve health, and aid even more in Cairo's SWM in the long run (DAC, 2014; SWEEPNET, 2014).

Although SWM has been the city's foremost sustainability concern, Cairo is also exploring other options as part of its sustainable development. There has been a push for green architecture and urban rooftop agriculture as ways to both improve the urban quality of life and their carbon footprint. The American University of Cairo (AUC) is pioneering green architecture in both Cairo and in Egypt as a whole with the first ever green rooftop on the AUC faculty housing building. The same building is currently pursuing Leadership in Energy and Environmental Design certification from the US Green Building Council and is a model of green architecture for the entire country with its green roof for urban rooftop farming, solar water heaters that account for 100% of the building's hot water, non-chlorofluorocarbon refrigerants for air-conditioning, and light-emitting diode bulbs. The faculty building was built not only to be environmentally conscientious but also to be an educative tool that is socially and economically responsible. Given the government's focus on culture and heritage in its Vision 2030, efforts are also being made to research the possibility of historic preservation mixed with green architecture within the city. Of course, green architecture is not without challenges. AUC Assistant Professor of Sustainable Design, Khaled Tarabieh believes the major challenges to be twofold: education and affordability.

> *People are resistant because they lack education. They want to do the same thing they have always done, or they don't know about the technology or they don't think they can maintain it.*
>
> Khaled Tarabieh, AUC (2015).

Tarabieh believes that education is the best way to enhance public opinion and support of green architecture. The issue of affordability, on the other hand, is a little more challenging. Green architecture, like renewable energy installations, has a high initial cost. However, as seen with distributed generation and solar panels, the materials and technologies are available, which has helped costs to decrease. Tarabieh believes that as green building techniques become more mainstream, innovation will be incentivized and the costs will fall even more (AUC, 2015).

Along with sustainable architecture, Cairo has held public panel discussions regarding the use of urban rooftop farming as an option to help alleviate poverty and naturally cool city buildings. The most recent discussion in

November 2015 showcased a rooftop pilot project in the area of Ezbet El-Nasr. The pilot project included six households and a total area of 90 m^2 and resulted in some very useful findings. First, the farmers asserted that the rooms below the rooftop gardens were noticeably cooler. They also found that income generation was the primary motive behind participation in the project and that the expected payoff seemed to be lower than expected. Even so, the farmers of the pilot project maintained farming without outside support after the project had ended and neighbors of participants actively expressed their wish to participate. Furthermore, women seemed to be very engaged in the project, showing it as a possible route for women's education and empowerment. That being said, it was noted that improved integration of marginalized groups (i.e., women and youth) is still needed and should be included in the overall planning process, as well as stronger involvement of public authorities to enhance capacity building and promote the project. Overall the Ezbet El-Nasr was a successful pilot project that has self-perpetuated and is looking to be scaled up in that area to 100 rooftop farms. The success here could be indicative of the possible success of these projects elsewhere in Egypt, including urban centers such as Cairo and Alexandria (Cairo Climate Talks, 2015).

The urban center of Alexandria has been focusing on other routes to become more sustainable. The primary focus for sustainable projects has been on accessibility and urban planning. Alexandria is strategically positioned along the Mediterranean Sea and is essentially the port of Egypt, which means that increasing accessibility would benefit the city economically and socially. Approximately 99% of Egypt's import–export trade is seaborne, and the main port through which most goods travel is Alexandria-Dekheila. In 2005, 284 million tons of goods were imported through the Alexandria-Dekheila port, and as of 2012, the port handles about 70% of Egypt's total maritime trade, which shows just how vital this port is to the Egyptian economy (Donato and Pallini, 2012). In spite of its key role in the Egyptian economy and its strategic position, the port was plagued by inadequate infrastructure, bottlenecks, and prohibitively cumbersome procedures. Everything about the port was in disarray: the roads and warehouses were poorly maintained, there was always traffic inside the port, there were not enough berths for the amount of incoming cargo ships, the unloading time for the ships was unacceptably long, and the customs procedures were also excessively time consuming leading to extra costs for cargo companies due to the need for storage and physical product inspections. A study by the World Bank in 1998 estimated that the port's procedures delayed cargos anywhere between 5 and 20 days, which cost the Egyptian economy about $1 billion per year. Prime Minister Ahmed Ebied launched the 2001–17 master plan for the restructuring and modernization of all state-owned Egyptian ports, with a special focus on Alexandria. Phase I of the renovations began in 2002 and was completed in March 2007. Among the upgrades were deeper quays to allow for offloading of larger vessels; redesign and upgrade of storage, warehouses, and related infrastructure; installation and implementation of an automated management system and related infrastructure; as well as a 15,000-m^2 passenger/cruise ship terminal. Additionally, a railway station was constructed to increase accessibility from the port to the city itself and areas beyond. Privatization of many of the port's operations looks to increase the efficiency as well as expansion of the container terminal. The port is still undergoing renovations; however, the changes from Phase I have significantly improved the accessibility of the region and the efficiencies of the port (Craig, 2007).

In April 2017, the city of Alexandria has come out with its own sustainable development goals in conjunction with those laid out in Egypt's Vision 2030. In December of 2016 the United Nations Development Programme, which worked closely with the central government on Vision 2030, has signed a protocol with the Egyptian Ministry of International Cooperation and the Ministry of Housing, Utilities and Urban Communities to implement "Strategic Urban Planning for Alexandria 2032." The main goal is to sustainably accommodate the governorate's increasing population in such a way that improves the quality of life for the citizens and meet the goals laid out in Vision 2030. Three primary areas of concern for the city are transportation, agricultural land management, and water management. Information gathered in the first phase of the project showed the public calling for immediate solutions to the traffic congestion caused by urban overpopulation. There has been talk of bridges and tunnel systems to alleviate the congestion within the city centers. However, any purposed expansion of Alexandria is challenging because of the surrounding agricultural land. Proper water management is required given the city's population and large agricultural sector. Proposals suggest that legislation such as the Unified Building Law of 2008 could aid in collaboration between urban, suburban, and rural areas for metropolitan planning, which would result in sustainable land and water use. At this point in time there are no set proposals or projects to show how Alexandria will go about tackling these challenges, but it is not unreasonable to expect a strategic plan in the near future (Khaled, 2017).

Efforts to facilitate the goals of sustainable development have been made at both the national and local levels within Egypt. In spite of the efforts with regard to people, the planet, and the economy, Egypt still faces challenges to fully executing Vision 2030. Nationally, it is struggling with a high birth rate, which needs to be addressed to harness the demographic dividend. Perhaps even greater is the challenge of water scarcity, which is only further exacerbated by the high birth rate, as well as the water needs for industry. Climate change mitigation still needs to be addressed, as well as energy needs to meet the needs of the growing economy and population. Another large challenge is the quickly growing informal economy within Egypt. To achieve the SDGs, the informal sector would need to be formalized through an incentive structure, and although a strategy for formalization is already in place, it is a large and potentially expensive process. Women's empowerment remains a challenge as well, in spite of the significant success by way of access to education and female representation in the parliament. Last, in addition to the national challenges, regional instability caused by unrest in the neighboring states of Libya and Syria has negatively impacted the Egyptian economy and, therefore, the financial and social conditions within the country (UNDP, 2016).

Nigeria

Nigeria is located in West Africa between Benin and Cameroon, on the Gulf of Guinea. It occupies an area slightly smaller than that of Egypt, but it is the most populous country in Africa. A little less than half of the 186 million people in Nigeria live in the urban areas, and this number is growing alongside the population growth rate of 2.4%. The country itself is urbanizing at the rapid rate of 4.6% annually, and this is causing significant environmental issues including soil degradation, deforestation, and urban air and water pollution. Rapid urbanization is causing significant social issues as well. Between 2002 and 2011, the Nigerian economy grew around 6.2% annually; however, there seemed to be a disconnect between the growth and human development. About 70% of the Nigerian population lives below the poverty line, with the majority of those in poverty located in the rural areas. In addition, Nigeria boasts one of the highest levels of income inequality in the world. Nigeria's unemployment is also significantly high, around 24%. The 2010 Nigeria Education Data Survey shows some promising trends, primary school attendance ratios have increased, whereas pupil repetition and dropout rates are low. About 60% of children in urban areas attend secondary school; however, only 36% of children in rural areas attend secondary school, indicating that school attendance is still directly linked to household socioeconomic status, which poses a significant challenge considering the large impoverished population. Gender inequality and violence toward women also remains pervasive.

> *At present, several harmful practices against women are prevalent in Nigeria. Female genital mutilation is one of the most common of these. Various widowhood rites, denial of inheritance rights and access to land and other resources, early and forceful marriage, domestic violence and harmful childbirth practices are some of the several practices infringing on the human rights of the Nigerian female.*
>
> NG Gov (2012).

Health care poses a significant challenge in Nigeria as well. As of 2009, there was a physician density of 0.41 physicians per 1000 people. The health care system remains inadequate and inefficient due to the rapidly increasing population and challenges in infrastructure and skill development. All of this is leading to significantly low life expectancy throughout the country, 53.4 years as of 2016, as well as high infant mortality, and high prevalence of communicable disease including human immunodeficiency virus/acquired immunodeficiency syndrome (HIV/AIDS). As of 2014, Nigeria was ranked second in the world for people living with HIV/AIDS and first for HIV/AIDS-related deaths. Other challenges to sustainable development within the social dimension include water and sanitation (68% of the population has access to improved water resources, 29% of the population has access to improved sanitation facilities) and security, especially against crime and drug trafficking (NG Gov, 2012; CIA, 2017b).

The Nigerian economy is considered typical of a developing country. The three production activities, agriculture, mining, and quarrying (including crude oil and natural gas), accounted for the majority of gross domestic output and 80% of government revenue, as well as over 90% of foreign exchange earnings and 75% of employment in 2012. However, as of 2016, agriculture by way of crop production, fisheries, and animal husbandry, as well as mineral and hydrocarbon extraction, only represents 21.1% of the national GDP (Fig. 23.1).

However, the agriculture and extractive sectors are still vital to the Nigerian economy in that they employ over half of the population. Still, both sectors have challenges that the government must help overcome. The agricultural sector is largely reliant on natural rains; very few agricultural areas have irrigation capacity, which means they are vulnerable to climate variation. In addition, the total planted area as well as production efficiency must increase rapidly to meet the growing domestic demand for food and exportable commodities, which necessitates responsible and environmentally sound agricultural land use to ensure against desertification and soil degradation. The

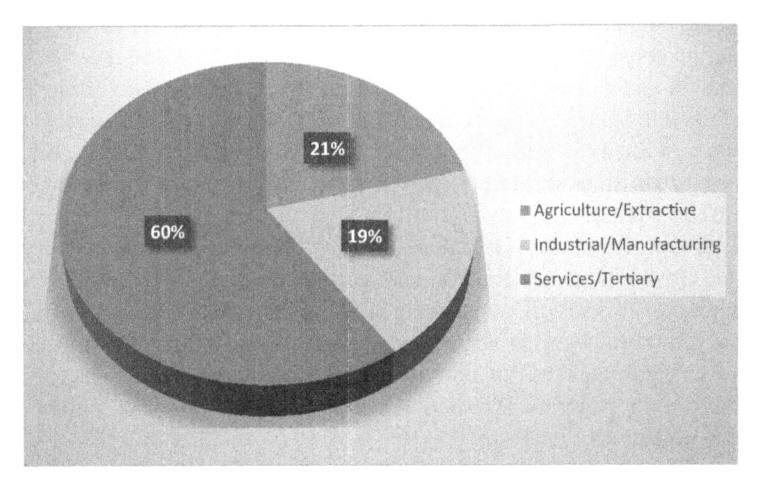

FIGURE 23.1 Economic sector shares in Nigeria as percentage of gross domestic product.

extractive sector faces its own challenges in relation to environmental protection, social concerns, and effective governance. In addition to addressing the sectoral challenges of the economy, the government must also consider the best ways to improve foreign and private sector investment, trade policies both domestically and internationally, and the reliability of the country's electricity supply.

Rapid population growth and urbanization have put significant strain on Nigeria's natural resource base. The environmental issues within the country can be broadly classified as land degradation and air and water pollution. Urbanization and industrialization are deteriorating the air and water quality.

> *Much of Nigeria's arable land is being sapped insidiously of its productive potential through overuse and inappropriate technologies. Rapid deforestation, resulting from unsustainable use of forest resources for human survival (e.g., fuel wood and energy, housing etc.) is a major contributing factor to land degradation. The end result of deforestation and other agricultural activities, including intensive grazing, over-plowing and over-cultivation, is severe land degradation, usually referred to as desertification, particularly in the northern part of the country.*
>
> NG Gov (2012).

It is not just agriculture that has exacerbated land degradation; indiscriminate and sometimes illegal mining has significantly reduced productivity of the land. Human activity has also been responsible for biodiversity loss. Deforestation has destroyed 43% of the forest ecosystem within the country; only about 2% of the original forest cover remains undisturbed. Destruction of Nigeria's unique ecosystems has led to an increase in the number of species considered threatened or endangered, and these problems afflict the coastal and marine ecosystems as well. The negative effects of natural disasters such as droughts and floods have been exacerbated by environmental destruction caused by humans as well as overall climate change. Waste management presents a problem as well; only between 30% and 50% of waste is collected and most urban areas still lack effective waste management systems (NG Gov, 2012; CIA, 2017b).

Given the magnitude of sustainability challenges faced by the country, the Federal Government of Nigeria has partnered with the United Nations Development Projects to put forth goals for sustainable development. The country has been working to implement Vision 20:2020 since it was first created in 2009, and the Vision still provides a framework for changes and projects within Nigeria. The initial vision consisted of three pillars with strategic objectives:

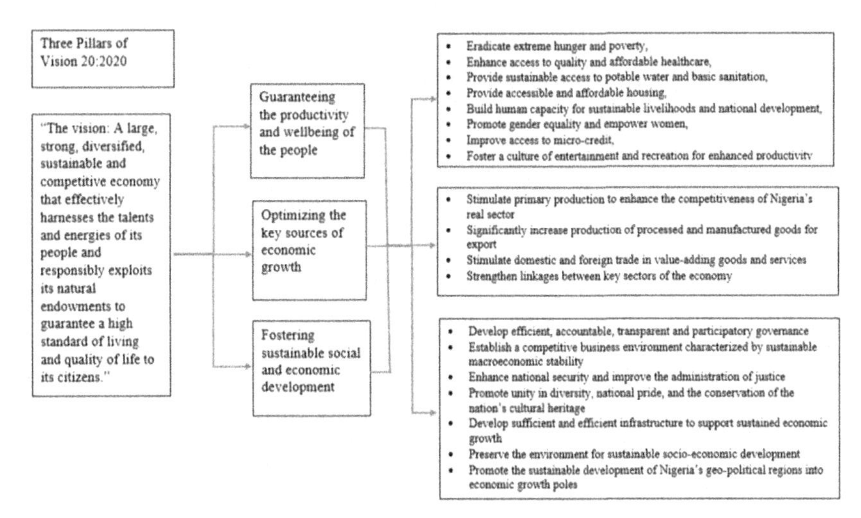

Reproduced from Government of Nigeria, 2009. Vision 20:2020. Economic Transformation Blueprint. Retrieved from: http://www.nationalplanning.gov.ng/index.php/national-plans/nv20-2020.

These would be achieved through a series of policy reforms to create a stable economic platform. The areas to be addressed included revenue allocation, corruption, investment in critical infrastructure, private-sector-powered nonoil growth for economic diversification, investment in human capital to enhance competitiveness, national security threats, and social equity from the national to the subnational level. The following table outlines the specific hindrances to sustainable growth and development that the government is looking to address in the Vision 20:2020.

Issue	Details
Poor and Decaying Infrastructure	• Transportation systems within Nigeria remain largely underdeveloped • The intermodal system has not been developed creating challenges moving goods and people • Telecommunications access remains sporadic
Epileptic Power Supply	• Inadequate generation • Inefficient transmission and distribution
Weak Fiscal and Monetary Policy Coordination	• Need to stop fiscal federalism and ways and means of financing
Fiscal Dominance	• Public sector borrowing crowds out of private sector participation due to hindered financing • Banks become more risk averse when lending
Pervasive Rent Seeking and Corruption	• Causes distortion of price signals that result in favoring of short-term and speculative investments • This is not conducive to long-term sector growth • Economic growth and poverty reduction cannot be attained in this environment

Weak Institutions and Regulatory Deficit	• Need proactive institutions with the authority to execute the rule of law and foster economic growth
Policy Reversals and Lack of Follow-Through	• Policy inconsistency is a significant hindrance to growth in Nigeria • Need measures to ensure policy sustainability and effective implementation
Dependence on the Oil Sector for Government Revenues	• The Nigerian government relies on crude oil revenue for over 80% of total revenue •Measures to diversify the economy are imperative
Disconnect between the Financial and Real Sectors	• The financial sector has been successful in trading government debt and foreign exchange, as well as financing the wholesale and retail trade sector • Financing of the real sector has been suboptimal • High interest rates are restrictive • Constraints on access to credit
Exchange Rate Instability	• Heavy reliance on imports opens the economy up to exchange rate risk, which contributes to: • Price volatility and inflation • Economic uncertainty that is detrimental to business planning and growth
Insecurity of Lives and Property	• Sustainable economic growth driven by the private sector requires a conducive environment characterized by security of lives and property, rule of law, sanctity of contracts, and respect for property rights • Internal impediments to this include ethnic/religious disturbances, kidnapping, armed robbery, and corruption. • External impediments include commodity price volatility, oil market boom and bust cycles, and intermittent droughts
Growth in a Depressed Global Economy	• Given the global recession and its potential to externally stifle Nigeria's economic growth, the government will focus more on addressing the internal constraints on growth, strengthening the internal institutions as well as incentivizing foreign direct investment and creating more functional cross-sectoral linkages
Climate Change and Environmental Degradation	• Without adequate action, it is predicted that African country's crop yields could be reduced up to 50% by 2020 • Nigeria will adopt environment-friendly practices to avoid the negative growth effects as well as capitalize on any competitive advantage opportunities as environmental issues put more pressure on international trade regulations
The Effects of Global Energy Transition	• Energy security is a critical issue for both emerging and developed countries • An energy diversification strategy that recognizes a transition away from fossil fuels is key

Reproduced from Government of Nigeria, 2009. Vision 20:2020. Economic Transformation Blueprint. Retrieved from: http://www.nationalplanning.gov.ng/index.php/national-plans/nv20-2020.

Around the same time, the government released the Nigeria a National Implementation Plan (NIP) for 2010–13 to outline the sectoral plans and programs that it would be undertaking to accomplish Vision 20:2020. This 256-page document outlines the goals, specific objectives, and priority projects of each sector to be reformed under the Vision 20:2020. Although it seems that environmental issues take a backseat to both people and profit in the rhetoric of the Vision, the NIP outlines many projects to combat environmental degradation. The government has planned three specific large-scale renewable energy projects while also planning for solar rural electrification projects and small and medium hydro implementation. The first of these was the 10-MW Katsina Wind Project, which would be the first wind installation in the country (NG NIP, 2010). Completion was anticipated to be in 2011; however, the project has been delayed significantly and it still pending completion. The second large-scale renewable energy project stated within the NIP is the 2600-MW Mambilla Hydroelectric Power Plant. Construction of the dam and power station along with the necessary transmission infrastructure will boost the availability of electricity to the area as well as irrigation. This project too has faced significant setbacks since bids were made in 2011, primarily due to heated negotiations between the Nigerian government and several Chinese developers. As of summer 2016 the government has indicated that an agreement has been reached and the project is set to start (ESI, 2016). The third project is the Zungeru Hydropower Project. The feasibility study returned an optimal capacity of 525 MW with expansion up to 700 MW. This project was handled jointly by the Chinese developers SINOHYDRO and China National Electric Engineering Corporation and is scheduled to begin generating electricity in 2019 in spite of multiple setbacks, including allegations of fraud and compensation for the local community. The government is also facilitating solar rural electrification projects in Cross River State, Ogun State, Bauchi State, and Katsina State as well as small and medium-scale hydro projects depending upon the results of feasibility studies (NG NIP, 2010).

In addition to renewable energy, the Nigerian government outlines several strategies to deal with the environmental dimension of sustainable development. With regard to land degradation and biodiversity loss, the efforts are classified as threefold: institutional policy, legal regulatory, and special initiatives. Institutionally, the Nigerian government has established the Federal Ministry of Environment, the National Oil Spillage Detection and Response Agency, and the National Environmental Standards and Regulations Enforcement Agency. Additionally, the country has developed national policies with regard to environmental sanitation, drought and desertification, flood and erosion control, and environmental impact assessment. Nigeria is also signatory to several treaties for the prevention of environmental degradation and biodiversity loss, including the Convention on Biological Diversity, RAMSAR Convention on Conservation of Wetlands, the United Nations Framework Convention for Climate Change, and the Kyoto Protocol.

A few years later the Nigerian government released "Nigeria's Path to Sustainable Development Through Green Economy," which builds on the goals, projects, and programs outlined in Vision 20:2020 through the concept of Green Economy. The Path to Sustainable Development was presented at the Rio +20 summit in 2012 and it very clearly has a more concerted focus on the environmental dimensions of sustainable development. A Green Economy "involves promoting growth and development while reducing pollution and greenhouse gas emissions, minimizing waste and inefficient use of natural resources, maintaining biodiversity and strengthening energy security (NG Gov, 2012)." At the operational level, the Green Economy is driven by investments that reduce carbon emissions and pollution, increase resource use efficiency, and prevent the loss of ecosystem services, while resulting in improved human well-being, social equity, and economic development.

Implementation of policy and projects for sustainable development are best seen at the local level, and because of the massive population in Nigeria, urban sustainability is imperative. Lagos is the largest urban center within the country and home to about 13 million people (CIA, 2017a,b,c). Lagos is already considered a megacity, since its population is already over 10 million, and its population is expected to reach 24 million by 2020. It is estimated that by the same year, half the Nigerians will live in urban centers and the country's megacities (Obia, 2016). The architecture, design, and urbanism firm NLE, founded by the Nigerian Kunle Adeyemi, has proposed several high-profile sustainable urban development projects for the city of Lagos and elsewhere in Nigeria. Some of these projects include a fourth Mainland Bridge and Masterplan for Lagos as an urbanization project to aid in pedestrian mobility as well as traffic flow of cars and boats. The multilevel bridge will accommodate vehicular traffic on its upper level, whereas the lower level will provide a space for pedestrians with social, cultural, and commercial activities. The design would create a roadway ring around the city that would provide alternative traffic routes and decrease congestion (NLE, 2008).

A second proposed urbanism project is the Lagos Water Communities Project. Given the pressures of climate change and the resulting rise in sea level and increased instances of heavy rainfall and flooding, the Lagos Water Communities Project aims to create low-cost floating infrastructure as a way to protect the urban city from water damage as well as to accommodate the ever-growing population (NLE, 2012a).

The Makoko Floating School was the pilot structure for the Lagos Water Communities Project. Completed in 2013, the building could accommodate 100 students and was kept afloat by 256 plastic drums filled with air. The electricity needs would be provided by the rooftop solar panels, while rainwater collection would help the indoor plumbing. The building itself was constructed from locally sourced wood, and construction was done by Makoko residents at a total build cost of $6250 (FutureLagos, 2014; NLE, 2012b).

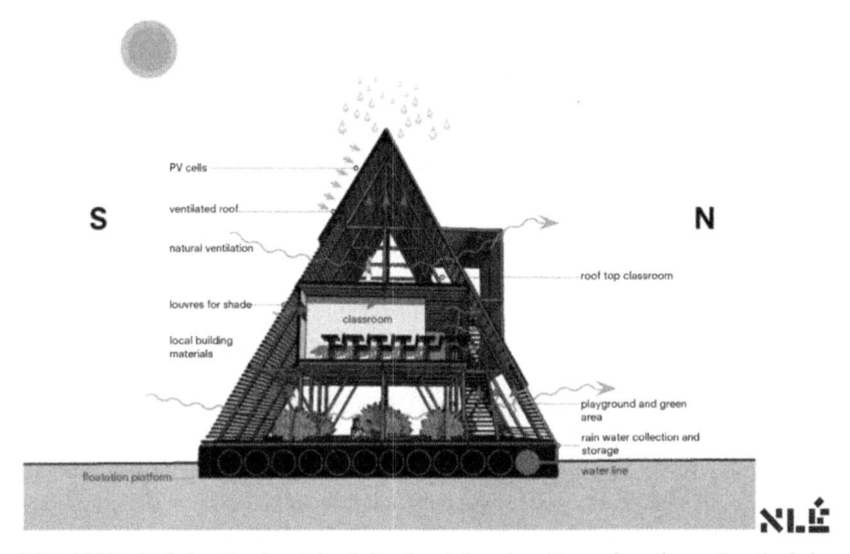

NLE, 2012b. Makoko Floating School. Retrieved from:http://www.nleworks.com/case/makoko-floating-school/.

Another project proposed by the NLE would make Bonny Kingdom one of the most important investment destinations in West Africa. The Bonny Kingdom Masterplan looks to change Bonny Kingdom from the Nigerian home of oil and gas into a diversified economy with low unemployment, strong infrastructure, and improved quality of life for millions of skilled workers. Bonny Kingdom is a coastal community in close proximity to Port Harcourt giving it a strategic location for water trade both internationally and internally. Bonny Island itself contains one of the country's most industrialized communities (Bonny Kingdom), mangrove forest, and some of the most fertile farmlands in West Africa. The plan, alternatively named "Green Heart City," looks toward four goals for the development of Bonny Kingdom:

1. Human capital development
2. Economic expansion and diversification
3. Infrastructure development
4. Environmental sustainability

To accomplish these, the plan outlines affordable social amenities such as schools; efficient commercial, residential, and industrial spaces; large use of green space for recreation; efficient and sustainable land use for agriculture; as well as the Bonny ring road that will circumnavigate the island, leaving at the center a "green heart"—a nature conservation area (NLE, 2014).

Nigeria is also tackling sustainable development with direct help from the UN, the Sahara Group, the Nigerian government, and three of the world's top chefs. The new pilot initiative in Kaduna called Food Africa is aimed at job creation, increasing farmer's revenue, improving farm productivity, enhancing nutrition, and reducing food waste. Agriculture employs 70% of the Nigerian population, although it only amounts to 22% of the total GDP, indicating potential for productivity gains. Additionally, lack of adequate storage infrastructure and market information leads to harvest loss of 50%–70%, whereas the World Bank estimates that climate change could reduce harvest outputs by an additional 30%. All these factors have led to the declaration of state of emergency in Kaduna due to tomato shortages and price spikes. The Food Africa project is meant to provide an integrated approach to food supply chain management through introduction of sustainable practices. One of the key pieces of the pilot program is an agroprocessing facility that will help eliminate food loss due to inadequate infrastructure as well as provide training on food safety, business planning, and product diversification. Eventually the facility and its programs will be maintained and managed by the local farmers. The Kaduna State Government will provide the land for the facility, as well as farmland and personnel, and work with local stakeholders on the improvements of necessary infrastructure such as roads. The UN will provide expertise pertaining to food production, labor and employment, international trade, and use of an early warning geographical information system. The Sahara Group will provide financing as well as oversight into the facility's viability and business operations. Last, the master chefs will showcase how local food production can be cultivated for new markets and consumers, as well as food preparation and cooking skills to help combat malnutrition. The pilot project is expected to directly improve the welfare of 5000 men and women in Kaduna by way of increased income, new job prospects, and valuable food industry skills, and the project will indirectly benefit around 500,000 local citizens (UN Food Africa, 2016).

Kenya

Located in East Africa, Kenya occupies an area 580,367 km^2, a little more than twice the size of Nevada. The country has a population of around 46 million people, significantly less than both Egypt and Nigeria, and the population is growing around 1.8% annually. Similarly to Nigeria, agriculture employs the majority of the labor force, 75%. However, unemployment is significantly higher at 40% and about 43% of the population of Kenya lives below the poverty line. The Kenyan economy remains dominated by the service sector (49% of GDP) and the agriculture sector (33% of GDP) (CIA, 2017c).

Health care and sanitation remain significant challenges in Kenya. Kenya is considered to be at a high risk for infectious diseases and is ranked among the top 10 countries for the number or people living with HIV/AIDS and annual deaths from HIV/AIDS. The CIA estimates 1 physician for every 5000 people and 1.4 hospital beds available for every 1000 people. Only 30% of the population both urban and rural has access to improved sanitation. However,

Kenya has made some strides with regard to social amenities. About 81.6% of the population has access to an improved drinking water source. About 78% of the total population is literate, although on average Kenyan children do not continue schooling past the age of 11 (CIA, 2017c).

As of 2013, only 20% of the total population of Kenya had access to electricity. This means approximately 35 million people are without electricity, and 60% of the urban areas were electrified, whereas only 7% of the rural areas were connected. Electrification, or lack thereof, has significant implications in the development of a country's economy as well as social and environmental welfare. Electricity generation by source is shown in the following figure:

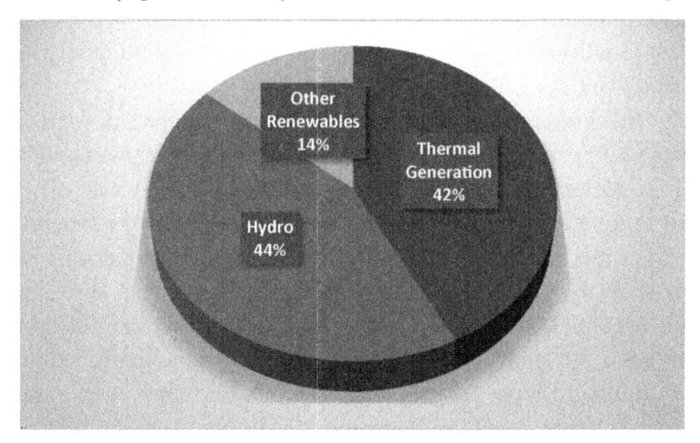

CIA World Fact Book, 2017a. Egypt. Retrieved from: https://www.cia.gov/library/publications/resources/the-world-factbook/geos/eg.html, CIA World Fact Book, 2017b. Nigeria. Retrieved from:https://www.cia.gov/library/publications/resources/the-world-factbook/geos/ni.html, and CIA World Fact Book, 2017c. Kenya. Retrieved from:https://www.cia.gov/library/publications/resources/the-world-factbook/geos/ke.html.

Kenya is a net electricity importer, getting the vast majority of its supply from its neighbor Uganda. Kenya is also an importer of crude oil as the country has no domestic fossil fuel production. The country is also home to one of the largest refineries in East Africa, which refines around 90,000 bbl/day of primarily Middle Eastern crude oil (IBP, 2013). Kenya does not consume, and therefore does not import, natural gas (CIA, 2017c).

Kenya has a significant historical framework for sustainable development with a primary focus on people, politics, and the economy. Kenya gained independence in 1963 and immediately began working on a development strategy to address poverty, hunger, illiteracy, and disease. The first edition of the Poverty Reduction Strategy Paper (PRSP) was released in 1965, and after a number of revisions and edits the World Bank and the International Monetary Fund approved the PRSP approach to poverty reduction. Kenya's PRSP in 2001 represented a short-term approach to the long-term poverty reduction goals set forth in the National Poverty Eradication Plan (NPEP). The NPEP was created in 1999 and aligned with the UN Millennium Development Goal (MDG) of reducing poverty by 50% by 2015. The PRSP aimed to facilitate economic growth and reduce poverty through linking public actions, donor support, and the outcomes of the UN's MDGs (UN KY, 2012).

During the first ever United Nations Conference on Environment and Development in 1992, Kenya adopted Agenda 21, an outline for sustainable development, as well as most of the treaties, international agreements, and protocols associated with this first conference. Furthermore, as a demonstration of its commitment to sustainable development and addressing climate change, Kenya hosted the second meeting of the parties to the Kyoto Protocol as well as the 12th session of the parties to the United Nations Framework Convention on Climate Change in 2006. Ratifying both Agenda 21 and the various other international agreements required Kenya to actively set up institutions and actions to address climate change, biodiversity loss, and desertification (UN KY, 2012).

In 2003, the government switched gears and began to focus more on Kenya's economic revival than the NPEP and PRSP, which were subsequently abandoned. A national economic recovery strategy (ERS) was a 5-year plan launched to combat poverty through wealth creation and employment. Agriculture, at the time, was the leading productive sector in Kenya, so the ERS focused on revival of the agricultural institutions as well as investment in agricultural research. Four pillars were created, from which policy would be created to spur economic recovery: (1) macroeconomic stability, (2) strengthening institutional governance, (3) rehabilitation and expansion of physical infrastructure, and (4) investment in human capital, especially the poor. After noticeable success, the ERS expired in 2007, and in 2008, Kenya launched Vision 2030 for sustainable development (UN KY, 2012).

Kenya's vision 2030 was launched with the goal of transforming the country into a globally competitive and prosperous nation with a high quality of life. The Vision took over where the ERS had left off in its first 5-year medium-term plan from 2008 to 2012. Kenya's Vision is based on three pillars:

The Economic Pillar	Improve prosperity of all regions of the country and all Kenyans by: • Achieving 10% GDP growth rate by 2012 • Expanding the six priority sectors that make up ~57% of Kenya's GDP and 50% of the country's formal employment: *Tourism, Agriculture, Wholesale and Retail Trade, Manufacturing, IT-Enabled Services/Business Process Off-shoring, and Financial Services*
The Social Pillar	Investing in the people of Kenya to improve quality of life through human and social welfare projects and programs with a specific focus on: • Education and training • Health • Housing and urbanization • Environment • Gender • Children and social development • Youth and sports

The Political Pillar	Focusing on national unity and a democratic system that is: *issue based, people centered, result oriented, and accountable to the public* Transforming Kenya's political governance across five strategic areas: • The rule of law—the Kenya Constitution 2010 electoral and political processes • Democracy and public service delivery • Transparency and accountability • Security • Peace building and conflict management

GDP, gross domestic product; *IT*, information technology.
Reproduced from United Nations, 2012. Sustainable Development in Kenya: Stocktaking in the Run up to Rio+20 and Government of Kenya, 2017. Kenya Vision 2030. Retrieved from: http://www.vision2030.go.ke/.

The government added a fourth pillar to Vision 2030. This pillar, Enablers and Macroeconomics, includes things such as environment, water, sanitation, energy, and infrastructure that are required as foundations for the reforms laid out in the first three pillars.

Enablers and Macroeconomics	Enablers and Macroeconomics can be broken down into key sectors that provide stable platforms for the three pillars and overall growth: • Macroeconomic stability for long-run growth • Infrastructure • Energy • Science, technology, and innovation • Land reform • Human resources development • Security • Public sector reforms • Environment, water, and sanitation

United Nations, 2012. Sustainable Development in Kenya: Stocktaking in the Run up to Rio+20 and Government of Kenya, 2017. Kenya Vision 2030. Retrieved from: http://www.vision2030.go.ke/.

To achieve the goals set out in the pillars of the Vision, the Kenyan government launched a significant number of flagship projects. Under the economic pillar, several initiatives were established to boost the aforementioned sectors, including development of resort cities on the coastline to boost tourism, a fertilizer cost reduction initiative and land-use master plan to boost the agricultural sector, establishment of a free trade port in Mombasa to connect Kenya to Dubai for the benefit of the wholesale and retail sectors, as well as issuance of benchmark sovereign bonds for the benefit of the financial sector. The flagship projects for the political pillar include establishment of the Kenya School of Government, judicial and legal reform, and security and policing reform, as well as building a nonpartisan professional research center to improve parliamentary law making. The social pillar was by far the focus of the flagship projects under Vision 2030. Projects and programs for this pillar include enacting of a housing bill, production of 200,000 housing units annually, providing financial support to female entrepreneurs, increasing women representation in all branches of government, building and equipping 560 new secondary schools, improving data collection with regard to the impoverished to identify their pressing needs, and establishing protection funds to benefit orphans, vulnerable children, persons with disabilities, and the elderly. Finally, to lay the foundation for the aforementioned reforms and projects, the government outlines the projects in the Macroeconomics and Enablers pillar, including Tana and Lake Victoria catchment management initiatives, securing wildlife corridors and migratory routes, relocation of the Dandora dumping site, improvement of sewage management, dredging/deepening of the Mombasa port, rural electrification programs, development of light rail and bus transit to improve mobility and traffic flow, increased use of solar power and cogeneration especially in agricultural processing, and construction of pipeline infrastructure and liquefied petroleum gas handling facilities in Mombasa and Nairobi (KY Gov (B), 2017).

In 2012, both the Kenyan government and the UN took stock of the country's sustainable development programs and the progress made with respect to the three pillars of Vision 2030. The midterm review of initiatives and projects returned a fair deal of progress under the first three pillars. With regard to the economic pillar, through the first half of 2011, tourism was up 13.6%, peaking at 1.8 million tourists in 2010, indicating significant improvement. Initially the agricultural sectors recorded negative growth rates in 2008 and 2009, attributed to postelection unrest and inclement weather. The sector has since recovered, posting a +6.3% growth rate in 2010. The wholesale and retail sectors saw slower growth, decreasing from +11% through 2006, to +5.5% between 2008 and 2010. In contrast, the financial services sector was able to recover in full and has exhibited 8.8% growth between 2008 and 2010 (UN KY, 2012).

In conjunction with the UN's MDGs, Vision 2030 aims to meet the eight goals with regard to social sustainable development:

1. To eradicate extreme poverty and hunger
2. To achieve universal primary education
3. To promote gender equality and empower women
4. To reduce child mortality
5. To improve maternal health
6. To combat HIV/AIDS, malaria, and other diseases
7. To ensure environmental sustainability
8. To develop a global partnership for development

Kenya has been able to make significant strides toward achieving the MDGs. Continued commitment is being made to provide free primary schooling to Kenyan citizens as well as to provide the schools with the adequate resources to provide quality education. Primary school enrollment increased from 73.7% in 2000 to 91.4% in 2010. With regard to gender inequality, the new Constitution requires a minimum 30% female representation in parliament and 33% female representation minimum in government appointments. Additionally, the ratio of female to male students in primary school was 0.95% in 2012, which indicates that primary school attendance in Kenya will achieve gender parity. Children (younger than 5 years) mortality (per 1000 live births) has fallen from 114 in 2003 to 74 in 2009. Infant mortality (per 1000 live births) has fallen from 77 in 2003 to 52 in 2009, and as of 2016, it has fallen even further to 38.3. National HIV prevalence declined from 7.4% in 2007 to 6.3% in 2010, and as of 2015, it was down to 5.9%. Even with the significant progress toward meeting MDGs 2, 3, 4, and 6, Kenya was still struggling to meet MDGs 1, 5, 7, and 8. This is primarily attributed to inadequate resources and financing, the postelection unrest in 2007/2008 that resulted in a food and fuel crisis, as well as unfavorable international trade practices within Kenya (UN KY, 2012; CIA, 2017a,b,c).

Under the political pillar, the Vision saw the successful establishment of several commissions in charge of national cohesion, postelection violence, independent truth, justice, and reconciliation, and public complaints. Additionally, the promulgation of the Constitution in 2010 is another significant success for Vision 2030 and resulted in establishment of various oversight committees in charge of implementing the Constitution, as well as institutional offices including the Supreme Court, Commission of Revenue Allocation, and the Judicial Service Commission (UN KY, 2012).

In addition to the efforts being made by the Kenyan government, several international institutions including the World Bank and the Nordic Development Fund (NDF) have been working at the state and local levels to advance Kenya's sustainable development goals. The NDF with help from both international and local partners is involved in a significant number of projects throughout the country. For the purposes of this case study we will focus on three: (1) climate-resilient low-cost buildings in Marsabit County, (2) leveraging markets for climate friendly sustainable development in Laikipia, and (3) off-grid electrification using wind and solar energy.

1. Marsabit county is located in northern Kenya. With a hot and arid climate, it is one of the least developed regions in the country with scarce water and energy resources. Climate change is a significant threat to the citizens as they have been forced to adapt to increasingly harsh weather conditions. The majority of the population in the urban areas live in housing often constructed from iron sheets, which proves little protection from the harsh climate and generally fails to provide basic services such as plumbing or electricity. The NDF, along with its partners, has decided to launch a pilot project to address this issue. The project itself is expected to provide training to local citizens as well as showcase sustainable construction and energy-efficient technologies in both public and private sector buildings. Furthermore, partnership with local governments will facilitate the incorporation of energy efficiency (EE) and renewable energy (RE) measures

into building regulations. The project is in progress but the expected results include:

a. "Training and awareness-raising material developed on sustainable building, EE/RE and green business management

b. Skills on sustainable construction and EE/RE technologies of local masons, artisans and energy technicians, construction sector, local vocational centers and polytechnics developed. At least 400 people trained, with women making up at least 50%.

c. Awareness-raising on the environmental, social and economic benefits of sustainable building and EE/RE technologies raised among the community of Marsabit, public and private sector related to the construction and energy sector

d. Manual on sustainable building and EE/RE technologies appropriate for hot and arid areas developed

e. Local building regulations and bylaws influenced in favor of adopting sustainable building design and EE/RE technologies. Recommendations and enforcement plan delivered to the Government of Marsabit

f. A scaling-up plan developed to address sustainability and EE/RE in building in different counties and sectors – residential buildings, public facilities and commercial buildings

g. The potential green building markets to be mapped and financing models and marketing materials for green building business developed

h. To showcase the affordable and sustainable housing, the building of 100 housing units facilitated by introducing the EE/RE housing technologies: reduction of energy consumption in construction and in operation and maintenance, reduction of the use and demand for water from the mains supply, reduction of consumption of wood to heat water, reduction of the use of charcoal and/or wood for cooking and introduction of water saving technologies (NDF, 2017)."

2. Laikipia County is another area of Kenya that is considered extremely vulnerable to climate change. With a mostly agrarian population, the NDF's project aims to improve the livelihood of the area's marginalized population as well as sequester carbon as a way of reducing GHG emissions. This will be done through "introduction and scaling up of resilient and environmentally and socially appropriate conservation agriculture practices and introduction of sustainable marketing and supply chains (NDF, 2017)." Conservation agriculture and innovative rainwater collection, as well as land restoration practices, will increase the area's carbon sequestration and ultimately help mitigate the effects of climate change.

> *These gains, along with increased productivity and sustained access to profitable markets, will enhance the socio-economic well-being of local communities. Innovative low-cost technologies will help mitigate human-wildlife conflict to ensure functioning ecosystems support adaptation and mitigation.*
>
> NDF (2017).

Conservation agriculture will be introduced and scaled in 10 locations, which will be augmented by rainwater collection that is expected to provide 1 million liters of water annually. Additionally, over 530,000 ha of degraded land will undergo ecological restoration and reforestation. Overall the project aims to directly benefit the lives of 4000 farmers through access to reliable and profitable markets and indirectly benefit 20,000 local citizens with access to food and measures to mitigate climate change (NDF, 2017).

3. NDF has partnered with the World Bank to increase capacity, efficiency, and quality of electricity supply to all areas of Kenya. The objective of the project is to focus on off-grid electrification through renewable energies to reduce the use of fossil fuels, in particular, diesel power stations. The project will also promote the use of solar energy for the purposes of rural electrification. The NDF has pledged to directly finance two specific pieces of the project:
   a. Assistance of Kenya Power in design and installation of hybrid energy systems to replace use of diesel generation
   b. Provision of basic energy services to rural schools and households through solar facilities as well as solar charging stations to allow for use of portable lanterns in both schools and homes.

The total project cost is estimated to be around 4 million EUR, with financing coming from the local areas, the Kenyan government, and international agencies such as the International Development Association and the Japanese International Cooperation Agency. This project will have significant impacts on education, literacy, and poverty alleviation in the rural areas of Kenya (NDF, 2017).

At the local level, communities are contributing to realizing Kenya's Vision 2030. One classic example is the urban center Nairobi. Upon adoption of Vision 2030, Nairobi and its metropolitan area adopted Nairobi Metro 2030 to follow the nation's goals of development and growth. The growing urban center working to create a rapid transit system and increase mobility while decreasing congestion, but other options are being proposed to contribute to Nairobi's overall sustainability. First is the use of recycled tires and plastics to build sidewalks. In 2010, an estimated 34,000 tons of tires were either burned or reused in ways that polluted the air, soil, and water in Kenya, and this number is expected to increase with the infrastructure improvements under Vision 2030. Use of the tires for sidewalks provides a number of benefits:

- Flexibility: rubber sidewalks can move with both shifting soil and tree roots
- Safety: the surface is softer for pedestrian traffic
- Porousness: rubber allows rain water to seep through and reach the soil underneath, aiding in drainage
- Low maintenance

- Environment friendly: It is an alternative to burning the discarded tires or throwing them in a landfill.

A German firm GIZ has partnered with Bamburi Cement Ltd. as well as other local businesses to form a group called Waste Tyre Management Kenya to help the Kenyan government adopt and implement updated waste tire regulations. While the cement industry is expected to be the main user of waste tires, the growing demand for urban mobility infrastructure, such as sidewalks, means that the use of waste tires may not be out of the question (NPI, 2016).

Electricity-generating sidewalks are another innovation proposed for Nairobi. A company called Pavegen has developed flooring tiles that use footsteps to generate electricity while also collecting data on pedestrian and consumer behavior. The technology is currently in use in conjunction with solar panels to power a football pitch in Lagos, Nigeria, as well as along the route of the Paris marathon (seen the following figure).

Runner in the Paris Marathon. *Pavegen, 2017. Nigeria Football Pitch. Retrieved from: http://www.pavegen.com/shell-nigeria.*

In Nairobi, about 50% of the citizens walk and electricity supply consistency remains a challenge. Use of this technology, especially in shopping centers and areas with high foot traffic, could help with security of electricity supply in a sustainable and environment-friendly way (NPI, 2016; Pavegen, 2017).

With the help of international institutions, local community support, and the broad range of reforms pushed by the government, Kenya's Vision 2030 has been quite successful. The GDP per capita has increased from US $895 in

2007 to US $1434 in 2015, which is attributed to economic stability and high economic performance. Kenya's average annual economic growth between 2008 and 2016 was 4.9%, higher than the average of both sub-Saharan Africa and the world. Macroeconomic stability has aided Kenya's robust economic performance. The Kenya Shilling remains strong and stable against world currencies, and inflation has been stable at ~9%. The agricultural, financial, and mining sectors have experienced the highest annual growth: ~21%. In 2015, Nairobi was rated the most attractive city in Africa for foreign direct investment and the Kenyan financial sector has experienced exponential growth in commercial banking allowing them to play a significant role in the economy.

In addition to the great strides in reducing infant/child mortality, maternal mortality has fallen from 488 deaths per 100,000 live births in 2008 to 362 in 2015. Significant increases in primary school enrollment due to access to free primary school programs have resulted in an increase in the primary-to-secondary school transition rate. Only 60% of primary school students continued on to secondary school in 2007; however, as of 2015, 82.3% of students continue on to secondary school. Gross secondary school enrollment jumped to 48% in 2015 from 38% in 2007. In addition, educational resources have improved. As of 2016, 2000 Kenyan primary schools have access to computers, compared with 0 schools in 2007.

Mobility has improved as road construction and maintenance has had significant investment: US $1.3 billion in 2015. In addition to paving previously unpaved roads, new roadways connecting Kenya and Ethiopia have been constructed. The Mombasa–Nairobi Standard Gauge Railway is slated to be complete in June 2017, reducing travel time between the two urban centers to 4 h as opposed to the previous travel time of 12 h. Improvements have been made to security through resources available to the police force, which include forensic laboratories, multiple stations, helicopters, and other vehicles allowing for easy and efficient mobilization.

Last, Kenya's energy sector has improved significantly. Kenya plans to begin exporting crude oil in 2017. With current reserves estimated around 750 million barrels, the government estimates that the revenues will total around US $40 billion if prices remain near $50/bbl. Moreover, from 2007 to 2015 Kenya's installed electricity generating capacity doubled from 1197 to 2334 MW. Geothermal capacity has increased fivefold in the same period, contributing 627 MW as on 2015. Connection prices have decreased significantly, as have the consumer electricity prices; from $0.19/kWh in 2007 to $0.12/kWh in 2015. Customers connected via rural electrification programs have increased fourfold since 2007, providing electricity to 703,190 people in rural areas. Total electricity access has increased from 20% in 2013 to 57% in 2016 with a target of 100% by 2020.

REFERENCES

Cairo Climate Talks, 2015. Greener Cairo: Sustainability through Urban Agriculture? Retrieved from. http://cairoclimatetalks.net/events/greener-cairo-sustainability-through-urban-agriculture.

CIA World Fact Book, 2017a. Egypt. Retrieved from. https://www.cia.gov/library/publications/resources/the-world-factbook/geos/eg.html.

CIA World Fact Book, 2017b. Nigeria. Retrieved from. https://www.cia.gov/library/publications/resources/the-world-factbook/geos/ni.html.

CIA World Fact Book, 2017c. Kenya. Retrieved from. https://www.cia.gov/library/publications/resources/the-world-factbook/geos/ke.html.

Craig, G., 2007. In Depth-Alexandria Port Shines After Renovation. American Chamber of Commerce in Egypt, Retrieved from. www.amcham.org.eg/publications/business-montly/issues/91/July-2007/495/.

Danish Architecture Centre, 2014. Cairo: finding its own way in waste collection. Sustainable Cities. Retrieved from. http://www.dac.dk/en/dac-cities/sustainable-cities/all-cases/waste/cairo-finding-its-own-way-in-waste-collection/.

Donato, V., Pallini, C., 2012. Projects for accessibility and "sustainable" planning in Alexandria (Egypt): a case study. Journal of Civil Engineering and Architecture 6 (6), 756–767.

Egypt Vision 2030, 2016. Sustainable Development Strategy: Egypt Vision 2030. Retrieved from. http://sdsegypt2030.com/?lang=en.

El Gamal, M., 2012. Municipal Solid Waste Management in Egypt-Focus on Cairo, Retrieved from. https://www.academia.edu/4805143/MUNICIPAL_SOLID_WASTE_MANAGEMENT_IN_EGYPT_-_Focus_on_Cairo.

ESI Africa, 2015. Egypt: 200MW Gulf of El-zayt Wind Farm Inaugurated. Retrieved from. https://www.esi-africa.com/news/egypt-200mw-gulf-of-el-zayt-wind-farm-inaugurated/.

ESI Africa, 2016. Nigeria Mambilla Hydropower Project Back on the Cards, Says Minister. Retrieved from. https://www.esi-africa.com/news/nigeria-mambilla-hydropower-project-back-cards-says-minister/.

European Union, 2017. EU Results: The Gulf of El Zayt Wind Farm Project. Retrieved from. https://ec.europa.eu/budget/euprojects/gulf-el-zayt-wind-farm-project_en.

Fulbright, N.R., 2013. Renewable Energy in Egypt: Hydro, Solar and Wind. Retrieved from. http://www.nortonrosefulbright.com/knowledge/publications/74735/renewable-energy-in-egypt-hydro-solar-and-wind.

FutureLagos, 2014. 10 examples of 'Green' Architecture in Africa. Retrieved from. http://futurecapetown.com/2014/04/the-move-to-green-architecture/.

IBP Inc, 2013. Kenya energy policy. Laws and Regulation Handbook 1, 40.

Al Jazeera, 2016. Saudi Arabia, Egypt Agree to Build Bridge over Red Sea. Retrieved at. http://www.aljazeera.com/news/2016/04/saudi-arabia-egypt-announce-bridge-red-sea-160409032158790.html.

Government of Kenya (B), 2017. Flagship Projects Summary. Retrieved from. http://www.vision2030.go.ke/flagship-projects/.

Government of Kenya, 2017. Kenya Vision 2030. Retrieved from. http://www.vision2030.go.ke/.

Khaled, F., 2017. Alexandria Accelerates towards Sustainability in 2032. Investigate. Retrieved from. http://invest-gate.me/features/alexandria-accelerates-towards-sustainability-in-2032/.

Mansour, A., Mansour, H., 2016. Sustainable youth community development in Egypt. Alexandria Engineering Journal 55, 2721–2728.

Nairobi Planning Innovations, 2016. Nairobi: Innovative Technologies for a More Sustainable Future. Retrieved from. https://nairobiplanninginnovations.com/2016/07/30/nairobi-innovative-technologies-for-a-more-sustainable-future/.

New and Renewable Energy Authority, 2016. Government of Egypt. Retrieved from. http://www.nrea.gov.eg/english1.html.

Government of Nigeria, 2010. The first national implementation plan for NV20:2020. In: Sectoral Plans and Programmes. Economic Transformation Blue Print, vol. II. Retrieved from. http://www.nationalplanning.gov.ng/index.php/national-plans/nv20-2020.

Government of Nigeria, 2009. Vision 20:2020. Economic Transformation Blueprint. Retrieved from. http://www.nationalplanning.gov.ng/index.php/national-plans/nv20-2020.

Government of Nigeria, 2012. Nigeria's Path to Sustainable Development through Green Economy: Country Report to the Rio+20 Summit. Retrieved from. http://www.nationalplanning.gov.ng/index.php/national-plans/nv20-2020.

NLE, 2008. 4th Mainland Bridge Masterplan. Retrieved from. http://www.nleworks.com/case/4th-mainland-bridge-masterplan/.

NLE, 2012a. Lagos Water Communities. Retrieved from. http://www.nleworks.com/case/lagos-water-communities-project/.

NLE, 2012b. Makoko Floating School. Retrieved from. http://www.nleworks.com/case/makoko-floating-school/.

NLE, 2014. Bonny Kingdom Masterplan. Retrieved from. http://www.nleworks.com/case/bonny-kingdom-masterplan/.

Nordic Development Fund, 2017. Projects: Kenya. Retrieved from. http://www.ndf.fi/projects/kenya.

Obia, A.E., 2016. Emerging Nigerian megacities and sustainable development: case study of Lagos and Ajuba. Journal of Sustainable Development 9, 2.

Pavegen, 2017. Nigeria Football Pitch. Retrieved from. http://www.pavegen.com/shell-nigeria.

SWEEPNET, 2014. Country Report on the Solid Waste Management in Egypt. Retrieved from. www.sweep-net.org/sites/default/files/EGYPT%20RA%20ANG%2014_1.pdf.

The American University in Cairo, 2015. Khaled Tarabieh: AUC is Pioneering Green Architecture in Egypt. Retrieved from. News@AUC. http://www.aucegypt.edu/news/stories/khaled-tarabieh-auc-pioneering-green-architecture-egypt.

United Nations, 2012. Sustainable Development in Kenya: Stocktaking in the Run up to Rio+20.

United Nations Department of Economic, and Social Affairs (UNDESA), 2016. Sustainable Development Knowledge Platform: Egypt. Retrieved from. https://sustainabledevelopment.un.org/hlpf/2016/egypt.

United Nations Development Programme, 2016. UNDP in Egypt: Sustainable Development Goals (SDGs). Retrieved from. http://www.eg.undp.org/content/egypt/en/home/sdgoverview/post-2015-development-agenda.html.

United Nations Information Center-Lagos, 2016. New Pilot Initiative in Kaduna Will Boost Inclusive Growth in African Food Industry. Retrieved from. http://lagos.sites.unicnetwork.org/2016/07/14/new-pilot-initiative-in-kaduna-will-boost-inclusive-growth-in-african-food-industry/#more-1928.

Wahaab, R.A., 2003. Sustainable development and environmental impact assessment in Egypt: historical assessment. The Environmentalist 23, 49–70. Kluwer Academic Publishers.

Walker, K., 2013. Egypt-Saudi Bridge: Is the Government Sacrificing Natural Resources for Short-term Gains? Egypt Independent. Retrieved from. http://www.egyptindependent.com/news/egypt-saudi-bridge-government-sacrificing-natural-resources-short-term-gains.

Chapter 24

Sustainable Agriculture: The Food Chain

Attilio Coletta
Università degli Studi della Tuscia, Viterbo, Italy

Chapter Outline

INTRODUCTION

Sustainability is a polysemic and ubiquitous term: there no longer exists a sector wherein sustainability is not sought, certified, and too often boasted. This is why a correct approach requires, prior to any further analysis, a statement that can be referred to while discussing about the sustainability concept proposed.

This is even more important in the agricultural sector that deals with the production of food (both from vegetables and animals) by means of environmental factors (water, soil, landscape, etc.) mainly behaving like nonprivate goods, and other producing factors traditionally considered private goods (labor, equipment, fertilizers, seeds, fuels, pesticides, etc.).

The shift from the traditional production approach, related to the production of commodities, toward a more comprehensive role of agriculture implied a shift of attention from outputs to inputs. This is particularly true for those producing factors that induce externalities.

Basically, sustainability could be defined as the level of resource consumption that allows a natural regeneration of the resource itself (Godfray et al., 2010). This approach is very close to the point of view of those who focus on the resilience of the ecosystems in which human activities take place and where environmental implications of the food

Sustainable Cities and Communities Design Handbook. https://doi.org/10.1016/B978-0-12-813964-6.00024-0

chain are among the priorities, as we will see in the chapter devoted to these topics.

A different point of view, focusing on human beings' survival, refers to the sustainability in terms of food security: sufficient food for a growing population, as pointed out at the international level by FAO (2010), whose concern is regarding recent estimates that indicate a world population expected to grow up to 9 billion in 2050 (Godfray et al., 2010).

Obviously, these two approaches overlap when considering the consequences of a lack of sustainability and focusing on the causes.

For a better organization of the present contribution, sustainability will be analyzed in relation to social, economic, and environmental frameworks. The first and the second focus on human beings needs, whereas the last takes into consideration a more comprehensive approach that refers generally to ecosystems.

SOCIAL IMPLICATIONS

From a social point of view, sustainability deals with food security and food safety, as mentioned earlier. Summarizing the concern, we still face the problem of how to feed the entire world population with enough and safe food (see FAO, 2010; Godfray et al., 2010).

Despite the concern on the expected increase in number, the attention of many organizations focuses on the problem of human feeding that is already present now: 15% of the population cannot access a sufficient protein and energy resource in terms of available food (with respect to their purchase power). Even more people seem to be currently affected by malnourishment. Therefore sustainability in terms of food security (at global level) deals with how it is possible to sufficiently feed the entire population of the earth at present and in future.

In most wealthy countries, despite a lower growth rate of population, the higher income available pushes up the demand for high-value food like processed food, meat, and fish. Such demand pattern (like the one that China and India are showing since the recent past) is characterized by a strong consumers' preference toward food requiring a greater use of natural resources and a lower efficiency in land use (animal husbandry has a conversion of around 10% of grassland food into meat). Such products seem to give to consumers a higher level of satisfaction (utility achieved from the consumption of goods), and accordingly their consumption grows as the income rises.

This is why emerging countries show a development path characterized by a strict positive correlation between the per capita income and demand for high-protein diet, like the Western countries show, although healthy diet is adopted only by the richest and smallest part of the population, able to value the health implications of an unbalanced diet. The result is a frightening dynamics: hunger (among the poorest), obesity, and protein excess, due to the consumption of cheap food with high energy content (sugar and alcohol mostly) in the largest part of population, which have low wage levels; high-value food (i.e., preprocessed foods like frozen food) for high-income people; and a well-balanced healthy diet only for a few most fortunate people, who are well educated and rich. In such situations there is a clear "health gap[1]" among countries with different average per capita income level.

Food prices are part of the problem, making poor people able to buy only unhealthy food, and allowing them to satiate hunger often with the sole alternative of obesity due to the consumption of a higher energy content only via food available at convenient price. The frightening obesity is definitely coming back in countries deeply affected by the economic crisis (Lock et al., 2010) wherein food choice is strongly driven by a lower available income with respect to the past.

In such critical context, food security is a target yet to be reached, mostly in those countries in which there is an unsatisfactory level of enforcing population rights with respect to accessing resources, mainly land and food, like in less developed areas of the world (i.e., land grabbing, grazing rights on common land, access to public water resource; Freibauer et al., 2011).

ECONOMIC IMPLICATIONS

By tradition, sustainability in food sector has been referred in terms of yield gap,[2] and accordingly scarcity has been related to natural resources: soil water, energy, phosphorous, and nitrogen. These are the basic producing factors involved in food production. However, recent dynamics show new emerging scarcities that stem from the development path of the modern society (Freibauer et al., 2011) (Fig. 24.1).

Beside traditional scarcities, climate change sheds light on the future availability of rain water for crops and progressive desertification of cultivated land that is claimed by many competing uses [food production, bioenergy crops, fodder, and urbanization; Beddington et al. (2011)] and therefore is becoming scarce. Referring to natural resources involved in food production, biodiversity is also becoming scarce. Intensive agriculture implies an increasing biodiversity loss with respect to both animal and vegetable genetic resources, dramatically reducing the opportunities for future plant and animal breeding. The economic response mechanism typical of free markets is unable to alleviate the scarcity effects of these resources solely by means of prices, like it is usually done for private goods, showing increasing negative externalities that end up in the so-called "tragedy of the commons" (Hardin, 1968).

---

1. See Joffe and Robertson (2001) and Friel et al. (2008).
2. The difference between the actual productivity and the best level achievable using current technologies available.

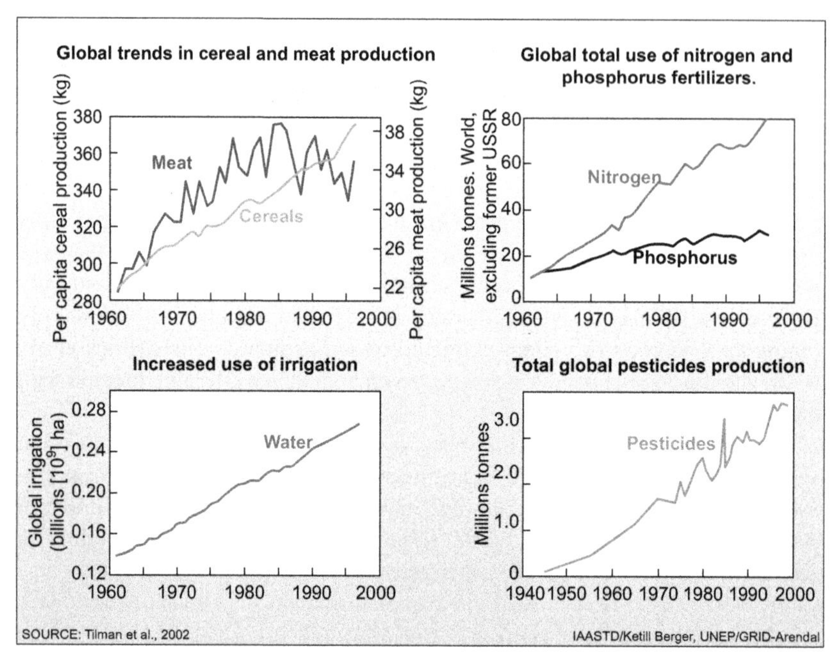

FIGURE 24.1 Trends in production and inputs use in agriculture. *Tilman, D., Cassman, K. G., Matson, P. A., Naylor, R., Polansky, S., 2002, Agricultural sustainability and intensive production practices, Nature 418 in UNEP http://www.grida.no/graphicslib/detail/global-trends-in-cereal-and-meat-production-total-use-of-nitrogen-and-phosphorus-fertilizers-increased-use-of-irrigation-total-global-pesticides-pr_ef80.*

Urbanization destroys most fertile areas, and economic framework offers room for oligopoly, monopoly, and distribution asymmetries, all reducing the development and investment opportunities in the whole economy in general, and in the food sector specifically. Therefore the challenge is a global approach that requires sustainability in the way food is produced, stored, processed, distributed, and assessed.

It has been discussed earlier that a higher income drives a higher demand for better food, which hence encourages new investment on the distribution side. This process leads to more industrial concentration in the retail sector, inducing in turn higher concentration on the whole supply chain.[3] Also, bigger enterprises usually adopt less sustainable industrial processes due to the intensive use of producing factors.

3. Concentration backward along the supply chain could be due to two different reasons. It might either be a countervailing reaction of producers to balance increased retailers' market power or a solely business evolution to satisfy growing logistic, organizational, and supplying needs of a concentrated retail sector.

In the Organisation for Economic Co-operation and Development countries in which sustainability might be threatened by such physiological industrial evolution of food chain, measures to mitigate the negative effects are often adopted. For instance, the European Union and United States have introduced and financed agricultural policy measures devoted to strengthen environment-friendly production (i.e., organic agriculture and low-impact agriculture, both characterized by lower input necessities and lower output level). However, the debate on the environmental effects of these agricultural practices displays different opinions regarding cost/benefit results. Nevertheless, there is no doubt that these policies have generated negative externalities on less developed countries: the curtailment of total production (the European Union and the United States play a key role in the global context of most commodity production) obviously causes a rise in prices hard to cope with especially for developing countries.

Hence, a small increase in environmental sustainability in a specific country may induce a loss of global sustainability.

Economic growth in Western countries allows private companies to adopt marketing strategies oriented to stimulate and increase the demand.[4] With respect to food sector, an increase in consumption often does not imply a growth in welfare, thus causing a double cost for the consumer: the cost for a higher demand for food and the recovery cost for the sanitary sector (due to food-related pathologies) as a tax payer. However, food processing industry, retail sector, and media unanimously play a key role in changing consumer habits toward a high-demand consumption style.

Indeed the consumer is encouraged by offers, price discounts, supersized portions, and strategic use of an anticipated "best before" date. Consumer is induced to buy an unnecessary amount of goods that inevitably becomes waste to be managed with additional resource consumption.[5]

On the contrary, in developing countries a similar amount of food is wasted due to the unfit storage and distribution sector. Insects and spoilage cause loss of one-third of the southeastern harvest, and 35%–40% of food harvested in India is lost due to the lack of cold storage systems in retail sector (Godfray et al., 2010). Overall, recent estimates indicate that one-third (equal to 1.3 billion of tons) of food produced for human consumption is ether lost or wasted along the global food system (Gustavsson et al., 2011).

4. The search of profit maximization strategy requires companies to adopt a behavior to increase revenues via an increase in the sold quantity, an increase in prices, or even both. Therefore a company's survival strictly depends on the increase in consumers' consumption.
5. Around 30% of food is wasted in developed countries (Segre and Gaiani, 2011).

ENVIRONMENTAL IMPLICATIONS

Because of the rapid population growth, the availability of freshwater per person has decreased from 15,900 m^3 in 1950 to 10,800 m^3 in 1970 and 8000 m^3 in 1990 down to 6500 m^3 in 2000. In 2010, the availability of freshwater has continued to decrease arriving at 5800 m^3, and it is expected to decrease in the future to 4400 m^3 per person in the year 2050 (UNEP, 2008).[6]

This quantity (overestimated because not all groundwater resources are technically/economically accessible) could be sufficient to satisfy the needs of the entire world population if it could be provided equally, but many countries of Africa, Middle East, and East Asia and some Eastern European countries much lower freshwater availability than the average and often under levels of subsistence. It is estimated that in 2025 about 3.5 billion people will fall into the category of "water scarcity" with an average yearly availability of 1700 m^3 per capita (UNEP, 2008).

It is important to remark that freshwater source distribution is not homogeneous on the whole planet; 60% of accessible water is concentrated in only nine countries: Brazil, Russia, China, Canada, Indonesia, United States, India, Columbia, and the Democratic Republic of Congo (WBDSC, 2009). Freshwater despite being a renewable resource, is not inexhaustible and is even subject to the risk to become scarce.

As regard water of lower quality, agriculture is the most important consumer, accounting for an average up to 70% of water withdrawn at the global level (FAO, 2002). For irrigation systems to be sustainable, there should be proper management (to avoid salinization) and there must not be more water used from the source than is naturally replenished. Otherwise, the water source effectively becomes a nonrenewable resource.

Irrigated agriculture plays a relevant and increasing role in producing food. The crop output level obtained from irrigation is more than double the level achievable by means of solely rain water contribution. A minimal irrigation in critical times makes it possible to obtain much better results than those achievable merely by rain water, even if with a good amount of rain. This is the result of the advantage of controlling the quantity of water absorbed by the plant roots when they need it.

This explains why irrigation contributes to the increase of crop productivity mainly in areas characterized by arid, semiarid, and subhumid climates. Irrigation is often associated with excessive water use of both surface and underground resources, therefore causing an overexploitation associated not only with a quantitative depletion but also with a qualitative degradation (e.g.,

6. These data refer to the share of renewable resources (potentially usable rainfall amounted to 40,000 km^3—The FAO State of Food and Agriculture, 1993); however, since the rains are not all used because of their distribution, it has been estimated that the share actually usable is 12,500–14,000 km^3 (UNEP, 2008).

chemical and biological contamination of underground pure water). In addition, the creation of such negative externalities is not internalized in current irrigation cost (often set on the simple need to cover the service and delivery costs).

Moreover, evidence of the negative consequences of high intensive irrigation is the increase of marshlands and soil salinization. It is estimated that about 30% of irrigated land is affected by these two problems in a more or less extensive way; progressive salinization of irrigated areas is provoking the reduction of usable land with an increase of 1%–2% per year (FAO, 2002).

The effect of water on the soil is strictly influenced by water management actions. Increasing erosion effects are evident in many areas, and erosion is one of the major factor causing loss of both organic matter and fertile soil. Estimates indicate a yearly loss of 12 million hectares of agricultural land due to soil degradation, which equals a loss of 20 million tons of grain (UNCCD, 2011).

Erosion is still one of the most critical concern with respect to climate change due to the intensification of heavy rain phenomena. Some of the world's most important agricultural districts are located in megadeltas where salt water intrusion and rising level of seas will become a severe threat for food production (Beddington et al., 2011).

Climate change is, moreover, influenced by human agricultural activities, and among others animal husbandry is the most critical. Agriculture as a whole generates approximately 20%–35% of greenhouse gas (GHG) emissions (IPPC, 2001) and 40%–50% of the total anthropogenic emissions of CH_4 and N_2O (up to more than 80% according to some authors; Smith et al., 2008).

Leaving aside the solutions that rely on abatement of GHG emissions by means of a reduction of livestock and those related to research effects in the field of animal diet and breeding that are yet to give consistent outcomes, the most effective managing option to be adopted is the concentration of husbandries. This may allow to manage the housing and storage activities in a closed environment and apply "end-of-pipes strategies" to treat wastes and mitigate the emissions.

This is the most reliable choice if considering the consequences of a change in the diet. Given the current genetic value of animals, alteration of feeding protocols would determine a lower production per capita; accordingly a larger number of animals would serve to satisfy the market demand, and eventually the production of GHG would increase.

Different strategies to limit the production of GHG also present strong implications on other relevant aspects of agriculture. One of the most invoked solutions with respect to food safety and environment-friendly cultivation practice is organic agriculture. However, organic agriculture requires a higher use of manure and slurry, which on the other hand increases the soil emissions of CH_4 and N_2O. Extensive agriculture, in contrast to intensive agriculture, can contribute to lowering the emission of CH_4 (thanks to the grassland diet) but at the same time leads to an increase of N_2O leaching in soil and water bodies, caused by grazing livestock excreta. Extensive grazing, in addition, lowers the soil productivity (yield gap) and requires a higher amount of land to reach the same output level.

Intensive husbandry allows the efficient treatment of manure at farmyard level (e.g., biogas production); however, it implies intensive agriculture as well on the fields surrounding. This may lead to an increase in N_2O leaching in the soil due to an increase in fertilizer use and pesticide pollution (which conversely are not used in grazing systems).

GHG reduction may be easily combined only with intensive farming contributing to a renewable energy production and a lower emission caused by the digested mass spread on soil.

Despite the fact that human demand for protein foods may be satisfied with vegetable protein too (using legumes not requiring fertilization since they are self-sufficient with respect of nitrogen) instead of animal products (meat, eggs, and milk products), demand for meat is expected to grow rapidly due to the increase in consumption caused by the expected rise of per capita income level generally already observed all over the world in the past years (Figs. 24.2 and 24.3).

Energy production from manure and slurry is important as well as the potential energy produced by food chain wastes. Companies' packaging strategies drive consumers to dispose a growing amount of wastes both in terms of food unused and package not eatable. In Western countries nowadays food packaging often have the same importance as the content in terms of

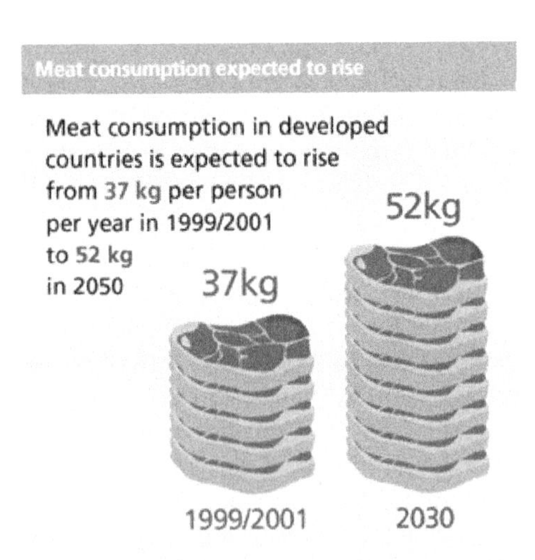

FIGURE 24.2 Meat consumption in Western countries. *http://www.unwater.org/statistics_sec.html.*

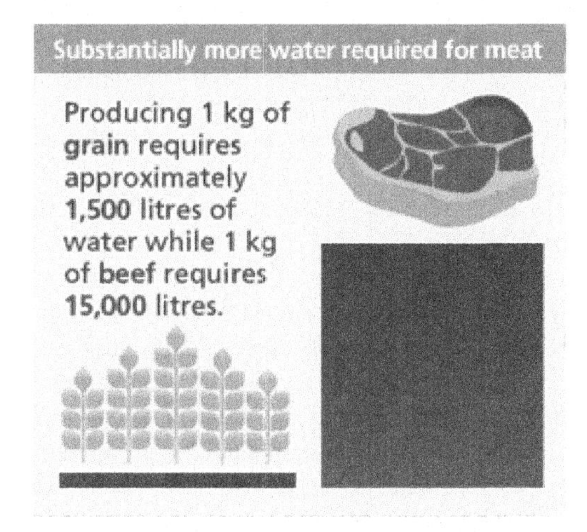

FIGURE 24.3 Resources involved in the production of meat and cereals. *http://www.unwater.org/statistics_sec.html.*

energy consumption and economic value. Reuse of packaging and eventually energy production is a keystone for sustainability of food consumption. Per capita food waste by consumers in Europe and North America is estimated at 95–115 kg/year, whereas this figure in sub-Saharan Africa and South/Southeast Asia is only 6–11 kg/year (FAO, 2009).

A 50% reduction of food losses and waste at the global level would save 1350 km^3 of water (for comparison, the mean annual rainfall over Spain is 350 km^3, water passing Bonn in Die Rhine is around 60 km^3 per year, and the storage capacity of Lake Nasser is nearly 85 km^3) (Lundqvist, 2011).

DEVELOPING NEW SOLUTIONS

The search for solutions follow two different approaches. The first is devoted to find mitigation practices to manage the actual negative effects of developing the food chain. The second focuses on prevention measures by means of research and developing long-term solutions.

Furthermore, all considerations mentioned earlier can be analyzed at different levels with regard to geographical scale: sustainability has to be assessed referring to a specific region or to the ecosystem in which specific activities take place, or even to the whole planet. Development paths usually show a growing concern about sustainability only after certain thresholds in terms of well-being have been trespassed, but measures devoted to increase in sustainability are mostly restricted to a local dimension. Induced externalities in neighbor countries or foreign market are too often not dealt with, since decision-making processes are built at government scale and international cooperation on such topics is difficult to achieve due to divergence of interests between the actors involved in the decision (as international agreements on environmental matters are often show). However, the environmental and human diversity forbids the identification of a unique solution for the whole planet.

With respect to food scarcity, research aims to reduce the yield gap by increasing the productivity of factors and/or reducing the use of scarce resources, and among these those more influenced by climate change.

Earth carrying capacity might be increased by means of plant breeding and genetic improvements that can help the cultivation of marginal soils increasing cultivated crops tolerance with respect to salt, dry, and pests.

Investments devoted to the reduction of food production impacts can help, mostly in mitigating the effects of by-products. The production of energy from renewable resources (as production wastes are with respect to agriculture) in the short term can lower the consumption of fossil fuels and support the reach of CO_2 emissions level agreed at international level.

Investments, however, at whatever level considered (agricultural research, biotechnology, recycling technologies, etc.) are constrained by the uncertainty related to future scenarios and the rights ownership framework. One of the most debated fields, nowadays, concerns property rights on biotechnology research results. Since the legal framework is still not clear, there may be a delay in investments and adoption of research results.

As a last remark, sustainable development implications usually imply different effects at different levels, often with opposite results, both quantitative and qualitative, which require the identification of optimal thresholds between different goals. Hence a coherent theoretical framework that takes into account different sources of information (like multicriteria models allow) seems to be preferred.

REFERENCES

Beddington, J., Asaduzzaman, M., Fernandez, A., Clark, M., Guillou, M., Jahn, M., Erda, L., Mamo, T., Van Bo, N., Nobre, C.A., Scholes, R., Sharma, R., Wakhungu, J., 2011. Achieving Food Security in the Face of Climate Change: Summary for Policy Makers from the Commission on Sustainable Agriculture and Climate Change. CGIAR Research Program on Climate Change, Agriculture and Food Security (CCAFS), Copenhagen, Denmark.

FAO, 2009. The state of agricultural commodities markets. In: High Food Prices and the Food Crises – Experiences and Lessons Learned, Rome.

FAO, 2010. A Conceptual Framework for Progressing towards Sustainability in the Agricultural and Food Sector (Discussion Paper).

FAO, 2002. Acqua per le colture. Corporate Document Repository.

Freibauer, A., Mathijs, E., Brunori, G., Damianova, Z., Faroult, E., Gomis, J.G., O'Brien, L., Treyer, S., 2011. Sustainable food consumption and production in a resource-constrained world. In: 3rd SCAR Foresight Exercise, Standing Committee on Agricultural Research (SCAR). European Commission, Brussels.

Friel, S., Marmot, M., McMichael, A.J., Kjellestrom, T., Vagero, D., 2008. Global health equity and climate stabilisation: a common agenda. The Lancet 372, 1677–1683.

Godfray, H.C.J., Beddington, J.R., Crute, I.R., Haddad, L., Lawrence, D., Muir, J.F., Pretty, J., Robinson, S., Thomas, S.M., Toulmin, C., 2010. Food security: the challenge of feeding 9 billion people". Science.

Gustavsson, J., et al., 2011. Global Food Losses and Food Waste. FAO, Rome, Italy.

Hardin, G., 1968. The tragedy of the commons. Science 162 (3859), 1243–1248. New Series.

IPCC, 2001. In: Houghton, J.T., et al. (Eds.), Climate Change 2001: The Scientific Background. Cambridge University Press, UK.

Joffe, M., Robertson, A., 2001. The potential contribution of increased vegetable and fruit consumption to health gain in the European Union. Public Health Nutrition 4, 893–901.

Lock, K., Smith, R.D., Dangour, A.D., Keogh-Brown, M., Pigatto, G., Hawkes, C., Mara Fisberg, R., Chalabi, Z., 2010. Health, agricultural, and economic effects of adoption of healthy diet recommendations. The Lancet 376, 1699–1709.

Lundqvist, J. (Ed.), 2011. On the Water Front: Selections from the 2011 World Water Week in Stockholm. Stockholm International Water Institute (SIWI), Stockholm.

Segre, A., Gaiani, S., 2011. Transforming Food Waste into a Resource. Royal Society of Chemistry (RSC), UK.

Smith, P., et al., 2008. Greenhouse gas mitigation in agriculture. Philosophical Transactions of the Royal Society 363, 789–813.

Tilman, D., Cassman, K.G., Matson, P.A., Naylor, R., Polansky, S., 2002. Agricultural sustainability and intensive production practices. Nature 418.

UNEP, 2008. An Overview of the State of the World's Fresh and Marine Waters, second ed.

United Nations Convention to Combat Desertification, 2011. Desertification: a visual synthesis. UNCCD Secretariat, Bonn (Germany).

World Business Council for Sustainable Development ed. (WBDSC), 2009. Facts and Trends. Water (Version 2), Geneva.

FURTHER READING

Baldos, U.L.C., Hertel, T., 2012. Economics of global yield gaps: a spatial analysis. In: Paper Presented at the 2012 AAEA Annual Meeting, August 12–14, Seattle, USA.

European Commission, 2010. The CAP towards 2020: Meeting the Food, Natural Resources and Territorial Challenges of the Future. European Commission, Brussels.

European Commission, 2012. Innovating for Sustainable Growth: A Bioeconomy for Europe. COM(2012)60 final, Brussels.

Iglesias, A., Quiroga, S., Diz, A., 2011. Looking into the future of agriculture in a changing climate. European Review of Agricultural Economics 38 (3), 427–447.

Isenburg, T., 2006. La cultura dell'acqua. In: L'acqua fra tecnologia e ambiente. Università di Firenze.

Monteny, G.J., Bannink, A., Chadwick, D., 2006. Greenhouse gas abatement strategies for animal husbandry. Agriculture Ecosystems and Environment 112, 163–170 (Elsevier).

OECD-FAO, 2012. Agricultural Outlook 2012–2021 (Paris/Rome).

Schmidhuber, J., 2007. Does the EU need a food security policy? In: Presented at the Seminar Food Security in the EU, Issues and Outlook, 28 February 2009, Brussels.

Chapter 25

Insights on Establishing a Cohesive and Enduring Campus Sustainability Initiative

Sierra Flanigan, Talia Arnow
Coalesce Accelerator, Boston, MA, United States

Chapter Outline

Sustainable Cities and Communities Design Handbook. https://doi.org/10.1016/B978-0-12-813964-6.00025-2

INSIGHT 1: MANY CAMPUSES STILL NEED A COLLECTIVE AND SHARED DEFINITION OF SUSTAINABILITY, A VISION FOR SUSTAINABILITY, AND A ROADMAP WITH CLEAR GOALS AND METRICS

Coalesce's Campus Sustainability Pyramid.

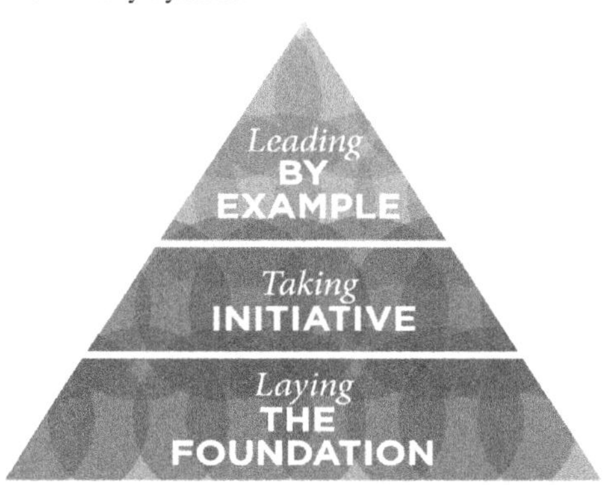

Community-generated definition and vision for sustainability are foundational elements of a strong initiative. The institutional adoption of this definition and vision bring clarity, direction, shared language, and consensus. With these key ingredients a campus is ready to outline a roadmap to achieve its vision. This roadmap or "action plan" articulates campus-wide commitments, goals, strategies, and metrics. This equips a school to execute effectively with purpose.

INSIGHT 2: SUSTAINABILITY IS BEING ESTABLISHED AS AN INSTITUTIONAL PRIORITY BECAUSE IT IS AN ECONOMIC AND STRATEGIC IMPERATIVE FOR ORGANIZATIONAL GROWTH, COMPETITIVENESS, AND LONG-TERM STAKEHOLDER VALUE CREATION

Millbrook School's Solar System.

As historical institutions, many campuses were built in the past defined by a different set of challenges. In order for campus systems to reflect 21st century realities, institutions must evolve as adaptive organizations to mitigate risk, stay competitive, and carry out their mission in preparing the next generation for success. Academic institutions are recognizing the significant risk of relying on production of products from fossil fuels and experiencing that there are a growing number of students, parents, and trustees who have expectations for schools to be ethical, environmental stewards. Schools are realizing cost savings from investing in energy efficiency and renewables and rebranding themselves to attract/retain talent and increase revenue. Leading schools are weaving sustainability as an organizing principle throughout their Strategic Plans and Master Plans and investing in sustainability initiatives because it bears directly on bottom-line business performance. Many parallels can be drawn from research conducted with corporate companies. See McKinsey and Company: Profits with purpose: How organizing for sustainability can benefit the bottom line.

INSIGHT 3: THE OPPORTUNITY SET FOR FINANCING SUSTAINABILITY PROJECTS IS EXPANDING AND THE ECONOMIC RETURNS ARE ATTRACTIVE

There is an array of financing mechanisms worth exploring to fund sustainability projects. Third-party financing through Power Purchase Agreements and Energy Service Agreements and working with energy savings companies are fitting for some schools, whereas other schools are taking advantage of utility rebates and incentives. More internal approaches are beings taken through establishing Green Revolving Funds and Green Community Funds, investing in campus infrastructure through cash reserves, and fundraising for sustainability projects from benefactors. See the Greening the Bottom Line report written by the Sustainable Endowments Institute (Fig. 25.1).

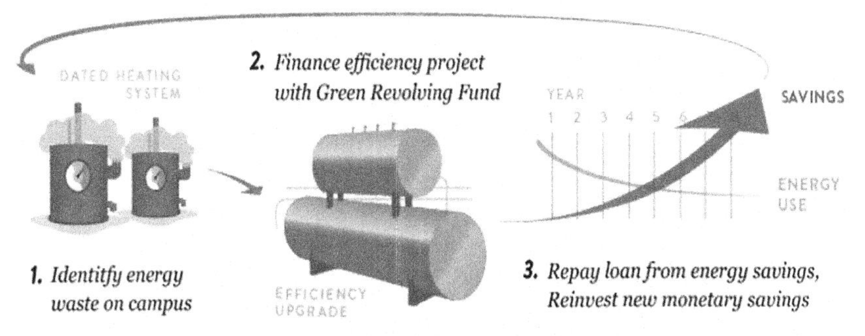

FIGURE 25.1 How Green Revolving Funds work. *Image from Greening the Bottom Line Report, Sustainable Endowments Institute.*

INSIGHT 4: COLLABORATION BETWEEN CAMPUSES AND COMMUNITIES IS FUNDAMENTAL TO MITIGATE AND ADAPT TO CLIMATE CHANGE AT THE LOCAL AND REGIONAL SCALE

Hotchkiss School's Biomass Facility.

A campus is nested within a community, and on a larger scale, is embedded within a region. There is an inherent economic, environmental, and social relationship between a campus, its local community, and its region, whether it be through attracting talent and increasing intellectual capital of the place, to influencing transportation demands, to water availability. Collaborative and collective planning between city stakeholders and campus stakeholders is fundamental to determining meaningful and aligned sustainability and resilience measures. A growing number of college and university presidents are signing Second Nature's resilience commitment. See Second Nature's Resilience Commitment here.

INSIGHT 5: DEEP ENGAGEMENT AND PARTICIPATION OF STAKEHOLDERS IS VITAL FOR EMBEDDING SUSTAINABILITY INTO THE CULTURE OF A SCHOOL

There is a significant difference between add-on approaches to sustainability solutions and built-in, community-generated approaches to sustainability solutions. After years of critical analysis of successful campus sustainability initiatives, Coalesce recognizes that consistent stakeholder engagement and capacity building drives the adoption of sustainability commitments. Increasing stakeholder participation broadens ownership, which propels visibility and communication and empowers champions. See Sustainable Universities - a study of critical success factors for participatory approaches research paper here.

FOUR CORE COMPETENCIES TO CONSIDER FOR ENSURING A SUCCESSFUL CAMPUS SUSTAINABILITY INITIATIVE

- Assessment and Benchmarking
  - Discover what has been accomplished, understand diverse stakeholder perspectives, realize priorities and next steps, and compare your campus' practices to peer efforts.
- Governance and Leadership
  - Integrate sustainability into decision-making structures as an economic and strategic imperative to mitigate risks, seize opportunities, and institutionalize commitments.
- Planning and Measurement
  - Chart a course of action that articulates goals, strategies, and meaningful metrics to meet commitments and measure progress over time.
- Capacity and Execution
  - Identify the bandwidth and expertise needed to accomplish goals either internally or through forming external partnerships.

The goal is to enhance your school's capacity to model behavior, prepare students, and align values with practices

Reasons Why It Is Important to Establish a Cohesive and Enduring Campus Sustainability Initiative.

Sharpen Competitive Edge	Align Strategic Priorities	Educate the Community
Foster innovation	Steward the environment	Engage stakeholders across campus
Attract top talent	Realize savings	Equip members to practice sustainability
Enhance reputation	Improve infrastructure	Increase employee satisfaction
Build school brand	Advance educational outcomes	Retain talent

Check out Coalesce's website for additional resources on Campus Sustainability: www.coalesce.earth

Chapter 26

The Power of Sustainability: The Story of Kent, Ohio

Myra Moss
Ohio State University Extension, Columbus, OH, United States

Chapter Outline

BACKGROUND

The City of Kent and the Kent State University (KSU) have long been dependent upon each other for public services, employment, and income. Despite this dependence and impact, there was little more than necessary—bordering on begrudging—interaction between the city and the university. Even more confounding than this detachment is the proximity of the two—Kent's downtown and the campus were within a block of each other, separated by a neighborhood with student rental housing and a state highway named Haymaker Parkway.

Sustainable Cities and Communities Design Handbook. https://doi.org/10.1016/B978-0-12-813964-6.00026-4

Haymaker Parkway, a four-lane road located between the KSU Campus and Downtown Kent, served to create a cultural and physical barrier between the campus and the downtown. One side of Haymaker was access to campus, and at the other side was the Kent central business district, without much interaction between the two. Haymaker was designated as a limited access highway by the Ohio Department of Transportation. The few available crosswalks were inconvenient and far apart. Traffic tended to speed on this four-lane road, so crossing at places other than those few designated crosswalks was taking your life in your hands.

Even if Haymaker did not present a physical and cultural barrier, there was not much for the students to do in the downtown area, other than to meet friends at local bars or eat at one of the few student-friendly restaurants. Students admitted that they did not feel welcomed by local shopkeepers, who they felt watched them closely, anticipating shoplifting. Kent's merchants and service providers did not take advantage of the "captive" resource literally in their backyard; in fact, they often treated them with disdain.

Another sore point, creating resentment among some Kent leaders, was that KSU faculty and administrators choose not to live, or shop, in the city. Several "up-scale" lakefront developments in the adjacent township were preferred. Kent's housing stock largely reflected the city's blue-collar roots, so the type of housing sought by professional families was not easy to find within the city limits.

Town—gown relationships were further eroded on May 4, 1970. An anti-Vietnam war student demonstration on the college green, following student riots in downtown Kent 2 days earlier, erupted in violence. The Ohio National Guard, sent to Kent to quell the riots, shot 13 unarmed students, killing 4. *The New York Times* summed up the relationship between the city and the university as follows:

> *Though it is home to the second largest campus in Ohio's state university system by enrollment, this small Cuyahoga River city spent much of the last four decades neglecting, if not deliberately retreating from, its history as a college town and its place in the annals of the Vietnam War era.*
>
> Schneider (2013).

Kent's residents have long supported the importance of a healthy environment. It is one of the few cities with not one, but two environmental entities. In 1970, residents created the Kent Environmental Council (https://kentenvironment.org), and in 1995, the city established the Kent Environmental Commission (http://www.kentohio.org/dep/enviro_com.asp). The citizen-run Kent Environmental Council initiated the city's first recycling program in 1970, and from 1979 to 1989 it became the first comprehensive and self-supporting recycling program in Ohio. The Kent Environmental

Commission was tasked by the city to help monitor local environmental initiatives and, most recently, evaluate the outcomes and accomplishments of the City's Comprehensive Plan (KSU Special Collections and Archives).

In 2002, the city recognized the need to update its comprehensive plan and assigned this task to the Community Development Department and its director and staff. The director was aware of an innovative approach to community development and planning, i.e., sustainable development, that was beginning to emerge in the United States and had gained a foothold internationally. He decided that approaching their comprehensive plan update from a sustainability perspective would be desirable. Partnering with a retired sociology professor at the KSU, Kent's Community Development Director began to promote the concept to city officials and to leadership at the KSU, gaining support and buy-in.

CREATION AND IMPLEMENTATION OF THE KENT SUSTAINABLE PLANNING APPROACH

Sustainable comprehensive planning was a newly emerging approach to the development of a community land use plan, and the city recognized the need to find a "consultant" who could help to design and provide guidance throughout the process. Faculty at Ohio State University (OSU) Extension had just finished piloting such a planning approach in a rural southeast county and established a website to present their process and results. Upon finding the website, the City Community Development Director and the KSU Professor contacted OSU Extension to secure their assistance. Kent would become the first city in the State of Ohio to create a Sustainable Comprehensive Plan.

Ohio State University Extension: Sustainable Planning Model

In 1998, the Extension faculty at OSU formed the Sustainable Development Team. The purpose of this initiative was to help communities address their planning efforts from the perspective of sustainability. The most commonly used definition at that time was articulated by the United Nations "Bruntdland Commission": "Sustainable development is development that meets the needs of the present generation without compromising the ability of future generations to meet their own needs" (World Commission on Environment and Development, 1987, p. 8). Core to the many definitions of sustainability that followed were a number of key concepts: intergenerational equity, balance of social, environmental and economic needs, and inclusion of diverse populations in determining the long-term future.

OSU Extension's planning model (Moss, 2016) reflected the core concepts of sustainable development. The process was guided by a shared community vision and incorporated four cornerstones of sustainable community development:

- Inclusion: Creation of a Guidance Team (steering committee) that broadly represented the community; discovering the community's shared vision through an inclusionary visioning process
- Long-term: Using a planning window of two generations, or 50 years
- Balancing and linking social, environmental, and economic aspirations of the community in an intentional way
- Multidimensional: Creating multidimensional goals that would insure sustainable outcomes

Under this model, the role of residents/citizens was to identify their shared, desired future (vision); the role of elected and appointed officials then became to help the community achieve their vision though targeted planning efforts and the allocation of resources. The process was "grass roots, bottom-up," and resulted in greater ownership and buy-in from community residents who could see their place in the development and implementation of the plan. Examples of linkages, which are so critical to understanding and implementing sustainability, are depicted in the Fig. 26.1

In 2002 Kent began the creation of the Kent Bicentennial Plan based on the principles of sustainability (http://kentohio.org/reports/bicentennial.pdf).

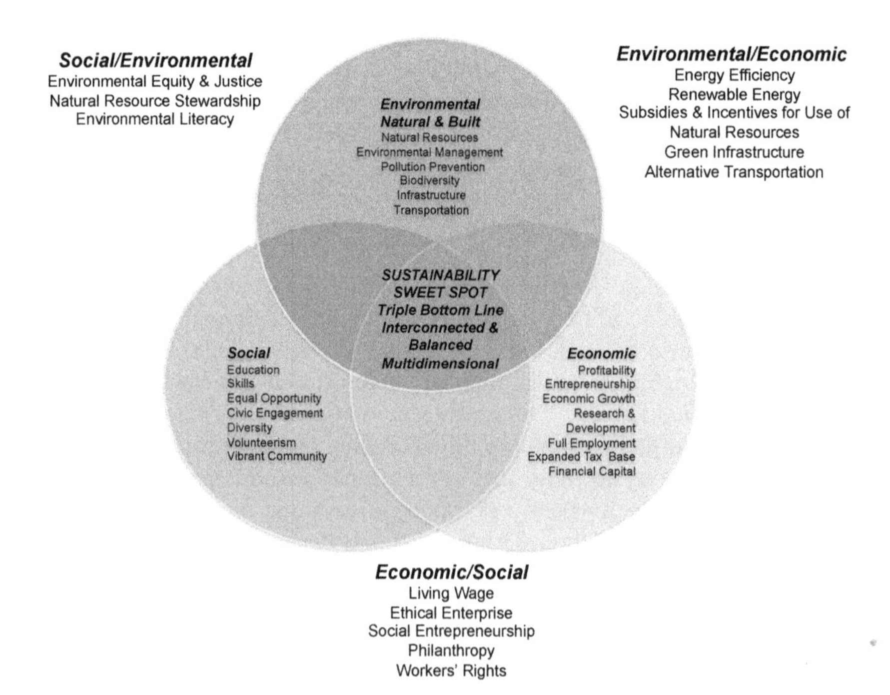

FIGURE 26.1 The three spheres of sustainability. *Adapted from U.S. EPA Framework.*

The plan's name was chosen in honor of Ohio's Bicentennial in 2003, the City of Kent's Bicentennial in 2005, and KSU's Centennial in 2010. The plan took over 2 years to complete and on November 4, 2004, it was approved and adopted by the Kent City Council.

STEPS IN THE CREATION OF THE BICENTENNIAL PLAN

Step 1: Building a Partnership

Through subsequent meetings between the City of Kent, Community Development Department, OSU, and KSU, an agreement was reached to use the OSU model and guidance, with technical assistance and support from KSU faculty and management by the Kent Community Development staff. Dr. Carol Cartwright, the President of KSU, tasked a team of faculty and administrators to provide expertise to the city in the development of the plan. Also, KSU committed the assistance of the Kent Urban Design Center (now known as the Cleveland Urban Design Collaborative), home of the urban design graduate program at KSU, and the public sector outreach efforts of the College of Architecture and Environmental Design (http://www.cudc.kent.edu/index.html). The Design Center provided technical design assistance and graphic renderings. They were critical in helping to interpret and conceptualize, in drawings and site plans, the resident's input into Kent's special planning areas.

The city, through previous planning efforts and conversations over the years, had identified three "special planning areas" that warranted special attention in the Bicentennial Plan. The Kent Design Center was tasked with providing master planning guidance for these three areas, including facilitation of community input sessions and drafting renderings that interpreted the resident's visions.

Step 2: Establishing Planning Governance

Two entities, an inclusive Steering Committee and a Design Team of experts, were created to guide the planning process and provide subject-matter expertise throughout the development of the plan. The Steering Committee represented a broad base of social, economic, and environmental sectors of the Kent community. The Design Team included faculty from KSU and OSU, the City Community Development Department staff, and experts from the KSU Urban Design Center.

Bicentennial Plan Steering Committee

A 46-member inclusive Steering Committee with representation from all sectors of the community was created; volunteers were sought from neighborhood associations, churches, schools, business organizations, Boards and

Commissions, community groups, and KSU administrators, faculty, and students. Volunteers spanned environmental, social, and economic sectors of the community.

The Steering Committee participated in an orientation, conducted by the OSU Team, to sustainable community development, the planning process built on sustainability principles, and their roles and responsibilities included:

- Championing the Bicentennial Plan and urging community involvement
- Helping to identify where and with whom vision sessions could be held to ensure that all interests in the community were encouraged to participate
- Facilitating visioning sessions to be held throughout the community
- Promoting attendance at the visioning sessions
- Recording results and determining themes that emerged from the sessions
- Providing guidance throughout the development and writing of the final plan

Bicentennial Plan Design Team

The Steering Committee was totally composed of a broad base of Kent residents, whereas the 19-member Design Team was created to provide technical and process assistance to this group and to the city. Design Team members were experts in specific disciplines that was anticipated would be useful in the creation of the plan. Examples include:

- Process guidance and expertise in community sustainability: OSU Extension Team
- Process management and coordination; knowledge of the community: Kent Community Development Office
- Interpretation of resident's shared vision through conceptual drawings and plans: Kent Design Center
- Topical knowledge: KSU faculty and administrators with expertise in technology transfer and economic development, geology, water resources, public policy, biodiversity, communications and marketing, architecture, environmental design, facilities planning, and transportation services. Transportation issues and facilities planning university staff would help to identify and develop shared initiatives between the university and the city.

Step 3: Discovering the Community's Shared Vision

OSU's sustainability planning process was guided by discovering the community's shared vision of a desired Kent 50 years into the future. Input was solicited by holding vision sessions throughout the community in locations that were familiar and comfortable for residents, the design of which was to "go to where people gather to reduce barriers to participation." To organize this effort, the city was divided into eight districts. Four rounds of input sessions were conducted in each district, resulting in a total of 8 to 14 meetings per round. Additional sessions were conducted with targeted populations (homeless, African American neighborhoods, downtown retail merchants, for example) to ensure that all voices throughout the community were heard and input captured.

The Steering Committee was instrumental in identifying groups/areas to visit to ensure inclusion and secured locations where these groups felt comfortable to share their input openly. Volunteers from the Steering Committee facilitated the community visioning sessions along with OSU, City of Kent, and Kent Design Center. Facilitators received training from the OSU in the importance of remaining neutral and respecting all participants' views as well as methods to capture, record, and organize input for inclusion in the development of the plan. The following Table 26.1 is an overview of the four rounds and the purpose and goals of each.

TABLE 26.1 Community Visioning/Input Sessions

	Goals/Purpose
Round 1	• Define values and aspirations for the Kent community • Conductasset-based fact-finding meetings • Discover what community values and want to preserve. Could be a tangible or intangible asset (building, feature, social environment, etc.) • Discover what the community hopes it will be for future generations • Learn about the three Special Planning Areas
Round 2	• Residents prioritize the ideas, assets, and values shared in Round 1 and organize them into social, environmental, and economic columns • Prioritize top three items in each column • Create linkages between priority items in one column with other two to set the stage for sustainable, multidimensional goals • Review and respond to initial concept plans developed for Special Planning Areas by Urban Design Center
Round 3	• Focus, in depth, on three Special Planning Area Plans • Review architectural drawings incorporating citizen input generated during Round 2 • Solicit additional input from community that can be used to develop final draft of architectural drawings • Share final draft of drawings with Planning Commission, Board of Zoning Appeals, Environmental Commission, and the Architectural Advisory Review Board; prepare presentation to share with community
Round 4	• Receive feedback on the draft Bicentennial Plan • View final architectural renderings for the Special Planning Areas through a PowerPoint Presentation

In the fourth and final round of community input meetings, residents mostly agreed with and validated the plans as presented. Many raised one concern, i.e., that students' needs were not addressed adequately by the plan. Residents felt that students preferred to shop at "big-box" stores and that an economic development approach interspersing locally owned businesses with "big-box" stores would attract students to shop in the downtown and keep them in Kent to shop on weekends. Residents believed that these types of businesses could coexist with smaller retail and still maintain the "feel" of the downtown, particularly if the "big boxes" could be encourage to reuse and rehab, rather than raise, older main street buildings.

Step 4: Finalize and Adopt Plan

The Design Team analyzed the input gathered from the many community visioning and input sessions and translated common themes that emerged into overall goals, initiatives, and strategies. The Bicentennial Plan was presented to the City Council and approved in 2004.

IMPLEMENTING MULTIDIMENSIONAL, SUSTAINABLE GOALS

Implementing the sustainable goals of the Bicentennial Plan has resulted in the transformation of the Kent community. Oftentimes community plans do not reach stated goals, thwarted by changes in elected and appointed officials, administrations, and community partners. New administrators and officials who evolve into leadership positions might not share the same passion for the plan as previous administrations did. Since the Kent Bicentennial Plan was adopted the city has a new Mayor, City Manager, Council Members, Community Development Director, and Economic Development Director and the KSU has a new President. However, unlike some comprehensive plans, Kent's sustainable plan was a bottom-up process through which the entire community, including the university, set the overall vision and confirmed the goals. Residents articulated a shared vision of their desired, long-term future. They also identified what they felt was unique about their community, and they sought to preserve, as embodied in tangible and intangible assets.

The widely engaged community bought into the plan, seeing their place in its creation and implementation. Residents held city officials, community leaders, and university administrators responsible and accountable for the implementation of their plan, and on many critical occasions "held their feet to the fire." The broad-based community engagement in and support of the plan enabled the implementation of sustainable goals to transcend changes in administrations, public officials, and local leadership.

CENTRAL GATEWAY: TRANSFORMING KENT'S DOWNTOWN

Kent's downtown was important to the Kent community, and resident's desired it to be the economic/social/environmental focal point of the community. Kent resident's downtown of the future follows.

Shared Vision Themes

- A place to gather for families and individuals (alleys, places to shop, parks)
- Economically vibrant with a mix of locally owned small businesses and student-friendly retailers; downtown includes office space, services, and quality housing;
- A location for cultural activities and opportunities, arts and entertainment, and shopping. Places to live, eat, and play
- Offers recreational opportunities taking advantage of the Cuyahoga River
- An exciting location for entertainment and events based on the arts, music, and culture
- Preservation of Kent's history through reuse/use of older and historic buildings
- Eliminate disconnect between the university and the community, and between the campus and the downtown. Students shop downtown and parents can stay in town when they visit—place to stay, things to do (restaurants, shopping, entertainment)
- Physically and visually appealing with plantings, benches, art, and attractive streetscapes

These vision themes were interpreted into goals and strategies. Multidimensionality was built in at this stage, intentionally linking social, economic, and environmental features to each goal to ensure that they supported tenets of sustainability (balance and linkage). The following flow chart presents the planning process and development of multidimensional goals followed by examples (Fig. 26.2).

Examples of Multidimensional Goals From the Kent Plan:

Encourage and promote locally owned businesses

- Economic/social: Downtown merchants organize to improve commercial activity and sidewalk events.
- Economic/social/environmental (built): Kent will become a Main Street Program.
- Economic/social/environmental (built): Kent students have the opportunity to open shops in the downtown alleyways that attract students and residents.

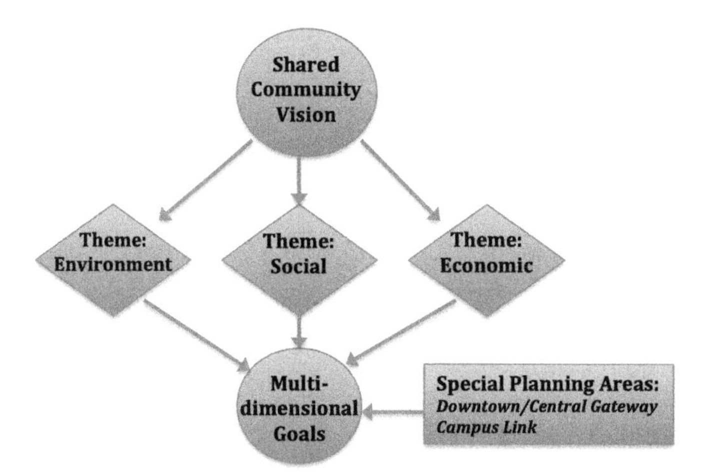

FIGURE 26.2 Process used to reach multidimensional goals.

Develop Kent's downtown as an economic focal point

- Environment (built)/economic: Create a comprehensive parking strategy for the downtown.
- Economic/social/environmental (built): Downtown Kent is a pedestrian-friendly area.
- Economic/social: Build a university hotel and convention center.

Promote traffic management and alternative forms of transportation

- Environmental (natural and built): Increase public transportation Portage Area Regional Transit Authority (PARTA) ridership by 2006.
- Environment/social/economic: Build a multimodal facility to increase the use of alternative forms of transportation.
- Environment (built)/social: Develop a 10-year plan for installation and repair of sidewalks.
- Environmental/social/economic: Eliminate Haymaker Parkway's barriers to pedestrian access from the campus to the downtown.

Use existing buildings for redevelopment

- Environment (built)/economic/social: Implement architectural standards for downtown by 2007.
- Economic/social/environmental: Develop a vibrant downtown by encouraging mixed uses in existing buildings.

Preserve natural resources for residents to enjoy

- Economic/social/environmental (natural): Cuyahoga River becomes a focal point and destination.

- Social/environmental (built/natural): Preserve and enhance significant historical features like the Main Street Dam on the Cuyahoga River.
- Environment/social/economic: Build a multi-modal facility to increase the use of alternative forms of transportation.

PLAN IMPLEMENTATION AND OUTCOMES

The City of Kent has accomplished much since presenting their plan to Council in 2004. Through public–private partnerships involving the city, private developers and investors, and the Kent State University Foundation and Alumni organization, the downtown revitalization and KSU connection as envisioned by community residents is taking place. The investment in Kent is projected to top $100 million and create up to 1600 jobs when completed. Much of the proposed investment in Fig. 26.3 below has been accomplished:

Development initiatives in Kent Central Gateway were designed to create a closer relationship between the university and the city, ease traffic and parking

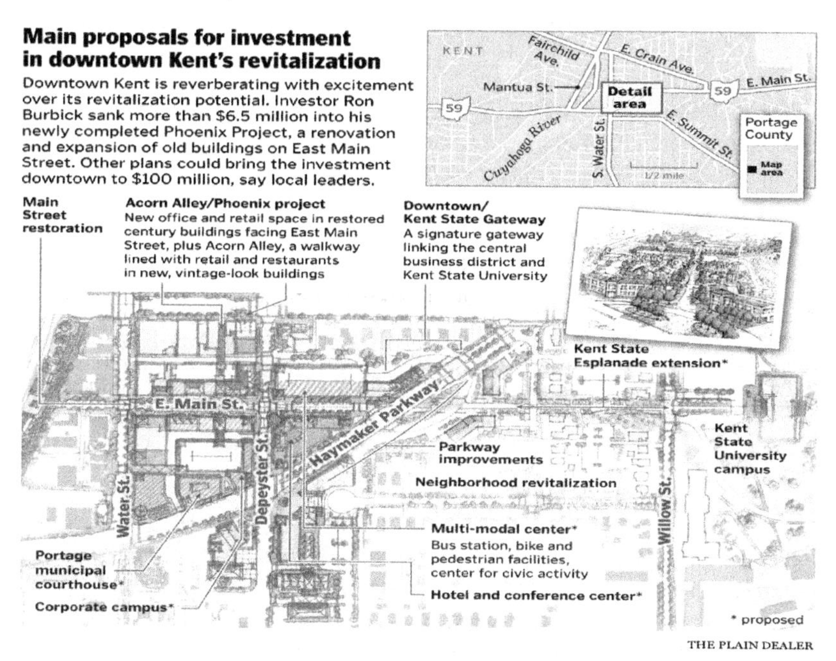

FIGURE 26.3 Proposed Central Gateway Projects. *Reproduced from ESRI; TeleAtlas; http://blog.cleveland.com/metro/2009/11/downtown_kent_ohio_rising_phoe.html.*

issues by reducing reliance on the automobile, and develop a multiuse downtown that was economically vital, socially diverse, and environmentally sensitive. Key initiatives included the following:

Promoting Traffic Management
PARTA Multimodal Center (left)

Opportunities for Parents to Stay and Shop in the Downtown
KSU Hotel/Conference Center (right)

The PARTA multimodal facility, an anchor for the Kent Central Gateway concept, was critical to Kent's vision of being a walkable, bikeable community with many alternative forms of transportation, reduced traffic congestion, and reduced need for parking in the downtown. It provided a transfer point for cars, busses, bikes, and pedestrians, as well as a visitor's center and retail shops. Commuters to Kent, many of whom are students, faculty, or employees in the local offices, can park their cars in the 300+ space center, access their bike stored in a rental locker, walk to the campus that is one block away, or catch a bus. The project was promoted jointly by the city; KSU; local, regional, and state political leaders; and the PARTA and funded through a $20-million TIGER Grant. PARTA is the ongoing manager of the facility.

Located directly across from the PARTA Multimodal Center, and another downtown anchor project, is the new Kent State University Hotel and Conference Center. Prior to this facility being built, there were no accommodations in the downtown area for parents and visitors and the University did not have an adequate conference and training facility. This location in the downtown provides easy access for visitors to shopping, entertainment, and dining. In addition, the KSU is recognized worldwide for its cutting-edge liquid crystal research. Kent state's 41,000 sq ft Centennial Research Park, located near the university, provides space to high-tech start-up enterprises that have been launched through KSU research activities. The Conference Center provides the university with a venue for events relating to this technology niche.

Tying Together the Campus and the Downtown: Haymaker Parkway and University Esplanade

*Improvements to **Haymaker Parkway** (left– looking toward campus from PARTA MultiOmodal)*

*KSU **University Esplanade** (below – looking from campus to PARTA and Conference Center)*

Haymaker presented an accessibility challenge and physical barrier between the campus and the downtown. The City Manager successfully negotiated with the Ohio Department of Transportation to have the Parkway's limited access highway status removed. Pedestrian-friendly and pedestrian-safe cross-walks were added to greatly improve access to the PARTA multi-modal facility, University Hotel and Convention Center shopping, and the main campus.

After the Bicentennial Plan was completed the University began addressing the need for additional parking on campus to ease congestion in city streets. Options included additional surface lots or a parking garage. KSU's Transportation Services Coordinator, who had been a member of the Bicentennial Plan's Design Team, partnered with the City of Kent to solicit input from community leaders and residents on options to improve parking and traffic congestion. These discussions, facilitated by OSU Extension and held over a number of months, resulted in a different proposal than was originally anticipated. Instead of adding more parking, which would require space and investment, the advisory group suggested improvements to enhance the use of alternative and public forms of transportation. The final initiative involved the extension of a KSU Esplanade that would provide dedicated pedestrian and bicycle access from the campus to the downtown, crossing Haymaker Parkway directly across from the PARTA Facility and the University Conference Center (http://thecollaborativeinc.com/project/esplanadeFextension#). The Esplanade was designed to facilitate bicycle access by tying into the Portage Bike and Hike Trail a 10-mile route from the City of Ravenna to Kent, with an offshoot that goes directly through the KSU campus.

Retail development in Acorn Alley with places to gather (historic Franklin Hotel in the background)

A place for residents and students to gather with access to shopping, entertainment, music, arts and culture

Entrance to Acorn Alley

Acorn Alley I and II, pedestrian-friendly multiuse downtown developments with shopping, entertainment, dining, and loft housing, is located in the block next to the PARTA multimodal facility and KSU Conference Center. A local developer/investor rehabbed and reused existing historic buildings for Acorn I and II, in keeping with the community's vision. The existing alleys and wide spaces between buildings were improved for small shops and places to sit and enjoy live music. The same investor substantially rehabbed the historic downtown Franklin Hotel, saving it from further deterioration and imminent demolition. Now known as Acorn Corner, this former hotel contains offices, a restaurant, and loft apartments.

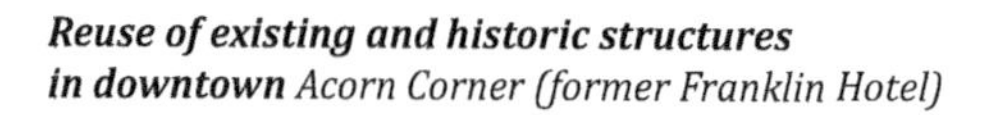

Reuse of existing and historic structures in downtown *Acorn Corner (former Franklin Hotel)*

Thanks to the efforts of local investors, along with the City of Kent administration and City Council, key players at Kent State University, Portage Area Regional Transportation Authority, and other community business leaders and stakeholders, there is a renaissance in downtown Kent. Their shared vision for a sense of community has infused more vitality and investment than the downtown has seen in a long time, restoring the pride and tradition of a vibrant downtown. Downtown Kent now boasts many points of interest uniting the old with the new by preserving the historical integrity of the past while creating a destination city of today.

About Acorn Alley, from http://acornalleykent.com/about/.

CONCLUSION

The Kent experience is a story of a community and its partners breaking down barriers and coalescing around a widely shared vision of the future, as articulated and implemented through a plan based on principles of sustainability.

The planning process itself generated collaborative approaches to shared problems and opportunities by bringing together citizens, community leaders, businesses, and technical experts in dialog with each other. The plan provided the catalyst to allow Kent to escape its past and work together for a common future. A clear indicator of success was the 2013 International Town-Gown Association's Larry Abernathy Award, given to Kent and KSU for their successful town–gown collaborative partnership (Record-Courier, 2013). The results that Kent has been able to achieve in a relatively short time—13 years—speak to the commitment of leadership to the resident's vision and hopes for the future, the residents' commitment to their sustainable community, and the power of sustainability.

REFERENCES

Moss, M.L., 2016. Comprehensive Planning Based on Sustainability: A Model for Ohio Communities. Ohio State University Extension, Columbus, Ohio.

Record-Courier, April 17, 2013. KSU, City of Kent Win Award From International Town-Gown Association. Retrieved from Schneider, K., February 5, 2013. A partnership seeks to transform Kent State and Kent. New York Times. Retrieved from: http://www.record-courier.com/news/20130417/ksu-city-of-kent-win-award-from-international-town-gown-association.

Schneider, K., February 5, 2013. A partnership seeks to transform kent state and kent. New York Times. Retrieved from: www.nytimes.com/2013/02/06/realestate/commercial/development-aims-to-bring-kent-state-and-its-city-closer.html.

World Commission on Environment, Development (Brundtland Commission), 1987. Our Common Future. Development and International Cooperation: Environment. Oxford University Press, Oxford.

FURTHER READING

City of Kent Bicentennial Plan (2005). City of Kent, Ohio. Retrieved from: http://www.kentohio.org/reports/BICENTENNIAL.pdf.

Kent Environmental Council, Records, 1970– Kent State University Special Collections and Archives. Kent State University. July 1997.

Chapter 27

The Los Angeles Community College District: Establishing a Net Zero Energy Campus

Calvin Lee Kwan[1], Andrew Hoffmann[2]
[1]*University of California, Los Angeles, CA, United States;* [2]*Independent Consultant, Chicago, IL, United States*

Chapter Outline

INTRODUCTION

With the distinct advantage of having "unique academic freedom, critical mass and a diversity of skills to develop new ideas" (Calhoun and Cortese, 2005), tertiary institutions are ideal locations to promote society changing technology, concepts, and practices. Industry, government, and private citizens have often contributed endowments, grants, and monetary gifts to support research and various functions of tertiary institutions. This is further aided by the fact that universities and colleges are essentially microcosms of society—miniature communities complete with power-generating facilities, transportation operations, residential and business functions, as well as food and waste services. Thus many of the challenges that the greater society faces can often be

Sustainable Cities and Communities Design Handbook. https://doi.org/10.1016/B978-0-12-813964-6.00027-6

observed or mimicked in the university setting, albeit in a more controlled environment. This provides an excellent opportunity and platform to develop, test and refine solutions aimed at addressing societal challenges, including those associated with climate change.

BACKGROUND

Net zero buildings—those that generate enough electricity from renewable energy sources to meet their total operational demand—Very few institutions in the United States—tertiary, secondary, or primary—have met more than 30% of its energy demand through renewable technologies. The best example to date is the small liberal arts campus located in Morris, Minnesota, part of the University of Minnesota system. The campus is able to meet 60% of its energy demand through its wind turbine farm.[1] The school is able to do this because of several factors, including the non-energy intensive nature of the programs offered, the small total campus population, and abundance of open land available for installing renewable energy resources. This combination of factors is not typical of a tertiary institution, let alone a city. Regardless, this is still a feat. However, the Los Angeles Community College District aims to set the standard even higher with the ambition of making all nine of its campuses operate at net zero energy. A net zero campus is defined as one which uses renewable energy technologies to generate enough electricity to meet its total consumption.

With over 180,000 students enrolled across nine colleges, the Los Angeles Community College District (LACCD) is the largest community college system in the United States. The LACCD has taken a major step along with the American Association of Sustainable Higher Education (AASHE) with its goal to advance the utilization of advanced energy technologies to make all college campuses "energy independent" and "climate neutral." This ambitious climate change policy was made feasible with the passing of Measure J in the November 2008 California State elections. This measure allocated $3.5 billion dollars to LACCD projects, including "modernization, renovation, improvement and new construction projects, constructing energy infrastructure improvements and upgrading technology systems.[2]" Of this funding, just over $200 million has been earmarked specifically toward making the District's goal of being energy independent a reality. What makes this goal especially unique is that the District has committed to actually produce the electricity by installing renewable energy technology on its campuses, as opposed to purchasing renewable energy certificates, which many other tertiary institutions currently do (AASHE). A net zero energy campus is defined as a campus where, over the course of a specified time, the amount of energy provided by

1. University of Minnesota, Morris Campus. http://www.morris.umn.edu/about/.
2. "**Measure J**: Community College Classroom Repair, Public Safety, Nursing & Job Training." http://www.smartvoter.org/2008/11/04/ca/la/meas/J/.

on-site renewable energy resources is equal to the amount of energy used. This is inherently different from an "off-the-grid" concept where at *any* point in time, *all* campus energy demand is met through renewable resources. Despite having strong financial support, whether the LACCD can actually realize its goal of developing a net zero energy school system has yet to be determined.

Using renewable energy resources to generate a portion of a campus energy demand is already difficult, let alone establishing a net zero energy operation. Currently, a number of viable renewable energy technologies exist, including solar photovoltaic (PV), wind, hydroelectric, and to a certain extent micro-turbines. However, one of the biggest concerns when considering installing renewable energy systems (RES)—and often times the deciding criteria of such projects—is the duration required before a return on investment (ROI) is observed. The limited number of case studies of tertiary institutions with RES installed and the relatively recent trend of installing such systems makes it difficult to accurately determine a payback period. Further complicating these calculations are government incentives, rebate programs, and initiatives that vary by state. As shown in Table 27.1, studies investigating small-scale (>1 MW) RES estimate ROI times anywhere between 3 and 30 years, although these are largely theoretical calculations. There are cases of deploying renewable technologies, but the approaches have largely been used to demonstrate technological capability rather than providing economic analyses.

TABLE 27.1 Summary of Estimated ROI Times for Various RES Projects

Study	Type of RES	Size of System[3]	Estimated ROI (Years)
Dalton et al. (2008)	Wind/diesel hybrid	1.8 MW	4.3
Rhoads-Weaver and Grove (2004)	Wind	100 kW	7
Yang (2004)	Solar PV		11.2
Smestad (2008)	Solar PV	1 GW	15
Edwards et al. (2004)	Wind	3.6 GW	20
	Solar PV	3 MW	30
	Biomass	1000 L of ethanol	11
Yue and Yang (2007)	Wind	10 kW	12
Forsyth et al. (2002)	Wind/diesel hybrid	1.8 MW	4.3

PV, photovoltaic; *RES*, renewable energy systems; *ROI*, return on investment.

3. MW = megawatt = 1,000,000 W; GW = gigawatt = 1,000,000,000 W.

As organizations, businesses, and nations begin focusing more attention on the role of RES in their energy portfolios, a greater understanding of the economics associated with RES projects is needed. This chapter aims to do just that by using the Los Angeles Community College District's net zero energy initiative as a case study.

For this study, the City College (LACC) campus of the LACCD was selected to determine whether a net zero energy campus can be achieved through a combination of renewable energy technologies and demand side management. City College was chosen specifically because it is situated in a dense urban area in the heart of Los Angeles. The campus itself mimics many cities—clusters of buildings up to 10 stories high situated in very developed areas with little open space available. This puts a unique challenge in identifying suitable locations for installation of various renewable energy technologies as well as identifying new types of technologies to implement in the project.

GOAL AND OBJECTIVES

The goal of this study is to evaluate whether the LACCD net zero energy campus concept is currently feasible both technologically and economically. This will be accomplished by meeting the following objectives:

1. Develop an understanding of the current energy demand of LACC including daily and annual fluctuations.
2. Complete a comprehensive energy audit and identify areas where energy demand can be reduced.
3. Identify the minimum PV array grid size to be built on LACC.
4. Examine the financial feasibility of the PV array taking into account the cost of materials, installation, operation, and maintenance, as well as Los Angeles Department of Water and Power (LADWP) rebates and government incentives.

IMPORTANCE OF THIS STUDY

Although installing renewable energy technologies is not a new concept, this study certainly adds new information to a rapidly growing field. These include:

1. Understanding of how government incentives, programs, and taxes can influence the construction of RES and whether such policies are actually needed or effective. This can help shape how government policies are geared toward promoting and developing a renewable energy infrastructure.
2. Establishing a clear guideline for organizations such as academia, towns, and cities to follow with the aim of becoming net zero energy consuming. Currently, many other universities, cities, and towns have made active commitments toward increasing their renewable energy portfolio. However, to this date, these efforts have largely involved purchasing renewable

energy credits, as opposed to actually using energy generated from renewable energy technologies. Schools and towns that have implemented renewable energy technologies have largely done so on a haphazard basis, installing randomly sized systems without the intention of fulfilling a particular percentage of their energy consumption or demand. Although the installed systems have yet to generate enough energy to completely fulfill an organization's total energy demand, this disorganized approach is inappropriate.

By providing information on demand side management and methods to calculate the size of renewable energy technology arrays, organizations will be able to take a more quantified approach to establishing a renewable energy infrastructure suitable for their operations. This work will also be valuable for large organizations and businesses, such as convention centers, shopping centers, and grocery stores, particularly information on demand side management used to minimize energy consumption.

3. Providing a model for cities to follow. The deliberate choice of City College for this study is significant. Instead of choosing other LACCD campuses that have more available open land, consequently making it easier to install enough solar PV to meet energy demands, City College's limited area requires innovative thinking to identify and install renewable energy technologies. Rooftops, walkways, and building sides will need to be utilized to produce enough energy, all without affecting the aesthetics of the campus. This is similar to the challenges that many towns and even other academic institutions face—a strong desire to increase renewable energy, but with limited space available. This study seeks to provide options for organizations to consider when trying to establish renewable energy programs.

CURRENT SITUATION

The City College campus is built on 49 acres of land. As of October 1, 2009, there were 21 existing buildings, 7 planned new building constructions, and 2 large parking lots. According to the LACC long-range development plan, eight buildings on campus were scheduled to undergo complete renovation over the next 5 years. A map of the campus can be found in Appendix 1.

Like many towns and cities, open, undeveloped land on the LACC campus is a premium commodity. Given the lack of open space and the presence of tall surrounding buildings that may impede wind exposure and solar radiation at ground level, it is difficult to install large RES on the ground. Therefore any RES installation will likely utilize existing rooftops and carports. Even so, not all rooftop areas are suitable for RES installation, as existing heavy equipment, building design, and orientation will affect the efficiency and effectiveness of an RES. An analysis of existing building rooftops, carports, and walkways was conducted to determine the total suitable area to install an RES and the results

TABLE 27.2 Summary of Total and Available Area for Installation of RES

Location	Total Space (m^2)	Available Space for Solar (m^2)
Rooftop	42,680	13,376
Carports		
Walkways		

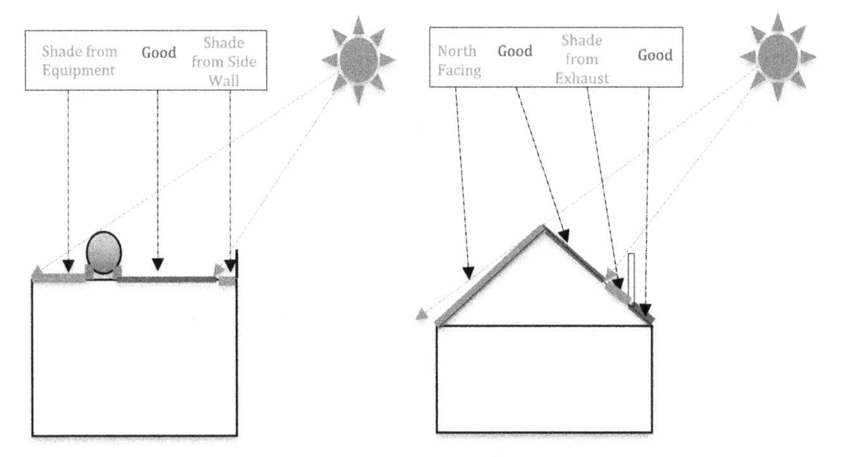

FIGURE 27.1 Illustration of available rooftop space for solar renewable energy systems installation on flat- and slope-topped buildings.

are summarized in Table 27.2. Fig. 27.1 illustrates how to identify rooftop areas that are suitable for solar RES installations.

Energy Demand Versus Energy Consumption

A key component of this project was analyzing and understanding LACC's electrical energy demand and electrical energy consumption. Although the two parameters are related, they are not the same. Most consumers will be familiar with electrical energy *consumption*, which is a measurement of total electrical energy that is used. The unit of this is kilowatt-hours (kWh) and is typically reported on electrical utility bills. Electrical energy *demand* is a measurement of the rate of consumption of total electrical energy and is measured in kilowatt-hours per hour, or simply, kilowatts (kW). For example, if a building consumed 24 kWh of energy over a 24-h period, its demand would be 1 kW. However, if the consumption occurred in 12 h, the demand would be 2 kW, likewise 3 kW if consumed over 8 h. Peak demand refers to the maximum electrical energy demand experienced at any point of the day.

Electric Utility Rates

The LACC campus is located within the service area of the LADWP, the largest municipal utility in the United States. Under the LADWP electrical tariff scheme, LACC is considered a Large General Service and is charged according to Schedules A-3A. Electrical rates differ depending on the time of year and time of day, varying between $0.02197/kWh up to as high as $0.04390/kWh. The calendar year is divided into high season starting from June 1 through September 30 and low season from October 1 to May 31. The daily rate period breakdown is summarized in Table 27.3. Full details of the tariff can be seen in Appendix 1.

By implementing a differential rate scheme, there is a financial incentive for LADWP consumers to minimize electrical energy consumption particularly during peak periods. It should be noted that installing RES is often used as a method to reduce energy consumption and demand during peak usage periods. This is known as peak shaving and is particularly useful if the RES is large enough to reduce electrical energy demand as opposed to consumption. However, the ability of an RES to maximize peak shaving depends heavily on its design.

City College Campus Energy Consumption and Demand

RES are not cheap and designing the system can be complicated—if it is too small, future growth may be compromised. Constructing larger systems that initially generate excess energy in anticipation of future growth may be financially unappealing and is even discouraged. Currently, there are two options for excess electrical energy; collect and store the electrical energy using batteries and energy storage systems, or feed the energy back into the grid. In the first option, batteries and storage devices are bulky and costly; finding additional land to build energy storage centers is also a whole other consideration. In the

TABLE 27.3 Summary of Daily Electric Rate Time Schedule

Charge	Time
TOU 1: High Rate	Monday to Friday
	13:00–16:59
TOU 2: Low Rate	Monday to Friday
	10:00–12:59, 17:00–19:59
TOU 3: Base Rate	Monday to Friday
	8:00–9:59
	Saturday and Sunday
	All day

second scenario, although many electric utility companies welcome feeding excess electrical energy back into the grid, they do not compensate the producers for the surplus electricity they send back into the grid. This discourages consumers from fully utilizing all spaces to install RES, and can even encourage consumers to waste energy to avoid giving it free to utility companies. Therefore, it is critical that a detailed understanding of the campus' energy demand is developed. By doing so, the correct sized RES can be identified and installed.

To understand the current energy demand and consumption of LACC, energy bills between January 1, 2007, and December 31, 2008, were collected and summarized and are shown in Table 27.4. In 2007 LACC spent just over $800,000 purchasing over 7,920,000 kWh of electricity from LADWP. Additional data for energy consumption by specific areas and buildings on campus were unavailable because LACC has only one electric meter.

The electrical energy consumption and demand were analyzed by collecting 15-min-interval utility energy consumption data and compiled to produce an average hourly energy demand profile for 2007 as shown in Fig. 27.2. As expected, the average hourly energy demand reached a peak of just under 1180 kW at noon each day, whereas the average daily minimum energy demand of City College was approximately 540 kW. A sudden spike in energy demand is observed at approximately 6 a.m. each day, and can be attributed to the start-up of campus equipment in preparation for daily operation.

A second graph was developed to identify how LACC's energy demand and consumption patterns fluctuated during the course of a year. These are summarized in Fig. 27.3. As expected, the electrical energy consumption reached a peak in July, with over 730,000 kWh used that month. This is likely due to the increased electricity needed to operate air conditioning during the summer months. The LACC electrical energy demand reached a maximum in August, drawing 2081 kW, although the demand in July was almost the same, at 2071 kW. The 2007 electrical energy demand of LACC ranged between 1447 and 2081 kW.

City College Campus Growth and Demand Side Management

Like other tertiary institutions, LACC is expected to grow and expand. This includes hiring more faculty and staff, enrolling more students, and retrofitting and constructing buildings. In support of this, since 2007, the LACCD has implemented two major programs that will significantly alter each campus' current hourly demand and consumption profiles. The first program consists of construction bond measures A/AA/J that are responsible for both campus retrofits and significant build out. From these measures, LACC is expected to construct seven new buildings and retrofit another eight. These projects will increase the total available building area of LACC from its current total of 721,609 ft^2 to 1,042,470 ft^2, an increase of over 44%. Once these building projects are completed, the electrical energy demand of the campus will also likely increase. However, it is uncertain exactly how much the energy consumption and demand will increase by, making it difficult to account for when

TABLE 27.4 LACC 2007 Monthly Energy Demand

Month	Peak Demand	TOU1 (kW)	TOU2 (kW)	TOU3 (kW)	Energy Cost ($)	Facility Cost ($)	Demand Charge ($)	Utility Bill ($)
1	1342.0	1342.0	1224.0	1230.0	39,861.74	5368.00	5293.00	50,672.74
2	1421.0	1421.0	1418.0	1239.0	40,368.10	5684.00	5609.00	51,811.10
3	1545.0	1532.0	1545.0	1391.0	47,287.64	6180.00	6053.00	59,670.64
4	1439.0	1439.0	1429.0	1220.0	43,198.64	5756.00	5681.00	54,785.64
5	1872.0	1872.0	1812.0	1490.0	51,615.08	7488.00	7413.00	66,666.08
6	1709.0	1709.0	1686.0	1569.0	48,139.72	6836.00	20,364.00	75,489.72
7	1805.0	1790.0	1805.0	1572.0	55,654.10	7220.00	21,450.00	84,474.10
8	1857.0	1836.0	1857.0	1648.0	56,341.74	7428.00	22,020.00	85,939.74
9	2142.0	2142.0	2100.0	1856.0	55,287.33	8568.00	25,503.00	89,508.33
10	1782.0	1782.0	1744.0	1436.0	56,698.48	7128.00	7053.00	71,029.48
11	1637.0	1637.0	1600.0	1325.0	46,748.25	6548.00	6473.00	59,919.25
12	1465.0	1465.0	1448.0	1276.0	40,909.59	5860.00	5785.00	52,704.59
Total					**582,110.41**			**802,671.41**

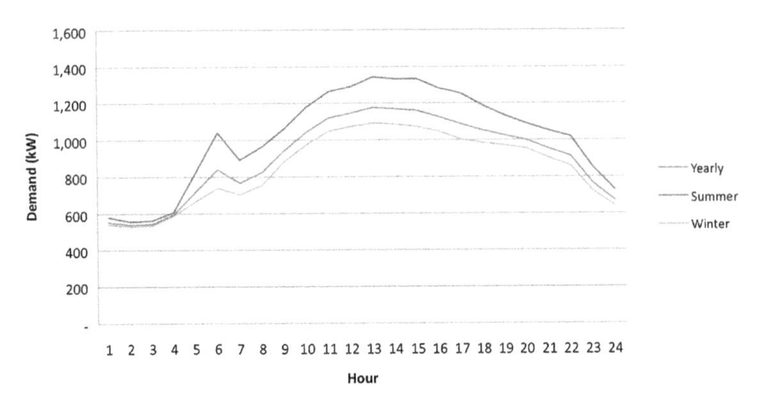

FIGURE 27.2 The 2007 average hourly energy demand profile for City College.

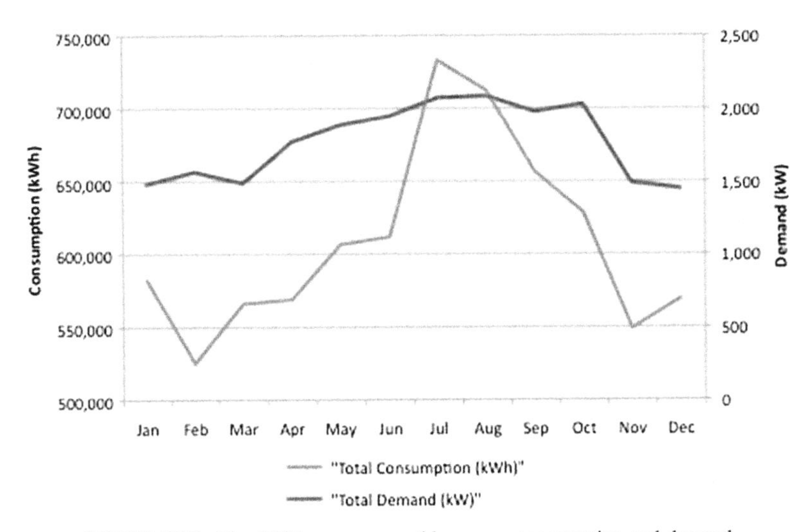

FIGURE 27.3 The 2007 average monthly energy consumption and demand.

designing the size of the RES. We can only estimate the future energy consumption and demand. The planned new buildings are similar to the existing buildings in that they are administrative rooms, classrooms, or computer laboratory rooms. Given this, it was assumed that energy consumption associated with the new buildings would result in a linear increase to LACC's energy demand proportional to the building square footage increase planned by the end of the bond programs. To do so, data from the California Energy Commission's "California commercial end-use survey,"[4] ASHRAE[5]'s "Pocket

4. California Energy Commission. "California Commerical End-Use Survey. http://www.energy.ca.gov/ceus/.
5. American Society of Heating, Refrigerating and Air-conditioning Engineers. http://www.ashrae.org.

guide," and existing LACC specific data were collected. Given this, it is estimated that within 5 years, LACC's electrical energy consumption will increase by almost 4,400,000 kWh, or approximately 13.63 kWh/ft^2.

The second program is a demand side management initiative aimed at reducing the current load burden, with the energy savings paying for capital investment of the energy-saving technologies. This involves examining the current energy infrastructure in place at LACC and making recommendations on how to reduce the energy demand. In Jun. 2009, a 3-month comprehensive energy analysis for LACC was completed. This involved evaluating each building on the LACC campus according to the following criteria:

- Age
- Condition
- Performance
  - Lighting system, including controls and fixtures
  - Building automation
  - Heating, ventilation, and air conditioning systems
  - Equipment upgrades
  - Ability to connect to a central plant
  - Structural upgrades

The results of the audit recommended a list of energy conservation measures with the cost, savings, and ROI. These are summarized in Table 27.5. By implementing these measures, it is estimated that LACC can reduce its current energy consumption by up to 18.5%. Simple payback periods were calculated,

TABLE 27.5 Summary of Proposed Electrical Energy Control Measures for LACC

Control Measure	Utility Savings ($)	Rebates ($)	Cost ($)	ROI (Years)
Interior lighting upgrades	72,896	67,590	122,722	15.8
Installation of lighting controls	14,347	7,340	167,028	11.1
Window glazing	10,535	6,723	156,482	14.2
HVAC fan and motor upgrades	2,275	1,759	35,308	14.7
Upgrade and integrate with EMS	250,551	27,141	1,996,901	7.9
Installation of utility submeters	0	0	498,416	n/a

EMS, environmental management system; *HVAC*, heating, ventilation, and air conditioning.

and all except upgrading and integrating and energy management system had ROIs of greater than 11 years.

SOLAR PV ARRAY AND SETUP

The renewable energy system to be installed at LACC consists of PV arrays, an inverter, a controller, and other essential cables and components. Solar inverters are a critical component to any solar power system. Inverters change DC (direct current) from solar panels, or the photovoltaic array, into AC (alternating current) for use in the campus and resale back to the utility grid.

Two key categories of solar inverters are used in solar energy systems. In systems that are not connected to the utility grid, the inverter takes DC from the PV array and converts it to AC powering a battery or series of batteries, from which power is drawn. In the case of this "off grid inverter" or "charge controller," the PV array and inverter are essentially charging batteries to keep power supplied to the building.

The second category ties into the utility grid, again converting DC to AC, but with this "grid-tie inverter" the power is first supplied to the building and the remaining energy is sent back to the LADWP grid.

Solar Energy

There will be three main configurations for PV arrays at City College: carport structures, roof-mounted structures, and ground-mounted arrays. The current standing design has a panel slope of 5° and an azimuth of 180°. For this study, it is assumed roof top units will be designed with a 20° and 180° panel slope and azimuth, respectively, and the ground-mounted units will have a slope equal to the latitude, approximately 34° with a 180° azimuth. If maximum winter production from the ground-mounted system is desired, the array tilt can be increased up to 15° past the location's latitude.

Hourly solar irradiance data for the year 2005 was downloaded from the National Renewable Energy Laboratory's (NREL) solar radiation database.[6] Data beyond 2005 were unavailable. A monitoring station located at the Downtown Los Angeles University of Southern California campus, the nearest station to the LACC campus, collected the 2005 data. The data were then analyzed using NREL PV Watts v2 program, which uses typical meteorological weather data for a selected location and determines the solar radiation incident of the PV array at the given array and azimuth tilt angles. To generate solar radiation data of hourly resolution, another NREL program HOMER v2.67 Beta[7] was employed to synthesize the desired profiles. HOMER uses an algorithm based on

6. National Renewable Energy Laboratory: Solar Radiation Database. http://rredc.nrel.gov/solar/old_data/nsrdb/.
7. National Renewable Energy Laboratory: HOMER https://analysis.nrel.gov/homer/.

the work of Graham and Hollands (1990). The algorithm produces synthetic solar data with certain statistical properties that result in a data sequence that has realistic day-to-day and hour-to-hour variability and autocorrelation.

The daily solar irradiance absorbed by an installed solar PV array will vary throughout the year. In Los Angeles, the maximum solar irradiance occurs in the summer months between June and August. In the winter months, the angle of the sun is lower, as well as the number of hours of sunlight each day, resulting in less daily solar irradiance. Average hourly solar irradiance was calculated for the entire year, winter months (December to February), and summer months (June to August). From Fig. 27.4 we see that in all three cases, hourly average solar irradiance is greatest at approximately 12:00 noon each day. The figure also shows that at midday, solar PV panels installed at LACC will generate on average 0.187 kW/m^2 more energy in the summer months than in winter.

The goal of this project is to examine whether City College can establish a net zero energy campus. To achieve this, we need to ensure that all energy demand and consumption during the course of the year is generated by the installed RES. The average amount of solar irradiance each day per square meter of solar PV was identified by calculating the area of each graph and summarized in Table 27.6.

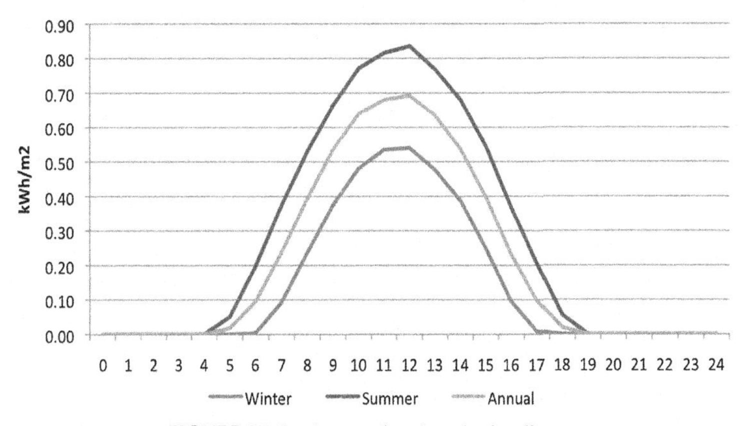

FIGURE 27.4 Average hourly solar irradiance.

TABLE 27.6 Average Daily Power Generated by 1 m^2 of Solar PV

Time Period	Average Daily Solar Irradiation (kW/m^2)
Annual	5.210
Summer	6.563
Winter	4.530

Using this information, we can determine the minimum size of the solar PV required to meet the maximum energy demand. Polycrystalline KD 205GX-LP cells manufactured by Kyocera were the choice for PV modules. They have a rated power of 205 W at a cell temperature of 25°C and solar irradiance of 1 kW/m2. The panels have dimensions of length 1.50 m and width 0.99 m, with a quoted cell efficiency of approximately 16%.

The following equation is used to calculate the output of the PV array [kW] at a given time step:

$$P_{PV} = Y_{PV} f_{PV} \left(\frac{G_T}{G_{T,STC}} \right) \left[1 + \alpha_P \left(T_c - T_{c,STC} \right) \right] \tag{27.1}$$

where Y_{PV} is the rated capacity of the PV array [kW]; f_{PV} is the derate factor [%]; G_T is the solar radiation incident on the PV array in the current time step [kW/m^2]; $G_{T,STC}$ is the incident radiation at standard test conditions [1 kW/m^2]; α_P is the temperature coefficient of power [%/°C]; T_c is the PV cell temperature in the current time step [°C]; $T_{c,STC}$ is the PV cell temperature under the standard test conditions [25°C].

The PV cell temperature, T_c, is given as follows:

Eq. (27.1) holds true if the PV system is equipped with a maximum power point tracker, which the modeled system does. A derate factor of 0.804 is assumed for the system. The effect of temperature on the PV array was neglected in this model.

The energy generated by the three PV arrays for hour t, $E_{G(t)}$, and can be expressed as follows:

$$E_{G(t)} = \sum_{i=1}^{3} \left(N_{PV} E_{PV(t)} \right)_i \tag{27.2}$$

where $E_{PV(t)}$ is the energy by PV array i during time step t [kWh]; N_{PV} is the number of PV modules in PV array i.

The minimum solar PV array size was calculated using the data in Table 27.5, solar PV specifications outlined earlier, and assuming that the campus needs to generate 8,000,000 kWh of energy each year. The results are summarized in Table 27.7.

TABLE 27.7 Summary of Minimum Photovoltaic (PV) Array Size Required to Generate 8,000,000 kWh of Energy

Time Period	Minimum PV Array (MW)	Minimum Area Required (m^2)
Annual	4.21	4210
Summer	3.34	3340
Winter	4.84	4840

CONCLUSION

It should be noted that, although the LACCD is truly a visionary in establishing a net zero energy goal, local government initiatives and incentives have also played a role in propagating this project. The question is how significant are these government efforts in promoting installation of RES, especially on such a large scale? An evaluation of the impact of California's renewable energy initiatives[8] on the LACCD net zero energy project would provide governments around the world with better information on how to develop similar effective programs. This is especially important as countries begin to roll out carbon trading initiatives and taxes.

Overall, the LACCD project has highlighted the technological and engineering feasibility of retrofitting an entire campus to become net zero energy. As existing renewable systems such as solar PV arrays become more efficient, wind turbines become smaller, and the maturity of other renewable technologies such as fuel cells evolves, it is clear that technology will be an enabler for organizations to meet net zero energy goals. In addition to this, efforts by the US Green Building Council to tighten Leadership in Energy and Environmental Design coupled with ongoing innovation in building design by architects will continually raise the benchmark in regards to building energy efficiency, thereby reducing energy demand. A key area this project has not addressed is the *impact of occupant behavior on the energy profile*. That is, we need to understand how occupants use energy in a building and anticipate how this will change over time. For example, in a mobile-enhanced society, there is a need to continually charge laptops, cell phones, and other devices. With the growth of the Internet of things, baseline energy demand will undoubtedly change. On an individual basis, these may not have a significant impact on an energy profile. However, when taken cumulatively, there may be impacts including increasing overall power demand and even *shifting* load profiles. This is even more likely with the rapid adoption of hybrid and all-electric vehicles that would require installation of more electric vehicle–charging stations and that can be operated 24 h a day. The LACCD vision is a good start. In the end, perhaps the biggest challenge we face in establishing net zero communities and truly sustainable cities may in fact be incentivizing and encouraging people to make sustainable lifestyle decisions.

APPENDIX 1: MAP OF LA CITY COLLEGE CAMPUS INDICATING PREVIOUS, CURRENT, AND PLANNED RENOVATIONS/CONSTRUCTION

8. See California Executive Order S-14-08. http://gov.ca.gov/executive-order/11072.

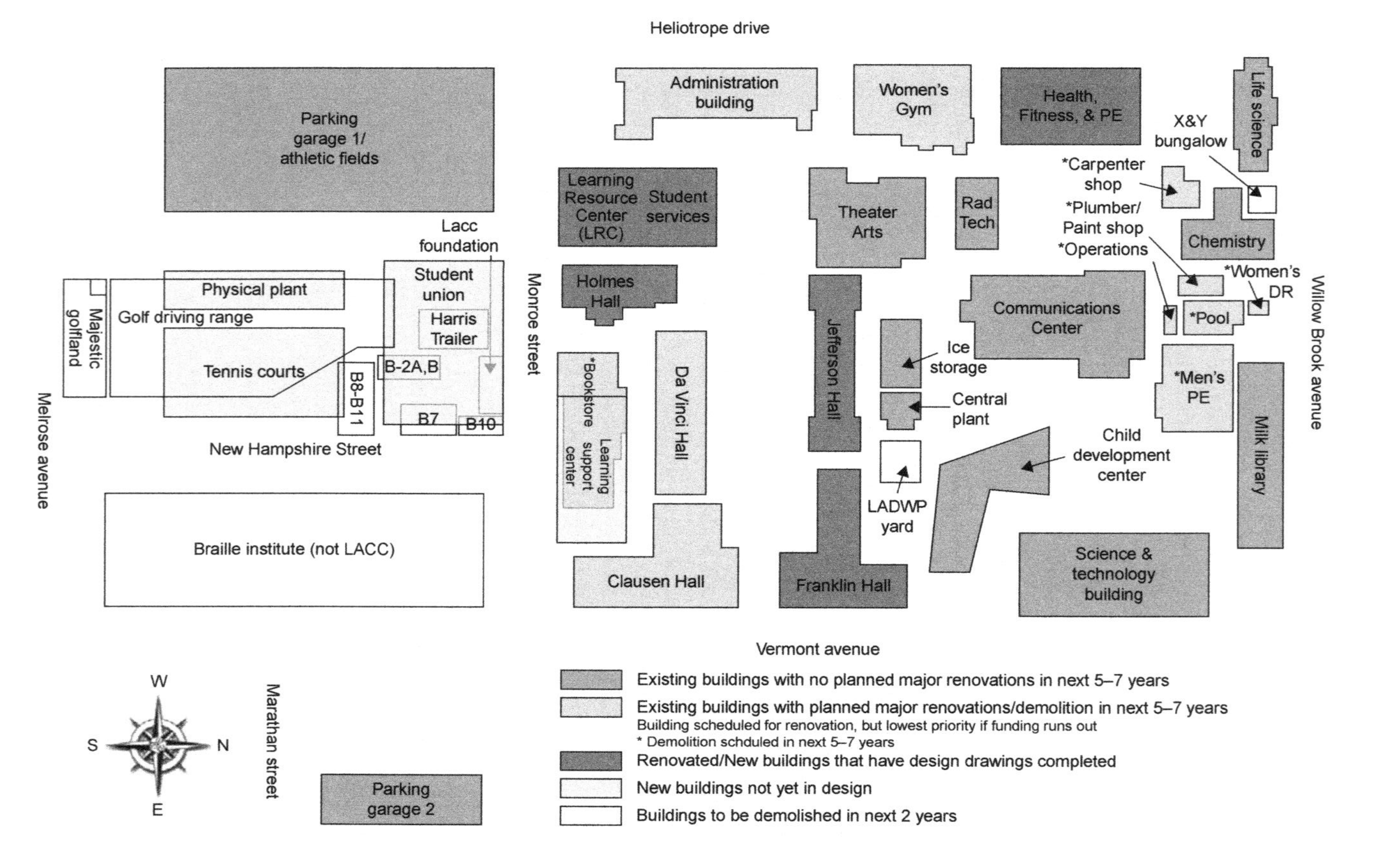
Heliotrope drive
Parking garage 1/ athletic fields
Administration building
Women's Gym
Health, Fitness, & PE
X&Y bungalow
Life science
*Carpenter shop
*Plumber/ Paint shop
*Operations
Chemistry
*Women's DR
*Pool
Learning Resource Center (LRC)
Student services
Theater Arts
Rad Tech
Lacc foundation
Student union
Harris Trailer
Monroe street
Holmes Hall
Communications Center
Physical plant
Golf driving range
Majestic golfland
Tennis courts
B-2A,B
B8-B11
B7
B10
*Bookstore
Learning support center
Da Vinci Hall
Jefferson Hall
Ice storage
Central plant
*Men's PE
Milk library
Willow Brook avenue
Melrose avenue
New Hampshire Street
Child development center
LADWP yard
Braille institute (not LACC)
Clausen Hall
Franklin Hall
Science & technology building
Vermont avenue
W
S
N
E
Marathan street
Parking garage 2
Existing buildings with no planned major renovations in next 5–7 years
Existing buildings with planned major renovations/demolition in next 5–7 years
Building scheduled for renovation, but lowest priority if funding runs out
* Demolition schduled in next 5–7 years
Renovated/New buildings that have design drawings completed
New buildings not yet in design
Buildings to be demolished in next 2 years

APPENDIX 2: LADWP ENERGY RATES AS OF OCTOBER 1, 2009—SPECIFIC FOR LACC OPERATIONS

Electric Tariff Structures			
Name of Utility:	*LADWP*		
Rate Schedule:	*A-3A—Large General Service*		
Component	**Charge**	**Unit**	**Description**
Customer Charge:	$75	Per Month	
Demand Charge:	$4.00	Per Monthly kW	Low Season, High Peak
	$3.00	Per Monthly kW	High Season, Low Peak
	$9.00	Per Monthly kW	High Season, High Peak
	$4.00	Per Billed kW	Facilities Charge
	$0.46	Per Billed kW	Electric Subsidy Adjustment (ESA)
	$0.96	Per Billed kW	Reliability Cost Adjustment (RCA)
Consumption Charge:	$0.03663	Per kWh	Low Season, Low Peak
	$0.03663	Per kWh	Low Season, High Peak
	$0.02197	Per kWh	Low Season, Base Peak
	$0.03764	Per kWh	High Season, Low Peak
	$0.04390	Per kWh	High Season, High Peak
	$0.01755	Per kWh	High Season, Base Peak
	$0.0499	Per Total kWh	Energy Cost Adjustment (ECA)
Power Factor Correction: 0.9 < pf < 0.949	$0.00059	Per kVARH	High Season, Base Peak
	$0.00113	Per kVARH	High Season, Low Peak

Continued

—cont'd

Electric Tariff Structures			
Name of Utility:	*LADWP*		
Rate Schedule:	*A-3A—Large General Service*		
Component	**Charge**	**Unit**	**Description**
	$0.00164	Per kVARH	High Season, High Peak
	$0.00073	Per kVARH	Low Season, Base Peak
	$0.00145	Per kVARH	Low Season, Low Peak
	$0.00145	Per kVARH	Low Season, High Peak
Other Charges:	$0.00022	Per Total kWh	State Energy Surcharge (SES)
Determination of Billed Demand:		kW	Maximum of greatest demand from last 12 months or 500 kW
Other Rate Details:	Current RCA charge is $0.51/kW; rate will change to $0.96/kW in July 2009		
	Current ECA charge is $0.0444/kWh; rate will change to $0.0454/kWh in July 2009; rate will change to $0.0499/kWh in October 2009		
	*High Season is from June 1 to September 30		
	*Low Season is from October 1 to May 31		
	*High Peak occurs Monday through Friday, 1:00 p.m. –4:59 p.m.		
	*Low Peak occurs Monday through Friday, 10:00 a.m. –12:59 p.m., 5:00 p.m.–7:59 p.m.		
	*Base Period occurs Monday through Friday, 8:00 p.m. –9:59 a.m. and Saturday through Sunday, all day		

Electric Tariff Structures (continued)			
Name of Utility:	*LADWP*		
Rate Schedule:	*A-2B—General Service*		
Component	**Charge**	**Unit**	**Description**
Customer Charge:	$23	Per Month	
Demand Charge:	$4.25	Per Monthly kW	Low Season, High Peak
	$3.25	Per Monthly kW	High Season, Low Peak
	$9.00	Per Monthly kW	High Season, High Peak
	$5.00	Per Billed kW	Facilities Change
Consumption Charge:	$0.46	Per Billed kW	Electric Subsidy Adjustment (ESA)
	$0.96	Per Billed kW	Reliability Cost Adjustment (RCA)
	$0.02252	Per kWh	Low Season, Base Peak
	$0.04045	Per kWh	Low Season, Low Peak
	$0.04045	Per kWh	Low Season, High Peak
	$0.01879	Per kWh	High Season, Base Peak
	$0.03952	Per kWh	High Season, Low Peak
	$0.04679	Per kWh	High Season, High Peak
	$0.0499	Per Total kWh	Energy Cost Adjustment (ECA)
Other Charges:	$0.00022	Per Total kWh	State Energy Surcharge (SES)
Determination of Billed Demand:		kW	Greatest demand from last 12 months
Other Rate Details:	Current RCA charge is $0.51/kW; rate will change to $0.96/kW in July 2009		
	Current ECA charge is $0.0444/kWh; rate will change to $0.0454/kWh in July 2DQ9: rate will change to $0.0499/kWh in October 2009		
	*High Season is from June 1 to September 30		

Continued

—cont'd

Electric Tariff Structures (continued)			
Name of Utility:	*LADWP*		
Rate Schedule:	*A-2B—General Service*		
Component	Charge	Unit	Description
	*Low Season is from October 1 to May 31		
	*High Peak occurs Monday through Friday, 1:00 p.m. −4:59 p.m.		
	*Low Peak occurs Monday through Friday, 10:00 a.m. −12:59 p.m., 5:00 p.m. to 7:59 p.m.		

REFERENCES

AASHE. Mandatory Student Fees for Renewable Energy and Energy Efficiency. From: http://www.aashe.org/resources/mandatory_energy_fees.php.

Calhoun, T., Cortese, A.D., 2005. We Rise to Play a Greater Part: Students, Faculty, Staff and Community Converge in Search of Leadership from the Top. Society for College and University Planning, Ann Arbor, Michigan. S. A. a. R. Panel.

Dalton, G.J., Lockington, D.A., et al., 2008. Feasibility analysis of stand-alone renewable energy supply options for a large hotel. Renewable Energy 33 (7), 1475−1490.

Edwards, J., Wiser, R., et al., 2004. Building a Market for Small Wind: The Break-even Turnkey Cost of Residential Wind Systems in the United States. Lawrence Berkeley National Laboratory. Lawrence Berkeley National Laboratory.

Forsyth, T., Tu, P., et al., 2002. Economics of Grid-Connected Small Wind Turbines in the Domestic Market. NREL.

Graham, V.A., Hollands, K.G.T., 1990. A method to generate synthetic hourly solar radiation globally. Solar Energy 44 (6), 333−341.

Rhoads-Weaver, H., Grove, J., 2004. Exploring Joint Green Tag Financing and Marketing Models for Energy Independence. Global Windpower.

Smestad, G., 2008. The Basic Economics of Photovoltaics. Optical Society of America, San Jose.

Yang, D., 2004. Local Photovoltaic (PV) Wind Hybrid System with Battery Storage or Grid Connection. NREL.

Yue, C.D., Yang, G.G.L., 2007. Decision support system for exploiting local renewable energy sources: a case study of the Chigu area of southwestern Taiwan. Energy Policy 35 (1), 383−394.

FURTHER READING

Borowy, B.S., Salameh, Z.M., 1996. Methodology for optimally sizing the combination of a battery bank and PV array in a wind/PV hybrid system. IEEE Transactions on Energy Conversion 11 (2), 367−375.

Chapter 28

Case Study: University of California, Irvine

Wil Nagel
University of California, Irvine, CA, United States

With more than 30,000 students, more than 15,000 full-time employees, and thousands of visitors on any given day, the University of California, Irvine (UCI), has a population that is comparable in size with the cities of Galveston, Texas; Coral Gables, Florida; and Aliso Viejo, a neighboring community in Orange County, California.[1] The main campus resembles a mid-sized city in other ways in that it has its own roads, parking structures and surface lots, bike paths, a large central park, athletic fields, a recreation center, housing, eateries, retail, classrooms, and medical clinics and office buildings.

The campus also houses numerous multistory research laboratories, which account for roughly two-thirds of the institution's energy use, and a highly efficient on-site combined heat and power plant, both of which play a significant role in the campus' internationally recognized energy-management program. Largely because of its energy-management practices, UCI is the only campus ranked among Sierra magazine's top 10 greenest campuses 8 years in a row, being placed first in 2014 and 2015. The campus was the highest ranked comprehensive university in Sierra's "Cool Schools" rankings and the leading UC school on the 2016 and 2017 lists. UCI has also appeared on The Princeton Review's Green Honor Roll for the past 5 years.

Officials in Denmark and Singapore and universities in Canada, Singapore, and the United Kingdom have sought UCI's counsel on the subject of "deep energy efficiency"—measures that reduce energy consumption and associated carbon emissions by half or more. These measures have been identified by the University of California system, with its 10 campuses and five medical centers, as the most immediate, cost-feasible strategy to effect a substantial reduction in UC's carbon footprint. Deep energy efficiency continues to play a major role in UC's commitment to be carbon neutral by 2025 and is fundamental to

1. https://factfinder.census.gov/faces/tableservices/jsf/pages/productview.xhtml?src=bkmk.

Sustainable Cities and Communities Design Handbook. https://doi.org/10.1016/B978-0-12-813964-6.00028-8

UCI's Smart Labs Initiative,[2] a proven program of safely reducing energy use in research laboratories by as much as 60%. The US Department of Energy (DOE) modeled its Smart Labs Accelerator[3] after UCI's comprehensive program, which is the basis of the campus' partnership in the DOE's Better Buildings Challenge.[4]

So, what defines a "smart lab"? Essentially, there are seven components, each of which needs to be addressed to attain the level of energy savings realized by UCI:

1. Fundamental platform of dynamic, digital control systems
2. Demand-based ventilation
3. Low power density, demand-based lighting
4. Exhaust fan discharge velocity optimization
5. Pressure drop optimization
6. Fume hood flow optimization
7. Commissioning with automated cross-platform fault detection

UCI began to seriously focus on energy savings in the early 1990s when it adopted a policy to outperform California's already stringent Title 24 Energy Code by 30% in new construction. By 2008, UCI's energy management team began to suspect that building energy systems had waste designed in, and by more than a few percent! The entrenched professional culture of overdesigning "margins of safety" and over-relying on what were then considered best practices appeared to be the cause of significant energy waste. Estimates of the potential for energy savings in this environment were understandably low and suppressed even further by the fact that early rollouts of "smart" buildings had been underdeveloped and oversold.

With a belief that energy savings of as much as 50% might be possible, UCI embarked on an overall program of redesigning laboratory building energy systems. A number of factors contributed to this pursuit, including the realization that buildings' energy systems had been, to that point, designed to waste more energy than realized for a variety of reasons: older, nondynamic systems operated at fixed volumes/levels/speeds and "worst case" parameters. Before digital controls and sensors, a margin of safety was needed. The energy needed to pump water and air is nonlinear, and reheating is excessive when air changes are high.

By focusing holistically on building energy system design (in retrofits and new construction), UCI concluded that the key to a "smart" building is to provide (1) just the right amount of energy, (2) at just the right place, (3) at just the right time. By challenging all accepted design practices and using a

2. https://www.ehs.uci.edu/programs/energy/UCISmartLabsInitiative_Feb222016.pdf.
3. https://energy.gov/eere/femp/smart-labs-accelerator.
4. https://betterbuildingssolutioncenter.energy.gov/showcase-projects/smart-labs-initiativenatural-sciences-ii.

comprehensive suite of software and sensors, UCI was able to demonstrate that 50% savings and more were, indeed, possible, without compromising occupant safety. The staff at the 52-year-old institution continue to improve on these findings and have since concluded that the "information layer" enabled by new sensors and software is just as important as the building control systems.

And people are just as key. A new generation of "digital savvy" tradespeople is increasingly essential to keep smart buildings smart. UCI has partnered with the State Employees Trades Council-United and Irvine Valley College to establish a rigorous apprenticeship program for next-generation and career technicians responsible for maintaining the increasingly complex computer-based systems in today's energy-efficient buildings.

Finally, there are a number of cobenefits to smart lab design: Many heating, ventilation, and air conditioning deferred maintenance problems are fixed and funded through energy savings. The availability of sophisticated information layers provides real-time commissioning and air quality control; the result is cleaner indoor air. Lighting quality has improved. Buildings are quieter inside and out. There is a longer service life for heat-producing and friction-producing building system components, and the resulting energy savings have eliminated the need for capital investments for generation, central plant chillers, and other infrastructure.

UCI Office of Sustainability.

Chapter 29

"Scrappy" Sustainability at Ohio Wesleyan University

Emily Howald, John Krygier
Ohio Wesleyan University, Delaware, OH, United States

Chapter Outline

A GRASSROOTS MODEL FOR SUSTAINABILITY IN HIGHER EDUCATION

There are colleges and universities with the expertise and financial resources to invest in large-scale, conspicuous sustainability efforts (such as large solar arrays, stylish LEED-certified buildings, and full-time sustainability staff) and there are those who do not. However, those without the funds for conspicuous sustainability are not necessarily excluded from substantive sustainability efforts. Indeed, we suggest that grassroots, "scrappy" sustainability efforts on college campuses and at other institutions may have certain benefits over top-down, high-investment sustainability.

THE CONTEXT OF SUSTAINABILITY AT OHIO WESLEYAN UNIVERSITY

Ohio Wesleyan University (OWU) is a small, private, liberal arts college in central Ohio that serves as a modest showcase for a relatively low-cost, grassroots, and distributed approach to sustainability (Fig. 29.1). The university neither has a sustainability coordinator position nor any other employee with distinct expertise in sustainability. None of the faculty have specializations in the field, and there are no classes taught on the subject. As of yet, there

Sustainable Cities and Communities Design Handbook. https://doi.org/10.1016/B978-0-12-813964-6.00029-X

FIGURE 29.1 Ohio Wesleyan Campus, Delaware Ohio. *Photo Credit: OWU, Office of Communications.*

is no official sustainability plan and there are neither funds nor donations set aside specifically for sustainability projects. OWU has, over the last decade, expanded its endowment, raised significant funds for student travel and research, and embarked on a substantial upgrade to campus student housing. These are all fundamentally important and easily justifiable priorities. Given this situation, it is easy for students, faculty, and staff to feel like not enough is being done to foster sustainability on campus. Instead of complaining about the lack of top-down, large-investment sustainability, a group of students, faculty, and staff have embarked on a grassroots effort to make sustainability work at OWU despite limited resources. Ultimately, we argue, sustainability efforts can succeed if those who believe in the value of sustainability actually do something, then persist in furthering the efforts until something takes hold, and then persist in keeping the efforts going. Successes with these smaller, "scrappy" efforts will, hopefully, lead to larger efforts, backed by a spreading culture of sustainability.

OWU has a rocky history with sustainability efforts. Many higher education institutions believe that they must be leaders in finding solutions to the environmental crisis by developing and promoting the knowledge, tools, and technologies needed to transition to a sustainable society. As the environmental movement emerged and developed in the 1960s and the 1970s, OWU established an Environmental Studies major, the first such program in an academic institution in Ohio. In its nearly 40-year existence, the program has produced hundreds of majors that have gone on to successful careers related to the environment. In 2009, a Sustainability Task Force was created to evaluate the President's Climate Commitment (PCC), which 80% of students voted to support. Despite the lack of any direct negative consequences for not meeting the PCC goals, the Task Force was concerned about the capital investments and employee time needed to implement and monitor the necessary energy efficiency upgrades to campus facilities, and recommended that a sustainability coordinator be hired (rather than signing the PCC). In 2011, an American Recovery and Reinvestment Act grant funded a 2-year sustainability coordinator position. The university hired Sean Kinghorn for the position, and his efforts generated significant rebate funds for the university, as well as energy-saving efforts and dozens of sustainability projects (many led by students). In 2013, Kinghorn's position ended, after the failure of several grants intended to acquire additional funds for the position. A student protest later that year demonstrated student commitment to the sustainability coordinator position. With the decision not to sign the PCC and the lack of funds to continue the sustainability coordinator position, one might expect the prospects for sustainability on campus to fade. At that point, the campus Sustainability Task Force set out on an effort to encourage grassroots sustainability efforts and create a campus sustainability plan, despite the setbacks.

The Sustainability Task Force is not an official campus committee; it is voluntary and open to all students, staff, and faculty. It coordinates and promotes sustainability on campus through the efforts of students, faculty, and staff working on projects in courses, as student-independent studies, through student organizations (such as the Tree House campus residence and the Environment & Wildlife Club), and as campus services (such as Buildings & Grounds and food service). Sustainability at OWU is one large, distributed, and voluntary collaboration. There are no experts and no one really in charge, but participation in sustainability efforts continues to grow, as do successful sustainability projects on campus.

Of course, this grassroots approach has its difficulties. There is a tendency for projects initiated by an individual student or a small group of students to work in the short term, until those students graduate and the project atrophies and eventually fails. OWU has many of these sustainability failures in its past. OWU students first developed a campus garden to grow food in the 1960s. Over the years there have been at least a half dozen such gardens. They are developed (often with student funding), exist for a few seasons, and then devolve into a large weedy eyesore. Our latest campus garden is currently in that weedy, decrepit stage, abandoned along with the special raised garden frames and portable greenhouse purchased with student funds. Another student received thousands of dollars to develop a campus bike share program in 2009. In 2011, the student graduated, and soon after all the bikes were abandoned, broken, or stolen; now the program is completely defunct. Funds from the regional waste authority were acquired in 2011 for campus composting. The effort required student volunteers to sort through campus dining services food waste to remove trash or contaminated materials. Students were initially excited to participate, but excitement faded fast: saving the Earth through sorting mounds of rotting food quickly lost its allure.

Given these experiences, grassroots campus sustainability may seem like a doomed cycle of "develop then fail." Such abjectly uncoordinated sustainability is just not sustainable. Yet the problem of no sustainability coordinator, no one on campus with expertise in sustainability, and very limited funds remained.

COORDINATING SUSTAINABILITY WITHOUT A SUSTAINABILITY COORDINATOR

Despite these setbacks there have been successful efforts, not the least of which is the university's newly proposed sustainability plan. Indeed, the cycles of failure of sustainability efforts on campus were a primary motivation for efforts by the campus Sustainability Task Force to develop a sustainability plan: significant effort and moderate funds were being put into perpetually failing projects. The plan was built on the foundation of efforts of the former sustainability coordinator; however, the development of the plan grew from the voluntary work of a grassroots group of students, faculty, and staff. The sustainability plan was created by students in Geography 499: Sustainability Practicum as overseen by the STF and the course instructor, Dr. John Krygier (the chair of the Environmental Studies program since 2010). Neither Krygier nor any of the students in the course had any clue about how to construct a sustainability plan when the course started. While initially rather disconcerting and even stressful, the students came to embrace their role: no one else was going to create a sustainability plan, so it was up to them.

Much thought was put into the reasons for the lack of successful sustainability efforts on campus. One key lesson learned from the failure to sign the PCC was that external, generic sustainability goals were simply not appropriate for our particular campus. Those creating the plan worked to make sure that all goals were appropriate for the institution, internally initiated rather than externally imposed. The students gathered information about hundreds of sustainability efforts on campus and began to shape what became a 40-page document. It became clear that this huge document was not really a plan, so the Sustainability Practicum was offered again and the effort focused on creating a much more succinct plan with short-, medium-, and long-term goals (see Proposed Sustainability Plan, above). Importantly, the goals were developed in consultation with students, faculty, staff, and administration. Student Emily Howald, as part of a course project and independent study, met with several academic committees, dozens of faculty, Buildings & Grounds, campus food service, student groups, and others for feedback on the plan. Concerns were considered and changes made. The plan was fine-tuned to the institution. Also important was the inclusion of a subset of campus sustainability projects that we could focus upon, semester after semester, in an attempt to stop the cyclic development and failure of sustainability projects on campus. This lent a level of coordination with a series of sustainability projects: course projects, student group projects, and efforts by the campus food service and Buildings & Grounds are focused on making this subset of projects work. As the Sustainability Task Force awaits official adoption of the Sustainability Plan (Fig. 29.2), positive outcomes are emerging from the slightly coordinated yet distributed and grassroots approach to sustainability at OWU.

"SCRAPPY SUSTAINABILITY" OUTCOMES

In 2012, OWU Environmental Studies student Sarah D'Alexander organized (as part of a class project) the first "May Move Out" at OWU. The goal was simple: to collect, rather than discard, usable materials left behind by students as they moved off campus at the end of the spring semester. The effort was successful in collecting tons of clothing, furniture, appliances, bikes, etc

Ohio Wesleyan University *March 29 2017*

Sustainability Plan

As a liberal arts higher education institution, Ohio Wesleyan University must strive to promote sustainability through educational, technical, and social means. The Ohio Wesleyan Sustainability Plan is intended to invigorate and expand a culture of sustainability on campus with an eye on the future of humans and the earth's environment.

Commit to a Sustainable Future on Campus and in the Community
- Strengthen Environmental Studies Program
- Fund sustainability coordinator with support & facilities by 2020
- 5% yearly growth in student sustainability involvement

Reduce Our Impact on Global Climate Change
- Commit to energy-efficient building investments
- Utilize 5% solar or renewable energy on campus by 2020
- Install solar panels on campus for learning purposes
- Annual student run energy reduction challenge initiated in 2017

Increase Our Health and Well-being
- Commit to a farm and food collaboration with Stratford Ecological Center and the Methodist Theological School's Seminary Hill Farm by 2017
- Purchase 15% of campus food from local and regional sources by 2020
- Develop a long-term plan for composting by 2018
- Maintain and grow student-run Food Recovery Network and related programs

Live Better on Campus and on Earth
- Develop Sustainable Living model residence option by 2019
- OWU activity courses on sustainable practices by 2017
- Maintain and expand May Move Out
- New campus bird, animal, insect and plant habitat enhancement each year
- Expand reusable container program for campus food by 20% each year

FIGURE 29.2 The Ohio Wesleyan University Proposed Sustainability Plan, page 1 (of 4). Pages 2 and 3 detail the four areas of focus outlined on page 1.

FIGURE 29.3 May Move Out storage container for student donations to Goodwill. *Photo Credit: John Krygier.*

(Fig. 29.3). The logistics were complicated: students had to move usable materials to several rooms on campus, and staff drove trucks full of materials to local social service providers. A significant amount of collected materials were also stored on campus, in an unused building, with a desire to open a free store the following fall. Ultimately, this model failed. It involved too much labor and organization. In addition, the planned free store never opened, and much of the stored material had to be discarded when a need arose for the building that housed the materials.

Instead of letting the May Move Out effort end, we encouraged students along with staff in Buildings & Grounds and Residential Life to rethink the May Move Out. A student, again part of a course project, came up with a simpler process: renting storage pods, which were located near dumpsters during the May Move Out period and used for donated items. The pods would then be emptied by our local Goodwill. This approach required minimal labor, but did incur costs for the pod rental, which was funded by a small grant from our local solid waste authority. The May Move Out in collaboration with Goodwill was a success in its first year: diverting over 10 tons of materials.

Alas, without the grant there were concerns about the cost of the storage pod rental. Buildings & Grounds foreman Jay Scheffel came up with a plan to reduce the number and size of trash dumpsters (thus reducing costs), given that tons of materials were being diverted. With Scheffel's plan in place, the reduced dumpster costs covered the cost of the pods. We are now able to divert over 10 tons of materials each May as donations to Goodwill without incurring additional costs. In addition, only a handful of volunteers are required. The moral of the story here is that persistence, experimentation, and collaboration between students, staff, and faculty over a number of years resulted in implementation of a low-cost successful sustainability effort on campus.

In the fall of 2014, another Environmental Studies student, Allie France, noticed the large amount of waste thrown away in our campus dining halls, especially the throw-away takeout containers, used by many students on campus, which could be seen filling many of the campus garbage receptacles after lunch and dinner. Why doesn't OWU use reusable carryout containers? This is the kind of question one often hears from environmentally-conscious students. The course instructor suggested Allie do something about the situation as a course project. Allie embarked on what she thought would be an easy task: convince our campus food service to offer reusable containers. However, such sustainability efforts are never easy.

Initially Allie was encouraged by the response from campus food service: they would love to offer students reusable carryout containers. Alas, soon afterward it became evident that our old campus industrial dishwasher could not handle the increased demand for washing reusable containers. Indeed, the shift to throw-away containers and utensils was in part the result of the inadequate dishwasher. All of a sudden Allie was faced with learning much more about industrial dishwashers than she ever imagined. What she suspected was that our old dishwasher was very inefficient, and a new dishwasher would quickly recoup costs due to energy savings alone while allowing the OWU food service to offer reusable food containers. At this point the semester was over, as was Allie's course project. In the spring, Allie continued to work on the project. She worked with Buildings & Grounds to develop a return on investment (ROI) analysis for a new dishwasher. She then had to go to the campus Finance Officer. There were many infrastructure projects on campus ahead of the dishwasher, and, indeed, it was not even on the radar. However, the short ROI (around 2 years) and the fact that Allie had drawn attention to the issue moved the effort forward, a new dishwasher was purchased and installed in the spring of 2015 and reusable food containers were offered in the fall of 2015.

The reusable food container initiative faced some significant hurdles, again illustrating how complicated initiating sustainability projects can be (Fig. 29.4). Unfortunately, despite each container having a bar code, our campus information system (used by food service, the library, for student records, and IDs, etc.) was old enough that there was no easy way to modify the code to allow students to "check out" reusable containers. Replacing the campuswide information system was also not feasible. Thus students paid $5 when they took a reusable container and were given $5 when they personally handed the containers back at certain food service locations on campus. A student project in the fall of 2016 surveyed students about the reusable containers. The additional effort involved in returning the containers proved too much for many students. In addition, some students indicated that carrying around the reusable container suggested the image of an "eco freak." These students were all for the reusable containers, but they did not seem themselves as part of the "ecological clique" on campus and felt uncomfortable using the

FIGURE 29.4 Forlorn reusable food container, discarded near a trash can on campus. *Photo Credit: John Krygier.*

containers because of the image they projected. Addressing these issues took some effort, and several more students, Izzy Sommerdorf and Sarah Hanes, took on the project in the spring of 2017, have increased the number of drop-off locations, and come up with a simple suggestion to encourage the use of reusable containers: provide *larger* reusable containers and *smaller* throw-away containers. The idea is to make up for the added hassle of returning the container by allowing students to pile more food into the reusable containers. While this tweak to the process has led to increased use of the reusable containers, it also may lead to more food waste: sustainability efforts are always complicated. Once again, the moral of the story is persistence, experimentation, and collaboration between students, staff, and faculty over a number of years to put in place a low-cost successful sustainability effort on campus.

These are not the only success stories from years of grassroots efforts. In addition to the creation of an institutional sustainability plan, many other accomplishments in sustainability have taken hold. New and renovated buildings on campus are now routinely upgraded for energy efficiency such as the new geothermal-regulated pool in Meek Aquatic Center (these building improvements are, indeed, one instance of a significant investment in sustainability by the university). The university has hosted several successful years of the Sagan National Colloquium with environmental topics such as Food, Waste, Water, and Climate Change, bringing experts from around the world to Delaware, Ohio, to share their insights. Campus dining halls now feature vegan options, many more local food options, and a general movement toward serving less meat. The recycling program has been successful for many years and has transitioned from a grassroots effort (begun in the 1980s) by students emptying recycling bins to having this task incorporated into housekeeping duties. Each year the students host Green Week, a collection of events and activities related to Earth Day and the environment. All campus printers are set to print double-sided pages as part of Information Services Print Green Initiative. Each of these efforts experienced similar troubles as those mentioned earlier, yet persistence and creativity led to success.

The university is expanding the number of filtered water hydration stations on campus, as an alternative to bottled water, rather than "banning the bottle." Student research determined that athletes were among the largest purchasers of bottled water, as there were no hydration stations in most of the campus athletic facilities. Hydration stations are being installed in six locations, almost all in athletic facilities, this fall, and a student was awarded $800 to buy OWU water bottles to promote the new hydration station to athletes.

There is also work on two related sustainable food issues. The first effort is to revive the campus garden and develop a means for sustaining it over time. To these ends, Environmental Studies student and Sustainability Development intern, Emily Howald, has developed a plan to offer campus "activity courses" (partial credit courses offered by the physical education program on campus, typically activities like yoga, running, and conditioning) that involve gardening. These courses will be offered in the second half of the spring semester (planting/harvesting early crops) and first half of the fall semester (planting/harvesting late crops) to take into account Ohio's growing season. In addition, students Maddie Coalmer and Larynn Cutshaw undertook a project to document a dozen out-of-the-way locations on campus to plant perennial crops (asparagus, mint, raspberries), which require minimal maintenance.

Second, due to increasing student interest in local foods, student Ellen Sizer undertook a project to get more local foods on campus. She developed a proposal for a "Hyper Local Salad Bar," which will be supplied by the nearby Seminary Hill organic farm, part of the Methodist Theological School of Ohio, managed by Tad Peterson and Noel Deehr. Tad and Noel have the capacity to provide many salad bar ingredients year round by using a greenhouse as well as a local food network (of organic farms) that they have developed.

Finally, OWU is expanding its sustainability and environmental vision beyond campus. During the fall of 2015 (and again during fall, 2017) we have offered a travel learning course focused on assessing environmental change, with a strong sustainability component, led by OWU Geography faculty member Nathan Amador Rowley. Students and faculty in the course work with Geoporter, a nongovernmental organization located in Bahia Ballena, Uvita, Costa Rica. Amy Work, a 2004 OWU Geography major, manages Geoporter. As residents of a coastal area in transition from a fishing economy to one based

FIGURE 29.5 OWU Faculty member Nathan Amador (left) and Amy Wok (OWU 2004, right) and various assistants hone in on a drone during an OWU Travel Learning trip to Costa Rica. *Photo Credit: John Krygier.*

on ecotourism, community members in Bahia Ballena, Uvita, are interested in understanding their natural environment and the potential impacts of global environmental change. Amy has been working with her community members to collect and map environmental information (including garbage, water quality, and whales) for several years, providing a solid basis in practice. OWU students learn the practice of data collection and mapping, but also, importantly, develop an understanding of the theories and concepts required to analyze and understand collected data (Fig. 29.5). Theories and concepts are put into practice in Costa Rica, the collaboration designed so that students and community members in Bahia Ballena, Uvita, will come to understand both the theory and practice of environmental change at a range of scales.

A NEW MODEL FOR SUSTAINABILITY?

The aforementioned examples illustrate the idea of grassroots, distributed (but not too grassroots and too distributed) kind of sustainability: students, staff, and faculty figure out how to make sustainability happen on campus with no full time staff and limited, devoted funds. Sustainability is not going to happen otherwise, at least in the short term. Upon reflection, there are some benefits to this approach to sustainability.

Most, if not all of these projects have required substantive collaboration between students, staff, and faculty. Creative and viable solutions arise from the cooperation of a diverse set of minds, all of whom can contribute some specific kind of expertise to the effort. In a way, this approach lends itself to more integration of sustainability across campus, and more active engagement, without depending on (or deferring to) one individual (a sustainability coordinator) for guidance and leadership. The engagement of an increasing number of students provides many excellent theory-into-practice experiences, a significant part of a student's education at OWU. Success after facing many challenges, but moving forward anyway, may be more meaningful given the persistence it requires. This persistence and creative engagement reveals dedication and commitment to environmental causes. Finally, this approach has put in place a strong foundation of sustainability upon which a sustainability coordinator, if one is hired in the future, can build.

The OWU Sustainability Task Force did not set out to develop a model for low-resource, high-engagement sustainability, but we have developed one, by experimentation, collaboration, and persistence. We are still learning and plotting new ways to get sustainability to work on campus, but we are making progress. We hope this model may help other colleges, businesses, and organizations in similar situations make sustainability move forward, as it inevitably must, despite the numerous obstacles to doing what is necessary and right.

Chapter 30

Afterword: A Sustainable Economic and Finance Proposal

Woodrow W. Clark, II
Clark Strategic Partners, Beverly Hills, CA, United States

Chapter Outline

THE PROBLEM

California Governor Brown has put forth a "Green Energy Job and Company Four Year Plan" (June 2010) when he was Attorney General and stated at the conference in Silicon Valley: "Over the next decade, the market for renewable energy will triple to more than $2 trillion," Brown proposed then an eight "point action plan" ranging from energy efficiency to renewable energy power generation for buildings and homes..." The key point or bottom line that Attorney General Brown made a year later as Governor is:

> *Probably the most significant reason people don't make their homes more efficient is the high up-front costs of major efficiency upgrades, even though they save money in the long run ...The State and utilities have to help local governments, businesses, homeowners finance the costs of efficiency upgrades.*

Sustainable Cities and Communities Design Handbook. https://doi.org/10.1016/B978-0-12-813964-6.00030-6

The conclusion of the book focuses on the critical issue: where is the money for homes and businesses to install renewable energy systems that save energy costs, directly require skilled job areas, and create entrepreneurial companies. Financing and resources, even with few tax or other government incentives, are limited at this critical time in the global economic environment.

Brownout and blackouts in California to Arizona and into Northern Mexico impact millions of people. The extent of the damage, loss of life, and health results are not all known yet. The message is that the current dependency on central grid energy generation and transmission (Clark and Bradshaw, 2004) need to be created differently now that the 21st century has arrived with concepts and ideas ranging from renewable energy to smart grids to sustainable communities (Clark, 2009a,b, 2010).

The extent and magnitude of the lessons learned from energy and environmental programs around the world have yet to be acknowledged and implemented in the United States. Such Global Energy Innovations (Clark and Cooke, 2011) stem directly from new economic theories and practices. Unfortunately, the United States has fallen behind, and even become entrenched in its traditional and conventional economic theories that have been proved to be wrong in the 2008 economic collapse as noted by Clark and Fast in *Qualitative Economics,* written before the economic collapse. The result is a need for "the next economics" (Clark, 2003a,b), which looks carefully and then redefines what government, business, and workers (including entrepreneurs) must do, as pointed out in Clark's book, *The Next Economics* (Springer Press, 2012), and other books since that, including solutions to climate change (Clark, 2010, 2014 and Clark and Cooke, 2016).

Above all, the book, *The Green Industrial Revolution* (GIR) in English in 2014 and in Mandarin in 2015, outlines the past as well as the near future. Many nations, such as the Nordic countries, China, Japan, and South Korea, which have created a new economic model labeled "civic capitalism" (Li and Clark, 2011; Clark and Li, 2009), are already in the GIR.

> *Basically the idea (social or civic capitalism) rests on the core concept that binds citizens and governments; people have a social contract with their governments. It is only that contract that allows governments to rule, provide services, and promote the common good. What is social capitalism? In part, it is the support of social concerns within any society: education, health, and infrastructure like roads, water, and waste systems. But it is also a societal concern with the environment and those systems that pollute it such, as vehicles and the energy sector.*
>
> Clark and Li (2004, p. 6)

With social–civic capitalism, the concern is making money, but also with doing what is best for people and society. One current element of the next economics as social capitalism is the concern for clean or green economics in terms of growth, jobs, and new ventures. As the Brookings Institute stated: "the clean economy remains an enigma: hard to assess. Not only do 'green' or 'clean' activities and jobs related to environmental aims pervade all sectors of the U.S. economy; they also remain tricky to define and isolate—and count." (Muro et al., 2011, p. 3).

The need is to outline various areas of focus that research proposals should consider. The context for these five areas is that the "world is round." Unlike some conventional economists, the point to this entire proposal is look at the world and understand, how more than ever, every region of the world is connected and dependent on another—be it through the environment, energy or economics, people, communities, and businesses depend and work with one another from around the world. Hence the five areas suggested directly reflect what are considered the most important.

Corporate Governance in an Age of Economic Globalization

Any action taken in one part of the world is very likely to be known quickly around the world. The current result of social networks in the Middle East are just one recent example. Other regions of the world will soon be engulfed in the same immediate awareness that can turn into new behaviors and trends. The results will range from violence to controlled governance. But the awareness will continue. Hence the economic view among many people now is that there is a growing caste society of the very rich (only a small percentage of any population) and the poor (most of the population). The emergence of a middle class is rapidly leaving the developed industrial world. Where are the leaders in the world for new economics that will recreate a middle class? The examples are coming from South America (Brazil), Asia (China), and Africa (Mauritius), along with other nations.

The role of government is critical and very different from what is now being done in the United States and parts of the European Union, especially like the United Kingdom and the Netherlands. What that new role of government is must be learned from newly developed countries that have successfully redefined both economics and governments. The best model appears to be from China, and also from the Nordic countries and some EU nations.

Political Economy of the State and Public Goods Provision

As we look into the "next economics," the concern is not only with making money but also the public good, hence "social capitalism." The Chinese have become the leading symbol of this merged approach from the extremes of capitalism to socialism. What is dynamic and even measurable so that results can be seen is their use of Five-Year Plans. Now into the 13th Five-Year Plan (implemented in March 2016), is presented in some of the Appendices and some Cases focus on the entire nation is on environmental issues and sustainable development.

The basic concern is to meet the middle class and growing business demands for energy but without further polluting the environment. In fact, the current Five-Year program provides trillions of US dollars for renewable energy and systems that will stem the use of fossil fuels and provide large renewable energy systems. Now with the aftermath of the Japanese earthquake and tsunami in Fukushima, the question of continued nuclear power plants is being reviewed and will be adjusted significantly.

Private and Public Sector Intergenerational Responsibility

The Chinese economic model raises other societal questions, which range from their control on birth (only one child per couple) to generational concerns for the older retiring workers and what to do about them. Where will these people live? How will they be supported? And can the nation grow with an aging population? The middle class in China, for example as defined by international economic standards, is about equal to the entire population of the United States. The fact that Chinese couples have decided over the last two to three decades to *not* keep female children because the parents want male on whom they can depend upon will be changing. With that there will be a significant change in global adoption of female Chinese babies as well as a different approach to China's economic and social development.

Political Economy of Income and Wealth Distribution

In addition to the aforementioned facts, the issues about how the global income and wealth distribution are changing focus on political and policy concerns around the world. The United States is not the only nation seeing a wider gap between the rich and the working class, while losing its middle class. The rise of the middle class in China is significant but not unusual there and with other "developing" nations. What is significant can be seen in the next economics in terms of how people have financial resources for their daily lives, future, and plans for retirement. The United States and other developed nations are facing the future now for "baby boomers" where many of them, born after World War II, cannot afford to retire at 65 years of age as they had expected. Instead, this generation, who will also be the largest consumers of government services from medical care to social security, must work beyond the age of 65 into their 70s. The political economics of this massive transformation in the United States and other developed nations projects into the issues that the developing nations may soon face as well.

Human Capability and Economic Development

Above all, the next economics is concerned with people. As described with the concept of social capitalism, concern for people, the environment, and future generations must be foremost in any economic paradigm.

ECONOMIC APPROACHES: OVERVIEW IN BRIEF

Feed-in Tariffs

Feed-in Tariffs (FiT), also known as standard offer contract or advanced renewable tariff, are policy mechanisms that are designed to accelerate investment in renewable energy systems and technologies. They achieve this by offering long-term contracts to renewable energy producers, typically based on the cost of generation of each different technology. Technologies like wind power, for instance, are awarded a lower per-kWh price, whereas technologies like solar, photovoltaic, and tidal or wave power are currently offered a higher price, reflecting their higher costs.

In addition, FiT often include "tariff degression," a mechanism according to which the price (or tariff) ratchets down over time. This is done to track and encourage technological cost reductions. The goal of FiT is ultimately to offer cost-based compensation to renewable energy producers, providing the price certainty and long-term contracts that help finance renewable energy investments (Gipe, 2011). The FiT has been tested and perfected since 1991 in Germany. Although suspended temporarily a few years ago in Germany, the FiT has been reinstituted as the results have been demonstrated and well documented in a Deutsche book report *The German Feed-in Tariff for PV* (Fulton and Mellquist, 2011) with an important economic subtitle, *Managing Volume Success With Price Response* (Fulton and Mellquist, 2011).

Carbon Incentive (Tax)

As in most countries, there needs to be a high tax on fossil fuels in the United States. When many people oppose the use of the word "tax," they then call it an "incentive," which is what was done in California when it took the national and then international lead on taxing cigarettes. The tax was explained as an incentive to stop people from smoking, and prevention of second-hand smoke was seen to prevent lung and other related diseases. The money collected was used to promote antismoking advertisements and materials as well as support health research in this problem. Now other nations are doing the same.

Master Contracts

The basic issue is always money. How do you pay for something that is needed but new. The costs are usually high. Some programs have been tried and have successfully addressed the economic issues. To start with, governments need new "green" buildings—company, campus, groups, etc. Some companies, for example, are now providing their employees solar panels and related systems at the same price that they bought them for their offices, facilities, and plants. Other companies are even supplying and encouraging subcontractors to purchase renewable energy systems, recycled products, and related environmentally sustainable products for their own use.

Solar Company Finance

Renewable energy companies finance their own products. In short, the program avoids both intermediate financial institutions and installers. Instead, there are some solar companies who will fund and finance solar systems. The interesting twist to this program, however, has come from Hanergy, which is the largest renewable energy company in China. The only requirement is that the solar and other products are from Hanergy or companies owned by or related to them.

The other renewable energy company strategy is the use of power purchase agreements (PPA), which was started about 6 years ago from Japanese companies. The requirement was that the purchaser of the solar system, for example, must do so from a finance company or bank. The PPA became a popular way to fund solar systems since 2006 because the price for solar panels was high. The long-term pay period made the solar systems available at lower costs. However, when the costs are added up, the total price for the customer or user of the solar panels was very high, and was payable over a long period of time.

Cap and Trade

The old economics has created a new version of the "market economy" through what it calls and promotes as "cap and trade." The implementation of Assembly Bill #32 in California (2012) after its passage under Governor Schwarzenegger was due to years of testimony and data gathering by the California Air Resources Board. The concept of cap and trade has been supported by members of both political parties since it creates credits for the savings of emissions that can then be traded to others. The critical issue is that there is a legal requirement to do so, hence creating a market for trading of carbon credits: cap is a legally set requirement either reverse emissions to lower levels such as 1990; and trade is the saving credit that is then traded on an climate exchange.

While cap and trade has gathered a lot of support from the neoclassical economists, it has created a "market" that is based on credits given to companies (and individuals) based on measurable environmental savings from energy efficiency technologies to renewable energy systems used primarily for buildings. A positive impact reversing climate change can happen, but primarily for the user of the technologies and systems. The negative problem is then trading that savings to some company or people who are not stopping climate change through reduction of emissions, etc. There are now trading markets and exchanges in the European Union as well as some established in the United States. There are serious questions about such cap and trade credits and exchanges. Much of it is due to the inability to regulate and set enforceable standards. Market exchanges have difficulties in doing this. Furthermore, there is a parallel similarity to this approach to climate change, similar to "deregulation" of the energy sectors around the world. In several US states, especially California, such deregulated markets were the basic problem with energy in the state that set off an energy crisis in 2000. Finally, environmental groups have filed legal briefs and opinions that argue and question the impact on the local environment of those companies and people who purchase carbon credits to avoid reducing their own local carbon emissions.

Investments (Equity)

The use of private investments ranging from angle capital to private equity to venture capital, which is often seen as becoming a public stock offering, is the key financial mechanism today for funding new technologies.

Property Assessed Clean Energy

The Property Assessed Clean Energy (PACE) program was started in Berkeley in 2007. Now 17 states and over 200 cities have started similar programs. PACE tackled some of the problems mentioned earlier. It pays for the energy project through property taxes levy. The property owner is cash-flow neutral as the energy savings would be equal to the tax increase for a period of say 5 years, at which point the project is paid off. Then all the energy savings reverts back to the borrower.

The City of Berkeley's pilot program, called Berkeley FIRST, was launched in November 2008 as a solar-only financing program. Berkeley FIRST was a small-scale program designed to test the idea of PACE-type financing in the real world. The program's 40 available slots were all claimed in 9 min. Once approved, property owners had 9 months to install their projects. In the end, 90% of participants installed or planned to install solar systems, 85% said the program was responsible for their decision to "go solar," and roughly a third of participants used Berkeley FIRST financing to pay for their new systems.

The PACE momentum in the commercial space seems to be building momentum, but there are still challenges. In terms of this research project, for example, it is not clear that a program with sufficient scalability has been built for small commercial properties, which is an important target segment for energy projects.

Tax Shifting

A very direct approach to controlling climate change and even reducing it has been through taxation. The carbon taxes on fossil fuels for transportation and buildings, especially in the European Union, have proved that when a carbon tax is set and raised, companies begin to find and implement technologies that either use less or no fossil fuels. The development of hybrid vehicles is primarily based on that hypothesis: carbon tax means the need to use less fossil fuels for driving; the tax funds are then often invested by governments in mass transportation.

California had a hybrid vehicle rebate program, but also special permits for users of hybrid cars in car pool lanes on highways, which helped create a demand for these passenger cars. In different countries similar strategies were implemented through the use of the tax funds to build train and light rail systems.

The taxation of off-shore oil (only permitted in one area of California, near Santa Barbara) could have netted the state from $4–5 billion a year under a statewide voter proposition that was defeated by a massive political campaign from the oil and gas sector. The voters decided by 55%–45% not vote for the proposition. California is the only state that does not tax off-shore drilling.

Another approach to creating funds for new mass transportation systems that do not use or limit fossil fuels is to "shift" taxes that already exist from one purpose to another. For example, some taxes that exist today were created decades ago and should be examined as to the use of the funds collected. There are taxes that could be still enforced, but the use of the funds is no longer needed at the levels originally set. Today those funds could still go for specific needed services or other purposes, but it needs to be examined whether the funds can be shifted to mass transit and systems that could help reverse emissions.

Bank Mortgages: (Long-Term Transferable Debt)

Identify a bank and/or target area, community, or company to start whereby there can be segmented sectors from existing property owners, buyers of properties, and new construction. Current mortgages where borrowers want to finance energy programs through their equity can be done through refinancing.

The issue is that from the value of property or loan to value (LTV), a building owner would have to have built up lendable equity in the property by either paying down the initial mortgage or experiencing property appreciation. There are numerous problems in the current environment.

From a loan structure perspective, the challenges will be: (1) getting lenders to either increase the LTVs or evaluate the property value as being increased by the retrofit project, and (2) getting lenders to incorporate the cash flow savings into the underwriting. If a borrower is borrowing as much as can be handled at the moment (banks apply what is called debt service coverage), then a lender will not lend the borrower anymore. A bank would have to allow for cash flow savings in their underwriting calculations due to prospective lower energy costs.

Usually lenders are reluctant to apply prospective cash flow savings in underwriting calculations, especially in this environment when many of the loan losses were due to pro forma cash flow predictions that were never realized. However, my understanding is that some of these energy retrofits are arranged by engineering or service companies that will guarantee the energy savings. In this case, there may be a flexibility here.

CONCLUSION NOW AND FOR TOMORROW

As Governor Brown put the concern and vision correctly, while serving as Attorney General

> *By 2020, California should produce 20,000 new megawatts (MW) of renewable electricity, and also accelerate the development of energy storage capacity. California can do this by aggressively developing renewables at all levels: small, onsite residential and business systems; intermediate-sized energy systems close to existing consumer loads and transmission lines; and large scale wind, solar and geothermal energy systems. At the same time, California should take bold steps to increase energy efficiency.*

And above all, as the Brookings Institute stated: "The clean economy has remained elusive in part because, in the absence of standard definitions and data, strikingly little is known about its nature, size, and growth at the critical regional level" (Muro et al., 2011, p. 3). Although Brookings did this somewhat in their report, they also noted that "Currently no comprehensive national database exists" (Muro et al., 2011, p. 3). This was one of the conclusions that Clark and Fast (2008) came up with in their study of "qualitative economics" as a new field of economics that would focus on definitions and meanings that ranged from numbers to symbols to make economics "a science." More studies and research must be done in this area since words like "clean" and "green" are very different.

In conclusion, the next economics is based on the following.

> *What the social capitalism paradigm argues, is that states or governments cannot be invisible. or leave certain societal areas and sectors open to .market forces.. Government must be active and even protective in certain areas that impact on all citizens, including businesses and new enterprises. Governments' role is to provide guidance through some regulation, oversight and investment stimulation policies and programs.*

Clark and Li (2004, p. 9)

The basic issues are using qualitative economics in outlining and discussing the "next economics," but such studies must also concern politics, governments, employment with new companies, and union.

REFERENCES

Clark II, W.W., 2009a. Sustainable Communities. Springer Press.

Clark II, W.W., June 2009b. Sustainable Communities Design Handbook. Elsevier Press, NY, NY, pp. 9–22.

Clark, W.W., 2003a. California's Next Economy. California Governor's Office Planning & Research.

Clark, W.W., 2003b. Southern California "LifeSciences Strategic Plan". Governor's Office of Planning and Research, Sacramento, CA.

Clark II, W.W., November 2012. The Next Economics. Springer Press.

Clark II, W.W., Bradshaw, T., 2004. Agile Energy Systems. Elsevier Press.

Clark II, W.W., June 2010. Sustainable Communities Design Handbook. Elsevier Press, NY.

Clark II, W.W., 2014. Global Sustainable Communities Handbook. Elsevier Press.

Clark II, W.W., Cooke, G., March 2016. Smart Green Cities. Routledge Press.

Clark II, W.W., Cooke, G., November 2011. Global Energy Innovations. Praeger Press.

Clark II, W.W., Fast, M., 2008. Qualitative Economics: Toward a Science of Economics. Coxmoor Press, Oxford, UK.

Clark, W.W., Li, X., 2004. Social capitalism: transfer of technology for developing nations. International Journal of Technology Transfer and Commercialization 3 (1). Inderscience, London, UK.

Clark, W.W., Crises, LiX., 2009. opportunities and alternatives globalization and the next economy: a theoretical and critical review, Chapter #4. In: LiX, Winther, G. (Eds.), Globalization and Transnational Capitalism. Aalborg University Press, Denmark.

Fulton, M., Mellquist, N., 2011. The German Feed-in Tariff for PV: Managing Volume Success with Price Response. DB Climate Change Advisor. Deutsche Bank Groups.

Gipe, P., 2011. Feed-in-Tariff Monthly Electronic Updates.

Li, X., Clark, W.W., January 2011. Energy concerns in China's policy-making calculations: from self-reliance, market-dependence to green energy. Special Issue Contemporary Economic Policy Journal. Western Economic Association.

Muro, M., Rothwell, J., Saha, D., August 2011. Sizing the Economy: A National and Regional Green Jobs Assessment. Brookings Institute, Washington DC.

FURTHER READING

Bailey, E.M., Wolfram, C., 2012. A whole different kind of innovation. The Journal Report: Innovations in Energy, Wall Street Journal. www.wsj.com.

Clark II, W.W., 2007. Eco-efficient Energy Infrastructure Initiative Paradigm. UNESCAP, Economic Social Council, Asia, Bangkok, Thailand.

Clark II, W.W., 2008. Public policy in the People's Republic of China: toward a model for planning and regulatory rules for the future.

Clark II, W.W., January 2011. The next economics: a green industrial revolution. Special Issue Contemporary Economic Policy Journal. Western Economic Association.

Clark, W.W. II, Cooke, G., The Third Industrial Revolution Chapter #2.

Clark II, W.W., Isherwood, W., 2009a. Leapfrogging energy infrastructure mistakes for Inner Mongolia. Utility Policy Journal, Special Issue.

Clark II, W.W., Isherwood, W., 2009b. Report on energy strategies for Inner Mongolia autonomous region. Utilities Policy. https://doi.org/10.1016/j.jup.2007.07.003.

Clark II, W.W., Intriligator, M., Winter, 2011. Global cases in energy, environment, and climate change: some challenges for the field of economics. Special Issue Contemporary Economic Policy Journal. WEAI, California.

Clark II, W.W., Isherwood, W., Winter 2010. Special issue on China: environmental and energy sustainable development. Utility Policy Journal.

Clark, W., Lund, H., 2008. Integrated technologies for sustainable stationary and mobile energy infrastructures. Utilities Policy 16 (2), 130–140.

Clark II, W.W., Lund, H., 2009. Analysis: 100% renewable energy. In: Renewable Energy Systems. Elsevier Press, pp. 129–159. Chapter 6.

Demirag, I., Khadaroo, I., Clark II, W.W., 2009. The institutionalization of public-private partnerships in the UK and the nation-state of California. International Journal of Public Policy 4 (3/4), 190–213.

Lund, H., Clark, W., June 2008. Sustainable energy and transportation systems introduction and overview. Utilities Policy 16 (2), 59–62.

Xing, L., Clark, W.W., 2009. Crises, opportunities and alternatives globalization and the next economy: a theoretical and critical review. In: Xing, L., Winther, G. (Eds.), Globalization and Transnational Capitalism. Aalborg University Press, Denmark. Chapter #4.

Contributing scientist to United Nations Intergovernmental Panel on Climate Change (UN IPCC); Lead author, IPCC (1995–1999); Lead Expert, Final IPCC Report Overview Approval (Nepal 1999); and Co-Author, Chapter #2, "Finance Mechanisms" and Co-Editor Chapter #3: "Legal/ Political Framework", UN, New York, NY: 2000. Nobel Peace Prize awarded to UN IPCC as co-winner with Al Gore, "An Inconvenient Truth." December 2007.

Index

'*Note:* Page numbers followed by "f" indicate figures, "t" indicate tables.'

B

D

F

G

H

I

J

M

N

O

P

S

T

U

Printed in the United States
By Bookmasters